Conceptual Perspectives in Quantum Chemistry

This book is a companion volume to
Conceptual Trends in Quantum Chemistry,
edited by E.S. Kryachko and J.L. Calais, ISBN 0-7923-2621-0
and
Structure and Dynamics of Atoms and Molecules: Conceptual Trends,
edited by J.L. Calais and E.S. Kryachko, ISBN 0-7923-3388-8

Conceptual Perspectives in Quantum Chemistry

edited by

Jean-Louis Calais†

and

Eugene Kryachko
*Bogoliubov Institute for Theoretical Physics,
Kiev, Ukraine*

SPRINGER SCIENCE+BUSINESS MEDIA, B.V.

A C.I.P. Catalogue record for this book is available from the Library of Congress.

DOI 10.1007/978-94-011-5572-4

Printed on acid-free paper

All Rights Reserved
© 1997 Springer Science+Business Media Dordrecht
Originally published by Kluwer Academic Publishers in 1997
MyCopy version of the original edition 1997
No part of the material protected by this copyright notice may be reproduced or
utilized in any form or by any means, electronic or mechanical,
including photocopying, recording or by any information storage and
retrieval system, without written permission from the copyright owner.
www.springer.com/mycopy

CONTENTS

Foreword   xv

Recent Developments In Multiple Scattering Theory And Density Functional Theory For Molecules And Solids

R. K. Nesbet   1

1. Introduction   1
2. Multiple Scattering Theory   3
   2.1. Full-Potential Theory   4
   2.2. Angular Momentum Representation   6
   2.3. Surface Matching Theorem   8
   2.4. Surface Integral Formalism   10
   2.5. Atomic Cell Orbitals   14
3. Variational Principles   14
   3.1. Kohn-Rostoker Variational Principle   15
   3.2. Schlosser-Marcus Variational Principle   16
4. Wave Functions at Cell Boundaries   19
   4.1. Wave Functions in the Near-Field Region   20
   4.2. Convergence of Internal Sums   22
5. Green Functions   26
   5.1. Definitions   26
   5.2. Properties of the Green Function   29
   5.3. Construction of the Green Function   30
6. Computational Theory   34
   6.1. Basis Functions   34
   6.2. Linearized Methods   36
   6.3. Canonical Scaling   40
   6.4. Poisson Equation   42
7. Density Functional Theory   44
   7.1. Reference-State Density Functional Theory   45
   7.2. Variation of Occupation Numbers   49
   7.3. Self-Interaction Corrections   52
Acknowledgments   55
References   55

Localized Atomic Hybrids: A General Theory

G. G. Hall and D. Rees   59

1. Introduction	59
2. Use of Symmetry	60
2.1. Tetrahedral Example	61
2.2. Use of Group Theory	62
3. Basis Space	64
4. Sample Space	66
5. Alias Functions	67
5.1. Operator Alias Functions	67
5.1.1. Square hybrids	68
5.2. Basis Alias Functions	69
6. Example of Hybrids	71
6.1. Octahedral Hybrids	71
6.2. The Cubic Hybrids	73
6.3. Other Hybrids	75
7. Localization	76
7.1. Properties of Localized Hybrids	77
8. Several Equivalent Sets	78
8.1. Radial Eigenfunctions	79
8.1.1. Five function example	80
8.1.2. Eleven function example	81
8.2. General Comments	84
9. Quadrature	86
10. Calculation of Molecular Properties	88
References	89

Quantum Electrodynamics And Molecular Structure
Molecular Electrodynamics

R. G. Woolley   91

1. Introduction   91

2. Quantum Mechanics of the Spectroscopic Experiment	94
3. The Electrodynamics of Molecules	97
3.1. Gauge Transformations	97
3.2. The General Hamiltonian	100
3.3. The Coulomb Gauge Theory	103
3.4. Polarization Fields	105
3.5. Scattering at High Energies	108
4. Electron-Molecule Scattering	109
5. Photon-Molecule Scattering; the Kramers-Heisenberg Formula	115
6. Discussion	119
7. *References*	123

Aspects Of The Chemical Bond 1996

J. F. Ogilvie   127

1. Introduction	127
2. Ionic Materials	127
3. Application of an Alternative Approach to Inorganic Structures	133
4. Mechanisms of Reactions	138
5. Conclusion	141
References	142

Lie Symmetries In Quantum Mechanics

D. R. Truax   145

1. Introduction	145
2. Lie Symmetries	147
2.1. Real Schrödinger Algebras, $(\mathcal{SA})_N$	147
2.2. Complexification of $(\mathcal{SA})_N$	151
3. Solutions to the Time-Dependent Schrödinger Equation	155
3.1. Systems in One Dimension	155
3.2. Systems in Two Dimensions	159
3.3. Comments	165
4. Coherent States	166

5. Examples	170
5.1. Harmonic Oscillator	171
5.1.1. One dimension	171
5.1.2. Two dimensions	173
5.2. Driven Harmonic Oscillator	176
5.2.1. One dimension	176
5.2.2. Two dimensions	178
6. Transformation of the Schrödinger Equation	180
6.1. A Velocity Dependent Potential	182
6.2. Spinless Particle in an Electromagnetic Field	184
Acknowledgement	187
Appendix I	187
References	191

The Interplay Between Quantum Chemistry And Molecular Dynamics Simulations

S. M. Kast, J. Brickmann, and R. S. Berry   195

1. Introduction	195
2. Molecular Dynamics Simulations	198
2.1. Principles	198
2.2. Constant Temperature Simulations	199
2.2.1. General outline	199
2.2.2. Finite-mass stochastic collision dynamics	200
2.3. Constant Pressure Stimulations	206
2.4. Forces by Electronic Structure Theory	207
2.4.1. *Ab initio* molecular dynamics	207
2.4.2. Hybrid QM/MM methods	209
3. Aspects of Parametrization	211
3.1. General Remarks	211
3.2. Optimal Simulated Annealing	212
3.3. Extended Systems Revisited	214
3.4. Examples	216
3.4.1. Lennard-Jones parameters from quantum chemical data	216
3.4.2. *Ab initio* data and thermal averages	218
4. Conclusions	219
5. Acknowledgements	220
6. *References*	220

The Permutation Group In Many-Electron Theory

S. Rettrup   225

1. Introduction	225
2. The Electronic Schrödinger Equation	226
2.1. Permutation Symmetry	227
2.2. The Pauli Principle	228
2.3. N-Electron Spin Eigenfunctions	229
2.4. Spin-Free Quantum Chemistry	230
3. Approximate N-Electron Functions	232
3.1. Configuration State Functions	233
3.2. Matrix Elements	235
4. Concluding Remarks	236
5. Acknowledgment	237
References	237

New Developments In Many Body Perturbation Theory And Coupled Cluster Theory

D. Cremer and Z. He   239

1. Introduction	239
2. Møller-Plesset Perturbation Theory	243
2.1. Derivation of the Møller-Plesset Correlation Energy at Lower Orders	244
2.2. Derivation of Møller-Plesset Correlation Energies in Terms of Two-Electron Integral Formulas	252
2.3. Correlation Effects Covered at Various Orders of Møller-Plesset Perturbation Theory	260
3. Møller-Plesset Perturbation Theory at Sixth Order	262
3.1. Derivation of a Sixth Order Energy Formula in Terms of Cluster Operators	264
3.2. Setting Up Two-Electron Integral Formulas	270
3.3. Implementation and Testing of a MP6 Computer Program	275

3.4. Comparison of MP6 and Full CI Correlation Energies	277
4. Coupled Cluster Theory	278
4.1. The Projection Coupled Cluster Approach	279
4.2. The Quadratic CI Approach - an Approximate Coupled Cluster Method	283
5. Analysis of Coupled Cluster Methods in Terms of Perturbation Theory	285
5.1. Expansion of CC Methods to Higher Orders of Perturbation Theory	288
5.2. Comparison of CCSD and QCISD	290
5.3. Comparison of CCSD(T) and QCISD(T)	295
6. Coupled Cluster Methods with Triple Excitations	297
6.1. Implementation of a Coupled Cluster Singles, Doubles, and Triples Method: CCSDT	298
6.2. Development of a QCI Method With Single, Double, and Triple Excitation: QCISDT	301
6.3. Analysis of CCSDT and QCISDT	307
6.4. Implementation and Application of QCISDT	310
7. Conclusions and Outlook	313
8. Acknowledgements	314
9. *References*	315

A Philosopher's Perspective On The "Problem" Of Molecular Shape

J. L. Ramsey 319

1. Introduction	319
2. A Physical Account of Shape	320
3. Interlude: Exactly Which Concept of Shape is Being Challenged?	322
4. Reduction and Explanation	323
5. Realism, Physicalism and Materialism	327
6. Conclusion	330
Acknowledgements	332
Endnotes	332
References	334

Van Der Waals Interactions From Density Functional Theories:
The He-CO System As A Case Study

F. A. Gianturco and F. Paesani	337
1. Introduction	337
2. An Outline of Interaction Contributions	339
3. The Density Functional Approach	342
4. An Overview of Earlier Studies on the He-CO Interaction	345
4.1. A Comparison of the Latest PES	348
5. DFT Calculations of the Interaction	350
5.1. The Effects of DFT Exchange and Correlation	352
5.2. The Post-Hartree-Fock Treatments	360
5.3. Comparing DFT and *ab Initio* TKD Results	364
5.4. DFT Results Versus a Multiproperty Potential	368
5.5. DFT Results Versus a Spectroscopic Potential	370
6. Summary and Conclusions	373
7. Acknowledgements	376
References	376

Different Legacies and Common Aims: Robert Mulliken, Linus Pauling And The Origins Of Quantum Chemistry

A. Simões and K. Gavroglu	383
Part I	384
Part II	392
Notes	406

Potential Energy Hypersurfaces For Hydrogen Bonded Clusters $(HF)_N$

M. Quack and M. A. Suhm   415

1. Introduction	415
2. Sampling and Representation Methodology	417
2.1. Freezing the Structure of the Flexible Monomers?	417
2.2. Local or Global Surfaces?	418
2.3. 'On-the-Fly' or Analytical Representations?	419
2.4. Compromise 'On-the-Fly' Strategies: Voronoi Step Representations	420
2.5. Many-Body Decomposition	422
2.6. Effective or True Pair Potentials?	425
2.7. *Ab Initio* and Empirically Adjusted Potentials	427
3. Approximations in *ab Initio* Electronic Structure Theory	428
3.1. Hartree-Fock Based Methods (Including Electron Correlation)	428
3.2. Density Functional Methods	429
3.3. Quantum Monte Carlo Methods	430
4. Potentials	430
4.1. Monomer Potential	430
4.2. Pairwise Interaction Potential	431
4.3. Three-Body Potential	436
5. Clusters	436
5.1. HF Dimer	436
5.2. HF Trimer	438
5.3. Larger HF Clusters - Structure and Energetics	440
5.4. Spectroscopic Properties	444
5.5. Beyond Monomer Integrity - Hydrogen Exchange	446
6. Conclusions and Outlook	448
Acknowledgement	450
References	450

One-Electron Pictures Of Electronic Structure: Propagator Calculations On Photoelectron Spectra Of Aromatic Molecules

J. V. Ortiz, V. G. Zakrzewski, and O. Dolgounitcheva   465

1. Introduction	465

2. The Electron Propagator	466
3. The Uncorrelated Case	467
4. Superoperator Theory	468
5. Projection and Partitioning Techniques	469
6. Relationships to Hilbert Space Theories	471
7. The Dyson Equation	473
8. Perturbative Self-Energies	474
9. Quasiparticle Methods	475
9.1. OVGF Methods	475
9.2. Partial Third Order Theory	476
10. Computational Techniques	478
10.1. Pole Search Methods	478
10.2. Semidirect Contractions	479
10.3. Integral Transformation and Storage	480
11. Interpretations of Photoelectron Spectra	482
11.1. Benzene	482
11.2. Chlorobenzene	484
11.3. Dichlorobenzenes	486
11.3.1. *Para*-Dichlorobenzene	486
11.3.2. *Meta*-Dichlorobenzene	488
11.3.3. *Ortho*-Dichlorobenzene	488
11.4. Azabenzenes	491
11.4.1. Pyridine	491
11.4.2. Pyridazine	493
11.4.3. Pyrimidine	495
11.4.4. Pyrazine	497
11.4.5. S-Triazine	499
11.4.6. S-Tetrazine	499
11.5. Polyacenes	502
11.5.1. Anthracene	502
11.5.2. Phenanthrene	505
11.5.3. Naphthacene	507
11.6. Borazine	509
11.7. TCNQ	511
12. Conclusions	514
13. Acknowledgments	514
References	514

Shape in Quantum Chemistry

P. G. Mezey   519

1. Introduction   519

2. Molecular Shape: The Shape Group Methods 523
3. The Shape of Potential Energy Hypersurfaces:
   The Fundamental Group of Reaction Mechanisms 528
4. An Analogy Between Shape Groups and
   The Fundamental Groups of Reaction Mechanisms 545
Summary 545
References 546

Index 551

FOREWORD

The rivers run into the sea, yet the sea is not full

Ecclesiastes

What is quantum chemistry? The straightforward answer is that it is what quantum chemists do. But it must be admitted, that in contrast to physicists and chemists, "quantum chemists" seem to be a rather ill-defined category of scientists. Quantum chemists are more or less physicists (basically theoreticians), more or less chemists, and by large, computationists. But first and foremost, we, quantum chemists, are conscious beings.

We may safely guess that quantum chemistry was one of the first areas in the natural sciences to lie on the boundaries of many disciplines. We may certainly claim that quantum chemists were the first to use computers for really large scale calculations. The scope of the problems which quantum chemistry wishes to answer and which, by its unique nature, only quantum chemistry can answer is growing daily. Retrospectively we may guess that many of those problems meet a daily need, or are say, technical in some sense. The rest are fundamental or conceptual. The daily life of most quantum chemists is usually filled with grasping the more or less technical problems. But it is at least as important to devote some time to the other kind of problems whose solution will open up new perspectives for both quantum chemistry itself and for the natural sciences in general.

When chemistry developed in the nineteenth century, its explanatory systems began to separate from those of physics, as peculiarly chemical theories of molecular structure and of reactivity were conceived. By the early twentieth century attempts were made to heal this rift and perhaps the most influential of these was the creation of the electronic theory of valency.

However it was not until the development of quantum mechanics in the late nineteen-twenties that it became clear how a proper reconciliation should be effected. It is in this context that quantum chemistry rises as a subject of its own and gets a very special place among the natural sciences. Quantum chemistry provides the conceptual apparatus by means of which chemical explanation is tied to the explanatory scheme of quantum mechanics, one of the most fundamental of all physical theories.

Until the nineteen-fifties - even though important numerical work was certainly carried out also before that time - the chief contribution of quantum chemistry to chemistry was in terms of concepts like atomic and molecular orbitals, resonance, hybridisation and many others. Since about nineteen-sixty the development of computers has led to a change in emphasis. The conceptual apparatus has remained largely that of previous generations but thanks to large scale sophisticated quantum chemical computations, quantitatively accurate results have made it possible to assist in an effective way the understanding of many phenomena in chemistry, physics, and biology. There can be no doubt about the tremendous quantitative successes that quantum chemistry has had in the last three decades. However, it may not be out of place to have a look retrospectively at different angles and to ask: Whether time might not be ripe for making a paradigm shift in the nature of the concepts that are used, for breaching the decorum to say what has not yet been said before because it has seemed too trivial to bother with or too embarassing to own up to, and for showing us something that had been there all the time that we had not seen for ourselves.

That is why it was our idea to launch "Conceptual Trends in Quantum Chemistry". The former collection of essays of a number of active quantum chemists appeared under this title in early 1994. The second one, "Structure and Dynamics of Atoms and Molecules : Conceptual Trends", was published in 1995. The present one has the same underlying idea, although it is somewhat broader. We hope that all these collections will stimulate a lively and fertile discussion. Our intention is to produce a new Volume sometime in 1998. We invite quantum chemists and other scientists working in neighbouring fields like theoretical physics and chemistry, philosophy of quantum mechanics to contribute to this series.

We are deeply indebted to the KLUWER Academic Publishers and, personally, to Mrs. Wil Bruins, Ms. Margaret Deignan, and Dr. David J. Larner for all their interest and assistance in the preparation of this series. We gratefully acknowledge Dr. Thomas Godsk Jørgensen for his kind help in preparing this Volume.

JEAN-LOUIS CALAIS EUGENE KRYACHKO
Uppsala Kiev

RECENT DEVELOPMENTS IN MULTIPLE SCATTERING THEORY AND DENSITY FUNCTIONAL THEORY FOR MOLECULES AND SOLIDS

ROBERT K. NESBET
IBM Almaden Research Center
650 Harry Road, San Jose, California 95120-6099, USA

Abstract. Density functional theory (DFT) and multiple scattering theory (MST) have been used extensively for calculations in condensed matter physics of the electronic structure of materials built from atomic species throughout the periodic table. As originally formulated for the muffin-tin model, MST in general gave disappointing results in applications to molecules, which discouraged use of this methodology. Recent developments in so-called 'full-potential' MST have removed these muffin-tin constraints while retaining the computational efficiency associated with these methods. Full-potential MST is reviewed here, with emphasis on the theoretical basis of computational methods that are viable for molecules. Several recent developments in the formalism of DFT are reviewed, in particular reference-state DFT, which frees the theory from the constraint that the electronic density in the variational model state must be identical to that of the true ground state, and places DFT in the context of the Fermi liquid theory of Landau. The theory of self-interaction corrections is reexamined from this point of view.

1. Introduction

Density-functional theory (DFT)(Hohenberg and Kohn, 1964; Kohn and Sham, 1965), which maps the many-electron Schrödinger equation onto an independent-electron model, has become the standard formalism for first-principles calculations of the electronic structure of extended systems. Most applications of DFT have been carried out in the local spin-density approximation (LSD), in which the exchange-correlation energy functional is a local function of the spin-indexed electron density. Of many reviews of this formalism and its applications, the author has found (Callaway and March, 1984; Parr and Yang, 1989; Jones and Gunnarsson, 1989) to be especially informative. The latter authors discuss applications to finite systems in detail, together with various practical extensions and modifications of LSD. In Kohn-Sham theory, the mean-field potential function is a local field. This makes it possible to subdivide a large molecule or solid into local atomic cells. Exact solutions computed independently in each cell can be combined into global eigenfunctions using multiple scattering theory (MST). The secular equation of MST varies nonlinearly with energy, which appears to preclude use of matrix diagonalization methods to determine eigenstates. However, energy-linearization techniques have been introduced that make MST a very efficient computational method. Very small basis sets are needed, composed of explicit solutions of the LSD Schrödinger (or modified Dirac) equation, obtained separately in each atomic cell and not used outside that cell. These functions can be evaluated on a coarse grid of fixed energies and interpolated in energy. Frozen inner shells and semirelativistic calculations for heavy atoms are a natural part of the formalism. This aspect has made solid-state DFT-MST calculations feasible for the entire periodic table.

Although the basic formal ideas of multiple scattering theory go back at least a century, to Lord Rayleigh(Korringa, 1994), modern applications are based on the KKR method(Korringa, 1947; Kohn and Rostoker, 1954), originally derived for the electronic structure of solids in the muffin-tin model. This assumes a spherical potential within each nonoverlapping muffin-tin atomic sphere, and a constant potential in the interstitial volume. The muffin-tin approximation is useful for close-packed metals but is inappropriate for the bonded structures of typical molecules. Extension of MST to three-dimensional potential functions without muffin-tin constraints, (the full-potential problem) in nonoverlapping, space-filling atomic cells has been considered for many years, but has only recently been proved to be valid. The present article will develop the relevant theory, with emphasis on methodology that can be applied to molecules as well as solids. Several new developments in the density functional theory itself will be presented, since their implications may be especially relevant to localized systems such as atoms and molecules and to localized states in solids. The recent developments presented here make it interesting to reconsider earlier proposals for applications of MST to large molecules and localized states(Johnson and Smith, 1971).

In the theory of Kohn and Sham, a model state is postulated that can be described by a single-determinant wave function. The electronic density of this model state is identical to that of the correlated N-electron system. Interpretation of properties of this model state is problematic, especially for atoms and molecules. Any optimized model state such as the Hartree-Fock function defines an electronic density function that is in general different from the exact electronic density of a correlated wave function. Although it has been shown that Kohn-Sham one-electron energies are derivatives with respect to occupation number of the DFT total energy(Janak, 1978), a property of quasiparticle energies in Fermi-liquid theory, the Kohn-Sham model state has no clear relationship to many-body theory. There is no systematic theory of corrections to approximate DFT functionals or theory that would allow DFT to be used as a first approximation for excitations from the ground state. It has been shown recently(Nesbet, 1996) that density-functional theory can be developed from an alternative postulate, using the density of a variationally defined reference state, rather than the density of the fully correlated system. This reference-state density functional theory (RDFT) will be presented here, and will be shown to be the ground-state counterpart of the more general parametrized Fermi-liquid theory of Landau. This makes the RDFT suitable as a starting-point for applications of many-body theory.

Formal multiple scattering theory is derived in Section 2, including full-potential theory, the angular momentum representation, the surface matching theorem, surface integral formalism, and atomic-cell orbitals. The Kohn-Rostoker and Schlosser-Marcus variational principles are considered in Section 3. The historically important issue of wave functions at cell boundaries and in the near-field region is discussed in Section 5, including the question of convergence of internal sums in the full-potential theory. Definitions, properties, and construction of the free-particle and full-potential Green functions are presented in Section 6. Aspects of computational theory are treated in Section 7, including basis functions, energy linearization, canonical scaling, and the use of MST methods to solve the Poisson equation.

Several relevant aspects of density functional theory are considered in Section 8, including reference-state theory (RDFT), the related issue of variation of occupation numbers, and a reformulation of the theory of self-interaction corrections in the light of these developments.

2. Multiple Scattering Theory

Standard muffin-tin multiple scattering theory (KKR) can be derived from the variational principle of Kohn and Rostoker (KR) (1954). Extension of this formalism to so-called full-potential theory has been a subject of considerable controversy. The original full-potential derivation(Williams and Morgan, 1972; Williams and Morgan, 1974) was criticized because of the apparent use of divergent series expansions, which could require 'near-field' corrections in the region between enclosed and enclosing spheres of each nonspherical atomic cell(Ziesche, 1974; Faulkner, 1979; Badralexe and Freeman, 1987), or otherwise invalidate the theory. Several proofs by different methods were eventually proposed, showing that the full-potential theory is indeed well-founded (Gonis, 1986; Nesbet, 1986a; Zeller, 1987; Nesbet, 1988; Gonis et al, 1988; Molenaar, 1988; Gonis et al, 1989; Zhang and Gonis, 1989; Nesbet, 1990a; Newton, 1990). Test calculations at first gave inconclusive results(Faulkner, 1986), essentially because of difficulties with slow convergence of series expansions. These issues were ultimately settled by showing that carefully constructed model studies did converge to correct results(Zeller, 1988; Butler and Nesbet, 1990; Butler and Zhang, 1991). Convergence of intermediate sums involving the matrix of structure constants remains a serious practical problem in full-potential MST. Several methods to alleviate this problem have been proposed(Nesbet, 1992b; Zhang and Butler, 1992b), including various applications of the Schlosser-Marcus variational principle, to be discussed below.

2.1. FULL-POTENTIAL THEORY

Expressed as a modified Helmholtz equation, the Schrödinger equation for energy κ^2 in Rydberg units is

$$(\nabla^2 + \kappa^2)\psi(\mathbf{x}) = V(\mathbf{x})\psi(\mathbf{x}). \tag{1}$$

A Helmholtz Green function with given boundary values satisfies

$$(\nabla^2 + \kappa^2)G_0(\mathbf{x}, \mathbf{x}') = \delta(\mathbf{x}, \mathbf{x}'). \tag{2}$$

A solution of Eq.1 throughout the range $\Re^3$ of coordinates x satisfies the Lippmann-Schwinger integral equation

$$\psi(\mathbf{x}) = \chi(\mathbf{x}) + \int_{\Re^3} G_0(\mathbf{x}, \mathbf{x}')V(\mathbf{x}')\psi(\mathbf{x}')d^3x', \tag{3}$$

where χ is a solution of the homogeneous Helmholtz equation. In a simplified operator notation, Eq.3 is expressed as

$$\psi = \chi + G_0 V \psi. \tag{4}$$

Eigenfunctions, determined by requiring the function χ to vanish, satisfy the same boundary conditions as G_0.

It is assumed that $\Re^3$ is subdivided into space-filling cells. A typical cell τ_μ has closed surface σ_μ. For a finite system, an external cell can be considered to be centered at an appropriately defined 'point' at infinity. Solutions of the homogeneous Helmholtz equation about the origin of cell μ define regular and irregular generalized solid harmonic functions, denoted here by J_L^μ and N_L^μ, respectively. Each solid harmonic function is the product of a radial factor and a spherical harmonic whose angular quantum numbers (ℓm) will be denoted here by the combined index L. For positive energies, the radial factors of J_L, N_L, respectively, are proportional to regular and irregular spherical Bessel functions. Standard normalization is assumed such that Wronskian surface integrals of solid harmonic functions define a Kronecker delta relation. Specifically, functions J,N satisfy the condition,

$$(N_L|W_\sigma|J_{L'}) = \delta_{L,L'}, \tag{5}$$

where Wronskian integrals over surface σ of cell τ are defined by

$$(\phi_1|W_\sigma|\phi_2) = \int_\sigma [\phi_1^* \nabla_n \phi_2 - (\nabla_n \phi_1)^* \phi_2] d\sigma. \tag{6}$$

Here ∇_n denotes an outward normal gradient. These integrals vanish for any two J-functions or any two N-functions.

The Helmholtz Green function can be expanded in local relative coordinates about the origin of a specified cell, in the form

$$G_0(\mathbf{r}, \mathbf{r}') = -\sum_L J_L(\mathbf{r}) N_L^*(\mathbf{r}'), \quad r < r'. \tag{7}$$

If coordinate $\mathbf{x} = \mathbf{X}_\mu + \mathbf{r}$ lies in cell τ_μ and coordinate $\mathbf{x}' = \mathbf{X}_\nu + \mathbf{r}'$ lies in a different cell τ_ν, it is convenient to represent $G_0(\mathbf{x}, \mathbf{x}')$ in the local coordinates $(\mathbf{r}, \mathbf{r}')$. $N_L^\mu(\mathbf{x}' - \mathbf{X}_\mu)$ is a regular solution of the Helmholtz equation in a sphere of radius $|\mathbf{X}_\mu - \mathbf{X}_\nu|$ about the origin of a displaced cell ν. This implies that

$$N_L^\mu(\mathbf{r}' + \mathbf{X}_\nu - \mathbf{X}_\mu) = -\sum_{L'} J_{L'}^\nu(\mathbf{r}') g_{L'',L'}^{\nu\mu} \tag{8}$$

for $r' < |\mathbf{X}_\mu - \mathbf{X}_\nu|$, if $\nu \neq \mu$. Substituting the Hermitian conjugate of Eq.8 into Eq.7,

$$G_0(\mathbf{X}_\mu + \mathbf{r}, \mathbf{X}_\nu + \mathbf{r}') = \sum_L \sum_{L'} J_L^\mu(\mathbf{r}) g_{L,L'}^{\mu\nu} J_{L'}^{\nu*}(\mathbf{r}'). \tag{9}$$

If the standard solid harmonic functions defined above are used in Eq.7, G_0 is a principal-value Green function, and the matrix $g_{L,L'}^{\mu\nu}$ is Hermitian. The elements of this matrix are the structure constants at the given energy. This two-center expansion converges in general for muffin-tin geometry, when the two coordinate values lie within nonoverlapping spheres. It is assumed here that cell coordinate origins are chosen to satisfy this geometric condition for some neighborhood of each cell origin. G_0 is a regular solution of the Helmholtz equation in any cell that does not contain both coordinate points.

2.2. ANGULAR MOMENTUM REPRESENTATION

For an eigenfunction determined by the same boundary conditions as the Green function G_0, the global function χ is set to zero, and Eq.3 takes the form

$$\psi = \int_{\Re^3} G_0 V \psi. \tag{10}$$

This homogeneous integral equation has regular solutions only at energy eigenvalues of the Schrödinger equation. To analyse a particular global eigenfunction ψ_α within a particular local cell τ, it is convenient to define an auxiliary function (Nesbet, 1990a)

$$\chi_\alpha = \psi_\alpha - \int_\tau G_0 V \psi_\alpha = \int_{\Re^3 - \tau} G_0 V \psi_\alpha. \tag{11}$$

by subdividing the integral in Eq.10. The function χ_α is a regular solution of the Helmholtz equation within cell τ. Hence χ_α can be represented in τ as a sum of regular solid harmonic functions J_L, excluding irregular solid harmonic functions N_L. This series representation of χ_α in cell τ is

$$\chi_\alpha(\mathbf{r}) = \sum_L J_L(\mathbf{r}) C_{L\alpha}. \tag{12}$$

As will be discussed below, such a sum can be assumed to converge within a muffin-tin sphere enclosed by the cell surface σ, but not outside this sphere unless the sum is finite. The function itself is defined in Eq.11 by an integral representation valid throughout cell τ. This provides a continuation of the series expansion, analogous to analytic continuation in complex variable theory, which extends a Taylor series beyond its range of convergence.

Assuming only that the series in Eq.12 converges in a neighborhood of the local cell origin, the coefficients in this series are determined in such a neighborhood. Since ψ_α is regular throughout $\Re^3$, it can be expanded in a series of regular generalized solid harmonic functions in an infinitesimal neighborhood of each cell origin. Denoting the coefficients in this expansion for cell τ_λ by $c_{L\alpha}^\lambda$, the coefficients in Eq.12 are determined by Eq.7 and the second member of Eq.11 to be

$$C_{L\alpha}^\lambda = c_{L\alpha}^\lambda + \int_{\tau_\lambda} N_L^* V \psi_\alpha. \tag{13}$$

The third member of Eq.11 provides an alternative representation of the regular function χ_α within cell τ_λ. Assuming cellular decomposition of

$\Re^3$, this expansion is

$$\chi_\alpha(\mathbf{r}) = \sum_{\mu \neq \lambda} \int_{\tau_\mu} G_0 V \psi_\alpha. \qquad (14)$$

If Eq.9 is substituted into Eq.14, it follows that

$$\chi_\alpha(\mathbf{r}) = -\sum_L J_L(\mathbf{r}) \sum_{L'} \sum_{\mu \neq \lambda} g_{L,L'}^{\lambda\mu} S_{L'\alpha}^{\mu}, \qquad (15)$$

where

$$S_{L\alpha}^{\mu} = -\int_{\tau_\mu} J_L^* V \psi_\alpha. \qquad (16)$$

In a neighborhood of the origin of cell τ_λ, Eqs.12 and 15 imply the consistency condition, for all L,

$$C_{L\alpha}^{\lambda} + \sum_{L'} \sum_{\mu \neq \lambda} g_{L,L'}^{\lambda\mu} S_{L'\alpha}^{\mu} = 0. \qquad (17)$$

Because the integrals in Eqs.13 and 16 extend only over a local cell, they can be expressed in terms of a local expansion of the wave function $\psi_\alpha = \sum_i \phi_i \gamma_{i\alpha}$ in a set of basis functions $\phi_i(\mathbf{r})$ that are exact solutions of the Schrödinger equation in the cell. Completeness of this basis set at a specified energy is defined by the ability to represent any regular solution of the Schrödinger equation within the cell and on the cell boundary. Basis functions $\phi_L(\mathbf{r})$ with these properties, each matched to a regular generalized solid harmonic $J_L(\mathbf{r})$ at the cell origin, can be computed by integrating the multichannel radial Schrödinger equation outwards. Nonspherical terms in the local potential function mix spherical harmonics into each ϕ_L as the radial integration proceeds. No boundary conditions are imposed at the cell boundary. Since these basis functions are used only within a local cell, their extension outside the cell is irrelevant (Nesbet, 1990a).

In this basis, Eq.17 becomes a set of linear equations, for all values of λ and L,

$$\sum_{\mu,L',L''} (\delta_{L,L''}^{\lambda\mu} C_{L'',L'}^{\mu} + g_{L,L''}^{\lambda\mu} S_{L'',L'}^{\mu}) \gamma_{L'\alpha}^{\mu} = 0. \qquad (18)$$

The matrix elements here are defined for each cell τ by

$$C_{L,L'} = \delta_{L,L'} + \int_\tau N_L^* V \phi_{L'}, \quad S_{L,L'} = -\int_\tau J_L^* V \phi_{L'}. \qquad (19)$$

If t is defined as the matrix $-SC^{-1}$, Eq.18 reduces to the generalized MST equation in the form $(I-gt)C\gamma = 0$, where all indices have been suppressed. Energy eigenvalues ϵ_α are determined by zeroes of the secular determinant of this equation. In a regular periodic solid, the sum over cells in Eq.17 for a Bloch wave ψ_k takes the form $\sum_{L'} g_{L,L'}(k) S_{L'k}$. The structure-constant matrix $g(k)$ is a Bloch sum over the second cell index of the structure-constant matrix $g^{\lambda\mu}$ and $S_{L'k}$ is the same for all translationally equivalent cells. For energy-band calculations on a regular periodic solid, Eq.18 is used in the form of the generalized KKR equation,

$$\{t^{-1}(\epsilon) - g(\epsilon; k)\} S(\epsilon) \gamma(\epsilon; k) = 0. \tag{20}$$

The secular matrix of this linear system is of the form $M(\epsilon; k) = t^{-1} - g$.

2.3. SURFACE MATCHING THEOREM

If a function ϕ and its normal gradient $\nabla_n \phi$ are specified on the closed surface σ of a cell τ, a unique solution of the Helmholtz equation is defined by numerical integration away from this surface either inside or outside the enclosed cell. Within the cell, unique regular solutions are determined separately by the function values on σ (classical Dirichlet problem) or by the normal gradients (classical Neumann problem). These two solutions are not in general compatible, implying that the general solution must include both regular solid harmonic functions J and irregular solid harmonic functions N within the enclosed cell. At positive energies, the external functions J and N are all bounded, so no conclusion can be drawn about divergence outside the cell.

If the given function can be fitted to a finite sum including both regular functions J and irregular functions N, then its extension by solution of the Helmholtz equation both inward and outward from σ is valid except at the coordinate origin where the irregular functions diverge. The coefficients in such a finite expansion are uniquely determined by Wronskian surface integrals, giving the expansion on σ

$$\phi(\mathbf{r}_\sigma) = \sum_L [J_L(N_L|W_\sigma|\phi) - N_L(J_L|W_\sigma|\phi)]. \tag{21}$$

The set of Wronskian integrals for any bounded function ϕ with bounded normal gradient on a closed surface σ define an infinite series expansion of this form. It has been proven(Nesbet, 1992b) that this is a valid representation of any such function on σ, in the sense that if the combined series converges, its value and normal gradient on σ equal the value and normal gradient of the given function. If the series does not converge, the given function is the sum of two functions that are each uniquely defined as solutions of the Helmholtz equation. The indicated Wronskian integrals define a surface δ-function.

Any bounded function ϕ on closed surface σ whose normal gradient is bounded defines two discontinuous functions that are regular solutions of the Helmholtz equation in the interior τ and exterior $\Re^3 - \tau$, respectively, of the cell enclosed by σ. Expressed in terms of the Green function G_0 of the Helmholtz equation, these functions are

$$\chi(\mathbf{r}) = -(G_0(\mathbf{r}, \mathbf{r}_\sigma)|W_\sigma|\phi), \qquad (22)$$

for $\mathbf{r} \in \tau$, and

$$\eta(\mathbf{r}) = (G_0(\mathbf{r}, \mathbf{r}_\sigma)|W_\sigma|\phi), \qquad (23)$$

for $\mathbf{r} \notin \tau$. Because the singularity of G_0 is weaker than the area element $d\sigma$, the two functions defined here approach finite limits on the bounding surface, which closes their disjoint regions of definition. The Wronskian integral and its normal gradient are discontinuous at the surface σ. It can be proved(Nesbet, 1992b) that the function ϕ and its normal gradient on σ are equal to the values of these discontinuities. The expansion given by Eq.21 represents this limit in the form

$$lim_{\mathbf{r} \to \mathbf{r}_\sigma -}\chi + lim_{\mathbf{r} \to \mathbf{r}_\sigma +}\eta \equiv \phi(\mathbf{r}_\sigma), \qquad (24)$$

where $\equiv$ is used to indicate that both function value and normal gradient approach their respective indicated limits.

The proof of this theorem uses the fact that any function Φ and its normal gradient must both vanish on surface σ if the surface Wronskian integrals defined above vanish for all functions N and J. This lemma is proved by *reductio ad absurdum*. By Green's theorem, Wronskian surface integrals between solutions of the Helmholtz equation are independent of the surface of evaluation so long as no singularities occur between a set of nested surfaces. Any nonvanishing function with zero Wronskian integrals on surface σ could be extended by integration to both the enclosed sphere (radius r_0) and the enclosing sphere (radius r_1) as a solution of the Helmholtz equation. Because the Wronskian integrals are the same for either of these spheres and for the intermediate surface σ, the extended function and normal gradient must have all zero coefficients in a spherical harmonic expansion over either of these spheres. Hence the extended function vanishes on both spheres, using the completeness of the spherical harmonics. The classical uniqueness theorem for the Helmholtz equation requires this function to vanish identically between these spheres, in contradiction to the hypothesis of nonzero values of Φ or of its normal gradient on σ.

Expansion of the Green function G_0 gives series representations valid, respectively, within the enclosed sphere and outside the enclosing sphere,

$$\chi(\mathbf{r}) = \sum_L J_L(N_L|W_\sigma|\phi) \qquad (25)$$

for $r < r_0$, and

$$\eta(\mathbf{r}) = -\sum_L N_L (J_L | W_\sigma | \phi) \qquad (26)$$

for $r > r_1$. The indicated series must converge because they are bounded integrals over the convergent expansion of G_0. By Green's theorem, Wronskian integrals between functions J or N and χ, evaluated on any sphere with $r < r_0$, have the same value on the limiting surface σ. Similarly, integrals of this kind with η, evaluated on any sphere with $r > r_1$, have the same value on σ. It follows from the values of the expansion coefficients in Eqs.25 and 26 that the difference function $\Phi = \phi - \chi - \eta$ has vanishing Wronskian surface integrals on surface σ with all functions J and N. This satisfies the conditions of the lemma proved above, and implies the theorem stated in Eq.24. Eq.21 follows as a corollary if the combined series terminates or converges. In any case the functions χ and η and their sum on surface σ are uniquely determined by the Wronskian surface integrals that appear as coefficients in Eq.21. The combined series may converge even though the two separate series in N_L and J_L both diverge, because the singularity of the Green function cancels exactly as the two functions χ and η approach a common boundary point $\mathbf{r}_\sigma$ in Eq.24.

As an additional corollary, it can be proven that for any two basis solutions of the Schrödinger equation at specified energy within cell τ,

$$\sum_L (\phi_i | W_\sigma | J_L)(N_L | W_\sigma | \phi_j) - \sum_L (\phi_i | W_\sigma | N_L)(J_L | W_\sigma | \phi_j) = 0. \qquad (27)$$

This equation follows immediately from the expansion of the two functions ϕ_i and ϕ_j on surface σ, using Eq.21. Because both functions are regular solutions of the same Schrödinger equation, the Wronskian surface integral $(\phi_i | W_\sigma | \phi_j)$ must vanish. If both functions are expanded as in Eq.21, the resulting expression for this integral reduces to Eq.27. This argument holds even if the combined series diverges for either function ϕ_i or ϕ_j, because the argument depends only on values of Wronskian surface integrals.

2.4. SURFACE INTEGRAL FORMALISM

A function Φ and its normal gradient on the surface σ of a closed local cell τ can be matched to a regular solution of the Schrödinger equation in this cell at a given energy ϵ if and only if it satisfies the linear integral equation,

$$\Phi(\mathbf{r}_\sigma) = \int_\sigma \Re(\epsilon; \mathbf{r}_\sigma, \mathbf{r}'_\sigma) \nabla_n \Phi(\mathbf{r}'_\sigma) d\mathbf{r}'_\sigma. \qquad (28)$$

$\Re$ is an Hermitian linear integral operator, defined over σ, that characterizes boundary matching on this surface(Nesbet, 1984). Any linearly independent set of basis functions that are regular in cell τ defines a variational construction of $\Re$ that becomes exact in the limit of completeness. If basis functions ϕ_i are used that are regular solutions of the Schrödinger equation at the given energy ϵ, Eq.28 is equivalent to the Wronskian integral condition(Nesbet, 1986a; Nesbet, 1990a)

$$(\phi_i|W_\sigma|\Phi) = 0, \qquad (29)$$

for all i. The t-matrix of multiple scattering theory is determined so that each function of the form $\Phi = J - Nt$ on surface σ matches a regular solution of the Schrödinger equation within cell τ. Eq.28 must hold for each such function Φ_L and reduces to a set of equations derived from Eq.29 in any particular basis of functions ϕ_i that are themselves regular solutions in τ,

$$(\phi_i|W|J_L) - \sum_{L'}(\phi_i|W|N_{L'})t_{L',L} = 0, \qquad (30).$$

for all i, L. Eq.27 implies that the t-matrix defined by Eq.30 is Hermitian. This matrix is related to the reactance matrix (K-matrix) of standard scattering theory, (Nesbet, 1980) for example. Since only the Kronecker-delta Wronskian relation between functions N and J is assumed here, specific asymptotic forms must be chosen in order to give the K-matrix directly. If for positive energies k^2 these forms are defined using orthonormal spherical harmonics Y_L by $J_L = r^{-1}j(r)Y_L$ where $j(r) \sim k^{-1/2}\sin(kr - \ell\pi/2)$ and by $N_L = r^{-1}n(r)Y_L$ where $n(r) \sim k^{-1/2}\cos(kr - \ell\pi/2)$, then $K = -t$. The asymptotic radial factor in N_L must decay exponentially for negative energies. In the muffin-tin model, with spherical local potentials, the K-matrix reduces to the diagonal form $\tan\eta$, where η is a partial-wave phase shift.

It is postulated here that a set of basis functions ϕ_L can be constructed that are local solutions of the Schrödinger equation in each atomic cell τ. From the surface expansion theorem proved in Section 2.3, any bounded function ϕ with bounded normal gradient on the surface σ of cell τ determines a unique expansion of the form

$$\phi = \sum_L [J_L(N_L|W_\sigma|\phi) - N_L(J_L|W_\sigma|\phi)], \qquad (31)$$

for points on surface σ. A standard notation is used for the coefficients in this expansion:

$$C_{L,L'} = (N_L|W_\sigma|\phi_{L'}), \quad S_{L,L'} = -(J_L|W_\sigma|\phi_{L'}). \qquad (32)$$

Using these coefficient matrices, Eq.31 implies

$$\phi_L = \sum_{L'}(J_{L'}C_{L',L} + N_{L'}S_{L',L}), \qquad (33)$$

for points on the local cell boundary σ. The indicated combined series may in general be semiconvergent or divergent. In practice, it is truncated as a finite series. If not truncated, the combined series represents a function uniquely defined on σ by the indicated Wronskian surface integrals.

A global solution of the Schrödinger equation for a multicellular system must satisfy Eq.28 on each cell boundary. In the muffin-tin model, a global matching function Φ is defined as a solution of the Helmholtz equation (constant potential) in the interstitial volume between the local cells. Since there are no singularitites of the potential function in the interstitial volume, the matching function can be expanded in the form $\Phi = \sum_{\mu,L} N_L^\mu \beta_L^\mu$, where the coefficients are determined by Eq.29 on all cell boundaries. The standard multiple-scattering theory (MST) equations follow on expanding the irregular functions N for external cells in the basis of regular functions J on the boundary of a particular cell, using the structure constants defined by Eq.8.

When muffin-tin cells are expanded to fill space, the interstitial volume shrinks to a limiting separating surface in the form of a honeycomb lattice(Nesbet, 1986a; Nesbet, 1990a). It is convenient to retain the global matching function Φ in this limit, and to replace Eq.29 by two sets of equations derived from the conditions that a global eigenfunction ψ, expanded in a local basis set ϕ_L^μ for each cell τ_μ, should match Φ in both value and normal gradient on each surface σ_μ. This requires that Wronskian integrals of the difference function $\psi - \Phi$ with both regular and irregular solid-harmonic functions should vanish on each cell surface. These conditions provide the equations needed to determine both the coefficients β that define the global matching function Φ and the coefficients γ in the local expansion of ψ in each cell(Nesbet, 1988; Nesbet, 1990a). The first matching equation is

$$(J_L^\mu | W_\sigma^\mu | \psi - \Phi) = 0, \tag{34}$$

for all (μ, L). Assuming that the enclosing sphere of each cell excludes all other cell centers, all contributions to the matching function Φ on σ_μ coming from other cells can be expanded in regular solid harmonics and have vanishing Wronskian integrals with J_L^μ. Hence Eq.34 reduces to an explicit formula $\beta = S\gamma$ for each cell τ_μ. The second matching equation is

$$(N_L^\mu | W_\sigma^\mu | \psi - \Phi) = 0, \tag{35}$$

for all (μ, L). Only the part of Φ due to other cells contributes to this equation for cell μ, and can be evaluated in terms of structure constants given by Eq.8. Substituting $\beta = S\gamma$ for the coefficients in Φ and using Eq.5, Eq.35 reduces to a set of equations identical in form to Eqs.18, above, except that here the coefficient matrices are given by Eqs.32. The MST t-matrix given by these equations is

$$t = (J|W_\sigma|\phi)(N|W_\sigma|\phi)^{-1}, \tag{36}$$

suppressing all indices. Given that the t-matrix is Hermitian, Eq.30 is consistent with this formula.

The matrix elements S and C defined by Eqs.19 in the conventional theory are volume integrals over a local cell τ. Using the fact that the basis functions ϕ_L are solutions of the Schrödinger equation, while the functions J,N are solutions of the Helmholtz equation, these volume integrals can be converted by partial integration to Wronskian integrals over the cell surface σ. In detail,

$$S_{L,L'} = -\int_\tau J_L^*(\nabla^2 + \kappa^2)\phi_{L'} = -(J_L|W_\sigma|\phi_{L'}). \tag{37}$$

Similarly,

$$C_{L,L'} = \delta_{L,L'} + \int_\tau N_L^*(\nabla^2 + \kappa^2)\phi_{L'} = (N_L|W_\sigma|\phi_{L'}). \tag{38}$$

Because N_L is singular at the cell origin, integration by parts in Eq.38 produces an inner surface integral, over an infinitisimal sphere about the origin, that exactly cancels the Kronecker delta if the solid harmonic functions satisfy Eq.5. The derived equations are identical to Eqs.32.

In a regular periodic lattice, the global matching function is a linear combination of functions $\tilde{N}_L(\mathbf{k}; \mathbf{r})$, each constructed as a Bloch sum of irregular generalized solid harmonics N_L centered at cell origins $\mathbf{X}_\mu$ displaced by lattice translations. The set of all such functions for each atomic cell in a translational unit cell has the required properties of linear independence and of defining bounded solutions of the Helmholtz equation outside the atomic cells. For space-filling atomic cells, the functions $\tilde{N}$ are used on the honeycomb lattice of cell interfaces. In coordinates whose origin is the center of a reference cell τ_λ, in a monatomic translational unit cell, the function $\tilde{N}_L$ is

$$\tilde{N}_L(\mathbf{k}; \mathbf{r}) = N_L(\mathbf{r}) + \sum_{\mu \neq \lambda} N_L(\mathbf{r} - \mathbf{X}_\mu)\exp(i\mathbf{k} \cdot \mathbf{X}_\mu). \tag{39}$$

Substitution of Eq.8 into Eq.39 gives the expansion needed on the surface of the reference cell

$$\tilde{N}_L(\mathbf{k}; \mathbf{r}) = N_L(\mathbf{r}) - \sum_{L'} J_{L'}(\mathbf{r})g_{L',L}(\mathbf{k}). \tag{40}$$

An important point in the present development is that the convergence condition for Eq.8 is much less restrictive than the conditions for Eqs.7

or 9. The convergence condition for Eq.8 can easily be satisfied either for Wigner-Seitz cells as usually defined or by judicious subdivision of cells in cases of distorted local geometry, since the only requirement is that no adjacent nucleus should lie within the enclosing sphere of a given local cell.

2.5. ATOMIC CELL ORBITALS

The concept of 'muffin-tin orbital' (MTO)(Andersen, 1971; Andersen and Kasowski, 1971; Kasowski and Andersen, 1972), used in developing efficient computational and scaling methods for local spherical potentials, can be directly generalized to full-potential multiple-scattering theory. Given the functions χ and η defined in Section 2.3, an 'atomic-cell orbital' (ACO), defined as $\phi - \chi$ within an atomic cell, matches the external function η on the cell boundary(Nesbet, 1986a; Nesbet, 1990a; Zhang and Butler, 1992b). From Eq.33, $\chi_L = \Sigma_{L'} J_{L'} C_{L',L}$ and $\eta_L = \Sigma_{L'} N_{L'} S_{L',L}$, where $C_{L,L'}$ and $S_{L,L'}$ are the matrices defined by Eqs.32. It was shown in Section 2.3 that these series expansions of χ and η represent well-defined local solutions of the Helmholtz equation even if the indicated series diverge. These functions should be used rather than the series expansions in an explicit representation of wave functions. With this proviso, the MTO or ACO functions are valid basis functions for the global Schrödinger equation. By construction, they are regular within the cell of definition, smooth at its boundary, and bounded in the exterior region. If 'canonical' basis functions are constructed so that matrix C is a unit matrix, then the corresponding ACO functions are of the form $\phi - J$, clearly well-defined in the local cell τ. These functions match to $-Nt$ on the cell boundary, where t is the MST t-matrix.

The MST secular equation is derived, following Andersen, by considering 'tail-cancellation' in the interior of each atomic cell. This concept is simply that the solid-harmonic parts of the MTO or ACO basis must cancel exactly in any variational linear combination of these basis functions that represents an exact solution of the Schrödinger equation. If the coefficients in the local ACO expansion of a global Schrödinger eigenfunction are denoted by γ_L, the coefficient of J_L in cell τ is $-(C+gS)\gamma$. Setting this coefficient to zero gives the general MST secular equation, as in Eq.18.

3. Variational principles

In full-potential theory, the KR variational principle(Kohn and Rostoker, 1954), from which full-potential MST equations will be derived here, requires an intermediate sum over a complete set of solid-harmonic functions. The characteristic matrices S and C of multiple scattering theory are usually considered to be square matrices, but to satisfy this completeness condition they must be extended as rectangular matrices to large or infinite order in the solid-harmonic index. These intermediate sums are avoided in an alternative variational principle based on the cellular model for a space lattice with a monatomic translational cell(Kohn, 1952), derived by adding surface integrals to the usual Rayleigh-Schrödinger functional. This was generalized to the SM variational principle(Schlosser and Marcus, 1963), formulated in terms of Wronskian integrals over the interfaces of adjacent local cells. SM variational solutions satisfy correct continuity and boundary conditions in the limit of variational completeness. These two variational principles are equivalent in the limit of completeness of the internal sum over solid-harmonic functions required in the KKR formalism(Nesbet, 1992a). A version of the SM variational principle valid for basis functions computed at a specified energy is used by Leite and collaborators in the variational cellular method (VCM)(Ferreira and Leite, 1978). Spurious solutions inherent in the VCM must be excluded by special procedures.

3.1. KOHN-ROSTOKER VARIATIONAL PRINCIPLE

Standard KKR theory for muffin-tin potentials is derived(Kohn and Rostoker, 1954) by varying the functional

$$\Lambda = \int_{\Re^3} \psi^* V(\psi - \int_{\Re^3} G_0 V\psi), \qquad (41)$$

where G_0 is the Green function of the Helmholtz equation. Generalization to potential functions defined throughout $\Re^3$, not restricted to spatial periodicity, follows from the argument given here. If $\Re^3$ is subdivided into nonoverlapping space-filling cells τ_μ, variation of ψ in a particular cell τ produces

$$\delta\Lambda = \int_\tau \delta\psi^* V(\psi - \int_{\Re^3} G_0 V\psi) + hc, \qquad (42)$$

where 'hc' denotes the Hermitian conjugate. This can be expressed in the form

$$\delta\Lambda = \int_\tau \delta\psi^* V(\chi^{in} - \chi^{out}) + hc, \qquad (43)$$

defining auxiliary local regular solutions of the Helmholtz equation. In a reference cell τ_μ, for a particular global wave function ψ_α,

$$\chi_\alpha^{in} = \psi_\alpha - \int_{\tau_\mu} G_0 V\psi_\alpha = \sum_L J_L C_{L\alpha}^\mu, \qquad (44)$$

$$\chi_\alpha^{out} = \sum_{\nu\neq\mu}\int_{\tau_\nu} G_0 V\psi_\alpha = -\sum_{\nu\neq\mu}\sum_{L,L'} J_L g_{L,L'}^{\mu\nu} S_{L'\alpha}^\nu. \qquad (45)$$

The matrix elements $g_{L,L'}^{\mu\nu}$ are structure constants computed for the Helmholtz Green function G_0. The functions J_L, N_L here are, respectively, regular and irregular solid-harmonic solutions of the Helmholtz equation in the reference cell. Index L denotes angular quantum numbers. The coefficients C and S are defined by Eqs.13 and 16. The functions χ_α^{in} and χ_α^{out} are uniquely defined in the closed interior of cell τ_μ. The indicated series are assumed to converge in some neighborhood of the cell origin, and may be continued out to the cell boundary σ_μ by integration of the Helmholtz equation.

When ψ is expanded as a linear combination of fixed local basis functions ϕ_L^μ in cell τ_μ, with coefficients γ_L^μ, allowed variations are of the form $\delta\gamma^\mu$, independently for each cell. Continuity of the trial function between cells is not required. In a basis of local solutions of the Schrödinger equation, the expansions given in Eqs.44 and 45 convert Eq.43 to an expression involving only the standard coefficient matrices C and S of full-potential theory, defined by Eqs.19. On integration by parts these matrices reduce to the Wronskian surface integrals given in Eqs.32. In this basis, for variations in cell τ_μ,

$$-\delta\Lambda = \delta\gamma^{\mu\dagger} S^{\mu\dagger} \sum_\nu (\delta^{\mu\nu} C^\nu + g^{\mu\nu} S^\nu)\gamma^\nu + hc, \qquad (46)$$

in a notation suppressing all indices L and intermediate summations.

The implied variational principle, valid if sums over solid harmonic functions with indices L are carried to completeness, is that $\delta\Lambda$ vanishes for all cells τ_μ if and only if the Schrödinger equation is satisfied in all cells, with continuity of ψ and of its normal gradient across all cell boundaries. Eq.46 can vanish only if the determinant of the associated matrix vanishes, determining a null-vector γ up to normalization. This condition implies the standard MST secular equation if S and C are square matrices, and if S is not singular. However, there is no intrinsic reason to restrict the intermediate sums over generalized solid harmonics J_L, N_L to the same number of terms as the number of basis functions ϕ_L. Hence the matrices S and C can be considered to be rectangular. The variational trial function is continuous across cell boundaries in general only if the intermediate sum over solid harmonic functions is complete, requiring the long dimension of these matrices to be infinite for nonspherical local geometry.

3.2. SCHLOSSER-MARCUS VARIATIONAL PRINCIPLE

The Schlosser-Marcus variational functional Z is defined (Schlosser and Marcus, 1963; Nesbet, 1988) such that

$$\delta Z = \delta\gamma^{\mu\dagger} \sum_\nu Z^{\mu\nu} \gamma^\nu + hc. \qquad (47)$$

Site-diagonal elements $Z^{\mu\mu}$ are zero in the local basis at a given energy. Off-diagonal elements of Z are Wronskian integrals over the interfaces $\sigma_{\mu\nu}$ of adjacent cells. In a local basis within each cell, the matrix elements between a reference cell τ_μ and an adjacent cell τ_ν are

$$Z^{\mu\nu} = -1/2(\phi^\mu|W_{\mu\nu}|\phi^\nu), \quad Z^{\nu\mu} = Z^{\mu\nu\dagger}, \tag{48}$$

where L-indices are suppressed and normal gradients are outwards from τ_μ.

As used in the variational cellular method (VCM)(Ferreira and Leite, 1978), the Schlosser-Marcus variational principle leads to both 'true' and 'false' solutions of the global Schrödinger equation. False solutions are characterized by maximum mismatch across cell boundaries, and must be eliminated in the VCM by a special criterion. Only 'true' solutions will be considered here, obtained by starting from a construction in which trial functions on two sides of an internal matching surface are matched separately to a common global function Φ that is everywhere continuous, with continuous normal gradient across any dividing surface.

The consistency condition of multiple scattering theory can be derived (Nesbet, 1988) by fitting a linear combination of basis functions within each cell to a global matching function on the boundary of that cell. This construction will be used here to demonstrate the equivalence between the Schlosser-Marcus variational principle and the Kohn-Rostoker variational principle. For 'true' VCM solutions, the sum $\sum \phi^\mu \gamma_\mu$ is matched to the global function Φ on each interface $\sigma_{\mu\nu}$. In a local basis at specified energy, Eq.47 is equivalent within variational error to

$$\delta Z = -1/2\, \delta\gamma^{\mu\dagger}(\phi^\mu|W_\sigma|\Phi) + hc. \tag{49}$$

If the two expansions of Φ, inside and outside the surface σ, were identical, the Wronskian integrals in Eq.49 would vanish identically, because these integrals vanish between any two regular solutions of the Schrödinger equation in cell τ_μ. Hence Eq.47 expresses a true measure of the failure of continuity on surface σ_μ.

In ACO/KKR theory (Nesbet, 1986a; Nesbet, 1988), the global matching function Φ is

$$\Phi = \sum_{\mu L} N^\mu_L \beta_{\mu L} = \sum_L (N^\mu_L \beta^\mu_L - J^\mu_L \sum_{\nu \neq \mu} \sum_{L'} g^{\mu\nu}_{L,L'} \beta^\nu_{L'}), \tag{50}$$

when expanded on surface σ_μ using Eqs.31 and 8. Standard Wronskian normalization of functions J,N is assumed. For any particular cell τ, Φ and its normal gradient are to be matched on surface σ to the local trial function, $\sum \phi_L \gamma_L$. Comparing Eqs.50 and 33, the coefficients of functions N_L in Φ are determined by

$$\beta_L = -(J_L|W_\sigma|\Phi) = -\sum_{L'}(J_L|W_\sigma|\phi_{L'})\gamma_{L'} = \sum_{L'} S_{L,L'}\gamma_{L'}. \tag{51}$$

Analysis of Eq.49 makes use of Eq.27, valid for equal-energy regular solutions of the Schrödinger equation. This is equivalent to $S^\dagger C - C^\dagger S = 0$, where matrices S and C are defined by Eqs.32. Substituting these results into Eq.49, it reduces to the form

$$2\delta Z = \delta\gamma^{\mu\dagger}(\phi^\mu|W_\sigma|J^\mu)[(N^\mu|W_\sigma|\phi^\mu)\gamma^\mu - \sum_{\nu\neq\mu} g^{\mu\nu}(J^\nu|W_{\sigma_\nu}|\phi^\nu)\gamma^\nu] + hc. \quad (52)$$

This equation is identical to Eq.46 if the intermediate sums in both equations are complete. Hence for 'true' solutions of the VCM equations, the two variational principles are equivalent. Because the Schlosser-Marcus intercell matrix element is evaluated in closed form, as a sum of Wronskian surface integrals, this equivalence is strictly valid only if the the intermediate sums over solid harmonic functions in Eqs.46 and 52 are complete.

Because the Schlosser-Marcus variational principle is formulated without using structure constants, it can be used to derive closed forms of the internal sums over solid-harmonic indices L that are characteristic of KKR theory. An important conclusion from this analysis is that computation of the Schlosser-Marcus intercell terms by surface integration provides a closure formula for the internal sums over solid harmonic functions in Eq.46. In the ACO formalism, by construction, the term in parentheses in Eq.46 is set to zero up to the value of L represented in the basis set. Terms beyond this are ignored in the usual KKR methodology, which treats the matrices in Eq.46 as square matrices. The off-diagonal Schlosser-Marcus term replaces these sums over solid harmonics by a closed-form quadrature over the surface of each atomic cell. Hence, in the linearized ACO method(LACO), inclusion of the Schlosser-Marcus intercell matrix elements, evaluated over each cell interface with basis functions from the independent sets for the two adjacent cells, is equivalent to completing the intermediate sums over solid harmonic indices in the full-potential KKR formalism.

Eqs.47 are deduced from the original SM formalism (Schlosser and Marcus, 1963), which involves matrix elements $(\psi^{in}+\psi^{out}|W_\sigma|\psi^{in}-\psi^{out})$. The secular matrix used in the variational cellular method (VCM)(Ferreira and Leite, 1978) is obtained by eliminating cell-diagonal terms, which vanish for basis functions all evaluated at the same energy, when integrated over a closed surface. Retaining these terms, the SM/VCM variational equations, for all cells τ_μ, are

$$\sum_{\nu\neq\mu}(\phi_L^\mu|W_{\mu\nu}|\sum_{L'}\phi_{L'}^\mu\gamma_{L'}^\mu - \sum_{L'}\phi_{L'}^\nu\gamma_{L'}^\nu) = 0. \quad (53)$$

These equations can be made linear in the local basis functions by replacing the functions ϕ_L^μ on the left by either regular or irregular solid harmonic functions, J_L^μ or N_L^μ. The resulting sets of equations are, respectively

$$\sum_{\nu\neq\mu}(J_L^\mu|W_{\mu\nu}|\sum_{L'}\phi_{L'}^\mu\gamma_{L'}^\mu - \sum_{L'}\phi_{L'}^\nu\gamma_{L'}^\nu) = 0, \qquad (54)$$

and

$$\sum_{\nu\neq\mu}(N_L^\mu|W_{\mu\nu}|\sum_{L'}\phi_{L'}^\mu\gamma_{L'}^\mu - \sum_{L'}\phi_{L'}^\nu\gamma_{L'}^\nu) = 0, \qquad (55)$$

for all τ_μ and adjacent cells τ_ν. As in the VCM, Wronskian surface integrals are required for adjacent cells only. These equations correspond to restricting variational increments of basis functions to either J or N functions on each cell surface. They will be designated, respectively, as the JVCM or NVCM equations. The former choice is appropriate to MST methods for constructing the Schrödinger Green function and for solving the global Poisson equation, to be discussed below. The latter choice defines the Green function cellular method (GFCM)(Butler et al, 1992; Zhang and Butler, 1992a; Zhang et al, 1992).

The standard Kronecker-delta relations imply that the JVCM equations eliminate only the N-component of the functional discontinuity on each cell surface σ, while the NVCM or GFCM equations eliminate only the J-component. However, the corresponding boundary components η and χ, respectively, of a local wave function are related at specified energy by the t-matrix, which is itself determined by the Schrödinger equation in the local cell. Hence equations that explicitly determine only one of these components suffice to determine the full wave function, and either the JVCM or NVCM equations determine a global wave function from sets of local solutions at a given energy. Considered in terms of the number of conditions associated with each interface between cells, the number of conditions at each interface is effectively doubled, because Wronskian surface integrals are required to vanish for two independent sets of solid-harmonic functions defined separately for each of the two cells that share an interface.

4. Wave Functions at Cell Boundaries

Multiple scattering theory is based fundamentally on matching solutions of the Schrödinger equation, computed by exact numerical integration within a local cell τ, to an external trial wave function represented on the cell surface σ by a global matching function Φ. This function must be chosen to be compatible with external boundary conditions and must match the internal wave functions of all other cells. The required matching conditions are simplified in the muffin-tin model, in which the individual cells are nonoverlapping spheres, because the radial factors of each local spherical harmonic component of the wave function ψ can be set equal in value and derivative at the boundary radius to the radial factor of the corresponding component of Φ. These matching conditions are expressed by various versions of the KKR consistency equations.

Extending multiple scattering theory to space-filling cells requires expanding an enclosed muffin-tin sphere S_0, with radius r_0, out to the cell surface σ, itself enclosed by an outer sphere S_1, with radius r_1. In order to avoid irrelevant mathematical complexities, atomic or empty cells will be assumed here to be convex, except for a possible exterior cell bounding a finite cluster of atoms. The enclosed and enclosing spheres are assumed to be centered at the atomic nucleus in each cell, defining a coordinate origin such that singularities of the potential function $V(\mathbf{r})$ occur only at cell origins. The near-field region is defined by the spherical shell with $r_0 \leq r \leq r_1$.

Both the potential function $V(\mathbf{r})$ and the wave function $\psi(\mathbf{r})$ are defined independently of the arbitrary subdivision of space into local cells. The local basis functions ϕ_L are defined as solutions of the Schrödinger equation entirely by their limiting behavior at the local cell origin. Outward integration of the Lippmann-Schwinger equations for different nested sequences of convex surfaces must produce the same unique function, since this is simply a change of the coordinate system in which the Schrödinger equation is integrated. This uniqueness theorem is proved (Nesbet, 1990a) by showing that the difference between solutions of the Lippmann-Schwinger equations for two different cell geometries must vanish at common interior points. The simplest choice of geometry for integration is to use a reference system of nested concentric spheres, corresponding to integration in spherical polar coordinates, using the full potential $V(\mathbf{r})$ out to the enclosing sphere(Brown and Ciftan, 1983). Because the required S and C matrix elements are evaluated on the cell surface σ, integration beyond the enclosing sphere is unnecessary, and would in general produce singularities at the nuclei of neighboring atoms. Hence the basis functions ϕ_L are unique, but are properly defined only within a local cell or its enclosing sphere.

4.1. WAVE FUNCTIONS IN THE NEAR-FIELD REGION

At any given energy, the Lippmann-Schwinger equation, Eq.3 here, defines a regular solution χ of the Helmholtz equation associated with any global regular solution of the Schrödinger equation. Global eigenfunctions are defined by the condition that the associated function χ should vanish everywhere. However, any specific eigenfunction ψ_α defines a local function χ_α, given by Eq.11 in any cell τ. The second equality in Eq.11 has a direct interpretation as the 'tail-cancellation' condition of Andersen (Andersen, 1971), generalized to nonspherical geometry. The function χ_α defined within cell τ must exactly cancel the corresponding function fitted on the surface σ to the external wave function. This latter function is equal to the third member of Eq.11, as can be seen directly by applying Eq.3 to the potential defined by setting $V(\mathbf{r})$ to zero inside cell τ without changing it elsewhere. When both functions are expanded in a neighborhood of the cell origin as sums of regular solid harmonic functions J_L, term-by-term cancellation of the coefficients implies Eq.17. The full-potential MST equations, Eq.18, follow when ψ_α is expanded in the local basis for each cell.

For further analysis of the variational trial function it is convenient to define a set of regular solid harmonic functions χ_L in τ, defined for each local basis function by the first equality in Eq.11, or equivalently, by χ in Eq.22. The functions ϕ_L and χ_L are used on the cell surface σ but not outside τ. Although each ϕ_L is unique in τ regardless of cell shape, this is not true of χ_L. It follows from Eq.25 that χ_L can be expanded in the form $\sum J_{L'} C_{L',L}$, where the coefficient matrix elements C are defined by Eqs.32. In the absence of unusual singularities it is postulated here that this sum converges in some neighborhood of the cell origin.

If χ_L and its normal gradient are represented by a convergent sum of functions J_L on any sphere about the cell origin, it follows from classical theory for the Helmholtz equation that χ_L is uniquely defined by outward integration. Hence the series representation is valid out to the enclosing sphere S_1, in the sense either that the series terminates, or that along each radial ray the series defines a unique function by analytic continuation. Green's theorem implies that Wronskian surface integrals between χ and any of the solid harmonic functions J_L, N_L take the same value on any set of nested surfaces enclosing the cell origin. Hence, using Eq.5, the coefficients C, which are determined by values of ϕ_L and its normal gradient on the surface σ, are equal to invariant Wronskian surface integrals $(N|W|\chi)$ over any nested surface, and uniquely determine χ_L in cell τ and on surface σ. A second function, which can be denoted by η_L, is defined outside τ by Eqs.23 and 26 as a sum of irregular solid harmonics $\sum N_{L'} S_{L',L}$, where the coefficients S are given by Eqs.32. This series is postulated to converge at some sufficiently large radius, and is continued uniquely inwards by integrating the Helmholtz equation. Because the external N-functions are bounded, η_L is uniquely defined outside τ and on σ.

It follows from Eq.5 and the surface matching theorem that ϕ_L and the sum $\chi_L + \eta_L$ have equal values and equal normal gradients on σ, represented on σ by

$$\phi_L = \sum_{L'}(J_{L'}C_{L',L} + N_{L'}S_{L',L}). \qquad (56)$$

This combined series cannot be assumed to converge unless it reduces to a finite sum, but it represents a function that is uniquely determined, together with its normal gradient, by its Wronskian surface integrals with the solid harmonic functions J,N, following the construction of functions χ_L and η_L described above. Eq.24 provides a rigorous statement of the relationship given by Eq.56.

The discussion above shows that the function η_L is the unique continuation outside cell τ of the difference function $\phi_L - \chi_L$, defined within τ. This difference function defines an 'atomic-cell orbital' (ACO) in full-potential theory (Nesbet, 1986a), as the generalization to nonspherical geometry of the 'muffin-tin orbital' (MTO) concept of Andersen (Andersen, 1971; Andersen and Kasowski, 1971). Since the asymptotic form of η is a sum of N-functions, it follows that the correct form of the global matching function Φ is the sum of such functions for all atomic cells, given in Eq.50. The variational trial function can be considered to be a sum of ACO's. By tail-cancellation, the continuation inside τ of the sum of contributions to Φ from all other cells removes the local χ_L component of the variational trial function. By construction, the continuation outside τ of the residual sum of canonical basis functions ϕ_L matches the contribution to Φ from the local cell τ. Thus, if all summations are complete, the global sum of ACO's is a local solution of the Schrödinger equation in each cell, and continues smoothly onto the global matching function on each cell boundary.

In formal scattering theory, the single-site scattering operator t_{op} is defined such that, for $V(\mathbf{r})$ in cell τ,

$$t_{op}\chi_L = V\phi_L. \tag{57}$$

Since χ_L can be expanded in the form $\sum JC$ in the enclosed sphere S_0, Eq.57 implies

$$t_{op}J_L = \sum_{L'} V\phi_{L'}(C^{-1})_{L',L}. \tag{58}$$

Hence matrix elements of t_{op} in the linear space of regular solid harmonic functions in τ are given by

$$(J_L|t_{op}|J_{L'}) = \sum_{L''}(J_L|V|\phi_{L''})(C^{-1})_{L'',L'} = -SC^{-1}, \tag{59}$$

using Eq.37 to evaluate $(J_L|V|\phi_{L''})$. This result is consistent with Eq.36 for the on-site t-matrix.

4.2. CONVERGENCE OF INTERNAL SUMS

It follows from Eq.46, for the Kohn-Rostoker (KR) variational principle, that global eigenfunctions of the Schrödinger equation are determined by the variational equations

$$S^{\mu\dagger}\sum_{\nu}(\delta^{\mu\nu}C^{\nu} + g^{\mu\nu}S^{\nu})\gamma^{\nu} = 0, \tag{60}$$

for all cells τ_μ. All summation indices L are suppressed in the notation here. Eq.60 can vanish only if the determinant of the associated matrix vanishes, determining a null-vector γ up to normalization. If matrix S is not singular, and if C and S are both square matrices, this condition implies the standard MST/KKR consistency condition, as in Eq.18. However, the S and C matrices, defined by Eqs.19 or 32, are indexed by two different linear function spaces, solid harmonic functions on the left and local basis functions on the right. There is no compelling reason for using equal numbers of these two distinct classes of functions. In order to ensure smooth matching of the global trial function across cell boundaries, the internal sums over solid-harmonic indices in the KR variational equations must be carried to completion.

It was found in full-potential calculations (Butler and Nesbet, 1990), for a square lattice with constant potential, that efficient convergence requires using a much larger number of solid-harmonic functions than of local basis functions. The general prescription is that internal sums over a solid-harmonic index should be carried to convergence for each specified set of local basis functions. When this was done, smooth and accurate convergence was demonstrated using relatively small values of the maximum basis index ℓ_{max}. The cutoff index ℓ_{cut} for internal sums over solid harmonic functions was required to be several times larger. The most rapid convergence was achieved by replacing the structure-constant expansion of displaced functions N_L^μ for nearest neighbor cells, as in the third member of Eq.50 here, by direct evaluation on the surface of a local reference cell, as implied by the second member of this equation. This study identifies the source of slow convergence as the attempt to use the expansion of Eq.8 at or near the limit of its domain of convergence.

This convergence behavior implies that accurate calculations should use rectangular S and C matrices, where the long dimension corresponds to the solid-harmonic index. As indicated above, this is consistent with direct use of the Kohn-Rostoker variational principle, but it casts doubt on the practical usefulness of full-potential multiple scattering theory if in analogy to the original muffin-tin KKR theory it must be based on the use of square matrices. The specific issues are presented by Eq.36, which expresses the on-site scattering matrix as $t = -SC^{-1}$, and Eq.20, which uses the matrix $t^{-1} = -CS^{-1}$. Both expressions are undefined in principle for rectangular column-matrices S or C, since the required right-inverse corresponds to the solution of an overdetermined set of linear equations. Expansion of these matrices to full square matrices by artificially extending ℓ_{max} to the value of ℓ_{cut} may require matrix dimensions so large that the whole method loses practical value. Moreover, the model study cited above indicates that optimal convergence is achieved by specifically using a sequence of rectangular matrices whose long dimension is several times ℓ_{max}, rather than a sequence of square matrices as ℓ is increased.

Many applications of multiple scattering theory are based on standard KKR consistency equations of the form

$$(t^\mu)^{-1}\beta_\mu - \sum_{\nu \neq \mu} g^{\mu\nu}\beta_\nu = 0, \qquad (61)$$

where the coefficients β are determined by $\beta = S\gamma$ as in Eq.51. These equations follow from Eq.60 if the site-diagonal matrices S and C are truncated to square matrices, if S has an inverse, and if matrix t is defined by $t^{-1} = -CS^{-1}$. The secular matrix in Eq.61 is $M = t^{-1} - g$, which clearly separates geometrical structural information, in the matrix of structure constants g, from the potential-dependent single-site matrix t^{-1}.

Empty-lattice model studies(Zeller, 1987; Zeller, 1988) for a 2D square lattice show a remarkable asymmetry between results based on truncating matrix t and results obtained by truncating matrix t^{-1}. Zeller found rapid exponential convergence with respect to ℓ_{max} in a full-potential formalism using the secular matrix $(I - gt)$. This behavior contrasts dramatically with the very slow convergence found in earlier studies of the same model, by Faulkner (Faulkner, 1986), who used several versions of full-potential KKR theory, essentially equivalent to using the secular matrix $(t^{-1} - g)$. Zeller showed that this slow convergence could be simulated by truncating the matrix t^{-1} rather than t.

This asymmetry can be understood from the definition of matrices S and C as Wronskian surface integrals, in Eqs.32. As ℓ_{max} is increased, the centrifugal potential in any cell τ must eventually dominate any bounded potential function $V(\mathbf{r})$. In scattering theory, this limit justifies the partial-wave Born approximation, in which basis functions ϕ_L are approximated by their high-ℓ forms, which are just the corresponding regular solid-harmonic functions J_L. It follows from the Kronecker-delta character of Wronskian surface integrals for the solid-harmonic functions, Eq.5, that the integrals S and C, in this limit, reduce to a null matrix and a unit matrix respectively. As the maximum value of ℓ increases for any bounded potential function, matrix S is augmented by null elements and becomes progressively more singular. In contrast, matrix C is augmented by elements of a unit matrix, and its inverse remains well-defined. This argument implies that elements of t^{-1} grow without bounds as maximum ℓ increases, consistent with the convergence difficulties found by Faulkner and verified by Zeller, while the higher order elements of t are well-defined and approach a null matrix.

If the S-matrix is rectangular, the relation $\beta = S\gamma$, given by Eqs.51, implies that the coefficient vectors β and γ cannot be used interchangeably. The elements of vector γ are the variational coefficients of local basis functions ϕ. Elements of vector β are coefficients of irregular solid harmonic functions in the global matching function Φ defined by Eqs.50. For a local cell embedded in an extended medium, the part of the global matching function defined by irregular solid-harmonic functions at all other sites and their coefficients β carries all geometrical and dynamical information describing the embedding medium. Coefficients γ are determined by solution of a local boundary-value problem defined by this medium. A consistent set of local β coefficients must satisfy the site-diagonal relation $\beta = S\gamma$. Equations that exhibit this relationship between the coefficient vectors β and γ can be derived from the Kohn-Rostoker variational principle.

Closed-form expressions can be introduced for two of the internal sums over solid-harmonic indices in Eq.60(Nesbet, 1992c). A site-nondiagonal generalization of the C-matrix is defined for displaced N-functions by

$$\tilde{C}^{\nu\mu}_{L,L'} = (N^{\nu}_L | W_{\sigma_\mu} | \phi^{\mu}_{L'}). \tag{62}$$

for $\nu \neq \mu$. Using Eq.8 for the sum over structure constants, the Hermitian conjugate matrix $\tilde{C}^\dagger$ is a closed form of the matrix product $S^\dagger g$. The site-diagonal matrix $S^\dagger C$ has the closed form

$$\sum_{L''} S^{\mu\dagger}_{L,L''} C^{\mu}_{L'',L'} = -\int_{\tau_\mu} \phi^{\mu*}_L V (\phi^{\mu}_{L'} - \int_{\tau_\mu} G_0 V \phi^{\mu}_{L'}), \tag{63}$$

evaluated by partial integration, defining χ_L from ϕ_L using the first equality in Eq.11, and using Eq.12 to expand χ_L in the enclosed sphere. From this closed form, it is obvious that the site-diagonal matrices $S^\dagger C$ and $C^\dagger S$ are Hermitian and equal if G_0 is chosen to be the Hermitian principal-value Green function.

Substituting these results, Eq.60 implies,

$$\gamma_\mu = -(S^{\mu\dagger} C^\mu)^{-1} \sum_{\nu \neq \mu} (\tilde{C}^\dagger)^{\mu\nu} \beta_\nu, \tag{64}$$

for all cells τ_μ. This equation determines the variational solution at site μ for an atomic cell embedded in a medium defined by the vector of coefficients β for all other cells. As indicated above, a consistent set of β coefficients must satisfy the site-diagonal equation $\beta = S\gamma$.

The Hermitian matrix $S^\dagger C$ is well-defined for rectangular matrices S and C. Since the long dimension is contracted, this matrix is of full rank in general, determined by the number of local basis functions, and will be singular only for isolated energy values. Hence its inverse is defined at arbitrary energies. If Eq.60 is multiplied on the left by the site-diagonal matrix product $S(S^\dagger C)^{-1}$, it takes the form $\Xi\beta = 0$ for a secular matrix defined for all values of μ by

$$\Xi^{\mu\nu} = \delta_{\mu\nu} + S^\mu (S^{\mu\dagger} C^\mu)^{-1} (\tilde{C}^\dagger)^{\mu\nu}, \tag{65}$$

where $\nu \neq \mu$ in the last term. Eq.65 can be compared directly with standard theory. If $\tilde{C}^\dagger$ is replaced by the expanded matrix product $S^\dagger g$, the secular matrix reduces to the form $I - tg$ if the site-diagonal scattering matrix t is defined by the Hermitian form

$$t^\mu = -S^\mu(S^{\mu\dagger}C^\mu)^{-1}S^{\mu\dagger}. \tag{66}$$

This expression reduces to $t = -SC^{-1}$ if matrices S and C are not singular. Because the matrices g and t are Hermitian, matrix Ξ given by Eq.65 is the Hermitian conjugate of the usual MST secular matrix $(I - gt)$. Energy eigenvalues must be the same for both matrices.

5. Green Functions

Although electronic structure calculations are usually concerned with finding the eigenfunctions $\psi_i(x)$ and eigenvalues ϵ_i of the Schrödinger equation, the corresponding Green function is required for many classes of applications. Problems dealing with impurities, transport, disordered systems, photoemission, etc. are most naturally solved using the Green function. In principle, the Green function can be constructed from eigenfunctions, using the spectral representation, $G(x, x') = \sum_i \psi_i(x)(\epsilon - \epsilon_i)^{-1}\psi_i^*(x')$ This representation converges very slowly in general, and an alternative method is needed. Direct methods for computing the Green function are considered here, using boundary-matching methods of multiple scattering theory.

5.1. DEFINITIONS

For energy κ^2 in Rydberg units, the Green function of the Helmholtz equation, with given boundary conditions, satisfies

$$(\nabla^2 + \kappa^2)G_0(x, x') = \delta(x, x'), \tag{67}$$

while the corresponding Schrödinger Green function satisfies

$$(\nabla^2 + \kappa^2 - V(x))G(x, x') = \delta(x, x'). \tag{68}$$

These two Green functions are related by the Lippmann-Schwinger integral equation

$$G(x, x') = G_0(x, x') + \int_{\Re^3} G_0(x, y)V(y)G(y, x')d^3y. \tag{69}$$

The wave function of a perturbed system defined by $\Delta V(x)$ is

$$\tilde{\psi}(x) = \psi(x) + \int_{\Re^3} G(x, x')\Delta V(x')\tilde{\psi}(x')d^3x'. \tag{70}$$

Eq.70 can be applied to a wide class of problems including the electronic structure of vacancies, impurities, and of other localized perturbations of solids and atomic clusters. The methods considered here make it possible to construct $G(x, x')$ for any potential function $V(x)$ defined throughout the coordinate space $\Re^3$.

It is assumed that $\Re^3$ is subdivided into space-filling cells. A typical cell τ_μ has closed surface σ_μ. An external empty cell can be included to fill space for a finite system. Solutions of the homogeneous Helmholtz equation about the origin of cell μ define regular and irregular generalized solid-harmonic functions, denoted here by J_L^μ and N_L^μ, respectively. Each generalized solid harmonic function is the product of a radial factor and a spherical harmonic function, defined here by a combined index L that denotes the quantum numbers (ℓm). Normalization is assumed to be consistent with the Kronecker delta relation, Eq.5.

The Helmholtz Green function, expanded in local coordinates about the origin of a specified cell, is given by Eq.7. This expansion can be verified by using Eq.5. Consider $G_0(r, r')$ for a fixed point r'. For $r < r'$, this function is a regular solution of the Helmholtz equation and must have an expansion of the form

$$G_0(r, r') = \sum_L J_L(r)(N_L|W_0|G_0), \qquad (71)$$

where the Wronskian integral is over the surface of any sphere centered at the cell origin with radius $r_0 < r'$. For $r > r'$, this function is a bounded solution of the Helmholtz equation and must have an expansion of the form

$$G_0(r, r') = -\sum_L N_L(r)(J_L|W_1|G_0), \qquad (72)$$

where the Wronskian integral is over the surface of any sphere centered at the cell origin with radius $r_1 > r'$. If Eq.67 is multiplied by $N_L^*(r)$, and the corresponding null term with the Helmholtz operator applied to N_L^* is subtracted, on integrating over the spherical shell between radii r_0 and r_1, such that $r_0 < r' < r_1$, this integral reduces to Wronskian surface integrals, implying

$$N_L^*(r') = (N_L|W_1|G_0) - (N_L|W_0|G_0). \qquad (73)$$

Since $r_1 > r'$, Eq.72 implies that the first integral on the right-hand side vanishes. The second integral gives the coefficient in Eq.71, which reduces to Eq.7. Similarly, if Eq.67 is multiplied by $J_L^*(r)$ and integrated over the spherical shell between radii r_0 and r_1,

$$J_L^*(r') = (J_L|W_1|G_0) - (J_L|W_0|G_0). \qquad (74)$$

In this case, the second integral on the right-hand side vanishes and the first integral gives the coefficient in Eq.72. This provides a formula valid for

$r > r'$, complementary to Eq.7. Comparison of these two formulas shows that the linear operator G_0 is Hermitian for real energies.

If coordinate $x = X_\mu + r$ lies in cell τ_μ and coordinate $x' = X_\nu + r'$ lies in cell τ_ν, it is convenient to represent $G_0(x, x')$ in the local coordinates (r, r'). When coordinates x and x' lie in different cells, a two-center expansion of the Green function is needed. $N_L^\mu(x' - X_\mu)$ is a regular solution of the Helmholtz equation in a sphere of radius $|X_\mu - X_\nu|$ about the origin of a displaced cell ν. This implies that

$$N_L^\mu(r' + X_\nu - X_\mu) = N_L^\mu(r')\delta_{\mu\nu} - \sum_{L'} J_{L'}^\nu(r')g_{L',L}^{\nu\mu}, \qquad (75)$$

for $r' < |X_\mu - X_\nu|$. The expansion coefficients $g_{L',L}^{\nu\mu}$ are structure constants at the given energy. This matrix is Hermitian for real energies. Using the above expansion, the Helmholtz Green function can be written in the form,

$$G_0(X_\mu + r, X_\nu + r') = \sum_{LL'} J_L^\mu(r)g_{LL'}^{\mu\nu}J_{L'}^{\nu*}(r') - \sum_L J_L^\mu(r)N_L^{\nu*}(r')\delta_{\mu\nu}, \qquad (76)$$

where $r < r'$ within any single cell μ. The two-center expansion converges in general for muffin-tin geometry, when the two coordinate values lie within nonoverlapping spheres. It is assumed here that cell coordinate origins are chosen to satisfy this geometric condition for some neighborhood of each cell origin.

These definitions can be generalized to give an expansion of the Green function $G(x, x')$ in terms of local solutions of the Schrödinger equation in each cell τ. Local regular solutions ϕ_L and local irregular solutions ξ_L can be defined by matching to functions J_L and N_L, respectively, on an infinitesimal sphere about the cell origin, or, alternatively, by requiring similar conditions on a local cell boundary σ. This construction imposes the Wronskian condition

$$(\xi_L|W_\sigma|\phi_{L'}) = \delta_{L,L'}. \qquad (77)$$

These integrals vanish between any two ϕ or two ξ functions. Green's theorem implies that this Kronecker-delta formula is valid for Wronskian integrals over any closed surface that encloses the local cell center and excludes all other singular points of the potential function. If singularities of the potential function occur only at cell centers, these functions can be assumed to be have regular extensions out to the nearest neighbor cell center, but in general no further. For this reason, the primitive function ξ must be modified by the addition of a sum of regular functions ϕ in its cell of origin to define a modified irregular function $\tilde{\xi}$ that is regular everywhere except at the defining cell center. With these definitions(Butler et al, 1992;

Zhang and Butler, 1992a; Zhang et al, 1992), by analogy to the Helmholtz Green function,

$$G(\mathbf{r},\mathbf{r}') = -\sum_L \phi_L(\mathbf{r})\tilde{\xi}_L^*(\mathbf{r}'), \quad r < r', \tag{78}$$

$$\tilde{\xi}_L^\mu(\mathbf{r}' + \mathbf{X}_\nu - \mathbf{X}_\mu) = \xi_L^\mu(\mathbf{r}')\delta_{\mu\nu} - \sum_{L'} \phi_{L'}^\nu(\mathbf{r}')G_{L',L}^{\nu\mu}, \tag{79}$$

for $r' < |\mathbf{X}_\mu - \mathbf{X}_\nu|$, and

$$G(\mathbf{X}_\mu+\mathbf{r},\mathbf{X}_\nu+\mathbf{r}') = \sum_L \sum_{L'} \phi_L^\mu(\mathbf{r})G_{L,L'}^{\mu\nu}\phi_{L'}^{\nu*}(\mathbf{r}') - \sum_L \phi_L^\mu(\mathbf{r})\xi_L^{\nu*}(\mathbf{r}')\delta_{\mu\nu}, \tag{80}$$

valid for $r < r'$ within a single cell, or for coordinates in two nonoverlapping spheres. The coefficients $G_{L,L'}^{\mu\nu}$ define the structural matrix of the Schrödinger Green function. For $\nu = \mu$, $\tilde{\xi}^\mu - \xi^\mu$ is a sum of regular functions ϕ^μ with coefficients $-G_{L,L'}^{\mu\mu}$.

The definition of irregular functions ξ depends on boundary conditions imposed on the Green function. Any variation of the set of functions ξ defined by adding linear combinations of the regular functions ϕ with coefficients that constitute an Hermitian matrix simply moves the corresponding term between the two summations in Eq.80, leaving the net sum invariant, and preserving the Kronecker-delta conditions of Eq.77. The present derivations are covariant with respect to the group of such transformations. For particular applications, the representation can be chosen for the greatest convenience, subject only to the Kronecker-delta condition.

5.2. PROPERTIES OF THE GREEN FUNCTION

The definition of the Schrödinger Green function can be extended to a complex-valued energy parameter z. Then Eq.68 can be written

$$(z - \mathcal{H}(\mathbf{x}))G(z;\mathbf{x},\mathbf{x}') = \delta(\mathbf{x},\mathbf{x}'). \tag{81}$$

The linear integral operator $G(z)$ is the resolvent operator of the Schrödinger equation. $G(z) = (z - \mathcal{H})^{-1}$ is an analytic operator function in the upper half of the complex z-plane. The singularities are poles or a continuum branch cut corresponding to energy eigenvalues. If $\mathcal{H}$ is Hermitian, the eigenvalues lie on the real energy axis. The corresponding orthonormal set of eigenfunctions ψ_i defines a spectral representation

$$G(z;\mathbf{x},\mathbf{x}') = \sum_i \psi_i(\mathbf{x})(z - \epsilon_i)^{-1}\psi_i^*(\mathbf{x}'). \tag{82}$$

This formula indicates that the residue of $G(z)$ at an eigenvalue pole is just the density matrix, in the coordinate representation, summed over all eigenfunctions with this eigenvalue. This residue can be extracted conveniently by using Dirac's formula after displacing each eigenvalue into the lower complex plane,

$$\lim_{\delta \to 0} \frac{1}{z - \epsilon_i + i\delta} = \mathcal{P}\frac{1}{z - \epsilon_i} - i\pi\delta(z - \epsilon_i). \tag{83}$$

Specializing to the electronic density for electrons of one spin in the local coordinates of a particular cell, the local density per unit energy is

$$n(\epsilon; \mathbf{r}) = -\frac{1}{\pi}\Im G(\epsilon; \mathbf{r}, \mathbf{r}). \tag{84}$$

Summing over all eigenenergies below a chemical potential or Fermi level μ, the local density function is defined by a contour integral passing above all poles on the real axis,

$$\rho(\mathbf{r}) = -\frac{1}{\pi}\int_{-\infty}^{\mu} \Im G(z; \mathbf{r}, \mathbf{r})dz. \tag{85}$$

In methodology based on computation of the Green function, this formula replaces the usual sum over eigenfunctions. When integrated over coordinate space, Eq.84 gives the density of states per unit energy,

$$n(\epsilon) = -\frac{1}{\pi}\int_{\mathcal{R}^3} \Im G(\epsilon; \mathbf{r}, \mathbf{r})d\tau. \tag{86}$$

These formulas must be summed over a spin index to give the corresponding total densities. In multiple scattering theory, the Green function as given by Eq.80 is subdivided into terms valid in separate atomic cells. The elements with $\mathbf{r}' = \mathbf{r}$ are single sums over the cells. Hence the density of states of given spin in a particular cell τ_μ is

$$n^\mu(\epsilon) = -\frac{1}{\pi}\int_{\tau_\mu} \Im G^\mu(\epsilon; \mathbf{r}, \mathbf{r}), \tag{87}$$

where G^μ denotes the terms with $\nu = \mu$ in Eq.80.

5.3. CONSTRUCTION OF THE GREEN FUNCTION

Eq.69 suggests that the structural matrix $G^{\mu\nu}_{L,L'}$ of the Schrödinger Green function at specified energy should have a simple relationship to that of the Helmholtz Green function, which is the matrix of structure constants $g^{\mu\nu}_{L,L'}$. The basis set of regular local solutions of the Schrödinger equation can be constructed so that matrix G depends only on g and the t-matrix. In deriving Eqs.18, 'primitive' basis functions were defined by outward integration, starting from specified regular solid-harmonic functions J_L at the origin in each atomic cell. A 'canonical' basis set is defined by transforming the primitive basis set by matrix C^{-1}, where the matrix $C_{L,L'}$ is defined by Eq.32. Hence C is a unit matrix in the canonical basis. The matching equations on the local cell boundary depend only on the t-matrix, because each basis function matches to a boundary function $J - Nt$. These basis functions are in one-to-one correspondence with the regular solid harmonic functions J_L and can still be indexed by L. Corresponding to the canonical regular basis function ϕ_L, which matches onto $J - Nt$ on cell surface σ, a paired canonical irregular function ξ_L can be computed by integrating the Schrödinger equation inwards from σ, starting from a single irregular function N_L on this surface. The full set of canonical basis functions satisfies Eq.77.

The structural matrix G can be derived in this canonical basis from the matrix g of structure constants, using the surface-matching theorem, as in Eqs.34 and 35. For a modified irregular function $\tilde{\xi}^\lambda$ and the corresponding global matching function $\Phi = \Sigma_\mu N^\mu \beta^\mu$, the matching conditions on the boundary of any cell τ_μ are

$$(J_L^\mu | W_\sigma^\mu | \tilde{\xi}^\lambda - \Phi) = 0, \quad (N_L^\mu | W_\sigma^\mu | \tilde{\xi}^\lambda - \Phi) = 0. \tag{88}$$

Expanding the difference function $\tilde{\xi}^\lambda - \xi^\lambda$ in the form $-\phi^\mu G^{\mu\lambda}$ in cell τ_μ, its representation in the canonical basis on surface σ_μ is $-(J^\mu - N^\mu t^\mu)G^{\mu\lambda}$. Similarly, ξ^λ is $N^\mu \delta^{\mu\lambda}$ on σ_μ and Φ is $N^\mu \beta^\mu - \Sigma_\nu J^\mu g^{\mu\nu} \beta^\nu$. Then from Eqs.88 the coefficients in functions Φ and $\tilde{\xi}^\lambda - \xi^\lambda$, respectively, are given by

$$\beta^\mu = \delta^{\mu\lambda} + t^\mu G^{\mu\lambda}, \quad G^{\mu\lambda} = \sum_\nu g^{\mu\nu} \beta^\nu, \tag{89}$$

omitting indices L. Combining these equations, for each cell τ_μ,

$$G^{\mu\lambda} = g^{\mu\lambda} + \sum_\nu g^{\mu\nu} t^\nu G^{\nu\lambda}, \tag{90}$$

again omitting L-indices and summations. This is a Dyson equation in the form of a matrix representation $G = g + gtG$ of the Lippmann-Schwinger equation, Eq.69. This equation can also be derived by using the idea of atomic-cell orbitals (ACO), generalized to irregular functions. An irregular ACO is defined in a reference cell τ_λ as the canonical function ξ^λ. In all other cells τ_μ it is the local expansion of N^λ, expressed in terms of the structure constants $g^{\mu\lambda}$. Similarly, a regular ACO takes the form $\phi - J$ in its indexed cell and Jgt in all other cells. The ACO form of $\tilde{\xi}^\lambda$ in cell τ_μ is

$$\tilde{\xi}^\lambda = \xi^\mu \delta^{\mu\lambda} - \phi^\mu G^{\mu\lambda} + J^\mu [\sum_\nu (\delta^{\mu\nu} - g^{\mu\nu} t^\nu) G^{\nu\lambda} - g^{\mu\lambda}], \tag{91}$$

omitting indices L. Following tail-cancellation logic(Andersen, 1971), the coefficients of functions J must vanish in all cells. This implies Eq.90.

The structural matrix $G_{L,L'}^{\mu\nu}$ can be evaluated by direct matching across adjacent cell interfaces, without using the matrix of structure constants. This construction was originally derived (unpublished notes) from the Green function cellular method (GFCM)(Butler et al, 1992; Zhang and Butler, 1992a; Zhang et al, 1992). The GFCM itself can be derived(Nesbet, 1992d) as a variant of the variational cellular method (VCM)(Ferreira and Leite, 1978; Ferraz et al, 1984; Nesbet, 1988), by restricting the form of allowed variations on cell surfaces in the Schlosser-Marcus variational principle(Schlosser and Marcus, 1963). The resulting NVCM or GFCM equations are given above as Eqs.55.

The structure constants or structural matrix for the Helmholtz Green function, defined by Eq.67, can be determined by applying Eqs.55 to the global solution of the Helmholtz equation given by Eq.75, with a specified singularity in cell τ_λ. For this application, the potential function vanishes everywhere except for an implied singular potential in cell τ_λ, designed to specify the irregular solid harmonic function N_L^λ as an exact solution in this cell. Then the local basis in each cell $\nu \neq \lambda$ is the set of regular solid harmonic functions J_L^ν. Using these local expansions, Eqs.55 become a set of inhomogeneous linear equations for column λ of the matrix of structure constants. Suppressing L-indices, these equations take the form,

$$\sum_{\nu \neq \mu}(N^\mu|W_{\mu\nu}|(N^\mu\delta^{\mu\lambda} - J^\mu g^{\mu\lambda}) - (N^\nu\delta^{\nu\lambda} - J^\nu g^{\nu\lambda})) = 0, \qquad (92)$$

for all cells τ_μ. Simplifying by use of Eq.5,

$$\sum_{\nu \neq \mu}(N^\mu|W_{\mu\nu}|J^\nu)g^{\nu\lambda} - g^{\mu\lambda} = \sum_{\nu \neq \mu}(N^\mu|W_{\mu\nu}|N^\nu)\delta^{\nu\lambda}, \qquad (93)$$

While these equations are consistent with the GFCM, their relationship to the VCM variational equations suggests that the alternative formalism defined by the JVCM might be more appropriate. In the global solution sought in the present case, the irregular term is fixed, and all variational increments are linear combinations of regular functions J_L. Hence variation of coefficients gives the JVCM Eqs.54. A similar situation occurs for the Poisson equation, since a particular solution of the inhomogeneous equation is fixed, and all incremental functions are regular solid harmonics. Recent calculations comparing these methods for the Poisson equation(Zhang et al, 1994) show improved convergence in the JVCM formalism. It was shown that extension of the Schlosser-Marcus variational principle to the Poisson equation implies the JVCM equations(Zhang et al, 1994). In the present case, Eqs.54 take the form

$$\sum_{\nu\neq\mu}(J^\mu|W_{\mu\nu}|(N^\mu\delta^{\mu\lambda} - J^\mu g^{\mu\lambda}) - (N^\nu\delta^{\nu\lambda} - J^\nu g^{\nu\lambda})) = 0, \quad (94)$$

for all cells τ_μ. Simplifying by use of Eq.5,

$$\sum_{\nu\neq\mu}(J^\mu|W_{\mu\nu}|J^\nu)g^{\nu\lambda} = \sum_{\nu\neq\mu}(J^\mu|W_{\mu\nu}|N^\nu)\delta^{\nu\lambda} + \delta^{\mu\lambda}. \quad (95)$$

Eq.95 is generalized to the Schrödinger Green function by using local regular functions ϕ on the left and by using Eq.79 to define an exact solution of the Schrödinger equation in each cell, with a specified singularity in cell τ_λ. The resulting linear equations take the form, suppressing L-indices,

$$\sum_{\nu\neq\mu}(\phi^\mu|W_{\mu\nu}|\phi^\nu)G^{\nu\lambda} = \sum_{\nu\neq\mu}(\phi^\mu|W_{\mu\nu}|\xi^\nu)\delta^{\nu\lambda} + \delta^{\mu\lambda}, \quad (96)$$

for all values of μ and λ. These equations determine the structural matrix G for any set of space-filling cells.

Eqs.96 simplify if the system has point-group or translational symmetry. Then local basis functions of equivalent cells are related to those of a smaller number of generating cells by phase factors and elementary rotations. Considering only translational symmetry, it is convenient to index cells in a reference translational cell by indices μ, etc. and to index translated equivalent cells by the corresponding indices $\tilde{\mu}$, etc., such that the relative displacement of cell $\tau_{\tilde{\mu}}$ with respect to cell τ_μ is a lattice translational vector $d_{\tilde{\mu}\mu}$. Using this notation, Eqs.96 separate into equations for each k-vector in the reduced Brillouin zone, in the form

$$\sum_{\tilde{\nu}\neq\mu}(\phi^\mu|W_{\mu\tilde{\nu}}|\phi^{\tilde{\nu}})exp(i\mathbf{k}\cdot\mathbf{d}_{\tilde{\nu}\nu})G^{\nu\lambda}(\mathbf{k}) =$$

$$\sum_{\tilde{\nu}\neq\mu}(\phi^\mu|W_{\mu\tilde{\nu}}|\xi^{\tilde{\nu}})exp(i\mathbf{k}\cdot\mathbf{d}_{\tilde{\nu}\nu})\delta^{\nu\lambda} + \delta^{\mu\lambda}, \quad (97)$$

for all values of μ and λ that index cells in the reference translational cell. The corresponding equations derived from Eq.93 have been verified by computing structure constants for an fcc lattice along the $\Gamma - K$ line in the reduced Brillouin zone, checked against a published structure-constant program STR(Skriver, 1984). Eqs.97 have not been tested, but are expected to show improved convergence because they are variationally correct for this problem.

6. Computational Theory

6.1. BASIS FUNCTIONS

Primitive basis functions are defined in an atomic cell τ as local solutions of the Schrödinger equation. Each basis function ϕ_L is matched to a regular solid harmonic J_L at the cell origin, and must satisfy a Lippmann-Schwinger integral equation

$$\phi_L(\mathbf{r}) = \chi_L(\mathbf{r}) + \int_\tau G_0(\mathbf{r}, \mathbf{r}') V(\mathbf{r}') \phi_L(\mathbf{r}') d^3r', \qquad (98)$$

where χ_L is a regular solution of the Helmholtz equation. The potential function $V(\mathbf{r}')$ is unrestricted in form, except that it can be singular only at the cell origin. Eq.98 indicates that the extension of V outside a local cell cannot affect the primitive basis functions in that cell. A rigorous proof of this uniqueness theorem is given below. The function χ_L is a linear combination of functions J_L with coefficients chosen to satisfy the boundary condition at the origin. Since ϕ_L satisfies Eq.1, it follows from Eqs.11, 13 and 19 that the coefficients in the expansion of χ_L are the Wronskian surface integrals $C_{L,L'}$ given by Eqs.32.

In a coordinate system adapted to the geometry of the local cell, the expansion given by Eq.21 could be used on a set of convex surfaces nested between the cell origin and its surface σ, parametrized by a generalized coordinate u such that $0 \leq u \leq 1$. On the surface indexed by u,

$$\phi_L(\mathbf{r}_u) = \sum_{L'} [J_{L'}(\mathbf{r}_u)(N_{L'}|W_u|\phi_L) - N_{L'}(\mathbf{r}_u)(J_{L'}|W_u|\phi_L)]. \qquad (99)$$

As discussed in Section 2.3, this double series represents a well-defined function whether or not it converges. For an infinitesimal outward displacement du, the differential increment of each of the Wronskian surface integrals here is converted by Green's theorem to a volume integral for the incremental volume, which is just a surface integral over the mean displaced surface. This analysis implies the differential equations

$$\frac{d}{du}(N_{L'}|W_u|\phi_L) = \int_{\sigma_u} N_{L'}^* V \phi_L d\sigma, \quad \frac{d}{du}(J_{L'}|W_u|\phi_L) = \int_{\sigma_u} J_{L'}^* V \phi_L d\sigma. \qquad (100)$$

These equations can be used to integrate the multichannel Schrödinger equation outwards from each cell origin, starting from initial values

$$(N_{L'}|W_u|\phi_L)_{u \to 0} = \delta_{L',L}, \quad (J_{L'}|W_u|\phi_L)_{u \to 0} = 0. \qquad (101)$$

From Eqs.32, the values of these integrals for $u = 1$ give the standard matrices $C_{L',L}$ and $-S_{L',L}$, respectively. In order to construct the Schrödinger

Green function, canonical regular basis functions are obtained by multiplying on the right by the inverse of matrix C. A canonical irregular function ξ_L can be obtained by integrating Eqs.100 inwards from surface σ, starting from boundary values

$$(N_{L'}|W_u|\xi_L)_{u \to 1} = 0, \quad (J_{L'}|W_u|\xi_L)_{u \to 1} = -\delta_{L',L}. \qquad (102)$$

In the original development of the full-potential theory(Williams and Morgan, 1972; Williams and Morgan, 1974), Eqs.100 were integrated using nested concentric spheres, while explicitly truncating the potential function on the cell boundary. This truncated potential function was expanded in a basis of spherical harmonics on each spherical shell, equivalent to expanding Heaviside cutoff functions in spherical polar coordinates. Such expansions converge slowly because of the abrupt cutoff. Given these expansions of the truncated potential, which were done by numerical quadrature on each nested sphere, Eqs.100 for concentric spheres become coupled radial differential equations(Williams and Morgan, 1974). The surface integrals in these equations reduce to radial functions times Gaunt coefficients, integrals of three spherical harmonics over a sphere. An alternative procedure, which avoids the use of Heaviside functions, is to represent the potential function by a spherical-harmonic expansion valid within the enclosing sphere(Brown and Ciftan, 1983). In order to evaluate the standard C and S matrices, the basis functions obtained by this procedure must be interpolated to the cell surface σ.

Despite the apparent difference of these integration methods in the near-field region (between enclosed and enclosing spheres)(Brown and Ciftan, 1983), the computed primitive basis functions are solutions of the same differential equation and for given index L are determined by the same boundary condition at the cell origin. It can be proven that the resulting functions are identical in the interior of the local cell(Nesbet, 1990a). Consider the alternative definitions of basis functions by WM(Williams and Morgan, 1974)and by BC(Brown and Ciftan, 1983). WM define basis functions ϕ_L^τ as solutions of the integral equation

$$\phi_L^\tau = \sum_{L'} J_{L'} C_{L',L}^\tau + \int_\tau G_0 V \phi_L^\tau, \qquad (103)$$

while BC define basis functions ϕ_L^S such that

$$\phi_L^S = \sum_{L'} J_{L'} C_{L',L}^S + \int_S G_0 V \phi_L^S, \qquad (104)$$

where S is the enclosing sphere of cell τ. Both ϕ_L^τ and ϕ_L^S approach J_L at the cell origin. For radii greater than the radius of the enclosed muffin-tin sphere, these two equations appear to give different results. This difference is illusory, as will be shown below by proving that ϕ_L^τ and ϕ_L^S are identical within cell τ.

To prove uniqueness, consider the difference function

$$\phi_L^S - \phi_L^\tau = \sum_{L'} J_{L'}(C_{L',L}^S - C_{L',L}^\tau) + \int_\tau G_0 V(\phi_L^S - \phi_L^\tau) + \int_{S-\tau} G_0 V \phi_L^S, \quad (105)$$

given by Eqs.103 and 104. When the operator $\nabla^2 + \kappa^2$ is applied to this equation, it follows from Eq.2 that for points within τ,

$$(\nabla^2 + \kappa^2)(\phi_L^S - \phi_L^\tau) = V(\phi_L^S - \phi_L^\tau). \quad (106)$$

Hence $\phi_L^S - \phi_L^\tau$ is a regular solution of the Schrödinger equation in τ. Since ϕ_L^S and ϕ_L^τ both approach J_L at the cell origin, the difference function vanishes with vanishing radial derivative on a small sphere centered at the origin, and outward integration produces a function that vanishes identically throughout τ. Hence ϕ_L^S and ϕ_L^τ are identical within τ. Since the shape of τ is arbitrary in this proof, generic primitive solutions ϕ_L are uniquely defined. Any particular choice of cell geometry produces functions ϕ_L^τ that agree with ϕ_L within their defined cell. The simplest and most easily computed representation of these functions is obviously that using the full sphere(Brown and Ciftan, 1983), but no correction is implied in the method using cutoff functions(Williams and Morgan, 1974).

Eq.105 does not imply that the coefficient matrices C^S and C^τ are equal. They are defined as Wronskian integrals over different surfaces. Expanded about the cell origin, the leading terms in Eq.105 imply that

$$C_{L',L}^S = C_{L',L}^\tau + \int_{S-\tau} N_{L'}^* V \phi_L^S. \quad (107)$$

These expansion coefficients are different because ϕ_L^τ is represented over a spherical shell of radius r as a spherical harmonic expansion whose sum must agree with the generic function ϕ_L on the portion of the shell inside cell τ, but in general differs outside τ. Since ϕ_L^τ and ϕ_L^S are not identical over the entire sphere, their spherical harmonic expansions must be different.

6.2. LINEARIZED METHODS

Because of the nonlinear energy dependence of the KKR secular equation, roots of the secular determinant of Eq.20 must be computed by a root-search procedure, requiring repeated evaluation of the secular matrix for each root. Despite the inherent inefficiency of this procedure, the simplifications implicit in the KKR-DFT theory for muffin-tin potential functions could be exploited. This methodology was used(Moruzzi et al, 1978) to compute self-consistent energy band structures for 32 elemental metals, ranging from H to In in the periodic table, including spin-polarized structures for the transistion metals. This reference provides a bibliography giving technical details of the analytical and computational methods used. In order to carry out self-consistent calculations on more complex materials with polyatomic translational cells, or to examine effects of lattice distortion, it was desirable to convert this formalism into a variational form in which energy eigenvalues could be computed from a linearized secular equation.

Starting from the idea of 'muffin-tin orbitals' (MTO)(Andersen, 1971), the linear muffin-tin orbital method (LMTO), which became one of the most widely-used computational methods in energy-band theory, was developed by Andersen and collaborators. A description of this method, including source listings of computer programs(Skriver, 1984), and a review of methodology and applications(Andersen et al, 1985) have been published. The LMTO method differs from the KKR muffin-tin model of multiple-scattering theory in three important aspects. In addition to conversion of the theory to an energy-linearized form, the LMTO method also replaces the original muffin-tin model by an atomic-sphere approximation (ASA), in which the local atomic potential function is extended from the enclosed muffin-tin sphere to a larger 'atomic' sphere, defined to have the same volume as the atomic Wigner-Seitz cell. The third element of difference from KKR is that structure constants are taken to be independent of energy. This is done by subtracting an energy shift from both sides of the Schrödinger equation,

$$(\nabla^2 + \kappa_0^2)\psi(\mathbf{x}) = (V(\mathbf{x}) - \kappa^2 + \kappa_0^2)\psi(\mathbf{x}). \qquad (108)$$

The Green function G_0 and solid-harmonic functions J_L and N_L are required only for the fixed energy κ_0^2, usually taken to be zero. This amounts to adding a linear energy term to the potential function in the interstitial region, justified in full-potential theory or the ASA because the net interstitial volume is reduced to zero. The practical effect is slower convergence of the angular momentum expansions. Although the ASA violates the original KKR conditions that ensure an exact solution of the muffin-tin model problem, the ASA model of overlapping atomic spheres is in many ways a more satisfactory approximation to the full-potential MST considered here. In particular, in the surface-integral formalism, approximating Wronskian integrals over the surface of a Wigner-Seitz cell by the corresponding integrals over an equivalent-volume sphere can be considered as a first approximation to a surface quadrature scheme for polyhedral cells. The full DFT local potential function can be used within the atomic sphere. The muffin-tin model, which assumes a constant potential in the near-field region, has no special justification as a computational approximation for atomic potentials.

In the usual ASA, the t-matrix is real and diagonal, but the elements have both zeroes and poles as functions of energy. The original proposal was to simplify KKR calculations by using constant-energy structure constants while fitting diagonal elements of t^{-1} in the standard secular equation, Eq.20 here, to rational functions of energy(Andersen, 1973). This also simplifies the energy dependence of muffin-tin orbitals, which can be used as basis functions in the standard Rayleigh-Schrödinger variational principle for the energy eigenvalues. The resulting LMTO formalism (called LCMTO in the original presentation) produces a linear eigenvalue problem that includes a 'combined correction' matrix representing the difference of the overlap matrix evaluated in atomic spheres versus atomic polyhedra together with a correction for basis functions of higher ℓ quantum numbers. LMTO energy-band calculations obtain all energy eigenvalues at a given k-point from a single matrix diagonalization. These eigenvalues are valid within an energy panel in which the rational fit to the t^{-1} matrix is sufficiently accurate.

In considering nonspherical local potentials, a simple rational fit to diagonal elements of t^{-1} is no longer valid. From the general definition of the t-matrix as $-SC^{-1}$ in terms of the standard MST matrices defined by Eqs.32, it is clear that for energy-independent solid-harmonic functions, linear energy expansion of the basis functions $\phi_L(\epsilon; \mathbf{r})$ is equivalent to fitting elements of the t-matrix by a simple rational formula. In the ASA this produces the parametrization proposed by Andersen. It was recognized at the same time(Andersen and Woolley, 1973) that a linear energy expansion of the basis functions also solves the practical problem in MTO theory that the regular solid-harmonic functions in the representation $\phi - J$ of canonical MTO's within a local cell are not orthogonal to inner-shell occupied orbitals. Replacing these functions by linear combinations of the energy-derivative functions, denoted here by $\dot{\phi}_L$, solves this problem. In the standard LMTO method, the basis functions are energy-independent MTO's that are linear combinations of functions ϕ_L and $\dot{\phi}_L$ computed at a fixed panel energy(Andersen, 1975; Skriver, 1984). Basis functions of this kind are used in alternative energy-linearized methods(Andersen, 1975; Williams et al, 1979) which will not be considered here.

The linearized atomic-cell orbital (LACO) method was developed as a full-potential analog of LMTO(Nesbet, 1986a). Energy-independent structure constants are computed using the LMTO program STR(Skriver, 1984). On a grid of energy values suitable for accurate interpolation, basis functions are computed by numerical integration within the enclosing sphere of each atomic cell. The basis functions and their normal gradients are interpolated to atomic-cell boundaries so that the standard MST C and S matrices given by Eqs.32 can be evaluated by two-dimensional quadrature over the surfaces of polyhedral atomic Wigner-Seitz cells. Energy-linearization is accomplished in two stages, each with several options as to the specific procedure. In the first stage, energy-independent atomic-cell orbitals are constructed in the form $\phi + \sum \dot\phi \omega$. Here, ϕ is a single primitive basis function ϕ_L^λ in a particular cell τ_λ. In this cell, the defining ACO form $\phi - \sum JC$ is modified by representing the regular solid-harmonic functions J as linear combinations of energy-derivative functions $\dot\phi_L$. In all other cells, the matched external function $\sum NS$ is expanded in the local $\dot\phi_L$ basis. The coefficient matrix $\omega_{L,L'}$ satisfies a set of inhomogeneous linear equations, obtained from one of the linear forms of full-potential MST equations, in the extended fixed-energy basis of functions ϕ_L and $\dot\phi_L$. In the second stage of LACO calculations, these energy-independent ACO functions are used in the standard Rayleigh-Schrödinger or Schlosser-Marcus variational equations to determine energy eigenvalues.

In an energy range or panel sufficiently narrow that linear energy interpolation of the basis functions is accurate, it can be shown that this general linearization procedure produces eigenfunctions identical to the null vectors of the MST secular equations. To prove this, for basis functions all at the same panel energy, consider right-eigenvectors of the complex, unsymmetric matrix ω, such that $\omega c_k = c_k \Delta \epsilon_k$, defining a displacement $\Delta \epsilon$ from a fixed panel energy. The definition of energy-independent ACO functions implies that the function corresponding to a given eigenvector is

$$\psi_k = \sum (\phi + \sum \dot\phi \omega) c_k = \sum (\phi + \dot\phi \Delta \epsilon_k) c_k = \sum (\phi(\epsilon_k)) c_k. \qquad (109)$$

This agrees with the form of an MST null-vector expanded in the energy-dependent basis at energy ϵ_k. Because of numerical approximations inherent in the methodology, it is generally inconvenient to diagonalize matrix ω directly. In particular the eigenvalues are complex numbers if there are any residual numerical errors. For this reason, a standard variational principle based on Hermitian matrices is used in the second stage of linearized calculations. The present analysis shows that the ϕ, $\dot\phi$ expansion must be used with caution unless all elements of the ω-matrix are small. A large energy shift $\Delta \epsilon$ may imply that basis functions fall outside the energy grid used in a particular calculation, so that the interpolated functions are not accurate. So-called 'ghost' bands can occur, especially for higher angular quantum numbers, when functions ϕ and $\dot\phi$ tend to become linearly dependent within a local cell(Andersen et al, 1985).

Of the available versions of MST equations for energy-independent ACO functions, the KR variational principle(Kohn and Rostoker, 1954) is one of the best justified. If the functional Λ of Eq.41 is evaluated in the extended basis (ϕ and $\dot\phi$), trial vectors can be defined in the form of energy-independent ACO functions ϕ^λ. Each vector has one fixed element, $\delta^{\mu\lambda}$ in the ϕ basis and one set $\omega^{\mu\lambda}$ of variable $\dot\phi$ coefficients. Varying each of these coefficients, using Eq.46, gives linear equations (suppressing L-indices),

$$\dot S^{\mu\dagger}\sum_\nu[(\delta^{\mu\nu}C^\nu + g^{\mu\nu}S^\nu)\delta^{\nu\lambda} + (\delta^{\mu\nu}\dot C^\nu + g^{\mu\nu}\dot S^\nu)\omega^{\nu\lambda}] = 0. \qquad (110)$$

Each subset of these equations for a given value of λ, L and all values of μ, L' determines a row of the ω-matrix. Matrices $C, S, \dot C$ and $\dot S$ here are to be considered as rectangular matrices. The internal sums over solid-harmonic L-indices should be carried to convergence. The L', L indices of matrix ω are basis function indices and may have a smaller range, treating ω as a square matrix. Empty-lattice calculations on an fcc space-lattice have been carried out to test this formalism(Nesbet, 1992b).

The SM variational principle(Schlosser and Marcus, 1963) provides an alternative that does not use structure constants. On substituting the expansion of an energy-independent ACO into the SM variational functional, varying the matrix elements of ω and using Eqs.53 gives

$$\sum_{\nu\neq\mu}(\dot\phi^\mu|W_{\mu\nu}|\sum(\phi^\mu\delta^{\mu\lambda} + \dot\phi^\mu\omega^{\mu\lambda}) - \sum(\phi^\nu\delta^{\nu\lambda} + \dot\phi^\nu\omega^{\nu\lambda})) = 0. \qquad (111)$$

As in Eqs.110, each subset of equations for given λ, L determines a row of the ω-matrix. Eqs.111 replace the internal summations in Eqs.110 by direct matching across the interfaces of adjacent cells. If Eqs.111 are used, the LACO method takes the form of an energy-linearized version of the variational cellular method (VCM)(Ferreira and Leite, 1978).

6.3. CANONICAL SCALING

Energy levels are roots of the determinant of the MST secular matrix in Eq.18. If the C-matrix has an inverse, this takes the form $I - gt$, or $t^{-1} - g$ if the S-matrix is not singular. The structure constant matrix g is purely geometrical, while all effects of the electronic potential are condensed into the t-matrix. In LMTO theory, g is independent of energy and energy band structures can be parametrized by the energy dependence of the t-matrix. This is expressed in terms of a 'potential function' $P_l(\epsilon)$, equivalent to the matrix t^{-1}, that is simply related to the logarithmic derivative of a radial wave function on the atomic sphere. The resulting canonical band theory(Skriver, 1984; Andersen et al, 1985) assumes that the electronic potential is spherical in each atomic cell. In a representation of the g-matrix diagonalized within each ℓ-block, each element $g_{\ell i}(\mathbf{k})$ defines a 'canonical ℓ-band' in k-space. This is mapped onto an energy-band structure through the scalar equation $P_\ell(\epsilon) = g_{\ell i}(\mathbf{k})$. The 'unhybridized' bands obtained from parametrized canonical ℓ-bands are a very useful approximation to true energy bands in elemental metals.

These ideas provide an estimate of changes in band structure due to changes of basis functions in a self-consistency iteration. The electronic density in each atomic cell is determined by the band structure, which depends on the functions P_ℓ defined by the boundary logarithmic derivatives of the basis functions. The canonical bands serve as an intermediate in estimating the ℓ-projected density of states $n_\ell(\epsilon)$ in each cell. The procedure of 'canonical scaling' is a partial self-consistency iteration carried out separately for each cell, given fixed input values of the energy moments of each ℓ-projected density of states. The zero-order moment, or number of occupied states, is a property of the canonical band structure, summed over all bands below a chemical potential or Fermi level. A scaling function $\hat{\epsilon}(\epsilon)$ is defined for each ℓ-value such that the function P_ℓ computed for the updated basis at $\hat{\epsilon}$ has the same value as P_ℓ computed for the input basis at ϵ. Thus the two energies correspond to the same points on the unhybridized canonical bands. For small changes, this defines a linear energy mapping.

Canonical scaling adjusts the local density function in each cell so that basis functions computed from it have boundary values and normal gradients consistent with the energy level structure assumed in constructing the density of states function. Because the fully self-consistent local density must have this property, enforcing this condition within each outer-loop iteration accelerates outer-loop convergence. The computational cost is relatively small because the inner-loop iterations are carried to convergence one atom at a time, and are required only for atoms that are inequivalent with respect to geometrical symmetries.

In the self-consistency iteration of canonical scaling, given energy-moments of the projected density of states about a fixed panel energy ϵ_0 are modified by energy scaling such that

$$\int \hat{n}(\hat{\epsilon})d\hat{\epsilon} = \int n(\epsilon)d\epsilon. \qquad (112)$$

$$\int (\hat{\epsilon} - \hat{\epsilon}_0)\hat{n}(\hat{\epsilon})d\hat{\epsilon} = \int (\epsilon - \epsilon_0)\frac{d\hat{\epsilon}}{d\epsilon}n(\epsilon)d\epsilon. \qquad (113)$$

$$\int (\hat{\epsilon} - \hat{\epsilon}_0)^2\hat{n}(\hat{\epsilon})d\hat{\epsilon} = \int (\epsilon - \epsilon_0)^2\left(\frac{d\hat{\epsilon}}{d\epsilon}\right)^2 n(\epsilon)d\epsilon. \qquad (114)$$

The local density function $\rho(\mathbf{r})$ is computed by combining products of the radial wave function at $\hat{\epsilon}_0$ and its first and second energy derivatives with these scaled energy-moments of the density-of-states function(Skriver, 1984). 'Hybridization', or interaction between different ℓ-states, must be computed by a full energy-band calculation in an outer-loop step. Because the upper limit of these integrals is changed by the mapping, the Fermi level must be recomputed in an outer-loop step that balances occupation numbers among the ℓ-projected densities.

In extending this procedure to the full-potential problem, a simple scalar 'potential' function of the energy no longer exists. In the LACO method(Nesbet and Sun, 1987), the matrices C and S of Eqs.18 and 32 are used directly in a representation of transformed spherical harmonics that belong to irreducible representations of the atomic-cell point group indexed by symmetry indices ζ, μ. The scaled energy $\hat{\epsilon}$ is determined by requiring $\hat{t}(\hat{\epsilon}) = t^\dagger(\epsilon)$ for the matrix $t = -SC^{-1}$. For rectangular matrices C, S this is expressed as

$$C^\dagger(\epsilon)\hat{S}(\hat{\epsilon}) - S^\dagger(\epsilon)\hat{C}(\hat{\epsilon}) = 0, \tag{115}$$

for diagonal elements indexed by ζ, which are independent of the subspecies index μ by symmetry. Matrices in the initial basis are denoted here by C, S and those in the updated basis by $\hat{C}, \hat{S}$. The internal sums in the matrix products in Eq.115 should be carried to convergence in the solid-harmonic indices ζ, μ. At self-consistency, the initial and updated basis functions are the same, and Eq.115 follows from Eq.27, which is just the statement that the t-matrix is Hermitian. In the current LACO program, Eq.115 is solved to determine $\hat{\epsilon}$ for three values of ϵ closely spaced about each panel energy $\epsilon_{0\zeta}$. This determines $\hat{\epsilon}_{0\zeta}$ and $d\hat{\epsilon}/d\epsilon$, needed in Eqs.113 and 114.

6.4. POISSON EQUATION

For a solid or molecule subdivided into atomic cells, the Poisson equation, given in Rydberg units by

$$\nabla^2 v = -8\pi\rho, \tag{116}$$

can be solved by full-potential multiple scattering theory methods. Here ρ is the local number density of electrons, and the negative electronic charge is factored out of both charge density and potential. Adopting the classical theory of inhomogeneous linear differential equations to a cellular model, particular solutions are first obtained for each atomic cell and then modified by the addition of regular solutions of the homogeneous equation to satisfy continuity and boundary conditions. The original full-potential MST(Williams and Morgan, 1974) was modified for the Poisson equation and applied to a periodic charge distribution on an fcc space lattice(Morgan, 1977). In this example all atomic cells are geometrically equivalent, so the actual calculation is for a single cell with periodic boundary conditions. The given charge distribution, with Heaviside cutoff factors at the polyhedral cell boundary, was expanded in spherical harmonics at radii up to that of the enclosing sphere. A local particular solution v_0 was obtained by integrating equations analogous to Eqs.100 in spherical polar coordinates, subject to the boundary condition that the particular solution should vanish at large r. The external continuation of v_0 is a sum of multipole potentials. Within a particular cell, the sum of these potentials due to all other cells defines a local potential Δv, expressed in terms of structure constants that are just the coefficients in the two-center expansion of the Coulomb interaction. Δv is added to v_0 to give the required global solution of the Poisson equation. The surface-integral MST method described below has been applied to the Poisson equation, and is incorporated in the current LACO program package(Nesbet, 1990b), still using structure constants. More recently, MST cellular methods, using direct matching theory with NVCM(GFCM) and JVCM equations rather than structure constants, were applied to this problem in a detailed numerical study(Zhang et al, 1994). This study finds that the JVCM is the most accurate and reliable of the methods considered, which included standard MST with structure constants.

In multiple scattering theory for the Poisson equation, the particular solution v_0 and complementary solution Δv within a local atomic cell τ are defined by subdividing the Green function solution such that

$$v = v_0 + \Delta v = -8\pi \int_\tau G_0 \rho - 8\pi \int_{R^3-\tau} G_0 \rho. \tag{117}$$

The Green function G_0 is that defined for the Laplace equation (zero-energy Helmholtz equation), such that the Coulomb potential is $2/r_{12} = -8\pi G_0$ in Rydberg units. The choice of solid-harmonic functions $J_L = r^\ell/(2\ell+1)Y_L$ and $N_L = r^{-\ell-1}Y_L$ gives the well-known one-center expansion of $1/r_{12}$ when substituted into Eq.7. It is assumed that the local density function is subdivided into spherical harmonic components as $\rho = \sum_L \rho_L$. Each of these components defines a particular solution within the enclosing sphere of the local cell. A very efficient numerical algorithm is available for solving the radial Poisson equation(Loucks, 1967), generalized to arbitrary ℓ-values in the current LACO program(Nesbet, 1990b). Denoting these primitive solutions by $\hat{v}_L$, the ACO construction can be used to define a particular solution v_{0L} for each ρ_L such that $v_{0L} = \hat{v}_L - \sum J\hat{c}$ in the local cell τ. v_{0L} matches onto an external function $\sum N\hat{s}$ on the cell surface σ. The coefficient matrices here are defined by

$$\hat{c}_{L',L} = (N_{L'}|W_\sigma|\hat{v}_L), \quad \hat{s}_{L',L} = -(J_{L'}|W_\sigma|\hat{v}_L). \tag{118}$$

The complementary potential Δv or generalized Madelung term is expanded in regular solid harmonics as

$$\Delta v_L^\mu = -\sum_{L'} J_{L'}^\mu \Delta c_{L',L}^\mu. \tag{119}$$

The coefficients here are

$$\Delta c^\mu = -\sum_\nu g^{\mu\nu}(J^\nu|W_{\sigma\nu}|v_0^\nu), \qquad (120)$$

obtained by substituting the two-center expansion given by Eq.8 into the second term in Eq.117, or, equivalently, into $\sum N\hat{s}$.

Alternative equations for the coefficients Δc are given by direct matching at cellular interfaces, bypassing the need for structure constants. The two alternatives considered by (Zhang et al, 1994) are the NVCM or GFCM equations,

$$\sum_{\nu\neq\mu}(N^\mu|W_{\mu\nu}|J^\nu)\Delta c^\nu - \Delta c^\mu = \sum_{\nu\neq\mu}(N^\mu|W_{\mu\nu}|v_0^\nu), \qquad (121)$$

and the JVCM equations,

$$\sum_{\nu\neq\mu}(J^\mu|W_{\mu\nu}|J^\nu)\Delta c^\nu = \sum_{\nu\neq\mu}(J^\mu|W_{\mu\nu}|v_0^\nu) - (J^\mu|W_{\sigma\mu}|v_0^\mu). \qquad (122)$$

Because of the high efficiency of the integration algorithm for the radial Poisson equation(Loucks, 1967), MST surface-integral formalism(Nesbet, 1986a; Nesbet, 1990a) reduces the computation of multipole moments of cellular charge densities to the evaluation of Wronskian surface integrals. Electrostatic multipole moments Q_L are defined by the asymptotic external potential for each cell τ. For the particular choice of solid-harmonic functions given above,

$$\lim_{r\to\infty} v_{0L} = \sum N_{L'}\hat{s}_{L',L} = -8\pi \sum_L N_L Q_L. \qquad (123)$$

A factor -2 included in the last term here compensates for the use of Rydberg units and for the omission of the negative electronic charge in potential functions derived from Eq.117. Hence the electrostatic multipole moments of atomic cell τ_μ are

$$Q_L^\mu = -\frac{1}{8\pi}\sum_{L'}\hat{s}_{L,L'}^\mu = \frac{1}{8\pi}(J_L^\mu|W_{\sigma\mu}|\sum_{L'}\hat{v}_{L'}^\mu). \qquad (124)$$

This equation is used for $\ell = 0$ to evaluate normalization integrals in the current LACO program, avoiding numerical volume quadrature.

7. Density Functional Theory

A reformulation of density-functional theory is presented here in which the Kohn-Sham model state is replaced by a reference state determined by a condition that establishes a one-to-one relationship to the electron quasiparticles of many-body theory. This reformulated theory will be referred to as reference-state density functional theory or RDFT. Quasiparticle energies correspond to energy differences for the addition or removal of one electron from a many-electron system. In many-body theory it is convenient to define a reference state whose one-electron wave functions develop as their interaction is turned on into the dressed quasiparticles of the fully interacting system. In general the electronic density of the reference state differs from that of the correlated N-electron system. Individual terms in the RDFT functional and the resulting one-electron energies have well-defined counterparts in many-body theory. The quasiparticle energies can be directly related to a Landau free-energy functional, as postulated in Fermi liquid theory(Landau, 1956; Landau, 1957; Nozières, 1964). An important goal of RDFT is to validate use of DFT as an initial step in applications of many-body theory that describe excitations or perturbations of the ground state. Because the one-electron states of the present theory have a definite relationship to physical quasiparticles, their use as a basis of such developments is justified, in contrast to the standard theory.

7.1. REFERENCE-STATE DENSITY FUNCTIONAL THEORY

An N-electron wave function Ψ determines a reference-state Slater determinant Φ by the condition that the projection of Φ on Ψ is maximized. If Φ is not unique in some special case, it will be assumed here that one such function is selected by some criterion. It has been proved(Brenig, 1957) that this maximal property implies that Φ has no one-electron matrix elements with the orthogonal remainder of Ψ. The set of orthonormal occupied one-electron orbital wave functions of Φ is determined by this condition up to a unitary transformation(Brueckner and Wada, 1956; Nesbet, 1958). The normalization convention to be assumed here, $(\Phi|\Psi) = (\Phi|\Phi) = 1$, implies, by a second basic theorem(Nesbet, 1965; Nesbet, 1969), that the unsymmetric formula $E = (\Phi|H|\Psi)$ is exact if Ψ is an eigenfunction of H with eigenvalue E.

If T and V are the kinetic energy and external potential operators, and U is the two-electron Coulomb interaction, the full N-electron Hamiltonian is $H = T + V + U$. For any one-electron Hamiltonian operator $H1$, $(\Phi|H1|\Psi) = (\Phi|H1|\Phi)$ by the Brenig theorem. The unsymmetric energy formula implies that the contributions of the one-electron operators T and V to N-electron energy eigenvalues can be evaluated as mean values in the reference state. This has the profound consequence in density-functional theory that the external potential function can be taken to interact with the electronic density of the reference state Φ rather than with that of the corresponding eigenstate Ψ. These two density functions are in general not identical. An alternative theory (RDFT), based on the reference-state density, is developed here.

To avoid purely mathematical difficulties, it is assumed that variational trial functions Ψ are of the correct form for eigenfunctions of a Hamiltonian $H_v = T + V + U$, where V is the sum over all electrons of some physically realizable one-electron potential function v. Thus any Ψ here determines a potential function v. The ground-state energy is a functional $E[v]$. The electron density of a correlated wave function will be denoted by $\hat{\rho}$ such that $\Psi \to \hat{\rho}$ and the reference-state electron density by ρ such that $\Psi \to \Phi \to \rho$. A ground state eigenfunction whose correlated density is $\hat{\rho}$ defines an expression of the form

$$\hat{F}[\hat{\rho}] = (\Psi|T + U|\Psi)/(\Psi|\Psi). \tag{125}$$

Hohenberg-Kohn theory proves this to be a functional of $\hat{\rho}$, universal in the sense that it does not depend explicitly on v. A constrained-search argument(Levy, 1979a; Levy, 1979b) shows that $\hat{F}$ is the minimum value of the indicated Rayleigh quotient, for trial functions constrained to have the density $\hat{\rho}$. The universal functional $\hat{F}$ is related to the ground-state energy $E[v]$ for a potential function v determined by the spin-indexed scalar field $\hat{\rho}$ as a spin-indexed scalar field of Lagrange multipliers. Explicitly, for unconstrained trial functions Ψ_t,

$$\hat{F}_v[\hat{\rho}] = \min_{\Psi_t}[(\Psi_t|T + U|\Psi_t)/(\Psi_t|\Psi_t) + \int v(\hat{\rho}_t - \hat{\rho})d^3\mathbf{r}.], \tag{126}$$

If the Lagrange multiplier field v is chosen so that $\hat{\rho}_v = \hat{\rho}$ for the minimizing trial function Ψ_v, then $\hat{F}_v = \hat{F}$. If the constant integral $\int v\hat{\rho}d^3r$ is added to both sides of this equation, the left-hand side is the Hohenberg-Kohn energy functional

$$\hat{E}[\hat{\rho}] = \hat{F}[\hat{\rho}] + \int v\hat{\rho}d^3\mathbf{r}, \tag{127}$$

and the right-hand side is the variational expression for the ground-state energy $E[v]$.

A derivation of an analogous theory based on the reference-state electron density follows from the same logic, if the Rayleigh quotient is replaced throughout by the unsymmetric formula $(\Phi|T+U|\Psi)$ and $\hat{\rho}$ is replaced by ρ. Because the unsymmetric formula is not itself a variational expression, but rather a formula for a stationary value, the argument must be modified in some points, which will be indicated here. A universal functional is defined by

$$F[\rho] = (\Phi|T + U|\Psi), \tag{128}$$

where Ψ is a ground-state eigenfunction such that $\Psi \to \Phi \to \rho$. If ρ is specified, this notation implies that an N-electron wave function Ψ is to be selected whose reference-state density is ρ. The one-electron potential function v_ρ can be derived as a spin-indexed Lagrange multiplier field from a variational expression

$$F_v[\rho] = \min_{\Psi_t}[(\Phi_t|T+U|\Psi_t) + \int v(\rho_t - \rho)d^3\mathbf{r}], \tag{129}$$

where Ψ_t determines Φ_t and ρ_t. If the constant integral $\int v\rho d^3\mathbf{r}$ is added to both sides of Eq.129, the right-hand side is the minimal value of $(\Phi|H_v|\Psi)$ for the potential function v_ρ, for trial functions as defined below. The left-hand side defines an energy functional

$$E[\rho] = F[\rho] + \int v_\rho \rho d^3\mathbf{r}. \tag{130}$$

Because $(\Phi|H_v|\Psi)$ is not in itself a variational expression, its minimum value for trial functions constrained only to have a given reference-state density ρ is not simply related to an eigenstate of the Hamiltonian H_v defined by v in Eq.(129), whereas Eq.(128) defines $F[\rho]$ only for such eigenstates. If the minimizing trial function in Eq.(129) were not an eigenfunction of H_v, then for some subset of trial functions, using the Brueckner-Brenig condition and defining $\Delta = \Psi - \Phi$,

$$(\Phi|H_v|\Psi) = (\Phi|H_v|\Phi) + (\Phi|U|\Delta) \leq E[v]. \tag{131}$$

It should be noted that $(\Phi|H_v|\Phi) \geq E[v]$, and $(\Phi|U|\Delta)$ is a generalized correlation energy. In order to exclude trial energy values below the ground state, and to ensure that the variational functional has a lower bound, trial functions Ψ_t whose reference state is Φ must satisfy an additional constraint. Any function $\Psi \to \Phi$ defines a linear manifold of functions that have the same Φ and ρ. This contrasts with the Levy constrained search, in which $\tilde{\rho}$ varies with the norm of Ψ. An arbitrary trial function Ψ_t in this manifold can be expressed in the form $\Phi + c\Delta$, where $\Delta = \Psi - \Phi$. If the coefficient c is determined by diagonalizing the 2-by-2 matrix of $T + U + V$ in the orthogonal basis (Φ, Δ), this fixes the norm of Ψ_t. Since the unsymmetric energy expression is valid for a matrix eigenvalue, it also follows from this construction that $(\Phi_t|T + U + V|\Psi_t)$ in Eq.129 is bounded below by the ground-state energy of H_v. This construction is assumed here as a constraint condition on trial functions Ψ_t. It retains the form of Ψ, modifying only the coefficient c, and retains Φ as reference state. An eigenstate remains unchanged, and the ground state is in the trial set. Hence the minimum in Eq.(129) for the modified set of trial functions corresponds to the ground state of H_v.

The wave function and reference function defined by this minimization condition, Ψ_ρ and by Φ_ρ, respectively, are functionals of ρ. The kinetic energy functional $T[\rho]$ is defined simply by $(\Phi_\rho|T|\Phi_\rho)$. Denoting the Coulomb part of $(\Phi_\rho|U|\Phi_\rho)$ by $U[\rho]$, a universal exchange-correlation energy functional is defined by

$$E_{xc}[\rho] = F[\rho] - T[\rho] - U[\rho] = (\Phi_\rho|U|\Psi_\rho) - U[\rho]. \quad (132)$$

Comparing RDFT with standard theory for the same potential function v_ρ, the universal ground state functionals are related by

$$F[\rho] = \hat{F}[\hat{\rho}] + \int v_\rho(\hat{\rho} - \rho)d^3\mathbf{r}. \quad (133)$$

The constrained-search procedure for $\hat{F}[\hat{\rho}]$ defines a wave function Ψ that determines reference state Φ and reference-state density ρ as functionals of $\hat{\rho}$. The functional $E[\rho]$ defined by Eq.(130) is not a variational expression and is defined only for $v = v_\rho$. However, if $v \neq v_\rho$, Eqs.(130) and (133) define

$$E[\rho] = \hat{E}[\hat{\rho}] - \int (v - v_\rho)\hat{\rho}d^3\mathbf{r}, \quad (134)$$

as a functional of $\hat{\rho}$, in terms of the Hohenberg-Kohn energy functional. For variations about a ground state, with fixed external potential $v = v_\rho$ but variable number of electrons N, these equations imply

$$\int d^3\mathbf{r}[\delta F/\delta\rho + v - \mu]\delta\rho = \int d^3\mathbf{r}[\delta\hat{F}/\delta\hat{\rho} + v - \mu]\delta\hat{\rho}, \quad (135)$$

where $\mu = \partial E/\partial N$ and $\int \delta\rho = \int \delta\hat{\rho}$. The Hohenberg-Kohn theory implies (Callaway and March, 1984; Parr and Yang, 1989) that the density $\hat{\rho}$ for a correlated ground state wave function Ψ is determined by

$$\delta\hat{F}/\delta\hat{\rho} + v = \mu, \quad (136)$$

where μ is the chemical potential. Eq.(135) implies for fixed $v = v_\rho$ that the reference-state density and the external potential are related by

$$\delta F/\delta\rho + v = \mu. \quad (137)$$

If N is fixed, Eq.(137) implies that $E[\rho]$ is stationary for fixed $v = v_\rho$. The minimal value must correspond to the ground state. If Eq.(137) has multiple solutions ρ for given v, it must be assumed that the ground-state reference density ρ is selected.

As in the original theory (Kohn and Sham, 1965), an effective Schrödinger equation for the orbital wave functions of the reference state can be derived from functional derivatives of $E[\rho]$. From Eq.(135), the present theory is operationally equivalent to Kohn-Sham theory, except that the computed electronic density is no longer the same as the correlated electronic density. Eq.(133) implies that density functionals derived from the interacting electron gas, widely used in the local density approximation (LDA), are equally valid in the present formalism, since the electronic densities of the reference state and the N-electron ground state are identical for the uniform electron gas. The general success of LDA using these functionals also supports the present theory. However, if the density functional is deduced from accurate calculations on finite systems, or if density-gradient corrections are introduced, as in many recent refinements of LDA(Becke, 1988a; Becke, 1988b; Lee et al, 1988; Perdew et al, 1992), the theory presented here implies quantitative changes and modified methodology.

Unlike the model state postulated in Kohn-Sham theory, the reference state used here is well-defined in detailed CI or many-body theory. The ground-state functional $E_{xc}[\rho]$ has an explicit representation in CI theory(Nesbet, 1986b), using the Brueckner-Brenig basis $\{\phi\}$ and the CI coefficients c_{ij}^{ab} for two-particle virtual excitations of the reference state. Introducing occupation numbers for the ground state, this is

$$E_{xc} = -\frac{1}{2}\sum_{i,j} n_i n_j (ij|u|ji) + \sum_{i<j} n_i n_j \sum_{a<b}(1-n_a)(1-n_b)(ij|\bar{u}|ab)c_{ij}^{ab}, \quad (138)$$

where $\bar{u}$ indicates an antisymmetrized Coulomb-exchange matrix element. All quantities here for a ground state are functionals of the reference-state density. Excited states require correction terms, no longer functionals of the density, that are finite and nonuniversal. Following a self-consistent ground state calculation of orbital wave functions and quasiparticle energies, using a standard approximate density functional, the proposed procedure for strongly correlated or excited states is to compute selected CI coefficients by standard many-body theory and then to evaluate the implied incremental changes of the Landau energy functional and of the one-electron Hamiltonian in a modified quasiparticle basis.

7.2. VARIATION OF OCCUPATION NUMBERS

In the Landau theory of Fermi liquids, the energy functional is parametrized as a function of occupation numbers, denoted here by $\{n\}$, that are indexed to correspond to quasiparticles represented here by the orbital functions $\{\phi\}$ (Landau, 1956; Landau, 1957; Nozières, 1964). This theory is not restricted to ground states. Occupation numbers are considered to be continuously variable in the range $0 \leq n \leq 1$. To validate this postulate, it will be assumed here that atoms or molecules are located in noninteracting equivalent cells of an infinite space lattice. Although it may not be experimentally realizable, an infinite space lattice of weakly interacting fractionally charged atoms, neutralized by a constant distributed charge or by charges on an internal honeycomb surface, is a valid physical system that should be described correctly by quantum theory. Occupation numbers can take arbitrary rational values, as they do in energy-band theory. The alternative of considering a statistical ensemble of atoms or molecules in different charge states conflicts with the definition of an optimal reference function here and with use of the unsymmetric energy formula.

Given the external potential function, an exact N-electron wave function is specified by the occupation numbers of a reference state, by the orbital basis set, and by a set of coefficients $\{c\}$ that are elements of an eigenvector of the Hamiltonian matrix in the configuration-interaction (CI) representation generated by virtual excitations of the reference state. It has been shown(Nesbet, 1986b) that a Landau functional is implied if the Brenig theorem is used to determine an optimal reference state and the implied orbital basis set for an exact wave function. An energy eigenvalue is stationary with respect to the CI coefficients $\{c\}$ by solution of the matrix eigenvalue equation and stationary with respect to the set of orbital basis functions $\{\phi\}$ because of the Brenig theorem. To lowest order in infinitesimal variations, the only residual energy variation for fixed external potential is due to variations of the occupation numbers $\{n\}$, which become the free parameters in a resulting Landau functional. The unsymmetrical energy formula is valid for all energy eigenstates and is used as an interpolation formula for continuously variable occupation numbers. In the case of finite variations, the CI and Brueckner-Brenig calculations must be repeated over a range of occupation numbers. A finite energy difference is obtained by integrating over this range.

The universal exchange-correlation functional defined above implies an equivalent Landau functional in which the explicit dependence of the variational energy on the CI coefficients is replaced by dependence on the reference-state density function ρ. A universal functional is only defined for ground states, giving derivatives that are valid for infinitesimal variations about a ground state. Finite excitations, with finite changes of occupation numbers, require augmentation of the density functional theory by an appropriate form of many-body theory.

If the ground-state functional and corrections due to selective finite changes of occupation numbers are known to adequate accuracy, the resulting generalized Landau theory can be used as an extension of density-functional theory, without requiring full solution of the N-electron CI problem, or the equivalent many-body theory. Because the coefficients $\{c\}$ for a ground state are properties of the eigenfunction, they are determined by the constrained-search procedure for $F[\rho]$ as universal functionals of ρ. The orbital basis set $\{\phi\}$, defined by the Brenig theorem, is also a ground-state functional of ρ. It follows that for infinitesimal variations about the ground state, for fixed external potential,

$$E[\rho] = (\Phi_\rho|T+V|\Phi_\rho) + U[\rho] + E_{xc}[\rho] \quad (139)$$

acts as a Landau functional, providing an interpolation formula between N-electron eigenstates, for which the unsymmetric energy formula is valid. The explicit dependence on occupation numbers is given by expanding $\rho = \Sigma \phi n \phi^*$. Then

$$E[\rho] = \sum_i n_i(i|t+v|i) + \frac{1}{2}\sum_{i,j} n_i n_j(ij|u|ij) + E_{xc}[\rho], \quad (140)$$

where t is the one-electron kinetic energy operator, and u denotes the Coulomb interaction. Differention of this expression with respect to an occupation number n_i gives the Landau one-particle energy,

$$\epsilon_i = \partial E/\partial n_i = (i|t+v+v_{cl}+\mu_{xc}|i), \quad (141)$$

in analogy to Janak's theorem(Janak, 1978; Perdew and Zunger, 1981). Here v_{cl} is the classical Coulomb potential due to ρ and μ_{xc} is the functional derivative $\delta E_{xc}/\delta \rho$. The implied computational procedure, exactly as in Kohn-Sham theory, is to solve the eigenvalue problem for the effective one-electron Hamiltonian defined by Eq.(141) and to iterate the reference state spin-density to self-consistency. The chemical potential (Fermi level) and occupation numbers are defined as in Kohn-Sham theory.

Landau one-electron energies are derivatives with respect to occupation numbers, and for atoms or molecules are distinct from the Dyson one-electron energies associated with poles of the one-particle Green function, which correspond to removal or addition of one electron(Nesbet, 1986b). Changes of total energy corresponding to finite changes of occupation numbers in general require integrating these energy derivatives over a continuous range of occupation numbers. It has been shown(Perdew et al, 1982) that use of a statistical ensemble to produce nonintegral occupation numbers implies that the minimum DFT energy must have discontinuous slope as a function of electron number. This conclusion does not necessarily apply here. The present theory does not rely on a statistical argument, but assumes instead that a finite system is replicated on an extended space lattice. The number of electrons in a unit cell can take on arbitrary rational values. In the present derivation, it would be very difficult to account for unsmooth behavior of the universal functional $F[\rho]$ with respect to variation of the number of electrons.

7.3. SELF-INTERACTION CORRECTIONS

In Landau Fermi-liquid theory, an energy functional is defined as a function of quasiparticle occupation numbers n_i. For electron quasiparticles, the one-electron energies are the partial derivatives

$$\epsilon_i = \partial E/\partial n_i. \qquad (142)$$

This relationship follows from Janak's theorem in DFT. In general, ϵ_i is a function of the local spin-density, which depends on all occupation numbers. However, a Fermi particle cannot interact with itself, so the self-interaction defined by $\partial \epsilon_i/\partial n_i$ should vanish in a correct theory. This condition is not generally true in the usual parametrized forms of DFT. For the DFT energy functional defined by Eq.140,

$$\epsilon_i = \partial E/\partial n_i = (i|t+v|i) + \sum_j n_j (ij|u|ij) + (i|\mu_{xc}|i). \qquad (143)$$

The self-interaction is

$$\partial \epsilon_i/\partial n_i = (ii|u|ii) + (i|\partial \mu_{xc}/\partial n_i|i). \qquad (144)$$

When this does not vanish, the theory requires self-interaction corrections (SIC)(Perdew and Zunger, 1981). Since individual Coulomb self-interaction integrals $(ii|u|ii)$ are small or vanish for delocalized wave functions in a solid or large molecule, it might appear that the self-interaction can be neglected. However, this conclusion depends on the degree of localization, and SIC terms can lead to self-consistent spontaneous localization(Perdew and Zunger, 1981; Jones and Gunnarsson, 1989). A computationally feasible formulation of the theory is desirable.

Any parametrized ground-state exchange-correlation energy formula in RDFT is an approximation to the exact expression from many-body theory given by Eq.138. If Coulomb and exchange terms are combined into the antisymmetrized interaction $\bar{u}$, the one-electron energies evaluated according to Eq.142 are

$$\epsilon_i = (i|t+v|i) + \sum_{j \neq i} n_j (ij|\bar{u}|ij) + \sum_{j \neq i} n_j \sum_{a<b} (1-n_a)(1-n_b)(ij|\bar{u}|ab)c_{ij}^{ab}$$

$$- \sum_{k<j} n_k n_j \sum_b (1-n_b)(kj|\bar{u}|ib)c_{kj}^{ib}. \qquad (145)$$

Because of the antisymmetry of $\bar{u}$, only the final summation here can contribute to the self-energy

$$\partial \epsilon_i/\partial n_i = -\sum_j n_j \sum_b (1-n_b)(ij|\bar{u}|ib)c_{ij}^{ib}. \qquad (146)$$

However, because the coefficient c_{ij}^{ib} with a repeated index i represents a single-particle virtual excitation, it must vanish by the Brueckner-Brenig theorem when Eq.138 is valid. Hence, as expected in many-body theory, there is no residual self-interaction.

The general conclusion from this argument is that any parametrized exchange-correlation energy functional should be modified so that the net residual self-interaction of Coulomb, exchange and correlation terms vanishes. The SIC correction is summed over occupied one-electron states, using orbital index i to include a spin index, in the form

$$\Delta E = -\sum_i \{\frac{1}{2}n_i^2(ii|u|ii) + E_{xc,i}^{SI}\}, \qquad (147)$$

where $E_{xc,i}^{SI}$ must be defined so that

$$\frac{\partial}{\partial n_j}\frac{\partial}{\partial n_i}(E_{xc} - \sum_i E_{xc,i}^{SI}) = 0. \qquad (148)$$

This is required for consistency in the RDFT because the density functional is a Landau functional for the ground state. It implies a choice of $E_{xc,i}^{SI}$ that differs in detail from the usual definition(Perdew and Zunger, 1981), and will be discussed below. Because $E_{xc,i}^{SI}$ must vanish when $n_i = 0$ it takes the form

$$E_{xc,i}^{SI} = (i|\epsilon_{xc}[\rho] - \epsilon_{xc}[\rho - n_i\rho_i]|i), \qquad (149)$$

neglecting the effect of other second derivatives $\frac{\partial}{\partial n_j}\frac{\partial}{\partial n_i}E$ on $(i|\epsilon_{xc}|i)$, and neglecting any change of the wave functions. Because one-electron energies are defined as Landau energies, the SI part of ϵ_i is $\partial E^{SI}/\partial n$. Including the Coulomb self-energy,

$$\epsilon_i^{SI} = n_i(ii|u|ii) + (i|\mu_{xc}[\rho] - \mu_{xc}[\rho - n_i\rho_i]|i)$$

$$= n_i(ii|u|ii) + \int_0^{n_i}(i|\frac{\partial \mu_{xc}}{\partial n_i}|i)_{n_i=n}dn. \qquad (150)$$

consistent with Eqs.144 and 148. From Eq.144, the incremental SIC potential function, defined so that $\Delta\epsilon_i = (i|\Delta v_i|i)$, is

$$\Delta v_i = -n_i(i|u|i) - \int_0^{n_i}(\frac{\partial \mu_{xc}}{\partial n_i})_{n_i=n}dn = -n_i(i|u|i) - \mu_{xc}[\rho] + \mu_{xc}[\rho - n_i\rho_i]. \qquad (151)$$

The usual definition of $E_{xc,i}^{SI}$(Perdew and Zunger, 1981), for $n_i = 1$, is

$$E_{xc,i}^{SI} = (i|\epsilon_{xc}[\rho_i]|i), \qquad (152)$$

which does not take into account the nonlinear dependence of the exchange-correlation energy on the total spin-density. By identifying the DFT one-electron states with Landau quasiparticles, the RDFT formalism implies that the self-consistent interaction with all other occupied electronic states must be included in one-electron energies and their derivatives about a ground state. This implies the present Eq.149. Detailed calculations are needed to see if this has a significant effect in particular cases.

A systematic study of SIC effects was carried out for neutral atoms, negative ions, solid rare gases, and transition metals(Perdew and Zunger, 1981). Significant improvements in detailed results were found in general compared with local spin-density (LSD) calculations. In particuar, binding energies of atomic negative ions H^-, O^-, F^-, and Cl^- were found to be in good agreement with experimental numbers (within 0.2 eV in all cases), in contrast with LSD binding energies, which have the wrong sign for H^- and O^-. Band gaps in insulating solids are generally underestimated by LSD calculations. The SIC was found to remove this discrepancy almost completely for rare-gas solids. Two technical aspects of this formalism have been improved in subsequent work. The first problem is that the SIC potential function Δv_i is different for each wave function, so that eigenfunctions of the SIC effective Schrödinger equation are not orthogonal. This problem is resolved by defining a 'unified' Hamiltonian, incorporating projection operators, such that orthogonal eigenvectors are obtained by diagonalizing its matrix(Harrison et al, 1983; Heaton et al, 1983). The second problem is that the SIC is representation-dependent. In a solid, SI effects are largest in a localized representation of the wave functions. This motivates the idea that a representation should be chosen by a variational criterion of optimal localization, selecting the unitary transformation of occupied orbital functions that minimizes the total energy including the SIC term ΔE(Pedersen et al, 1984; Pedersen and Lin, 1988). This leads to an auxiliary condition for orbital wave functions indexed by i and j,

$$(\psi_i|\Delta v_i - \Delta v_j|\psi_j) = 0. \tag{153}$$

It might seem counterintuitive that the most correct computation of an error such as the residual self-interaction in parametrized DFT should be defined by choosing a representation in which this error is maximized, but that is precisely the logical structure of optimal localization. It is very well justified by the results obtained in the calculations summarized below.

In a series of calculations(Svane and Gunnarsson, 1988; Svane and Gunnarsson, 1990; Svane, 1992; Svane, 1994), electronic localization in solids was examined using a self-consistent SIC method. It was found for the one-dimensional Hubbard model that the local spin moment, a measure of localization in an antiferromagnetic structure, is substantially in agreement with the known exact solution, and qualitatively different from uncorrected LSD. Similar results were obtained for the series of transition metal oxides, for which standard LSD calculations incorrectly give small or vanishing band gaps, while the real materials are antiferromagnetic insulators. The computed SIC-LSD band gaps are in reasonable agreement with experiment, and the exceptional member of this series, VO, is correctly computed to be a nonmagnetic metal(Svane and Gunnarsson, 1990). Similar calculations on the transition metal oxides(Szotek et al, 1993), using the unified Hamiltonian method with self-consistent localization, confirmed these results. Subsequent calculations were carried out on La_2CuO_4, the undoped precursor of high-T_c superconductors, and on metallic Ce, by both groups(Svane, 1992; Svane, 1994; Temmerman et al, 1993; Szotek et al, 1994). Results in both cases are in qualitative agreement with physical properties given incorrectly by the uncorrected LSD method.

Acknowledgments

The author is grateful to W.H. Butler for many enlightening discussions about the formal and practical problems of full-potential multiple scattering theory, and to O.K. Andersen for encouraging the development of the surface-integral formalism and the LACO method in its early stages. Discussion of issues considered here with A. Gonis and X.-G. Zhang (MST) and with W. Kohn, J.P. Perdew, M. Levy, P. Ziesche, and A. Savin (DFT) was a vital part of the author's education in these subjects.

References

Andersen, O.K. (1971) Comments on the KKR wavefunctions; extension of the spherical wave expansion beyond the muffin tins. In *Computational Methods in Band Theory* ed. P.M. Marcus, J.F. Janak, A.R. Williams, Plenum, New York,178-182.
Andersen, O.K. (1973) Simple approach to the band-structure problem, *Solid State Commun.* 13,133-136.
Andersen, O.K. (1975) Linear methods in band theory, *Phys.Rev.B* 12,3060-3083.
Andersen, O.K., Jepsen, O. and Glötzel (1985) Canonical Description of the Band Structures of Metals. In *Highlights of Condensed-Matter Theory, Soc.Ital.Fis.* Corso 89,59-176.
Andersen, O.K. and Kasowski, R.V. (1971) Electronic states as linear combinations of muffin-tin orbitals, *Phys.Rev.B* 4,1064-1069.
Andersen, O.K. and Woolley, R.G. (1973) Muffin-tin orbitals and molecular calculations: General formalism, *Mol.Phys.* 26,905-927.
Badralexe, E. and Freeman, A.J. (1987) Energy-band equation for a general potential, *Phys.Rev.B* 36,1378-1388.
Becke, A.D. (1988a) Density-functional exchange-energy approximation with correct asymptotic behavior, *Phys. Rev. A* 38,3098-3100.
Becke, A.D. (1988b) Correlation energy of an inhomogeneous electron gas: A coordinate-space model, *J.Chem.Phys.* 88,1053-1082.
Brenig, W. (1957) Zweiteilchennäherungen des Mehrkörperproblems I, *Nucl.Phys.* 4,363-

374.

Brown, R. G. and Ciftan, M. (1983) Generalized non-muffin-tin band theory, *Phys.Rev.B* 27,4564-4579.

Brueckner, K.A. and Wada, W. (1956) Nuclear saturation and two-body forces: Self-consistent solutions and the effects of the exclusion principle, *Phys.Rev.* 103,1008-1016.

Butler, W.H. and Nesbet, R.K. (1990) Validity, accuracy, and efficiency of multiple-scattering theory for space-filling scatterers. *Phys.Rev.B* 42,1518-1525.

Butler, W.H. and Zhang, X.-G. (1991) Accuracy and convergence properties of multiple-scattering theory in three dimensions. *Phys.Rev.B* 44,969-983.

Butler, W.H., Zhang, X.-G. and Gonis, A. (1992) The Green Function Cellular Method and its Relation to Multiple Scattering Theory, *Mat.Res.Symp.Proc.* 253,205-210.

Callaway, J. and March, N.H. (1984) Density Functional Methods: Theory and Applications, *Solid State Physics* 38,135-221.

Faulkner, J.S. (1979) Multiple-scattering approach to band theory, *Phys.Rev.B* 19,6186-6206.

Faulkner, J.S. (1986) Non-muffin-tin potentials in multiple-scattering theory, *Phys.Rev.B* 34,5931-5934.

Ferreira, L.G. and Leite, J.R. (1978) General formulation of the variational cellular method for molecules and crystals, *Phys.Rev.A* 18,335-343.

Ferraz, A. C., Chagas, M.I.T., Takahashi, E.K. and Leite, J.R. (1984) Variational cellular model of the energy bands of diamond and silicon, *Phys.Rev.B* 29,7003-7006.

Gonis, A. (1986) Multiple-scattering theory for clusters of nonoverlapping potentials of arbitrary shape, *Phys.Rev.B* 33,5914-5916.

Gonis, A., Zhang, X.-G. and Nicholson, D.M. (1988) Electronic-structure method for general space-filling cell potentials, *Phys.Rev.B* 38,3564-3567.

Gonis, A., Zhang, X.-G. and Nicholson, D.M. (1989) Multiple-scattering Green-function method for space-filling potentials, *Phys.Rev.B* 40,947-965.

Harrison, J.G., Heaton, R.A. and Lin, C.C. (1983) Self-interaction correction to the local density Hartree-Fock atomic calculations of excited and ground states, *J.Phys.B* 16,2079-2091.

Heaton, R.A., Harrison, J.G., and Lin, C.C. (1983) Self-interaction correction for density-functional theory of electronic energy bands of solids, *Phys.Rev.B* 28,5992-6007.

Hohenberg, P. and Kohn, W. (1964) Inhomogeneous electron gas, *Phys.Rev.* 136,B864-B871.

Janak, J.F. (1978) Proof that $\partial E/\partial n_i = \epsilon_i$ in density-functional theory, *Phys.Rev.* B18,7165-7168.

Johnson, K.H. and Smith, F.C.,Jr. (1971) Bands, Bonds, and Boundaries. In *Computational Methods in Band Theory* ed. P.M. Marcus, J.F. Janak, A.R. Williams, Plenum, New York,377-399.

Jones, R.O. and Gunnarsson, O. (1989) Density functional formalism, *Rev.Mod.Phys.* 61, 689-746.

Kasowski, R.V. and Andersen, O.K. (1972) Muffin tin orbitals in open structure, *Solid State Commun.* 11,799-802.

Kohn, W. (1952) Variational methods for periodic lattices, *Phys.Rev.* 87, 472-481.

Kohn, W. and Rostoker, N. (1954) Solution of the Schrödinger equation in periodic lattices with an application to metallic lithium, *Phys.Rev.* 94,1111-1120.

Kohn, W. and Sham, L.J. (1965) Self consistent equations including exchange and correlation effects, *Phys.Rev.* 140,A1133-A1138.

Korringa, J. (1947) On the calculation of the energy of a Bloch wave in a metal, *Physica* 13,392-400.

Korringa, J. (1994) Early History of Multiple Scattering Theory for Ordered Systems, *Physics Reports* 238,341-363.

Landau, L.D. (1956) The theory of a Fermi liquid, *Zh.Eksp.Teor.Fiz.* 30,1058-1064. [*Sov.Phys.JETP* 3,920-925 (1957)]

Landau, L.D. (1957) Oscillations in a Fermi liquid, *Zh.Eksp.Teor.Fiz.* **32**,59-66. [*Sov.Phys.JETP* **5**,101-108 (1957)]

Lee, C., Yang, W. and Parr, R.G. (1988) Development of the Colle-Salvetti correlation-energy formula into a functional of the electron density, *Phys.Rev.* **B37**,785-789.

Levy, M. (1979a) Universal functionals of the density and first-order density matrices, *Bull.Am.Phys.Soc.* **24**,626.

Levy, M. (1979b) Universal variational functionals of electron densities, first-order density matrices, and natural spin-orbitals and solution of the v-representability problem, *Proc.Natl.Acad.Sci.* **76**,6062-6065.

Loucks, T.L. (1967), pp 98-103. *Augmented Plane Wave Method*, Benjamin, New York.

Molenaar, J. (1988) Multiple-scattering theory beyond the muffin-tin approximation, *J.Phys.C* **21**,1455-1468.

Morgan, J. van W. (1977) Integration of Poisson's equation for a complex system with arbitrary geometry, *J.Phys.C* **10**,1181-1202.

Moruzzi, V.L., Janak, J.F. and Williams, A.R. (1978) *Calculated Electronic Properties of Metals*, Pergamon Press, New York.

Nesbet, R.K. (1958) Brueckner's Theory and the Method of Superposition of Configurations, *Phys.Rev.* **109**,1632-1638.

Nesbet, R.K. (1965) Electronic correlation in atoms and molecules, *Adv.Chem.Phys.* **9** 321-363.

Nesbet, R.K. (1969) Atomic Bethe-Goldstone equations, *Adv.Chem.Phys.* **14** 1-34.

Nesbet, R.K. (1980) *Variational Methods in Electron-Atom Scattering Theory*, Plenum, New York.

Nesbet, R.K. (1984) R-matrix formalism for local cells of arbitrary geometry, *Phys.Rev.B* **30**,4230-4234.

Nesbet, R.K. (1986a) Linearized atomic-cell orbital method for energy-band calculations, *Phys.Rev.B* **33**,8027-8034.

Nesbet, R.K. (1986b) Nonperturbative theory of exchange and correlation in one-electron quasiparticle states, *Phys.Rev.B* **34**,1526-1538.

Nesbet, R.K. (1988) Variational methods for cellular models, *Phys.Rev.A* **38**,4955-4960.

Nesbet, R.K. (1990a) Full-potential multiple scattering theory, *Phys.Rev.B* **41**,4948-4952.

Nesbet, R.K. (1990b) Atomic Cell Method for Total Energy Calculations, *Bull.Am.Phys.Soc.* **35**,418.

Nesbet, R.K. (1992a) Variational principles for full-potential multiple scattering theory, *Mat.Res.Symp.Proc.* **253**,153-158.

Nesbet, R.K. (1992b) Internal sums in full-potential multiple scattering theory, *Phys.Rev.B* **45**,11491-11495.

Nesbet, R.K. (1992c) Full-potential revision of coherent-potential- approximation alloy theory, *Phys.Rev.B* **45**,13234-13238.

Nesbet, R.K. (1992d) Full-potential multiple-scattering theory without structure constants, *Phys.Rev.B* **46**,9935-9939.

Nesbet, R.K. (1996) Alternative density functional theory for atoms and molecules. *J.Phys.B* **29**,L173-L179.

Nesbet, R.K. and Sun, T. (1987) Self-consistent calculations using canonical scaling in the linearized atomic-cell orbital method: Energy bands of fcc Cu, *Phys.Rev.B* **36**,6351-6355.

Newton, R. G. (1990) Korringa-Kohn-Rostoker Spectral-Band Theory for General Potentials, *Phys.Rev.Lett.* **65**,2031-2034.

Nozières, P. (1964) *Theory of Interacting Fermi Systems*, Benjamin, New York.

Parr, R.G. and Yang, W. (1989) *Density-Functional Theory of Atoms and Molecules*, Oxford University Press, New York.

Perdew, J.P., Chevary, J.A., Vosko, S.H., Jackson, K.A., Pederson, M.R., Singh, D.J. and Fiolhais, C. (1992) Atoms, molecules, solids, and surfaces: Applications of the generalized gradient approximation for exchange and correlation, *Phys.Rev.B* **46**,6671-6687.

Pedersen, M.R., Heaton, R.A. and Lin, C.C.(1984) Local-density Hartree-Fock theory of electronic states of molecules with self-interaction correction, *J.Chem.Phys* 80, 1972-1975.

Pedersen, M.R. and Lin, C.C. (1988) Localized and canonical atomic orbitals in self-interaction corrected local density functional approximation, *J.Chem.Phys.* 88,1807-1817.

Perdew, J.P., Parr, R.G., Levy, M. and Balduz, J.L. (1982) Density-functional theory for fractional particle number: derivative discontinuities of the energy, *Phys.Rev.Lett.* 49,1691-1694.

Perdew, J.P. and Zunger, A. (1981) Self-interaction correction to density-functional approximations for many-electron systems, *Phys.Rev.* B23,5048-5079.

Schlosser, H. and Marcus, P. (1963) Composite wave variational method for solution of the energy-band problem in solids, *Phys.Rev.* 131,2529-2546.

Skriver, H.L. (1984) *The LMTO Method*, Springer-Verlag, New York.

Svane, A. (1992) Electronic structure of La_2CuO_4 in the self-interaction corrected density functional formalism, *Phys.Rev.Lett.* 68,1900-1903.

Svane, A. (1994) Electronic structure of Cerium in the self-interaction corrected local spin density approximation, *Phys.Rev.Lett.* 72,1248-1251.

Svane, A. and Gunnarsson, O. (1988) Localization in the self-interaction corrected density-functional formalism, *Phys.Rev.B* 37,9919-9922.

Svane, A. and Gunnarsson, O. (1990) Transition-metal oxides in the self-interaction corrected density-functional formalism, *Phys.Rev.Lett.* 65,1148-1151.

Szotek, Z.,Temmerman, W.M. and Winter, H. (1993) Application of the self-interaction correction to transition metal oxides, *Phys.Rev.B* 47,4029-4032.

Szotek, Z.,Temmerman, W.M. and Winter, H. (1994) Self-interaction corrected, local spin density description of the $\gamma \to \alpha$ transition in Ce, *Phys.Rev.Lett.* 72,1244-1247.

Temmerman, W.M.,Szotek, Z. and Winter, H. (1993) Self-interaction corrected electronic structure of La_2CuO_4, *Phys.Rev.B* 47,11533-11536.

Williams, A.R., Kübler, K. and Gelatt, C.D. (1979) Cohesive properties of metallic compounds: Augmented-spherical-wave calculations, *Phys.Rev.B* 19,6094-6118.

Williams, A.R. and Morgan, J.van W. (1972) Multiple scattering by non-muffin-tin potentials, *J.Phys.C* 5,L293-L298.

Williams, A.R. and Morgan, J.van W. (1974) Multiple scattering by non-muffin-tin potentials: general formulation, *J.Phys.C* 7,37-60.

Zeller, R. (1987) Multiple-scattering solution of Schrödinger's equation for potentials of general shape, *J.Phys.C* 20,2347-2360.

Zeller, R. (1988) Empty-lattice test for non-muffin-tin multiple-scattering equations, *Phys.Rev.B* 38,5993-6002.

Zhang, X.-G. and Butler, W.H. (1992a) Simple cellular method for the exact solution of the one-electron Schrödinger equation, *Phys.Rev.Lett.* 68,3753-3756.

Zhang, X.-G. and Butler, W.H. (1992b) Multiple-scattering theory with a truncated basis set, *Phys.Rev.B* 46,7433-7447.

Zhang, X.-G. and Gonis, A. (1989) Secular equation of Korringa, Kohn, and Rostoker for the case of non-muffin-tin, space-filling potential cells, *Phys.Rev.B* 39,10373-10375.

Zhang, X.-G., Butler, W.H., Nicholson, D.M. and Nesbet, R.K. (1992) Green-function cellular method for the electronic structure of molecules and solids, *Phys.Rev.B* 46,15031-15039.

Zhang, X.-G., Butler, W.H., MacLaren, J.M. and van Ek, J. (1994) Cellular solutions for the Poisson equation in extended systems, *Phys.Rev.B* 49,13383-13393.

Ziesche, P. (1974) Multiple scattering within finite and infinite systems of generalized muffin-tin potentials. Generalizations of the cluster equations, the Lloyd formula and the KKR equations, *J.Phys.C* 7,1085-1097.

justified. Slater [3] and, later, Kimball [4] relied on the use of group theory for their definition. All of these discussed the angular factors of the orbitals and assumed that the radial factors were the same. (For further comments on earlier work and full references see Herman [5].) We need a new definition of hybrids which will be sufficiently general to apply to many different basis sets and to those with rather different radial factors. Our starting point will be the position operators x, y, and z. Throughout this work we have to distinguish between the use of x, y, z as operators, as basis functions and as variables even though, in some examples, they act in more than one way. Since these position operators obviously commute with one another as operators it is natural to ask that their matrix representatives in the required basis set should also commute. These matrices will then have a set of simultaneous eigenfunctions and these we will define as our localized hybrids.

In a recent series of papers [6, 7, 8, 9, 10, 11] we have devised methods of finding these matrices and developed the theory of their eigenfunctions. Examples have been given which are drawn from one, two and three dimensional situations. Emphasis has been placed on the angular behaviour of the orbital functions since this largely determines the results. There are several possible forms in use for the orbital radial factors and these need further investigation which is now in progress.

2. Use of symmetry

It is very convenient to follow Slater and Kimball and make the maximum use of symmetry in finding the localized hybrids. In a typical basis set for an atom we have the possibility of a transformation of the orbital basis functions into one or more sets of equivalent hybrids depending on the angular factors involved. An *equivalent set* of hybrids is one whose hybrids are permuted into one another by each element of the relevant symmetry group [12]. The combination of several sets raises some problems which will be discussed in Section 8 so, until then, we assume a single set of hybrids having their directions pointing towards the vertices of a regular polyhedron around the nucleus at the origin. We consider first an illustrative example and then develop the general theory.

LOCALIZED ATOMIC HYBRIDS:
A GENERAL THEORY

G. G. HALL
Shell Centre for Mathematical Education,
University of Nottingham, Nottingham, NG7 2RD
D. REES
Department of Mathematics,
University of Nottingham, Nottingham, NG7 2RD

1. Introduction

The combination of atomic orbitals into hybrids in order to achieve a better quantum chemical description of directed valence can be traced back to Pauling [1], Hultgren [2] and Slater [3] who derived and used the familiar tetrahedral hybrids to explain the directional characteristics of the tetrahedral carbon atom. Since their work is now seen as rather limited, because of the minimal basis sets which they used, and since directed valence is still needed for didactic and interpretative purposes, it is important to look at the topic afresh. In the context of modern computer calculations there are few advantages in the use of hybrid orbitals until we attempt to interpret the results. Hybrids can then play a major role in analysing meaningfully the structure of a molecule and relating it to its chemical analogues.

Pauling's discussion [1] started from a definition of the strength of a hybrid which may now seem arbitrary. He assigned to an s orbital the unit value and to p orbitals the values of $\sqrt{3}$. We will show later how these ideas may be made precise and

2.1. TETRAHEDRAL EXAMPLE

As an example of the group theoretical procedure [9] for calculating hybrids, we take a tetrahedral set consisting of the four tetrahedral vertices or grid points, labelled p_1, p_2, p_3, p_4 respectively, and situated at

$$\mathbf{r}_1=(1,1,1),\ \mathbf{r}_2=(1,-1,-1),\ \mathbf{r}_3=(-1,1,-1),\ \mathbf{r}_4=(-1,-1,1). \tag{1}$$

We note that these lie on the surface of a sphere of radius $\sqrt{3}$. These have the symmetry group T_d. We can take formal linear combinations of these points according to the irreducible representations of this group *(reps)*. The first of these combinations is simply their sum. It has the A_1 symmetry and is, after normalization,

$$P_1 = 1/2(p_1+p_2+p_3+p_4). \tag{2}$$

The next three have F_2 symmetry and are

$$\begin{align} P_2 &= 1/2(p_1+p_2-p_3-p_4), \\ P_3 &= 1/2(p_1-p_2+p_3-p_4), \\ P_4 &= 1/2(p_1-p_2-p_3+p_4). \end{align} \tag{3}$$

The various coefficients in these equations can be collected into a matrix U. The U matrix here is then

$$U = 1/2 \begin{pmatrix} 1 & 1 & 1 & 1 \\ 1 & 1 & -1 & -1 \\ 1 & -1 & 1 & -1 \\ 1 & -1 & -1 & 1 \end{pmatrix}. \tag{4}$$

This matrix is unitary and self-inverse. By using the coordinates of the points in turn as eigenvalues and the columns of U as the common eigenvectors, we can construct [9] the matrices which will be shown to represent the operators x, y, z and are given by:

$$X_{st} = \sum_v x_v U_{sv} U_{tv},\ Y_{st} = \sum_v y_v U_{sv} U_{tv},\ Z_{st} = \sum_v z_v U_{sv} U_{tv}, \tag{5}$$

where (x_v, y_v, z_v) are the coordinates of p_v. In this example, they will be

$$X = \begin{pmatrix} 0 & 1 & 0 & 0 \\ 1 & 0 & 0 & 0 \\ 0 & 0 & 0 & 1 \\ 0 & 0 & 1 & 0 \end{pmatrix},\ Y = \begin{pmatrix} 0 & 0 & 1 & 0 \\ 0 & 0 & 0 & 1 \\ 1 & 0 & 0 & 0 \\ 0 & 1 & 0 & 0 \end{pmatrix},\ Z = \begin{pmatrix} 0 & 0 & 0 & 1 \\ 0 & 0 & 1 & 0 \\ 0 & 1 & 0 & 0 \\ 1 & 0 & 0 & 0 \end{pmatrix}. \tag{6}$$

It follows from this definition that the columns of U are the required common eigenvectors.

The atomic orbitals of the basis set, whose angular factors are spherical harmonics, can be classified according to the various reps of T_d. By selecting an s orbital, which has the same symmetry as (2) and three p orbitals, which have the symmetry of the linear combinations (3), and applying U^T, the transpose of U (but here $U^T=U$), we can define four hybrids which will permute into one another under the T_d group. Thus from s, (p_x, p_y, p_z), spanning the A_1 and F_2 reps, we can produce four tetrahedral hybrids using the U matrix (4) viz.

$$h_1 = 1/2(s + p_x + p_y + p_z), \tag{7}$$
$$h_2 = 1/2(s + p_x - p_y - p_z), \tag{8}$$
$$h_3 = 1/2(s - p_x + p_y - p_z), \tag{9}$$
$$h_4 = 1/2(s - p_x - p_y + p_z). \tag{10}$$

The strength of a hybrid can be defined as the value of the function at its grid point, which is also its mean centre. If these orbitals are assumed to have the same radial factors then their directional properties are fixed by their angular factors only. Since the p functions have a normalization factor of $\sqrt{3}$ while s has 1, the strength of these hybrids will agree with the Pauling definitions. Each of these tetrahedral hybrids has a strength of 2. The shape of these hybrids must depend also on the radial factors of the s and p orbitals but their directions are fixed.

2.2. USE OF GROUP THEORY

The angular dependence of the basis functions in three dimensions can be discussed in general using group theory [9]. Around an isolated nucleus there is every reason to construct the hybrids as symmetrically as possible. To generate a set of equivalent hybrid functions we start from a given number, N, of equivalent points, with labels p_t located at r_t, t=1,..., N, no two of which coincide, and which admit a symmetry point group G. Only certain values of N are possible in three dimensions, N=1, 4, 6, 8, 12 and 24 being the most used. The origin of the position vectors of the points will be at the nucleus and their directions will be the directions of the hybrids. Equivalent means that there are permutations, belonging to G, which bring the vertices into exact coincidence with one another. When so acted on by any element of G, these points will permute among themselves and the effect is the same as that produced by a N×N permutation matrix, i.e. the points span a permutation representation of G. The vectors of these points, in the same set, must all have the same length so length can be used as one test of how many equivalent sets an arbitrary set of points will have.

For our present purpose it is helpful to assume that G is the octahedral group O_h or one of its subgroups. This group and the icosahedral group, I_h, are the two largest finite groups which are subgroups of $O(3)$, the group of orthogonal matrices in 3 dimensions. All the regular solids [13] have symmetry groups which are subgroups of one or other of these two. The vertices of such a solid will be a set of equivalent points. All these points will lie on the surface of a sphere. Thus the regular solids will determine geometrically the possible sets of equivalent hybrids that can be constructed for an atom. From these points, by applying the group projectors to any one vertex, elements of the group algebra, which are linear combinations of point labels, are deduced and these span various irreducible representations (reps) of the group. Each equivalent set contains the identity representation, A_{1g}, exactly once so the first combination will be the simple formal sum of all the points

$$P_1 = \frac{1}{\sqrt{N}} \sum_t p_t. \tag{11}$$

The factor $1/\sqrt{N}$ has been added for normalization. Note that the points are here treated as independent (as in an N-particle space). The complete set of coefficients in the various reps, formed from one equivalent set, will form a N×N unitary matrix U. If no rep occurs more than once in the set, then this procedure will produce a unique matrix U apart from a possible phase change in each combination.

In order that a set of orthonormal basis functions f_t, t=1,...N, that belong to a number of irreducible representations of G, can be transformed into equivalent hybrids, h_s, s=1,...,N, their irreducible representations must be those that occur in the combinations constructed above. For example, the identity representation must occur exactly once in each equivalent set, and its basis function will be taken as f_1. The unitary matrix U defines the orthonormal hybrids in terms of the basis functions spanning the given reps, i.e.:

$$h_s = \sum_{t=1}^{N} f_t U_{ts}. \tag{12}$$

These are the required functions.

3. Basis space

The finite set of orthonormal basis functions f_t, t=1, ..., N, will be taken as a basis defining a finite subspace of Hilbert space which we will call the basis space H. The hybrids h_t, as defined by (12), will be an alternative orthonormal basis set for H with the property, ensured by their construction, that they permute under every operation of the group. The function space of these functions will have a scalar product defined by integration. For functions defined on the surface of a sphere this integral is

$$<f\,g> = \frac{1}{4\pi}\int_0^\pi \sin\theta d\theta \int_0^{2\pi} f(\theta,\phi)g(\theta,\phi)d\phi. \qquad (13)$$

For functions with radial factors the scalar product integration is over all space and has the $1/4\pi$ factor. The functions will be normalised so that the basis set is orthonormal i.e.

$$<f_t\,f_s> = \delta_{ts}. \qquad (14)$$

The first angular function, f_1, in the basis set in H, which has A_{1g} symmetry, is chosen to be a constant. For f_1 to be normalized, $<f_1 f_1>=1$ and, since the scalar product has been so defined, $f_1=1$. The functions f_2, f_3, f_4, are defined as $f_2=x=\sqrt{3}\sin\theta\cos\phi$, $f_3=y=\sqrt{3}\sin\theta\sin\phi$, $f_4=z=\sqrt{3}\cos\theta$. We note that this is equivalent to taking the sphere radius as $\sqrt{3}$. Since $<x^2>=<y^2>=<z^2>$ by symmetry, it follows that the normalization $<f_2 f_2>=<f_3 f_3>=<f_4 f_4>=1$ is ensured by this choice for the radius of the sphere. If f_1 has a radial factor then it is convenient to take f_1^2 as a weight function in the integral and define a new $f_1 = 1$. If, further, the next basis function has the same radial factor i.e. $f_2=xf_1$, it can be redefined as $f_2=x$ and, similarly, $f_3=y$, $f_4=z$.

The commuting matrices representing x, y, and z have already been defined from the group theoretic results (5) but their relation to the basis functions must now be clarified. The representative matrices of quantum operators in H can be defined, in the usual quantum mechanical way, by integration of the operators over the different basis functions of H. The X matrix, representing the x operator, in contrast with the definition in (5), will now be:

$$X_{ij} = <f_i\,x\,f_j>, \qquad (15)$$

where f_i are the orthonormal basis functions having

$$\langle f_i f_j \rangle = \delta_{ij}. \tag{16}$$

Y and Z, representing the y and z operators, are similarly defined. Note that this definition and that of f_2, f_3, f_4, above also ensures that the matrix elements satisfy

$$X_{12}=\langle f_1 x f_2 \rangle = 1, Y_{13}=1, Z_{14}=1. \tag{17}$$

Furthermore, since xf_1 is given exactly by f_2, the first row and column in the matrix for the x operator has just this one non zero matrix element.

In a realistic calculation every basis function is the product of radial and angular factors. It has proved convenient to maintain this product form as far as possible and to normalize each factor separately. This calculation of matrices, using (15), representing the x, y, z operators, primarily concerns the angular factors but nevertheless includes the radial integration in the definition of the scalar product. The separation can be made more explicit if the radial factor in each cartesian operator is removed and the operator is normalized. The angular operators are defined as functions on the surface of a sphere of radius $\sqrt{3}$. Thus the operator x becomes $\sqrt{3}\sin\theta \cos\phi$ in the integrals and the operators satisfy $x^2+y^2+z^2=3$. Integration over the radial and angular factors then becomes independent. This allows us to discuss the matrices independently of the radial factors. The matrices resulting from this process are essentially angular and will be called the *angular matrices*. They will satisfy the relation $X^2+Y^2+Z^2=3I$ where I is the unit matrix. The resulting eigenvalues of X, Y, and Z will define the directions of the hybrid centres but not their radii. The radii can be recovered by replacing $\sqrt{3}$ by r and integrating over the radial variable.

Unfortunately, if the matrix elements of the operators x, y, z are defined in this way, the resulting matrices do not commute and do not agree with those defined group theoretically (5). To produce the same commuting matrices it is necessary to introduce the concept of *alias functions* [7] which supplement the operators and extend their action in the given finite function space. If A_x is an alias function for the x operator, then the matrix elements may now be defined as

$$X_{ij} = \langle f_i(x+A_x)f_j \rangle. \tag{18}$$

Alias functions A_y, A_z for the y and z operators are obtained by cyclic permutation of the variables. The alias functions for our tetrahedral example [9], using the s, p_x, p_y, p_z basis functions, are $(\lambda yz, \lambda zx, \lambda xy)$, again taking $r=\sqrt{3}$. These belong to the same rep (i.e. F_2) of the T_d group as the (x, y, z) functions but are orthogonal to them, have higher degree and are opposite in parity. The constant parameter λ is determined by the vanishing of the commutators and, in this example, λ=5/3. The matrices X, Y, Z, defined as in (18), can be proved to be identical with their group theory predecessors (5) and their simultaneous eigenfunctions are the familiar tetrahedral hybrids and their eigenvalues are the coordinates of their centres (1). Since the hybrids $h_s(\mathbf{r})$ are the eigenfunctions, the representation of the x operator in their basis will be the diagonal matrix

$$\langle h_s (x+A_x) h_t \rangle = x_s \delta_{st}. \tag{19}$$

4. Sample space

We now define [11] a new space, S, the sample space of H, in which each function of H is represented by an N-dimensional vector whose components are the values of the function at the given set of grid points, i.e. at the $\mathbf{r}_s$ of Section 2. Thus, for example, the continuous function $g(\mathbf{r})$ is represented by the vector g_s of its discrete values at the grid points with a weighting factor $1/\sqrt{N}$,

$$g_s = g(\mathbf{r}_s)/\sqrt{N}, \quad s=1,...,N. \tag{20}$$

Thus S will be an N-tuple vector space and will have a scalar product defined in the usual way as the sum over components, e.g. the product of the functions n and m is

$$\{nm\} = \Sigma n_s m_s = \Sigma n(\mathbf{r}_s)m(\mathbf{r}_s)/N. \tag{21}$$

The hybrids $h_s(\mathbf{r})$ have a special significance in S since they are represented by the unit basis vectors:

$$h_s = h_s(\mathbf{r}_t)/\sqrt{N} = \delta_{st}. \tag{22}$$

We have proved [11] that each function in H corresponds uniquely to a vector in S and, conversely, each vector in S corresponds to a function in H. Thus the function space H is represented, one to one, by the N-dimensional vector space S, the sample space of H.

In S, f_1=1 will be represented as a vector with all components equal to $1/\sqrt{N}$. This ensures that $\{f_1 f_1\}$=1 so that f_1 is normal both in S and in H. Since f_2 at the grid points takes the value of their x coordinates its normalization involves the sum Σx_s^2. Now, by O_h symmetry,

LOCALIZED ATOMIC HYBRIDS

$$\sum x_s^2 = \sum y_s^2 = \sum z_s^2. \qquad (23)$$

But every grid point lies on the sphere of radius $\sqrt{3}$, so each of these will sum to N giving

$$\{f_2 f_2\} = \{f_3 f_3\} = \{f_4 f_4\} = 1 \qquad (24)$$

so that these functions, also, are normalized in S and in H. This important property is not generally true for other basis functions.

5. Alias Functions

The idea of an alias function is of major importance for our theory. We have already introduced it above, see (18), in connection with the tetrahedral hybrids example. It is needed in every example where the matrix elements of a quantum operator are found by integration over a finite set of basis functions. It corrects for the finite nature of the basis set. The discussion takes two forms depending on whether the alias function is correcting an operator or a basis function.

5.1. OPERATOR ALIAS FUNCTIONS

The simplest example of an alias function is in the problem of N functions of an angle ϕ which point towards the vertices of a regular polygon [7]. The basis functions can be taken as the complex exponentials $e^{in\phi}$, $n=0,1,...(N-1)$, and $x=(e^{i\phi}+ e^{-i\phi})/2=\cos\phi$ is the x operator. When integrated over ϕ in $(0, 2\pi)$, these do not give the correct X matrix. Compare this result with the correct matrix i.e.

$$\frac{1}{2}\begin{pmatrix} 0 & 1 & \ldots & \\ 1 & 0 & 1 & \ldots \\ \cdot & \cdot & \cdot & \cdot \\ \cdot & \cdot & \cdot & 1 \\ \cdot & \cdot & 1 & 0 \end{pmatrix}, \quad X = \frac{1}{2}\begin{pmatrix} 0 & 1 & \ldots & 1 \\ 1 & 0 & 1 & \ldots \\ \cdot & \cdot & \cdot & \cdot \\ \cdot & \cdot & \cdot & 1 \\ 1 & \cdot & \cdot & 1 & 0 \end{pmatrix}, \qquad (25)$$

and note that the two matrix elements at the corners, which maintain the n-fold symmetry, are missing in the integrated form. By replacing $e^{i\phi}$ by $e^{i\phi}+ e^{-i(N-1)\phi}$ so that x becomes $x=\cos\phi+\cos(N-1)\phi$, these elements are added and the X matrix becomes correct. The operators $e^{-i(N-1)\phi}$ and $e^{i\phi}$ describe the same rotation of the N grid points in two ways depending on whether the rotation is clockwise or anticlockwise but they contribute to the integrals in a complementary way which builds up the whole of the correct matrix. Thus, in the integral definitions, the operator x is represented by $\cos\phi + \cos(N-1)\phi$ and y by $\sin\phi - \sin(N-1)\phi$. With these additions to the operators the matrices X and Y commute and become non-singular.

In general for a spherical surface, the basis functions will be spherical harmonics though they are usually written in cartesian form for convenience, i.e. they are solid harmonics. An operator alias function is also a spherical harmonic and has the same O_h symmetry as the operator it complements. Its spherical harmonics will have larger l values than any in the basis set but not more than twice that number. This difference in l means that the two angular functions will be orthogonal.

An important property, proved in [11] (see also Section 5.2.), is that the alias function takes values at the grid points which are proportional to those of the operator itself, i.e. it has the same representation in S space except possibly for a constant multiple. Its purpose is to correct the representation of the operator in H space. The choice of alias function is not, in general, unique. Each alias function will contain a number of parameters chosen to ensure that the resulting matrices X, Y, Z commute.

We have proved [11] that the angular alias functions for the x and y operators, when written in cartesian form, can be deduced by a cyclic permutation of the x, y, z variables from the alias function for the z operator. Furthermore, an analysis of all possible angular alias functions in terms of their angular momenta has been given [11]. We now illustrate the criteria for the choice of alias function by considering the example of the square hybrids in three dimensions and also later examples.

5.1.1. *Square hybrids*

The derivation of the square hybrids from an sp^2d set of atomic orbitals proceeds in a simple fashion [8]. When all r,θ,ϕ variables are introduced to describe realistic atomic orbitals the resulting basis functions, normalized by integration over all space, and using one Slater-type exponential form for all the radial functions, are

$$s = Ne^{-\zeta r},$$
$$p_x = Ne^{-\zeta r}\sqrt{3}\sin\theta\cos\phi,$$
$$p_y = Ne^{-\zeta r}\sqrt{3}\sin\theta\sin\phi. \qquad (26)$$
$$d = Ne^{-\zeta r}\sqrt{15/2}\sin^2\theta\cos 2\phi,$$

where $N^2=4\zeta^3$. Here, as suggested in Section 3, it is useful to regard $N^2 e^{-2\zeta r}$ as a weight function in the integrals and treat the basis functions as angular functions.

Since the Z matrix vanishes, the centres of the hybrids will be in the z=0 plane. The alias functions will be dominated by their dependence on ϕ. Since the square has four vertices, i.e. four grid points, a term containing $\cos 3\phi$ is expected in the x alias function and $-\sin 3\phi$ in the y alias function. The functions which are integrated to produce the matrices representing the operators x and y are then, respectively,

$$r\sin\theta(\cos\phi + \lambda\cos 3\phi), \quad r\sin\theta(\sin\phi - \lambda\sin 3\phi). \tag{27}$$

The value of λ is determined by making the matrices commute and is $\lambda=\sqrt{(5/3)}-1$. The matrices X and Y, using the basis functions in the order given in (26), are:

$$X = r°/\sqrt{3}\begin{pmatrix} 0 & 1 & 0 & 0 \\ 1 & 0 & 0 & 1 \\ 0 & 0 & 0 & 0 \\ 0 & 1 & 0 & 0 \end{pmatrix}, \quad Y = r°/\sqrt{3}\begin{pmatrix} 0 & 0 & 1 & 0 \\ 0 & 0 & 0 & 0 \\ 1 & 0 & 0 & -1 \\ 0 & 0 & -1 & 0 \end{pmatrix}, \tag{28}$$

where $r° = 3/(2\zeta)$ is the orbital radius. The normalized eigenfunctions and their grid points are:

$$\begin{aligned} T_1 &= 1/2(s+\sqrt{2}p_x+d), \quad r°\sqrt{(2/3)}\ (1,0,0), \\ T_2 &= 1/2(s+\sqrt{2}p_y-d), \quad r°\sqrt{(2/3)}\ (0,1,0), \\ T_3 &= 1/2(s-\sqrt{2}p_y-d), \quad r°\sqrt{(2/3)}\ (0,-1,0), \\ T_4 &= 1/2(s-\sqrt{2}p_x+d), \quad r°\sqrt{(2/3)}\ (-1,0,0). \end{aligned} \tag{29}$$

These are the familiar forms for the square hybrids. Their strength, the value of the angular factor of each hybrid at its grid point, is $1/2(1+\sqrt{6}+\sqrt{15/2})=2.693$ and this high value indicates their greater localization due to the mixing with the d function. They are directed along the positive and negative x and y axes around their respective grid points. These grid points all lie on a circle of radius $r°\sqrt{(2/3)}$.

5.2. BASIS ALIAS FUNCTIONS

We have shown already that, because of the definition of the scalar product, the basis functions 1, x, y, z have the property of being normal in both H and S. This makes it easy to use them to generate the matrix U. In general, basis functions are defined as normal in H but their sample values in S are not normal. For some purposes this is not a problem since the sample values can be renormalized but, to define the correct hybrids and examine their properties, the basis function has to be corrected. The introduction of a basis alias function, which mixes with the corresponding basis function, is a method of correcting the basis function so that it becomes normal in both spaces.

The centre piece of the argument is the alias theorem which we have proved [11]. This states that, if one basis function in H, for a single equivalent set, is replaced by another function from outside H with the same O_h symmetry, an alias function, then their representative vectors in S will be the same, apart from a possible constant scale factor. The theorem may fail when there are several equivalent sets unless the function is the only one in the basis set with its symmetry. Furthermore, since orthogonality of the basis functions in most examples is due to a difference in symmetry, we prove that functions with different symmetry are orthogonal in both H and S.

Balancing the norms of the functions in H and S can be achieved in this way. For each basis function f which is normal in H but not normal in S, we need to take a linear combination of it with another function F, a *basis alias function*, which has <FF>=1, is orthogonal to f and has the same symmetry. Retaining the symmetry means that those orthogonality relations, which are due to differences in symmetry, are not disturbed. Thus the new basis function is

$$g = \lambda f + \sqrt{(1-\lambda^2)} F, \qquad (30)$$

where λ ensures that $\{gg\}=1$. The effect of substituting g for f is to normalize the vector in S without changing its norm in H. In previous work [7,8,9,10,11] we have used alias functions to modify operators. These alias functions modify the basis functions rather than the operators but there are similarities since both are terms in the integrands that define the various matrices in H space. The modified basis functions will form a new N-dimensional space H°. Their purpose is to ensure the equality of the various scalar product integrals in H° with the corresponding scalar product sums in S. An important consequence of this normalization is that it can now be proved [11] that the values of these modified basis functions at the grid points are directly related to the elements of the U matrix, viz.

$$U_{st} = g_s(r_t)/\sqrt{N}. \qquad (31)$$

This result, therefore, establishes the uniqueness of the X, Y, Z matrices. Equation (31) also gives an alternative method of determining the constant λ in (30) by equating the value of the function at some grid point to the corresponding element in the group theoretical U matrix.

6. Examples of hybrids

To illustrate the power of our methods and show the complications due to the presence of alias functions we now outline the results of some spherical shell calculations [11]. The first example also shows how, by ordering the grid points and basis functions in a specific way, the inherent alternant symmetry of the situation is made explicit and the calculation simplified. This procedure can be applied to most of our examples.

6.1. OCTAHEDRAL HYBRIDS

The octahedral grid points are situated at equal distances along the three coordinate axes in both directions. We will now order these grid points and the basis functions in a special way. The six points are taken in two groups of three, at

$$(\sqrt{3},0,0), (0,\sqrt{3},0), (0,0,\sqrt{3});$$
$$(-\sqrt{3},0,0), (0,-\sqrt{3},0), (0,0,-\sqrt{3}), \quad (32)$$

where the second group are, respectively, the spatial inverses of those in the first group. The six basis functions are taken in the order A_{1g}, E_g, F_{1u}.

Because the g reps are taken before the u reps, the U matrix now becomes

$$U = \frac{1}{\sqrt{6}} \begin{pmatrix} 1 & 1 & 1 & 1 & 1 & 1 \\ \frac{\sqrt{3}}{\sqrt{2}} & -\frac{\sqrt{3}}{\sqrt{2}} & 0 & \frac{\sqrt{3}}{\sqrt{2}} & -\frac{\sqrt{3}}{\sqrt{2}} & 0 \\ -\frac{1}{\sqrt{2}} & -\frac{1}{\sqrt{2}} & \sqrt{2} & -\frac{1}{\sqrt{2}} & -\frac{1}{\sqrt{2}} & \sqrt{2} \\ \sqrt{3} & 0 & 0 & -\sqrt{3} & 0 & 0 \\ 0 & \sqrt{3} & 0 & 0 & -\sqrt{3} & 0 \\ 0 & 0 & \sqrt{3} & 0 & 0 & -\sqrt{3} \end{pmatrix}. \quad (33)$$

Note that the last three rows are formed from the x coordinates, y coordinates, and z coordinates of the grid points respectively. The form of this matrix follows from our systematic use of odd and evenness. The g reps, the first three rows, have the same values at each point and at its inverse point but the u reps, the last three rows, have opposite signs. The form of U now becomes

$$U = \begin{pmatrix} V & V \\ W & -W \end{pmatrix}. \quad (34)$$

From this U, the X, Y and Z matrices are derived (5). These are also simplified because of the ordering of the functions and the grid points. Because of the form of U, the matrices partition into the alternant form:

$$X = \begin{pmatrix} O & A \\ A^T & O \end{pmatrix}, \quad Y = \begin{pmatrix} O & B \\ B^T & O \end{pmatrix}, \quad Z = \begin{pmatrix} O & C \\ C^T & O \end{pmatrix}, \quad (35)$$

with

$$A = \begin{pmatrix} 1 & 0 & 0 \\ \frac{\sqrt{3}}{\sqrt{2}} & 0 & 0 \\ -\frac{1}{\sqrt{2}} & 0 & 0 \end{pmatrix}, \quad B = \begin{pmatrix} 0 & 1 & 0 \\ 0 & -\frac{\sqrt{3}}{\sqrt{2}} & 0 \\ 0 & -\frac{1}{\sqrt{2}} & 0 \end{pmatrix}, \quad C = \begin{pmatrix} 0 & 0 & 1 \\ 0 & 0 & 0 \\ 0 & 0 & \sqrt{2} \end{pmatrix}. \quad (36)$$

so that, for example,

$$Z = \begin{pmatrix} 0 & 0 & 0 & 0 & 0 & 1 \\ 0 & 0 & 0 & 0 & 0 & 0 \\ 0 & 0 & 0 & 0 & 0 & \sqrt{2} \\ 0 & 0 & 0 & 0 & 0 & 0 \\ 0 & 0 & 0 & 0 & 0 & 0 \\ 1 & 0 & \sqrt{2} & 0 & 0 & 0 \end{pmatrix}. \quad (37)$$

The form of the U matrix is typical of the set of eigencolumns of an alternant and the form of the operator matrices, with vanishing diagonal blocks, is the result of alternation (see [14]). The consequence is that the eigenvalues form pairs differing only in sign. Here it shows that Z has one such pair and 4 zero eigenvalues and this agrees with the z coordinates given in (32).

The orthonormal basis functions sd^2p^3, whose angular factors are spherical harmonics, have the correct O_h symmetries for octahedral hybrids and can be written in the above order as

$$s = 1, \quad d_z = \sqrt{15/6}(x^2-y^2), \quad d_h = \sqrt{5/6}(2z^2-x^2-y^2),$$
$$x \,(= \sqrt{3}\sin\theta\cos\phi), \quad y \,(= \sqrt{3}\sin\theta\sin\phi), \quad z \,(= \sqrt{3}\cos\theta). \quad (38)$$

Note that here $r=\sqrt{3}$, but that otherwise the radial factors for the basis functions do not enter into our discussion.

Because they belong to different reps, all six of these functions are orthogonal. The first and last three of these functions are normal in both H and S but the d functions, though normal in H, are not normal in S. This is a situation that requires the introduction of basis alias functions. We replace the d functions with the combinations

$$d°_z = \mu d_z + \sqrt{(1-\mu^2)} D_z, \quad d°_h = \lambda d_h + \sqrt{(1-\lambda^2)} D_h, \qquad (39)$$

where the normalized, harmonic alias functions are

$$D_z = \sqrt{5/12}(x^2-y^2)(6z^2-x^2-y^2),$$
$$D_h = \sqrt{15/36}(x^4+y^4-2z^4-12x^2y^2+6z^2x^2+6y^2z^2). \qquad (40)$$

These are the quartic functions with the symmetry of the two d functions. The two functions transform with the same coefficients since they are components of E_g, the same degenerate rep, i.e. the constants λ, μ are equal. Each is determined (c.f. equation (31)) so that the functions $d°_h$ and $d°_z$ have values at the grid points which agree with those in U (33). These conditions show that $\lambda=\mu$ and that their value is

$$\lambda = \mu = \sqrt{5(9+4\sqrt{2})}/35. \qquad (41)$$

The matrices representing the x, y and z operators, are formed by integrating these operators, modified by the operator alias functions, over these already modified basis functions. For the Z matrix this operator now becomes

$$z + \alpha z(2z^2-3x^2-3y^2), \qquad (42)$$

where the alias function is the next higher polynomial, cubic, with the symmetry, F_{1u}, of the z operator and α is determined by making this Z matrix agree with the group theoretical one above (37). This gives the value

$$\alpha = (162-75\sqrt{2})/238. \qquad (43)$$

The x and y operators are treated in the same way with alias functions obtained by cyclic permutation of the variables.

6.2 THE CUBIC HYBRIDS

The example of eight equivalent grid points, at the vertices of a cube, is discussed similarly. The points may be ordered so that the first set are at tetrahedral vertices (1) and the second set, their spatial inverses, form the opposite tetrahedron. We now order the basis functions, sp^3d^2f, so that those even under inversion are first, followed by those that are odd. This ordering leads to a partitioning for their U matrix similar to that above (34). The basis functions are

$$f_1 = 1,$$

$$f_2 = \sqrt{(5/3)}yz = \sqrt{15}\sin\theta\cos\theta\sin\phi,$$
$$f_3 = \sqrt{(5/3)}zx = \sqrt{15}\sin\theta\cos\theta\cos\phi,$$
$$f_4 = \sqrt{(5/3)}xy = \sqrt{15}\sin^2\theta\sin\phi\cos\phi, \quad (44)$$
$$f_5 = \sqrt{35/3}\,xyz = \sqrt{105}\sin^2\theta\cos\theta\sin\phi\cos\phi$$
$$f_6 = x = \sqrt{3}\sin\theta\cos\phi,$$
$$f_7 = y = \sqrt{3}\sin\theta\sin\phi,$$
$$f_8 = z = \sqrt{3}\cos\theta.$$

These functions are orthonormal in H and, therefore, orthogonal in S, but only the first and the last three are normal in S. To find a basis set which is orthonormal in both H and S, we take the linear combinations of the others with the normalized basis alias functions

$$F_2 = \sqrt{5/6}\,yz(y^2+z^2-6x^2),$$
$$F_3 = \sqrt{5/6}\,zx(z^2+x^2-6y^2), \quad (45)$$
$$F_4 = \sqrt{5/6}\,xy(x^2+y^2-6z^2),$$
$$F_5 = \sqrt{385/18}\,xyz(2z^2-x^2-y^2).$$

The combinations

$$g_2 = \lambda f_2+\sqrt{(1-\lambda^2)}F_2,\ g_3 = \lambda f_3+\sqrt{(1-\lambda^2)}F_3,$$
$$g_4 = \lambda f_4+\sqrt{(1-\lambda^2)}F_4,\ g_5 = \mu f_5+\sqrt{(1-\mu^2)}G_5, \quad (46)$$

are determined by fitting to the leading elements in the U matrix (48). The values of the constants are

$$\lambda = \sqrt{5}(3\sqrt{3}+2\sqrt{26})/35 = 0.9834988155,$$
$$\mu = 3/\sqrt{35} = 0.5070925528. \quad (47)$$

With these values, the functions g_t will be normal in both H and S as required.

The group theoretical U matrix is

$$U = \frac{1}{\sqrt{8}}\begin{pmatrix} 1 & 1 & 1 & 1 & 1 & 1 & 1 & 1 \\ 1 & 1 & -1 & -1 & 1 & 1 & -1 & -1 \\ 1 & -1 & 1 & -1 & 1 & -1 & 1 & -1 \\ 1 & -1 & -1 & 1 & 1 & -1 & -1 & 1 \\ 1 & 1 & 1 & 1 & -1 & -1 & -1 & -1 \\ 1 & 1 & -1 & -1 & -1 & -1 & 1 & 1 \\ 1 & -1 & 1 & -1 & -1 & 1 & -1 & 1 \\ 1 & -1 & -1 & 1 & -1 & 1 & 1 & -1 \end{pmatrix} \quad (48)$$

and this can be partitioned so that

$$U = \begin{pmatrix} A & A \\ A & -A \end{pmatrix}, \quad (49)$$

where

$$A = \begin{pmatrix} 1 & 1 & 1 & 1 \\ 1 & 1 & -1 & -1 \\ 1 & -1 & 1 & -1 \\ 1 & -1 & -1 & 1 \end{pmatrix} \quad (50)$$

In this example, since O_h is the direct product of T_d and the inversion, the eight basis functions are first considered as reps of T_d. The first four span the A_1 and F_2 reps and are g, i.e. even under inversion. The second four span the same reps in the same order but are u, i.e. odd under inversion. Because of this identity of reps, the two submatrices in the first column of (49) are the same and A is the same as the tetrahedral U matrix in (4). From this U, using (5), the matrices X, Y, and Z are obtained in alternant form. The Z matrix, for example, is

$$Z = \begin{pmatrix} 0 & B \\ B & 0 \end{pmatrix}, \text{ where } B = \begin{pmatrix} 0 & 0 & 0 & 1 \\ 0 & 0 & 1 & 0 \\ 0 & 1 & 0 & 0 \\ 1 & 0 & 0 & 0 \end{pmatrix}. \quad (51)$$

The alternative derivation of the X, Y, Z matrices by integration over the operators still requires operator alias functions. For Z, the z operator is modified by cubic and quintic functions of the same symmetry as z to become

$$z + \alpha z(2z^2 - 3x^2 - 3y^2) + \beta z(8z^4 - 40z^2(x^2+y^2) + 15(x^2+y^2)^2). \quad (52)$$

To fit the group theoretical Z matrix, we require the constants to have the values (these have been put into decimal form from their exact values for ease of understanding)

$$\alpha = -0.085421839,$$
$$\beta = -0.0646990013. \quad (53)$$

The X, Y, Z matrices will now agree with the group theory ones. Their eigenfunctions, which are the desired localized hybrids, are found by using the columns of U in turn as coefficients of the modified basis functions. Because of the basis alias functions, the hybrids are now guaranteed to have nodes at all grid points except their own where they reach the value $\sqrt{8}$.

6.3. OTHER HYBRIDS

The same methods have been applied to find hybrids for sets with 12 functions and 24 functions [11]. The first set of hybrids can be visualized as pointing towards the centres of the 12 edges of a cube. The second set have two grid points on each cube edge, spaced so that each face has a regular hexagon. The calculations involve the use of basis sets with more than one rep of the same symmetry. Because of these, orthogonality of the basis set in S is no longer produced by symmetry alone but has to be superimposed by ensuring that the functions are orthogonal in S as well as in H. It is then necessary for each basis function, whose symmetry matches one already used, to be mixed with another function, with the same symmetry but orthogonal to H, so that the combination is orthogonal in S to the function already included in the basis. They will then mix with one another, and may also mix with another alias function, so that the result is normal in H and S.

The one remaining possibility for an equivalent set within the O_h symmetry, one with 48 functions, has been considered too complicated to be worth calculating but all the methods for it are available. Similarly, the I_h equivalent sets, where the first one of interest has 60 basis functions, have not been attempted.

A number of examples have been calculated showing the effect of including the radial factors in the basis sets. These include two trigonal sets inside one another [8] and two tetrahedral sets inside one another [10]. The extra freedom of the radial variable has the advantage that more parameters can be introduced easily but the disadvantage that a large number of possibilities opens up for the radial factors and the results cannot easily be categorized.

7. Localization

One of the most used procedures for the localization of orbitals is that due to Boys [15]. He minimizes the sum of the variances of the occupied molecular orbitals about their centres with respect to a unitary transformation of these orbitals among themselves. The centre of the localized function b_s is defined by

$$x_s = <b_s\, x\, b_s>, \quad y_s = <b_s\, y\, b_s>, \quad z_s = <b_s\, z\, b_s>. \qquad (54)$$

Its variance is

$$<(xb_s-x_sb_s)^2> + <(yb_s-y_sb_s)^2> + <(zb_s-z_sb_s)^2>. \qquad (55)$$

The minimizing of the sum of these variances can be interpreted as a least squares fit of the functions $b_s(r)$, formed from the occupied orbitals by a unitary transformation, to the three simultaneous eigenvector equations.

$$xb_s(r) = x_s\, b_s(r),\; yb_s(r) = y_s\, b_s(r),\; zb_s(r) = z_s\, b_s(r). \qquad (56)$$

This Boys definition can be extended readily to give a transformation of any arbitrary basis set to localized form.

This definition and procedure can be considered as a practical approximation to our problem of finding common eigenfunctions for the x, y and z operators. However, since no alias functions are included and the basis sets are finite, the equations (56) have no exact eigensolution and the localized functions that result are approximations to the true localized eigenfunctions. In effect, the matrix elements due to the alias functions, which our work shows to be important, are absent and their absence will affect the results. We can, nevertheless, claim that our localized hybrids are localized in a sense close to that intended by Boys.

7.1. PROPERTIES OF LOCALIZED HYBRIDS

The localized hybrids h_t are formed from the columns of the U matrix, the eigenvectors, and the basis functions as

$$h_t = \sum_{s=1}^{N} f_s(r_t) f_s(r)/\sqrt{N}. \qquad (57)$$

The constant $\sqrt{N}$ ensures that the column of function values is normalized to 1. Several of the properties of hybrids can be deduced immediately from this definition. It is clear, for example, that h_t will vanish at the grid point r_w of any other hybrid since the sum (57) then becomes the scalar product of two different eigenvectors, which must vanish. This nodal property is one important aspect of localization. The centre of each hybrid is at its own grid point by definition of the eigenfunction.

Another important property is the strength, the value of the hybrid at its own grid point. For N equivalent functions in one set, this value will be $\sqrt{N}$. If the hybrid is normalized to 1, its value is a measure of the concentration of the localized function at that point. The greater the number of functions the greater the localization. This was illustrated in our paper [6] with an example in one dimension having N=31, but the result is true generally. In higher dimensions the effect of their normalization in H will ensure that the hybrids die away from their centres faster than in one-dimension.

8. Several equivalent sets

When an atom is described by a large basis set, it is often possible to associate these basis functions together into sets each of which can be transformed into localized hybrids in different directions and with a different radius for each set. The group theoretic restriction is that the reps of the basis functions should exactly match the reps of the number of equivalent sets. If necessary, a number of additional basis functions with specific symmetries can be included to make this possible. It is sometimes possible, where this restriction is not satisfied using the group G, to succeed by using a subgroup of G instead of G itself. The tetrahedral example, where T_d is used instead of O_h, illustrates this. Each of these sets can be treated as we have described for a single equivalent set provided that the sets do not interfere. This problem of removing interference must now be considered.

First we describe the nature of the problem. The matrix U continues to play a fundamental role in the theory. In general, when there are several equivalent sets, this unitary matrix, which transforms the basis functions into the final hybrids, is given a block structure with one diagonal block, such as (4), for each equivalent set; all off-diagonal blocks being zero. We note that, to achieve this, a preliminary adjustment of the basis set is required in which those basis functions in different sets that have the same symmetry are modified in order that each may have nodes at the radii of the other sets so that there is no interference in U. They also need to have zero matrix elements of X, Y, Z with each other in H. Satisfying these two conditions enables the various basis sets to achieve this block structure. All the basis functions with the A_{1g} symmetry, for example, must be given nodes at each other's centres and sufficient adjustable parameters to ensure that the matrix elements in the operator matrices outside the diagonal blocks will vanish. Other criteria, which arise from outside group theory, may also be applied. Our approaches differ in the practical means employed to make these adjustments. We outline here, in Section 8.1., one approach to this problem and, in Section 8.2., comment on a possible alternative approach.

8.1. RADIAL EIGENFUNCTIONS

A procedure for treating multiple shells using the notion of a direct product has been explored [10]. Each basis function is the product of a radial factor and an angular factor. The same radial factor is used for all the functions in each equivalent set. The radial factor then distinguishes the sets and the angular factor gives the distribution of hybrids within the set. The scalar product integration now includes the radial variable. The orthogonality of the radial factors ensures the orthogonality of the basis functions in H. It is helpful, in this approach, to introduce the R matrix as the matrix of the r operator:

$$R_{st} = <f_s \, r \, f_t>. \tag{58}$$

This matrix is not usually diagonal even when f_s are the common eigenfunctions of X, Y, Z. By imposing as a further condition on the basis functions, that R should be diagonal, we obtain unique hybrids. This means requiring the R matrix to commute with the angular X, Y and Z matrices. The angular factors of the basis functions ensure that many off-diagonal elements in R will vanish. The eigenvalues of these three angular matrices give the directions of the grid points, the centres of the eigenfunctions, and those of the R matrix give their radii.

While this solution is a practical one, it has some difficulties. It greatly restricts the radial factors of all the basis functions in each equivalent set by forcing them to be the same. Its description of the innermost hybrid, centred on the origin, tends to be rather different from the traditional 1s and there is an energy penalty to be paid for this. Because this approach requires use of the same radial factor for s and p functions, it gives a new justification for the use of the Slater nodeless 2s function, for the tetrahedral hybrids, rather than a 2s function having a radial node. The cost of this is that the 1s function then has a radial node at the radius of the sphere containing the centres of all the tetrahedral hybrids.

8.1.1. Five function example

We can obtain a realization of the five basis function example with two equivalent sets, a tetrahedral set and one hybrid at the origin. This reproduces all four commuting matrices and uses the following five orthonormal basis functions (see (13) and (14)):

$$1s = A\alpha^{3/2}(\alpha r - 5/2)(\alpha^2 r^2 + 6\alpha r - 20/3)e^{-2\alpha r},$$
$$2s = 2\alpha^{5/2} r e^{-\alpha r}/\sqrt{3},$$
$$p_x = 2\alpha^{5/2} r e^{-\alpha r}\sin\theta\cos\phi,$$
$$p_y = 2\alpha^{5/2} r e^{-\alpha r}\sin\theta\sin\phi, \tag{59}$$
$$p_z = 2\alpha^{5/2} r e^{-\alpha r}\cos\theta,$$

where $A = 24\sqrt{2}/\sqrt{1551}$ and α remains arbitrary. Note that the 2s has been given the same radial factor as the p functions and that the 1s has been adjusted to be orthogonal to 2s and to be an eigenfunction of R. The 1s has also been given a radial node at $5/2\alpha$, which is the radius of the 2s function. It also has nodes at $(3\pm\sqrt{(47/3)})/\alpha$. Each tetrahedral hybrid, formed from the 2s and the p functions, has a node at the origin because of its r factor. The R matrix, with the ordering as in (59), is diagonal i.e.

$$R = \begin{pmatrix} P & 0 & 0 & 0 & 0 \\ 0 & S & 0 & 0 & 0 \\ 0 & 0 & S & 0 & 0 \\ 0 & 0 & 0 & S & 0 \\ 0 & 0 & 0 & 0 & S \end{pmatrix} \tag{60}$$

where $P = 17301/(39808\alpha)$, $S = 5/(2\alpha)$. From this, the radius of the sphere containing the tetrahedral grid points is $5/(2\alpha)$, which is at a node of the 1s function. The small radius of the 1s function, P, shows the localized nature of this function. The U matrix now has the form

$$U = \begin{pmatrix} 1 & 0 & 0 & 0 & 0 \\ 0 & 1/2 & 1/2 & 1/2 & 1/2 \\ 0 & 1/2 & 1/2 & -1/2 & -1/2 \\ 0 & 1/2 & -1/2 & 1/2 & -1/2 \\ 0 & 1/2 & -1/2 & -1/2 & 1/2 \end{pmatrix}. \tag{61}$$

The angular matrix X has the same form as in Section 3 since the radial factors can be integrated out of the integral for each matrix element. Thus the angular operators and the angular factors are the same as before so that the operator alias function is also the same, viz.

$$A_x = \lambda yz = 3\lambda\sin\theta\cos\theta\sin\phi, \quad \lambda = 5/\sqrt{3}. \tag{62}$$

The angular matrices now have the corresponding block form, c.f. (6):

$$X = \begin{pmatrix} 0&0&0&0&0 \\ 0&0&1&0&0 \\ 0&1&0&0&0 \\ 0&0&0&0&1 \\ 0&0&0&1&0 \end{pmatrix}, Y = \begin{pmatrix} 0&0&0&0&0 \\ 0&0&0&1&0 \\ 0&0&0&0&1 \\ 0&1&0&0&0 \\ 0&0&1&0&0 \end{pmatrix}, Z = \begin{pmatrix} 0&0&0&0&0 \\ 0&0&0&0&1 \\ 0&0&0&1&0 \\ 0&0&1&0&0 \\ 0&1&0&0&0 \end{pmatrix}. \quad (63)$$

The strength of these hybrids is of interest. The 1s function has the strength:

$$\alpha_1 = 80\sqrt{10}/\sqrt{311}\ \alpha^{3/2}. \quad (64)$$

The tetrahedral hybrids each have the strength:

$$\alpha_2 = 10/\sqrt{3}\ e^{-5/2}\ \alpha^{3/2}. \quad (65)$$

8.1.2. Eleven function example

With several different sets of hybrids, the orientation of the sets may become a problem. It has sometimes been thought that hybrid sets such as a tetrahedral set inside an octahedral set were incompatible. We show in this example that this is not true. In the atom it is their relative orientation which is important. Their orientation in space will be modified in the molecular environment by the interaction effects due to the other atoms.

With 11 basis functions in the configuration $(1s)(2s)(2p)^3(3s)(3p)^3(3d)^2$ it is possible to have three equivalent sets since there are three s functions. The outer shell is octahedral, while the inner shell is tetrahedral and there is a hybrid at the origin. We take a as the radius of the tetrahedral shell and b as the radius of the octahedral shell. The basis functions (66) have the form:

$$\begin{aligned}
1S &= A(r-a)(r-b)(r^4-cr^3+dr^2+er-k)e^{-6r}, \\
2S &= Br(r-b)(r^2-hr+j)e^{-2r}, \\
2p_x &= 2S\ \sqrt{3}\ \sin\theta\cos\phi, \\
2p_y &= 2S\ \sqrt{3}\ \sin\theta\sin\phi, \\
2p_z &= 2S\ \sqrt{3}\ \cos\theta, \\
3S &= Cr^2(r-a)e^{-r}, \\
3p_x &= 3S\ \sqrt{3}\ \sin\theta\cos\phi, \\
3p_y &= 3S\ \sqrt{3}\ \sin\theta\sin\phi, \\
3p_z &= 3S\ \sqrt{3}\ \cos\theta. \\
d_{xx} &= 3S\ \sin\theta^2\cos2\phi\sqrt{15}/2, \\
d_{zz} &= 3S\ (3\cos\theta^2-1)\sqrt{5}/2,
\end{aligned} \quad (66)$$

where the normalization constants are

$$A = 188.280214, B = 1.604389, C = 0.132715, \qquad (67)$$

and the parameters needed to give the nodes and satisfy the various conditions in H space are

$$a = 0.611501, b = 4.674997, c = 3.228882, d = 2.409376, e = 0.144145,$$
$$k = 0.146707, h = 4.173047, j = 3.952782. \qquad (68)$$

Note that the 2S is the radial factor for all the tetrahedral functions and 3S for the octahedral ones. This 1S function has been forced to have nodes at a and b. Its high normalization coefficient shows how concentrated it is at the origin. The 2S function has nodes at 0 and at b. The 3S function has its nodes at 0 and a. The 1S and 2S functions also have "accidental" nodes not forced on them.

It then follows that, in the R matrix, the angular factors integrate to zero or one thus making R diagonal with four identical eigenvalues for the 2S shell and six for the 3S shell. The R matrix is, therefore:

$$R = \begin{pmatrix} e & 0 & 0 & 0 & 0 & 0 & 0 & 0 & 0 & 0 & 0 \\ 0 & f & 0 & 0 & 0 & 0 & 0 & 0 & 0 & 0 & 0 \\ 0 & 0 & f & 0 & 0 & 0 & 0 & 0 & 0 & 0 & 0 \\ 0 & 0 & 0 & f & 0 & 0 & 0 & 0 & 0 & 0 & 0 \\ 0 & 0 & 0 & 0 & f & 0 & 0 & 0 & 0 & 0 & 0 \\ 0 & 0 & 0 & 0 & 0 & g & 0 & 0 & 0 & 0 & 0 \\ 0 & 0 & 0 & 0 & 0 & 0 & g & 0 & 0 & 0 & 0 \\ 0 & 0 & 0 & 0 & 0 & 0 & 0 & g & 0 & 0 & 0 \\ 0 & 0 & 0 & 0 & 0 & 0 & 0 & 0 & g & 0 & 0 \\ 0 & 0 & 0 & 0 & 0 & 0 & 0 & 0 & 0 & g & 0 \\ 0 & 0 & 0 & 0 & 0 & 0 & 0 & 0 & 0 & 0 & g \end{pmatrix}, \qquad (69)$$

where e=0.135909, f=a, g=b and these are its eigenvalues. Thus the innermost hybrid 1S, with radius e, is close around the nucleus. The middle shell has four hybrids with centres on the sphere of radius a, and the four in the outer shell have radius b. The U matrix can be partitioned into blocks for the three shells, viz.

$$U = \begin{pmatrix} 1 & 0 & 0 \\ 0 & V & 0 \\ 0 & 0 & W \end{pmatrix}, \qquad (70)$$

and the two diagonal blocks, giving the tetrahedral and octahedral hybrids respectively, are:

$$V = \begin{pmatrix} m & m & m & m \\ m & m & -m & -m \\ m & -m & m & -m \\ m & -m & -m & m \end{pmatrix}, \quad W = \begin{pmatrix} n & n & n & n & n & n \\ r & -r & 0 & 0 & 0 & 0 \\ 0 & 0 & r & -r & 0 & 0 \\ 0 & 0 & 0 & 0 & r & -r \\ p & p & p & p & q & q \\ -m & -m & m & m & 0 & 0 \end{pmatrix}, \quad (71)$$

where the constants are

$$m=1/2,\ n=1/\sqrt{6},\ p=\sqrt{3}/6,\ q=-1/\sqrt{3},\ r=1/\sqrt{2}. \tag{72}$$

The centres of the hybrids are on the spherical shells, with radii a and b, mentioned above. The operator alias function for X has two parts since there are two shells. It is found to be

$$A_X = D(r-b)\sin\theta\cos\theta\sin\phi + E\sin\theta\cos\phi\,(1-5\sin\theta^2\cos\phi^2/3), \tag{73}$$

where $D = -0.710411$, $E = 6.022657$. The alias functions for Y and Z are found by cyclic permutation. Note that this alias function contains a term linear in r. This linear term vanishes when r=b, i.e. on the 3S outer shell, whose R eigenvalue is b, but it does contribute to the 2S inner shell. Since the radial factors are eigenfunctions of R, this linear term acts on the rest of the integrand as a constant in each diagonal block so that the structure of the block matrices is preserved. Because the radial functions are chosen to be orthogonal and also orthogonal with respect to r, the off-diagonal blocks continue to vanish. The final term contributes to the outer shell but has too high an angular momentum (l=3) to contribute to the inner shell. Note how the radial variable is used to give the alias function increased flexibility.

The angular matrices still show their block form and become:

$$X = \begin{pmatrix} 0 & 0 & 0 & 0 & 0 & 0 & 0 & 0 & 0 & 0 & 0 \\ 0 & 0 & 1 & 0 & 0 & 0 & 0 & 0 & 0 & 0 & 0 \\ 0 & 1 & 0 & 0 & 0 & 0 & 0 & 0 & 0 & 0 & 0 \\ 0 & 0 & 0 & 0 & 1 & 0 & 0 & 0 & 0 & 0 & 0 \\ 0 & 0 & 0 & 1 & 0 & 0 & 0 & 0 & 0 & 0 & 0 \\ 0 & 0 & 0 & 0 & 0 & 0 & 1 & 0 & 0 & 0 & 0 \\ 0 & 0 & 0 & 0 & 0 & 1 & 0 & 0 & 0 & r & s \\ 0 & 0 & 0 & 0 & 0 & 0 & 0 & 0 & 0 & 0 & 0 \\ 0 & 0 & 0 & 0 & 0 & 0 & 0 & 0 & 0 & 0 & 0 \\ 0 & 0 & 0 & 0 & 0 & 0 & r & 0 & 0 & 0 & 0 \\ 0 & 0 & 0 & 0 & 0 & 0 & s & 0 & 0 & 0 & 0 \end{pmatrix}, \tag{74}$$

where $r=1/\sqrt{2}$, $s=-\sqrt{3}/\sqrt{2}$.

$$Y = \begin{pmatrix} 0&0&0&0&0&0&0&0&0&0&0 \\ 0&0&0&1&0&0&0&0&0&0&0 \\ 0&0&0&0&1&0&0&0&0&0&0 \\ 0&1&0&0&0&0&0&0&0&0&0 \\ 0&0&1&0&0&0&0&0&0&0&0 \\ 0&0&0&0&0&0&0&1&0&0&0 \\ 0&0&0&0&0&0&0&0&0&0&0 \\ 0&0&0&0&0&1&0&0&0&r&t \\ 0&0&0&0&0&0&0&0&0&0&0 \\ 0&0&0&0&0&0&0&r&0&0&0 \\ 0&0&0&0&0&0&0&t&0&0&0 \end{pmatrix}, \tag{75}$$

where $r=1/\sqrt{2}$, $t=\sqrt{3}/\sqrt{2}$.

$$Z = \begin{pmatrix} 0&0&0&0&0&0&0&0&0&0&0 \\ 0&0&0&0&1&0&0&0&0&0&0 \\ 0&0&0&1&0&0&0&0&0&0&0 \\ 0&0&1&0&0&0&0&0&0&0&0 \\ 0&1&0&0&0&0&0&0&0&0&0 \\ 0&0&0&0&0&0&0&0&1&0&0 \\ 0&0&0&0&0&0&0&0&0&0&0 \\ 0&0&0&0&0&0&0&0&0&0&0 \\ 0&0&0&0&0&1&0&0&0&0&v \\ 0&0&0&0&0&0&0&0&0&0&0 \\ 0&0&0&0&0&0&0&0&v&0&0 \end{pmatrix}, \tag{76}$$

where $v=-\sqrt{2}$.

From these matrices the eigenvalues define the directions of the hybrids. The first set points towards the vertices (1,1,1), (1,-1,-1), (-1,1,-1), (-1,-1,1) and so defines the tetrahedron while the six outer ones point towards the vertices ($\sqrt{3}$,0,0), (-$\sqrt{3}$,0,0), (0,$\sqrt{3}$,0), (0,-$\sqrt{3}$,0), (0,0,$\sqrt{3}$), (0,0,-$\sqrt{3}$), the face centres of the cube that form the octahedron. The tetrahedral radius is f=a=0.611501 while the octahedral radius is g=b=4.674997. The central hybrid 1S is unchanged. The two sets of hybrids are oriented relative to each other in a rigid way.

8.2. GENERAL COMMENTS

The treatment offered above for the problems of several equivalent sets is restrictive. It excludes basis sets of the kinds in common use. A more general approach is needed. This is now being developed but is not yet complete. The block structure for the U matrix is still the ideal and each diagonal block can be determined using group theory. If the original basis functions can be modified to conform to these blocks, then they can be directly related to U. The column of basis function values at a grid point, $f_t(p)$, $t=1,...,N$, is, under certain conditions, a simultaneous eigenvector for the X, Y and Z matrices. The result is not usually normalized. We can define α_p, the strength of the hybrid at its grid point, by summing the squares of the function values there as:

$$\alpha_p^2 = \Sigma f_t(p)^2. \tag{77}$$

This eigencolumn can be normalized by dividing it by this α_p so that the value of the normalized hybrid at its grid point is α_p. Since the different eigenvectors are orthogonal, the resulting matrix of eigenvectors is unitary and is the U matrix for this basis set, viz.:

$$U_{st} = f_s(p_t)/\alpha_t, \tag{78}$$

where

$$\sum_s U_{st}U_{sv} = \delta_{tv}. \tag{79}$$

It follows that the sum of the squares of the row elements in each row of U is also normalized. The sums:

$$<f_sf_t> = \sum_v U_{sv}U_{tv} = \delta_{st} = \sum_v f_s(p_v)f_t(p_v)/\alpha_v^2. \tag{80}$$

are of great interest. The row sums over grid points, therefore, require each point to be weighted by $w_p = 1/\alpha_v^2$, the square reciprocal of its strength. These weights are those now required in any scalar product sum over the grid points in the S space. This is the generalization of the factor $1/\sqrt{N}$, which applies when there is only one equivalent set and when, obviously, all the points have equal weight. When the basis set contains the function $f_1=1$, the weights will add to unity because, if the integral over this basis function is taken, $<1> = 1$ and, as we have proved for this function, the integral is equal to the sum in S, which is just the sum of the weights.

If this result is applied to a single equivalent set then, since the weight $1/\sqrt{N}$ can be taken outside the summation, it shows that the rows of the function matrix must also be orthogonal and normalized to unity in the usual sense. Since this may not be true for the original basis functions, it reinforces the argument for introducing basis alias functions to achieve this.

This insight allows us to outline a practical procedure for dealing with these more complicated problems. We can divide the basis set into equivalent sets according to their angular symmetries and use them to calculate matrices for the x, y and z operators. We can then introduce alias functions for the operators and determine their constant parameters by the condition that the matrices commute. Once this has been achieved, the common eigenfunctions can be calculated. The eigenvalues will fall into the various equivalent sets, each having its own radius. The basis alias functions are then used to make the functions normal in S and to introduce the nodes needed to reinforce the block structure. To arrive at the block form for the U matrix, we use the modified basis functions instead of the originals. The building-in of the nodal properties of the eigenfunctions through the basis alias functions is essential in making the different sets independent of one another.

9. Quadrature

All this work is of considerable interest from a mathematical viewpoint. It is concerned with Gaussian quadrature - the exact evaluation of integrals by performing finite sums. This follows from the relation between the scalar products in the $H°$ and S spaces. The purpose of modifying the basis functions using the basis alias functions has been to ensure that the scalar product integrals

$$<f_s\, f_t>,\ <f_s(x+A_x)\, f_t>,\ <f_s(y+A_y)\, f_t>,\ <f_s(z+A_z)\, f_t> \qquad (81)$$

over the modified basis functions in $H°$ are exactly reproduced by the corresponding scalar product sums in S. The discussion of the alias functions shows that the function space $H°$ over which this result holds is not the original space. For N basis functions the integrand, which is a direct product of two basis functions and a three-dimensional vector, will have 2N+3 dimensions. Thus these integrals (81) are exactly reproduced by the sum over grid points of the value of the integrand at those points with their weights, as given in (80) for the first product in (81). It follows that any integrand which can be expressed as a linear combination of these integrands will be evaluated exactly from a similar weighted sum over the grid points of S.

The derivation of a quadrature formula from a given basis set does not require all the steps which we have described since the analytical form of the hybrids is not needed. A basis set sufficiently large and with several equivalent sets may be taken. The X, Y and Z matrices are formed and, using operator alias functions, made commutative. Their eigenvalues define the coordinates of the various grid points. The necessary renormalization of the eigenvectors determines the weighting factors needed in the quadrature formula. It should be noted that the weight of a grid point is the inverse square of the strength of the hybrid localized there. The greater the number of integrals which are exact the greater will be the accuracy of the quadrature for arbitrary functions. The fact that some basis functions have become more elaborate by mixing them with basis alias functions does not change the utility of the approximation and does not enter the quadrature.

The classical derivations of quadrature formulae have often started from an algebraic equation expressing the action of the operators on the basis functions as a linear combination of those basis functions but leaving a remainder. This is a generalization of the original Gauss discussion of the one-dimensional case. We have used this approach [6, 7] in our discussion of localized hybrids in one-dimension and on a circle. The simultaneous vanishing of the remainders at a grid point is the critical step in this approach. For several dimensions, this discussion becomes more complicated and only special examples are known. For example, Schmid [16] has discussed quadrature formulae for the unit square in the plane using polynomial basis sets. He uses ten basis functions (linear, quadratic and cubic polynomials) to derive ten grid points and their weights. Our alternative approach through the matrix eigenvectors is, in some respects, quite different and relies more heavily on the use of symmetry. However our current investigations show that we often obtain the same end results as Schmid and can trace parallels between our methods and those which he employs. In our methodology we often obtain more insight into the origin of difficulties and so can more easily avoid situations, such as those involving negative weights, which are unlikely to lead to useful results.

Clearly a normalized localized function, which has a large value at its centre, i.e. a large strength, can be approximated by a delta function at that point. An expansion in terms of such functions produces a sum formula for the integral which is the same as the S space sum if correctly weighted. A study of these localized hybrids is, therefore, needed for an understanding of the accuracy of quadrature formulae. Since we expect to use realistic basis sets, the quadratures derived from them will be designed to reproduce certain integrals exactly and others to good accuracy.

It is likely that, in the future, Gaussian quadrature methods will be applied to molecular calculations with results as accurate as existing methods but giving greater freedom in the choice of basis functions. We see our investigations as leading towards this goal.

10. Calculation of molecular properties

In all quantum calculations it has been assumed that properties of the static molecule such as its dipole, quadrupole moment, diamagnetic susceptability etc. can be approximated by performing the corresponding integrals as if the basis set were complete. What our calculations have shown, unexpectedly, is that this is often false. The group theoretical matrices for the x, y, z operators can claim to be a more accurate representation in that they commute as the operators themselves do so that the finite nature of the basis set has been compensated for. In our examples, almost half the matrix elements of the dipole moment, calculated by integration, are due to the alias functions which must be added to the operators to give commuting matrices. If these are ignored the results are poor. The well-known sensitivity of dipole calculations to small changes in the wavefunction are probably related to this effect. We conclude that, if the separate matrices for the x, y, z components of the dipole do not commute, then the significance of their average values is much reduced.

The relevance of this discussion to wider calculations has still to be seen. A dipole moment calculation is a test case for the effect of using a finite basis set and shows that, even within the limitations of a finite matrix, its integration definition does not give the best results. Clearly, if this also applies to the various other integrals in energy calculations, then most calculations of molecular properties will need to be revised.

References

1. Pauling, L. *Nature of the Chemical Bond* (Cornell University Press, Ithaca, New York, 1945); *J. Am. Chem. Soc.* **53**, 1367 (1931).

2. Hultgren, R. *Phys. Rev.* **40**, 891 (1932).

3. Slater, J. C. *Phys Rev.* **37**, 481 (1931).

4. Kimball, G. E. *J. Chem Phys* **8**, 188 (1940).

5. Herman, Z. S. *Int. J. Quant. Chem.* **23**, 921 (1983).

6. Hall, G. G. and Rees, D. *Int. J. Quant. Chem.* **53**, 189 (1995).

7. Rees, D. and Hall, G. G. *Int. J. Quant. Chem.* **54**, 351 (1995).

8. Rees, D. and Hall, G. G. *Int. J. Quant. Chem.* **54**, 361 (1995).

9. Rees, D. and Hall, G. G. *Int. J. Quant. Chem.* (1996) In press.

10. Rees, D. and Hall, G. G. *Mol. Phys* (1996) In press.

11. Hall, G. G. and Rees, D., *Int. J. Quant. Chem.* Submitted.

12. Hall, G. G. *Proc. Roy. Soc.* **A202**, 336 (1950).

13. Toth, L. F. *Regular Figures* (Pergamon Press, Oxford, 1964).

14. Hall, G. G. *Proc Roy. Soc.* **A229**, 251 (1955).

15. Boys, S. F. in *Quantum Theory of Atoms, Molecules and the Solid State* (Academic Press, New York, 1966).

16. Schmid, H. J. *Numer. Math.* **31**, 281 (1978).

QUANTUM ELECTRODYNAMICS AND MOLECULAR STRUCTURE

Molecular Electrodynamics

R G WOOLLEY
*Department of Chemistry and Physics, Nottingham Trent University,
Clifton Lane, Nottingham NG11 8NS, U.K.*

1. Introduction

Molecular structure is the cornerstone of chemistry; it is the remarkable synthesizing concept that makes chemistry intelligible. Why then should it appear as a topic in a volume concerned with `conceptual trends´ ? For most of us, molecular structure just *is*! We know enough about it to feel comfortable with it as the central idea in chemistry, and there is its extraordinary success in applications that is apt to shut off any motivation for further enquiry. All well and good, and yet from a modern theoretical perspective based on quantum theory, molecular structure *is* problematic. Quantum chemistry claims to account for the existence of chemical isomers (the key chemical question that molecular structure `solves´) in terms of states associated with the minima of potential energy surfaces. From the point of view of quantum theory this is at best an engineering solution, on two counts at least. Firstly, potential energy surfaces are classical constructions and beg the question as to what they can mean within quantum theory; secondly, granted the conventional quantum states of molecular species associated with individual minima of PE surfaces through the usual product forms of electronic, vibrational, rotational functions etc. why not form *superpositions* of states associated with *different* minima ? A pedestrian answer recognizes that the Rayleigh-Ritz variational principle guarantees a lowering in energy for the superposition, and that quest has motivated much of the history of quantum chemistry. More seriously the quantum formalism positively invites the construction of such superpositions which correspond to physically possible states. Why then are such superpositions not encountered in actual experiments ? The answer may reside in the part of modern quantum theory concerned with `decoherence´ and the emergence of classical properties. So far however there are only suggestive simple models, and nothing to predict, for example, the number of chemical isomers to be associated with any particular chemical formula.

The above remarks may seem to be of an entirely formal nature; they are concerned with the quantum theory of the customary molecular Hamiltonian. Perhaps one should rather focus on the quantum theory of specific experiments concerned with the determination of molecular structures. First however we should recall that molecular structure in its original sense has nothing to do with physics. When it was introduced to rationalize the developing body of information about chemical experiments, the only known forces that might hold atoms together were the electromagnetic and gravitational forces, but both of these were seen to be useless for chemistry. Chemists therefore turned to a structural rather than a mechanical (dynamical) account of their subject. One cannot be but impressed by the agreement over the molecular structure of practically any substance one cares to name that emerges from different 'structural techniques', particularly when one realises that, as just noted, the idea of molecular structure arose from a quite different tradition, namely the classical chemistry of the nineteenth century. It may be helpful therefore to begin by reviewing some specifics of the logical structure of classical chemistry in which molecular structure is the key concept.

There is the familiar classification of bulk matter in terms of the distinctions between mixtures and pure substances, and then elements and compounds. These have a microscopic structure in terms of molecules, which in turn are made of atoms, the basic 'building-blocks' of matter as far as chemical experiments are concerned. A molecule in classical chemistry is seen as a semi-rigid collection of atoms held together by chemical bonds. To each pure substance there corresponds a structural molecular formula, and conversely, to each molecular formula there corresponds a unique pure substance. Thus molecules are built up using atoms like the letters of an alphabet; the 'laws' that govern the relative dispositions of the atoms in ordinary 3-dimensional space are the classical valency rules which therefore provide the syntax of chemical structural formulae. Valency is a constitutive property of the atom. Of especial importance is the local structure in a molecule involving a few atoms coordinated to a specified centre for this results in the characteristic chemical notion of a *functional group*; the presence of such groups in a molecule confers specific properties to the corresponding substance (e.g. acid, base, oxidant etc.). Moreover each pure substance can be referred to one or several categories of chemical reactivity, and can be transformed into other substances which successively fall in other categories. Thus the classical structural formula of a molecule summarizes (represents) the connection between the spatial organization of the atoms and a given sequence of chemical reactions that the corresponding substance participates in, notably (but not only) the reactions required for its analysis and its synthesis.

A fundamental shift in perspective occurred during the years either side of the Second World War [1]. By and large the historical approach to molecular structure was highly successful for organic chemistry even though there were puzzles and anomalies that had to be regarded as 'special cases' (e.g. concerning the structural formulae for polycyclic hydrocarbons such as anthracene); it was much less successful for inorganic compounds. For this reason the systematic use of *physical* methods of structure determination, especially X-ray and electron diffraction techniques, in organic chemistry textbooks was much delayed with respect to those of inorganic chemistry, and did not become widespread until the late 1950's. This change in methodology implied a fundamental revision in the notion of molecular structure from being a hypothesis that encoded the actual and potential chemistry of a substance (the set of chemical reactions a substance may participate in) to being an *experimental observable*, to be *measured* by a physical technique.

Physical techniques of structure determination (infra-red and microwave spectroscopy, n.m.r., diffraction experiments etc.) fall under the general heading of spectroscopy, that is they involve the monitoring of some kind of radiation that has previously interacted with the chemical substance. With the passage of time it has become evident that the experimental results derived from these techniques are quite generally interpreted in terms of classical molecular structure. It is widely believed that this change in orientation of the basis for molecular structure is an inevitable outcome of the development of modern physical theory applied to molecular systems. An alternative view is that a far-reaching *reinterpretation* of these experiments has been made for reasons that are largely independent of any requirements of physics (specifically quantum mechanics); in other words, rather than the seamless integration of chemical theory into physics (the strong reductionist belief), all that has happened is that the nineteenth century rupture between chemistry and physics has been patched over.

It is generally held that diffraction experiments give the most direct access to the geometric arrangement of atoms in materials so that if one wishes to investigate the concept of molecular structure from a rather fundamental point of view, one might look towards the theory of such experiments. In this chapter a unified approach to electron diffraction and X-ray diffraction in fluid phases is presented. Rather than adopting the conventional approach that analyses these experiments in terms of the diffraction of classical waves by a distribution of atomic scattering centres - an approach that starts with the assumption of structure built in - we are interested in the calculation of the appropriate cross-sections using quantum mechanics. Of course this is not a new concern; the application of the Born approximation (first-order perturbation theory) to the diffraction of electrons and of X-rays by an atom (Mott, Weller) must rank among the initial inspiring successes of quantum mechanics - one thinks in this context of Gamow's theory of the α-decay of a nucleus, the Heitler-London treatment of the H_2 molecule, among others. Nevertheless, in neither case is the Born approximation *gauge-invariant* - the result apparently depending on the assumption of the Coulomb gauge condition - and one may reasonably seek a stronger justification for the customary basis of diffraction methodology that requires a cross-section depending on only the momentum transfer vector **q** and expressible in terms of averages involving the particle density operator. This requires both a more powerful treatment of scattering theory and some consideration of what gauge invariance entails in this context. The relevant theory is the non-relativistic quantum mechanical scattering formalism developed from the 1940's onwards, and described comprehensively in a series of monographs published in the 1960's - see e.g. [2].

In the next two sections we give an outline summary of time-independent scattering theory that leads to the quantum-mechanical expression for a scattering cross-section, and of the non-relativistic quantum electrodynamics of molecules. We then apply this general theory to two particular cases, namely scattering of fast electrons (the electron diffraction experiment), and scattering of X-Ray photons, both for molecules in fluid phases; both experiments can be modelled as the collision of a probe beam particle with a collection of bound charges (a molecule). The kinematics of the collision is dealt with explicitly and no approximations are made with respect to the wavevectors of the scattered particles; this is what is meant by the expression *molecular electrodynamics*. The theory is dynamical in character but limited to a perturbation theory viewpoint. The matrix elements involved in the scattering cross-sections are shown to be expressible in terms of the electric and magnetic polarization fields for a molecule. Some attempt is made to develop the analysis of the cross-section in the particular case of elastic scattering, defined by the condition that the detected scattered radiation should have the same energy as the incident radiation; some progress at least is possible in the case of electron scattering. The general form for the Kramers-Heisenberg formula for photon scattering is however intractable. The circumstance that the probe beam particles have energies that greatly exceed the average binding energies of the molecular particles justifies the use of the *impulse approximation*, [2], and this permits a kinematical analysis of the cross-section which we also describe.

2 Quantum Mechanics of the Spectroscopic Experiment

In a spectroscopic experiment a probe beam of particles (P) interacts with a target (T) and the scattered particles are detected in a region remote from the target. The difference between an absorption experiment and a diffraction experiment is then just that, in the former, the detector monitors particles moving in the same direction as the incident particle beam whereas in diffraction, particles that have suffered a change in direction are monitored. The key idea of the theory is that the interaction is localized in space and time, so that originally non-interacting parts come together to interact and then finally separate, after which detection occurs, so that the non-interacting system of probe beam and target can be used as a reference system to describe the outcome of the experiment. The quantum-mechanical description is based on: a Hamiltonian describing the dynamics of the probe beam and the target, and their coupling interaction,

$$H = H_o + H_c \qquad (1a)$$

$$H_o = H_P + H_T, \qquad (1b)$$

a specification of the states of the probe beam and the target before they interact which can be expressed as a simple product of density matrices

$$\rho_o = \rho_P \otimes \rho_T, \qquad (2)$$

and an idealization of the detector, in terms of an operator D_ϕ which will be specified further below (ϕ is a label for the state of the non-interacting system P + T that is detected) [2,3]. It is usually the case that H_o, as well as the full Hamiltonian H, commutes with the operators that generate time translations, space translations and space rotations, and this leads to transition amplitudes that contain delta functions expressing the Law of Conservation of Energy, the Law of Conservation of Momentum and so on. There are other important symmetry operations which have analogous properties (space and time inversions for example) that can be exploited to characterize the transition amplitudes in terms of appropriate symmetry labels. One important operator that does not fit this pattern is the generator of *gauge transformations* (see §3.1) and so the proof of the gauge invariance of the scattering theory is much more difficult than for the kinematical symmetries.

The Schrödinger equation for the probe beam and the target provides a basis of eigenstates for the reference system

$$H_P|\alpha\rangle = E_\alpha|\alpha\rangle \qquad (3a)$$

$$H_T|N\rangle = E_N|N\rangle \qquad (3b)$$

so that a typical basis element of the Hilbert space of the composite system is

$$|n\rangle = |\alpha\rangle|N\rangle \equiv |\alpha;N\rangle \qquad (4)$$

with energy

$$E_n = E_\alpha + E_N. \qquad (5)$$

The time evolution of the density matrix is governed by the Liouville (von Neumann) equation,

$$i\hbar(\partial \rho/\partial t) = [H, \rho] \qquad (6)$$

which has a formal solution in terms of a unitary transformation induced by the Hamiltonian

$$\rho(t) = e^{iH(t-t')/\hbar} \rho(t') e^{-iH(t-t')/\hbar} \qquad (7)$$

The probability at time t that the composite system (P + T) will be found in the state ϕ after the target and the scattered particles have separated is given by

$$P(\phi, t) = \mathrm{Tr}[\rho(t) D_\phi] \qquad (8)$$

and the corresponding probability per unit time, or transition rate, for the process $\rho_o \to \phi$ occurring because of the coupling interaction H_c is

$$\tau(\phi, \rho_o)^{-1} = dP(\phi, t)/dt. \qquad (9)$$

Quantum mechanical scattering theory provides formal expressions for the transition rate which can be approximated by perturbation theory. The solution for the density matrix is chosen so that

$$\rho(-\infty) = \rho_o \qquad (10)$$

and

$$\rho(+\infty) = S \rho_o S^+, \quad S = I - T \qquad (11)$$

where S and T are the scattering and transition operators, respectively. The transition operator is strictly off-diagonal and so in terms of matrix elements we have

$$S_{fi} = \delta_{fi} - 2\pi i \, \delta(E_f - E_i) \, T_{fi} \qquad (12)$$

where the 'transition matrix' is related to the coupling operator H_c by

$$\langle f|T|i\rangle = \langle \psi^+_f|H_c|i\rangle \qquad (13)$$

and ψ_f^+ is an outgoing scattering solution of the full problem with the same energy as the reference state $|i\rangle$. The operator T has the formal representation

$$T = H_c + H_c G_o(E) T \qquad (14)$$

where $G_o(E)$ is the Greens function for the non-interacting system,

$$G_o(E) = [E - H_o + i\varepsilon]^{-1} \qquad (15)$$

from which the usual perturbation series may be generated by iteration,

$$T = H_c + H_c G_o(E) H_c + H_c G_o(E) H_c G_o(E) H_c + \ldots \qquad (16)$$

Retaining only the first term, $T \approx H_c$, is the first Born approximation.

In an idealized experiment we suppose that the detector picks out a particular state $|k\rangle$ of the non-interacting system (P + T), and so may be represented by a simple projection operator

$$D_k = |k\rangle\langle k| = |\beta, N\rangle\langle \beta, N| \qquad (17)$$

in terms of the states defined by equation (3). On the assumption that the initial density matrix is stationary, and so may be represented in the eigenstate basis as

$$\rho_o = \sum_n w_n |n\rangle\langle n| \qquad (18)$$

where the real coefficients $\{w_n\}$ are probabilities

$$\sum_n w_n = 1, \quad 0 \leq w_n \leq 1, \qquad (19)$$

the scattering theory yields the usual 'golden rule' formula for the transition rate to the state $|k\rangle$,

$$\tau(k,\rho_o)^{-1} = (2\pi/\hbar) \sum_n w_n |T_{kn}|^2 \delta(E_n - E_k) \qquad (20)$$

In order to extract information from τ_k^{-1} that can be compared with experiment we have to recognize that usually the experimental control and information available for the probe beam and the target are quite different. In the typical experiment we may suppose that the probe beam is produced with particles having reasonably well defined energy and direction specified by a wavevector $\mathbf{k}$, and in some specified polarization state, λ. Hence the typical particle state $|\beta\rangle$ can be expressed more concretely as $|\mathbf{k}, \lambda\rangle$. A target molecule can be described in terms of its centre-of-mass momentum and its internal state, and we may write the molecular states as $|N\rangle = |\mathbf{K},s\rangle$. However the target molecules are lumped together in a bulk sample, and the initial and final states of the particular molecule that scatters a given probe beam particle into the detector are neither under our control nor subject to our observation, and so the observables will be averages over these molecular states [4].

The transition rate for a probe beam particle of momentum $\hbar\mathbf{k}$ being scattered into a state with momentum $\hbar\mathbf{k}'$ is therefore

$$\tau(\mathbf{k}',\lambda',\mathbf{k})^{-1} = (2\pi/\hbar) \sum w_\lambda w_{K,s} |\langle \mathbf{k}',\lambda';\mathbf{K}',s'|T|s,\mathbf{K};\lambda,\mathbf{k}\rangle|^2$$
$$\times \delta(E_k + E_K + E_s - E_{k'} - E_{K'} - E_{s'}) \qquad (21)$$

where w_λ describes the polarization state of the probe beam, and $w_{K,s}$ specifies the state of the target molecule prior to the interaction; the sum in (21) is over all the quantum numbers that do not appear on the lhs. This transition rate must be combined with a density of final states factor $\rho_{p'}$ for the scattered particle (since the detector will record the arrival of particles with momenta in the interval $[\mathbf{p}', \mathbf{p}' + d^3\mathbf{p}']$), and divided by the incident flux of probe beam particles, J_p, to yield finally a differential cross-section for the scattering process

$$d\sigma = \tau(\mathbf{k}',\mathbf{k})^{-1} J_p^{-1} \rho_{p'} \qquad (22)$$

where

$$\rho_{p'} = V/(2\pi\hbar)^3 d^3\mathbf{p}' dE^{-1} \qquad (23)$$

where $E = E_{f}$. For non-relativistic particles of mass m, and using the customary normalization of one particle per unit volume, we have

$$J_p = p/mV; \quad \rho_{p'} = V/(2\pi\hbar)^3 mp' d\Omega_{p'} \qquad (24)$$

and for photons

$$J_p = n_k c/V; \quad \rho_{k'} = V/(2\pi)^3 (k^2/\hbar c) d\Omega_{k'} \qquad (25)$$

$\rho_{p'}$ is the number of final states per unit energy; if the final state contains N scattered particles dE^{-1} in (23) must be multiplied by a product of N phase space factors.

The Law of Conservation of Momentum requires that the T-matrix elements have the property

$$\langle \mathbf{k}',\lambda';\mathbf{K}',s'|T|s,\mathbf{K};\lambda,\mathbf{k}\rangle \propto \delta^3(\mathbf{K}' + \mathbf{k}' - \mathbf{K} - \mathbf{k}) \qquad (26)$$

so that the transition rate and cross-section contain a momentum delta function squared; this is to be interpreted according to the rule [2]

$$\left(\delta^3(\mathbf{K}' + \mathbf{k}' - \mathbf{K} - \mathbf{k})\right)^2 = V/(2\pi)^3 \delta^3(\mathbf{K}' + \mathbf{k}' - \mathbf{K} - \mathbf{k}) \qquad (27)$$

In all these formulae V is the quantization volume for the composite system; the transition to continuous values of momenta is made with the usual rule,

$$V^{-1} \sum_K \rightarrow \int d^3K/(2\pi)^3 \quad (28)$$

and with this replacement, the momentum conservation δ-function makes the average over the final momentum **K'** immediate. Further progress requires specific information about the T-matrix elements and the initial density matrix, so we now turn to a description of the electrodynamics of molecules and some particular applications of these scattering theory relations.

3 The Electrodynamics of Molecules

3.1 Gauge Transformations

The essential framework of the formalism of quantum electrodynamics as used in atomic and molecular physics is a Hamiltonian description with the Schrödinger equation, and perturbation theory [5,6]. Classically, a Hamiltonian scheme for the electromagnetic field can be based on its 'field-coordinate', the vector potential **A(x)**, and its associated canonically conjugate variable $\pi(\mathbf{x})$, which is essentially the electric field strength, **E(x)** ($\pi(\mathbf{x}) = -\varepsilon_o \mathbf{E}(\mathbf{x})$). The magnetic field **B(x)** is the curl of the vector potential, $\mathbf{B}(\mathbf{x}) = \nabla \wedge \mathbf{A}(\mathbf{x})$, from which it follows that a *gauge transformation* of the vector potential

$$\mathbf{A}'(\mathbf{x}) = \mathbf{A}(\mathbf{x}) - \nabla f(\mathbf{x}) \quad (29)$$

leaves the magnetic field unchanged provided that the functions f(x) are integrable (Curl Grad f(x) = 0). In the Schrödinger equation for a charged particle, the momentum **p** must be replaced by the gauge-invariant substitution, $\mathbf{p} \rightarrow \mathbf{p} - e\mathbf{A}(\mathbf{x})$, where **A(x)** is the vector potential of the field evaluated at the position **x** of the charge; the gauge invariance of the full Hamiltonian for interacting charges and electromagnetic fields, and of the associated equations of motion it generates, can be settled fairly readily. More delicate questions arise in a perturbation theory context, such as is required for the implementation of the theory described in §2, where a gauge-dependent splitting, $H = H_o + V$, is involved; the kinetic energy term, $p^2/2m$, leads to both first- and second-order perturbation terms (treating the charge e as the perturbation parameter),

$$V = -(e/2m)\{\mathbf{p}.\mathbf{A}(\mathbf{x}) + \mathbf{A}(\mathbf{x}).\mathbf{p}\} + (e^2/2m)\mathbf{A}(\mathbf{x}).\mathbf{A}(\mathbf{x}) \quad (30)$$

and so the two types of term enter in different ways in perturbation theory to a given order in e. Since the vector potential is not unique one has to determine whether changes in the **p.A** terms can be compensated for by changes in the **A.A** term to give a well-defined answer independent of any arbitrariness in the vector potential.

The existence of such transformations complicates the Hamiltonian description of the electromagnetic field and its quantization since the classical Hamiltonian contains the generator of gauge transformations [7,8]. In quantum electrodynamics such transformations can be implemented by a unitary operator Λ that commutes with the full Hamiltonian

$$\Lambda[f] = \exp\{i/\hbar \int d^3x (\nabla.\pi(\mathbf{x}) + \rho(\mathbf{x}))f(\mathbf{x})\} \quad (31)$$

(square brackets are used to denote *functional* dependence) where $\rho(\mathbf{x})$ is the charge density

$$\rho(\mathbf{x}) = \sum_\alpha e_\alpha \delta^3(\mathbf{x} - \mathbf{x}_\alpha). \quad (32)$$

The operators $\Lambda[f]$ provide a unitary representation of an infinite dimensional additive group, the 'gauge symmetry group', isomorphic to U(1). There are two possible responses to this situation; the first, and usual one in atomic and molecular physics, is to impose an additional condition on the vector potential, a so-called *gauge condition* which allows the Maxwell equation

$$\varepsilon_o \nabla \cdot \mathbf{E}(\mathbf{x}) = \rho(\mathbf{x}) \tag{33}$$

to be treated as an ordinary equation; thereby the gauge transformations are eliminated from the Hamiltonian and Λ is reduced to the identity. These remarks need to be understood with some care; the gauge condition is arbitrary and one therefore has to ensure that quantities calculated from the reduced Hamiltonian are independent of the choice of gauge condition i.e. *gauge invariant*. A fundamental proposition of the theory is that only gauge-invariant quantities can be *physical observables*.

The second approach [9] does not require a gauge for the vector potential to be specified and instead the Maxwell equation (33) is imposed as a subsidiary condition on the state vectors of the theory so as to select out the physical states $|\Psi\rangle$

$$\varepsilon_o \nabla \cdot \mathbf{E}(\mathbf{x})|\Psi\rangle = \rho(\mathbf{x})|\Psi\rangle. \tag{34}$$

Gauge-invariant wavefunctions can be constructed as follows; let U[P] be a unitary operator,

$$U[P] = \exp\{(i/\hbar)A[P]\} \tag{35}$$

where

$$A[P] = \int d^3x \; \mathbf{P}(\mathbf{x}) \cdot \mathbf{A}(\mathbf{x}) \tag{36}$$

for some vector field $\mathbf{P}(\mathbf{x})$ which satisfies a linear inhomogeneous differential equation at all points in space $\mathbf{x}$, [9],

$$\nabla \cdot \mathbf{P}(\mathbf{x}) = -\rho(\mathbf{x}) \tag{37}$$

$\mathbf{P}(\mathbf{x})$ is called the *electric polarization field*. The condition (37) fixes only the *longitudinal* component of $\mathbf{P}(\mathbf{x})$,

$$\mathbf{P}(\mathbf{x})^{\|} = \nabla \phi(\mathbf{x}), \quad \phi(\mathbf{x}) = \int d^3x' \; \rho(\mathbf{x}')/\{4\pi |\mathbf{x} - \mathbf{x}'|\} \tag{38}$$

since $\nabla^2 (1/r) = -4\pi \delta^3(\mathbf{r})$, where $r = |\mathbf{x} - \mathbf{x}'|$, and so $\mathbf{P}(\mathbf{x})$ can contain any solution $\mathbf{P}(\mathbf{x})^\perp$ of the homogeneous equation related to (37). Then treating (34) as a functional differential equation shows that the states $|\Psi\rangle$ must be representable in the form [9],

$$|\Psi[A]\rangle = U[P] \int^{A(x)} \delta A'(x) \cdot t(A'(x)) \equiv U[P] |\Phi\rangle \tag{39}$$

where t is a transverse vector *function* of $\mathbf{A}(\mathbf{x})$. In the special case that a state $|\Phi\rangle$ depends on $\mathbf{A}(\mathbf{x})$ through its Curl (i.e. through $\mathbf{B}(\mathbf{x})$), an elementary calculation shows that the unitary operator $\Lambda[f]$ acts as the identity operator on $|\Psi\rangle$. Such states are manifestly gauge invariant, but how far this result can be extended to characterize gauge invariance remains an open question requiring the representation theory of the gauge symmetry group.

The conservation of electric charge is expressed by the equation of continuity,

$$d\rho/dt + \nabla \cdot \mathbf{j} = 0 \tag{40}$$

where the time derivative is calculated as usual by commutation with the particle Hamiltonian and $\mathbf{j}(\mathbf{x})$ is the current density

$$\mathbf{j}(\mathbf{x}) = \sum_\alpha (e_\alpha/m_\alpha) \mathbf{p}_\alpha \delta^3(\mathbf{x} - \mathbf{x}_\alpha) \tag{41}$$

which must be symmetrized to make an hermitian quantum operator. Taking note of (37) and (40) the current density can be linked to the polarization field P(x) through a relation that introduces a magnetic polarization field M(x),

$$j(x) = dP(x)/dt + \nabla \wedge M(x); \quad (42)$$

note that pairs of polarization fields {P,M}, {P',M'} related by

$$P' = P + \nabla \wedge G \quad (43)$$
$$M' = M - dG/dt \quad (44)$$

for any differentiable vector field G(x) are fully equivalent because they represent the *same* charge and current densities.

The relationship between the two approaches to formal gauge invariance in the quantum theory is simply this; requiring the functional A[P] to have a prescribed value, conventionally taken to be zero, and so fixing the phase of the states (39), is simply a gauge condition i.e. for any vector field P(x) satisfying (37) there is a corresponding vector potential such that A[P] = 0. If, for example, we take the purely longitudinal polarization field in (38), then after an integration by parts we have

$$0 = A[P^l]$$

$$= \int d^3x \, (\nabla \phi(x)).A(x) = -\int d^3x \, \phi(x) \, (\nabla.A(x)) \quad (45)$$

and this is essentially the radiation (or 'Coulomb') gauge condition

$$\nabla.A(x) = 0. \quad (46)$$

More generally, if we divide the polarization field, P(x), into its transverse and longitudinal components, [10],

$$P(x) = P(x)^l + P(x)^\perp, \quad (47)$$

then the gauge condition A[P] = 0 becomes

$$\int d^3x \, P(x)^l.A(x) + \int d^3x \, P(x)^\perp.A(x) = 0; \quad (48)$$

thus the Coulomb gauge theory is characterized by both terms separately vanishing ($P(x)^\perp = 0$, $A(x)^l = 0$), while in general the two terms are non-zero and equal and opposite.

The representation in (38) which exhibits explicitly the functional dependence of $P(x)^l$ on the charge density $\rho(x)$ can be usefully extended to the full polarization field by setting

$$P(x) = \int d^3x' \, p(x;x') \, \rho(x') \quad (49)$$

where p(x;x') is any vector field such that

$$\nabla_x.p(x;x') = -\delta^3(x - x'). \quad (50)$$

Since (36) is linear in the polarization field we can use (49) to remove its dependence on the charge density (an unessential complication) and write the general gauge condition as

$$A[p] = \int d^3x' \, A(x').p(x';x) = 0. \quad (51)$$

In making this change we recognize that as far as the gauge condition is concerned the essential feature of the polarization field is its *non-locality* in space; the non-local vector field p(x;x') has no dependence on time. The charges {e_α} occupy positions {x_α} in 3-dimensional space [x] and their dynamics give rise to the time-dependence of the polarization field P(x) as required by (42). If we now put $A(x) = \nabla f(x)$ in the first term of (48), we can display its formal solution in terms of the generating function

(48), we can display its formal solution in terms of the generating function

$$f(x) = -\int d^3x' \, A(x')^\perp \cdot p(x';x) \tag{52}$$

so that *any* vector potential can be written as a functional of the transverse vector potential characteristic of the Coulomb gauge, and the arbitrary non-local vector-field $p(x';x)$

$$A(x) = A(x)^\perp - \nabla_x \int d^3x' \, A(x')^\perp \cdot p(x';x). \tag{53}$$

A unitary operator U[P], equation (35), with **A** chosen specifically as the transverse Coulomb gauge vector potential is the basis of the *Power-Zienau-Woolley Transformation* theory that results in an unitarily equivalent QED Hamiltonian expressed in terms of the electric and magnetic field strengths coupled to the molecular electric and magnetic polarization fields respectively [11,12]. The original formulation of this transformation was confined to the leading terms of a multipole expansion of **P(x)**,

$$P(x) \approx \{d + Q \cdot \nabla + \ldots\} \delta(x - R) \tag{54}$$

where **d** and **Q** are the molecular electric dipole and electric quadrupole moment operators [13]; this approach can be traced back to important early work of Göppert-Mayer [14]. The complete multipole series for **P(x)** can be summed into an integral representation, initially given in a parameterized form, but subsequently reinterpreted as a line integral [11],

$$P(x) = \sum_\alpha e_\alpha \int_{C_\alpha} dz \, \delta^3(z - x) \tag{55}$$

where the integrations are over paths from **R** to each of the particle coordinates $\{x_\alpha\}$ (the path end-point **R** is the arbitrary centre about which the multipole expansion in (54) is made). In this case the unitary operator U[P] takes the form

$$U[P] = \exp\{(i/\hbar)\sum_\alpha e_\alpha \int_{C_\alpha} dz \cdot A(z) \tag{56}$$

which has a close connection with the formalism used to describe the Bohm-Aharonov effect [11]; this form for U[P] can be traced back to early work of Dirac [15], Fock [16], and Peierls [17]. As long as (55) is well-defined, *any* set of paths connecting **R** to $\{x_\alpha\}$ is as valid as any other, and so the particular linear path related to the multipole series has no special significance.

3.2 The general Hamiltonian

We have recently shown how a general Hamiltonian for non-relativistic QED can be obtained when the gauge condition is presented in the form (51), and we quote only the final results which are valid for any vector field $p(x;x')$ that satisfies (50) [18,19]. In the non-relativistic approximation with particles described in terms of two parameters (charge e and mass m) and position and momentum variables ($\underline{x}$, $\underline{p}$), the reduced Hamiltonian for electrodynamics in some gauge can be written as,

$$H = \tfrac{1}{2}\sum_\alpha m_\alpha^{-1}(p_\alpha - e_\alpha A(x_\alpha))^2 + \tfrac{1}{2}\varepsilon_0 \int d^3x \, \{E(x)^2 + c^2 B(x)^2\}. \tag{57}$$

with the vector potential in the gauge specified by (53). The constraint equation involving the charge density then holds as an ordinary equation (cf (33))

$$\nabla \cdot \pi(x) + \rho(x) = 0 \tag{58}$$

and so becomes an operator equation after quantization in contrast to (34) which holds when no gauge is fixed. The non-zero Dirac brackets which determine the commutation relations of the particle and field variables after quantization ($i\hbar\{..,..\}^* \rightarrow [..,..]$) are:

$$\{x_\alpha^r, p_\beta^t\}^* = \delta_{rt}\delta_{\alpha\beta} \tag{59}$$

$$\{A(x)^r, \pi(x')^t\}^* = \delta_{rt}\delta^3(x-x') - \nabla_x^r p(x;x')^t \tag{60}$$

$$\{p_\alpha^r, \pi(x)^t\}^* = e_\alpha \nabla^r p(x;x_\alpha)^t \tag{61}$$

(where ∇^r acts on the coordinate x_α).

If we choose $p(x;x') = p(x;x')^l$ (i.e. $P(x)^\perp = 0$) we recover precisely the usual Coulomb gauge theory since (60) is then the transverse delta dyadic [10]. Similarly choosing a line integral form

$$p(x;x') = \int_C^{x'} dz\, \delta^3(z-x) \tag{62}$$

so that, from (49),

$$P(x) = \sum_\alpha e_\alpha \int_C^{x_\alpha} dz\, \delta^3(z-x) \tag{63}$$

recovers the 'multipolar' Hamiltonian theory [11]. The separation of the delta functions in (62) and (63) into their longitudinal and transverse components makes clear how the longitudinal and transverse components of $p(x;x')$ (and $P(x)$) are to be interpreted; notice that x and x' generally play asymmetrical roles in $p(x;x')$; for example, in the line integral form (62) it is the second argument (x') that denotes the path end-point.

The equation of motion for the variable $\pi(x)$ yields the field equation for the electric field $E(x)$ to within a factor of ε_0, and we make the identification $\pi(x) = -\varepsilon_0 E(x)$; (58) is then Maxwell's equation that relates the electric field to the charges. These facts together with (61) lead us to put $\pi(x) = \pi(x)_{Rad} + P(x)$, where $\pi(x)_{Rad}$ is that component of $\pi(x)$ that has a vanishing Dirac bracket with the particle variables, and can be associated with electromagnetic radiation. The energy of the electric field is then

$$\tfrac{1}{2}\varepsilon_0 \int d^3x\, E(x).E(x) = \tfrac{1}{2}\varepsilon_0 \int d^3x\, E(x)^\perp . E(x)^\perp - \int d^3x\, E(x)^\perp . P(x)$$

$$+ \tfrac{1}{2}\varepsilon_0^{-1}\int d^3x\, P(x).P(x) \tag{64}$$

The term quadratic in $E(x)^\perp$ combines with the squared magnetic field term in (57) to yield H_{Rad}, the radiation field Hamiltonian. The term linear in $E(x)^\perp$ can be put in the form

$$-\sum_\alpha e_\alpha \int d^3x\, E(x)^\perp . p(x;x_\alpha) \tag{65}$$

and so couples charges to the radiation field. Finally if we define

$$\wp(x';x'') = \int d^3x\, p(x;x').p(x;x'') \tag{66}$$

the last term in (65) takes the form

$$\tfrac{1}{2}\varepsilon_0^{-1}\int d^3x\, P(x).P(x) = \tfrac{1}{2}\varepsilon_0^{-1}\int d^3x'\int d^3x''\, \rho(x')\wp(x';x'')\rho(x'')$$

$$= \tfrac{1}{2}\varepsilon_0^{-1}\sum_{\alpha\beta} e_\alpha e_\beta\, \wp(x_\alpha;x_\beta) \tag{67}$$

and the Hamiltonian (57) becomes,

$$H = \tfrac{1}{2}\sum_\alpha m_\alpha^{-1}(\mathbf{p}_\alpha - e_\alpha \mathbf{A}(\mathbf{x}_\alpha))^2 - \sum_\alpha e_\alpha \int d^3x\, \mathbf{E}(\mathbf{x})^\perp \cdot \mathbf{p}(\mathbf{x};\mathbf{x}_\alpha)$$
$$+ \tfrac{1}{2}\varepsilon_0^{-1} \sum_{\alpha\beta} e_\alpha e_\beta\, \wp(\mathbf{x}_\alpha;\mathbf{x}_\beta) + H_{Rad} \qquad (68)$$

Since (67) is independent of electromagnetic field variables it makes a contribution to any amplitude in which the state of the electromagnetic field is left unchanged. Earlier studies showed that it appears in a self-energy calculation such as that required for the non-relativistic contribution to the Lamb shift [13, 20]; it also contributes in any process where two systems interact via the QED Hamiltonian, for example the coupled oscillator model of optical activity [21]. The *longitudinal* part of $\mathbf{p}(\mathbf{x};\mathbf{x}')$ is well-defined (compare (38) and (49)) and yields the Coulomb interaction between the charges in (67) since

$$\wp(\mathbf{x}';\mathbf{x}'')^{\parallel} = \int d^3x\, \mathbf{p}(\mathbf{x};\mathbf{x}')^{\parallel} \cdot \mathbf{p}(\mathbf{x};\mathbf{x}'')^{\parallel} = 1/\{4\pi|\mathbf{x}' - \mathbf{x}''|\}. \qquad (69)$$

It is conventional to take this contribution to H with the particle kinetic energy terms to give the "unperturbed" Hamiltonian for matter

$$H_{matter} = \sum_\alpha \mathbf{p}_\alpha^2/2m_\alpha + \sum_{\alpha\neq\beta} \{e_\alpha e_\beta\}/\{4\pi\varepsilon_0 |\mathbf{x}_\alpha - \mathbf{x}_\beta|\}; \qquad (70)$$

if this and H_{Rad} are then subtracted off from H, equation (68), what is left is regarded as the coupling operator between matter and electromagnetic radiation,

$$V = H - H_{matter} - H_{Rad} \equiv H - H_0. \qquad (71)$$

This separation of the full Hamiltonian, H, into an unperturbed part H_0 describing non-interacting matter and radiation, and their coupling, V, is not gauge-invariant since H_{matter} does not commute with the gauge transformation operator Λ, equation (31), and hence neither does V. The gauge invariance of the S and T-matrices, (§2), is not therefore a trivial matter since the states of the free reference system are gauge dependent. Notice also that, so far, the operator V *cannot* be assumed to be a small perturbation since the transverse part of $\mathbf{p}(\mathbf{x};\mathbf{x}')$ is completely arbitrary and so the corresponding scalar $\wp(\mathbf{x}';\mathbf{x}'')^\perp$ resulting from (66) is also arbitrary, as is (65); the line integral form of $\mathbf{p}(\mathbf{x};\mathbf{x}')$, (62), for example, yields a form for $\wp(\mathbf{x};\mathbf{x}')^\perp$ that cancels the Coulomb interaction in (69) for all $\mathbf{x},\mathbf{x}'$ except along the chosen path between $\mathbf{x}$ and $\mathbf{x}'$ [20].

There are several possible approaches to this situation; the most direct, and conventional, is to assume at once that $\mathbf{p}(\mathbf{x};\mathbf{x}')^\perp$ is zero so that V only contains the usual p.A and $\mathbf{A}^2$ terms in the expectation that any other choice for $\mathbf{p}(\mathbf{x};\mathbf{x}')^\perp$ would yield the same results. More formally, one can give an *adiabatic* argument which suggests that there are approximate eigenstates of the full QED Hamiltonian, H, of a simple product form

$$\Psi \approx \phi_{matter} \cdot \Phi_{Rad}.$$

We give a brief sketch of the idea which follows the well-known Born-Oppenheimer idea in quantum chemistry; first of all we imagine solving the Schrödinger equation for a reduced Hamiltonian in which the kinetic energy terms for the charges have been dropped. The corresponding eigenfunctions and eigenvalues then depend on the particle variables in a purely parametric fashion; the eigenvalues provide `potentials´ under which the particles move (like P.E.surfaces). The lowest energy potential is the field vacuum expectation value of an expression like (14) in which G_0 is to be understood as a resolvent operator involving only H_{Rad}. In second-order perturbation theory the contribution from $\wp(\mathbf{x}_\alpha;\mathbf{x}_\beta)^\perp$ is cancelled precisely by the terms arising from (65) involving the exchange of a virtual photon between pairs of charges (the calculation is essentially the same as that described in [20], [21] adapted to interparticle interactions in place of intermolecular ones). There are also terms like those displayed in the Breit Hamiltonian, and renormalization terms from self-interactions. These comments are valid irrespective of the choice made for $p(x;x')^\perp$ and so to second-order the arbitrary terms are shown to be irrelevant. The physical interpretation of such a calculation is that the lowest energy eigenstates of the full Hamiltonian can be represented in terms of the photon vacuum weakly perturbed by the presence of charged particles which have a mutual static Coulombic interaction. As with the Born-Oppenheimer description of a molecule based on P.E. surfaces, the calculation is obviously guided by a prior conception of what an approximate solution of the Schrödinger equation should look like.

To see what the status of such states are in QED one may contemplate a calculation analogous to the derivation of the relativistic bound-state equation for charges (the Bethe-Salpeter equation). This development was a triumph for quantum field theory based on the insight that particle bound states are directly related to the poles of appropriately defined scattering amplitudes [22]; in the relativistic context however the gauge invariance of the amplitudes can be linked directly to the Lorentz invariance of the theory, and this feature is missing from the non-relativistic theory described here. Thus the role, if any, of the arbitrary transverse polarization field $p(x;x')^\perp$ remains an open question in molecular electrodynamics. We defer further comment on this point to §6.

3.3 The Coulomb Gauge Theory

The Coulomb gauge theory arises if we put $p(x;x') \equiv p(x;x')^\perp$ in (68); equivalently if we impose the gauge condition (46) on the vector potential corresponding to the choice $p(x;x')^\perp = 0$, the second term in H, equation (57) decomposes into two parts, one containing only transverse field variables that describe radiation,

$$H_{Rad} = \tfrac{1}{2}\varepsilon_o \int d^3x \{E^\perp(x).E^\perp(x) + c^2 B(x).B(x)\} \tag{72}$$

and a purely longitudinal part that describes the instantaneous Coulombic interaction between the charges,

$$H_{Coul} = \tfrac{1}{2}\varepsilon_o \int d^3x \, E^\parallel(x).E^\parallel(x) \equiv \sum_{\alpha \neq \beta} \{e_\alpha e_\beta\} / \{4\pi\varepsilon_o |x_\alpha - x_\beta|\}. \tag{73}$$

The Coulomb energies in (73) may also be written in terms of the charge density ρ and the 'scalar potential' ϕ,

$$H_{Coul} = \int d^3x \, \rho(x) \, \phi(x), \tag{74}$$

where ϕ (equation (38)) only depends on the instantaneous particle coordinates and is not a dynamical variable for the field (as it would be in a Lorentz invariant theory).

The transverse vector potential can be conveniently given a Fourier expansion (referred to a box of volume V)

$$A(x) = \sum_{k,\lambda} \sqrt{\{\hbar/2\varepsilon_o V k c\}} \hat{e}_\lambda(k) \{a_\lambda(k)e^{ik.x} + a^+_\lambda(k)e^{-ik.x}\} \tag{75}$$

with the polarization index $\lambda = 1,2$, and $e_1(k)$, $\hat{e}_2(k)$ and $k/|k|$ forming a set of orthonormal vectors; $a_\lambda(k)$ and $a^+_\lambda(k)$ are annihilation and creation operators, respectively, for photons with polarization λ in the field mode with wavevector k, and have the commutation relation

$$[a_\lambda(k), a^+_\lambda(k')] = \delta_{\lambda\lambda'} \delta_{k,k'}. \tag{76}$$

In the continuum limit, $\delta_{k,k'}$ becomes a Dirac delta function, $\delta^3(k - k')$. $\pi(x)$, canonically conjugate to $A(x)$, and the magnetic field $B(x) = \nabla \wedge A(x)$, can both be expanded in terms of these photon operators, and H_{Rad} then takes the simple form,

$$H_{Rad} = \sum_{k,\lambda} \{N_{k,\lambda} + \tfrac{1}{2}\}\hbar\omega_k \tag{77}$$

where $N_{k,\lambda} = a^+_{k,\lambda}a_{k,\lambda}$ is the photon number operator, and $\hbar\omega_k = \hbar kc$ is the photon energy; thus in this representation the field appears as a collection of independent harmonic oscillators. The zero-point energy contribution is a (formally infinite) constant and can be dropped from H_{Rad}.

We now adapt the Coulomb gauge formalism to the particular case of molecules interacting with radiation, and with charged particle beams. Having fixed the gauge of course there is no possibility of testing for dependence of the calculations on the choice of gauge, and we leave this for later. It is as well to recall at this stage that the classical structural conception of molecules that motivates this discussion is based on combinations of **atoms**. Unfortunately we do not know how to write down the electrodynamics of such a composite system using only operators for atoms (and radiation) as the basic building blocks; we are therefore obliged to work at one remove from our goal with a characterization of molecules based on sub-atomic physics. In this (wholly conventional) view, molecules are composite systems of electrons and nuclei, particles that carry electric charge; their interactions with other charged particles and with radiation, as well as their internal interactions are electrodynamic in character. The N particles in a typical molecule will be described by the following quantities: charge $e_\alpha = eZ_\alpha$, mass m_α, position coordinate x_α, momentum p_α, $\alpha = 1,....N$; later on it will be convenient to refine this notation by explicitly recognizing the positively charged nuclei as distinct from the negatively charged electrons. Charged particles in a probe beam have an analogous description in terms of a charge e, mass m, coordinate x, and momentum p. In terms of these variables we can define charge densities,

$$\rho_e(r) = e\delta^3(x - r), \quad \rho_{Mol}(r) = \sum_\alpha e_\alpha \delta^3(x_\alpha - r) \tag{78}$$

and corresponding current densities,

$$j_e(r) = \{e/2m\}\{(p - eA(x))\delta^3(x - r) + \delta^3(x - r)p\} \tag{79}$$

$$j_{Mol}(r) = \sum_\alpha \{e_\alpha/2m_\alpha\}\{(p_\alpha - e_\alpha A(x_\alpha))\delta^3(x_\alpha - r) + \delta^3(x_\alpha - r)p_\alpha\} \tag{80}$$

where the latter are written in hermitian and gauge-invariant form.

The Hamiltonian for the photons has already been given (77); that for the particles can be expressed in the Coulomb gauge theory as

$$H_e = p^2/2m \tag{81}$$

$$H_{Mol} = \sum_\alpha p^2_\alpha/2m_\alpha + \sum_{\alpha \neq \beta}\{e_\alpha e_\beta\}/\{4\pi\varepsilon_0|x_\alpha - x_\beta|\}.$$

The interaction terms involving photons can be expressed in terms of the current densities and the vector potential for the field

$$H_{e,Rad} = - \int d^3r\, j_e(r).A(r) \tag{82}$$

$$H_{Mol,Rad} = - \int d^3r\, j_{Mol}(r).A(r). \tag{83}$$

The electrostatic potential at the point r due to a charge e located at x is,

$$\phi(r) = e/\{4\pi\varepsilon_0|x - r|\} \tag{84}$$

so that the Coulombic interaction energy between this charge and a molecule is
$$H_{e,Mol} = \int d^3r \, \rho_{Mol}(r) \, \phi(r) \quad (85)$$
In order to make the scattering kinematics apparent we separate out explicitly the molecular centre-of-mass motion from the internal dynamics and write
$$x_\alpha = R + r_\alpha, \quad p_\alpha = \{m_\alpha/M\}P + \pi_\alpha, \quad \alpha = 1,....N \quad (86)$$
where
$$R = M^{-1}\sum_\alpha m_\alpha x_\alpha, \quad P = \sum_\alpha p_\alpha \quad (87)$$
are the centre-of-mass variables. The internal variables (r_α, π_α) as defined are not linearly independent for
$$\sum_\alpha m_\alpha r_\alpha = 0, \quad \sum_\alpha \pi_\alpha = 0 \quad (88)$$
but independent internal variables can always be obtained from these. We can then write
$$H_{Mol} = H_{CM} + H_i \quad (89)$$
where
$$H_{CM} = P^2/2M \quad (90)$$
and H_i does not contain the centre-of-mass variables (R, P); its explicit form depends on the choice of internal variables [23,24]. We can now collect these results together and write the Hamiltonian for the non-interacting system as
$$H_o = H_e + H_{CM} + H_i + H_{Rad} \quad (91)$$
while the coupling interaction is
$$H_c = H_{e,Mol} + H_{e,Rad} + H_{Mol,Rad}. \quad (92)$$
The stationary state eigenfunctions of the operators that make up H_o can be used as a set of reference states for the free system. It is simpler to work with plane-waves rather than physically more realistic wave-packets (it can be shown that the same results emerge) and we write
$$H_e|k\rangle = E_k|k\rangle, \quad E_k = \hbar^2 k^2/2m \quad (93)$$
$$H_{CM}|K\rangle = E_K|K\rangle, \quad E_K = \hbar^2 K^2/2M \quad (94)$$
$$H_i|s\rangle = E_s|s\rangle \quad (95)$$
$$H_{Rad}|\lambda[1_p]\rangle = \hbar\omega_p|\lambda[1_p]\rangle \quad (96)$$
where $|\lambda[1_p]\rangle$ is a 1-photon state (wavevector p, polarization λ); the photon vacuum $|0\rangle$ is annihilated by H_{Rad} of course if the zero-point energy constant is omitted. In a coordinate representation we can take
$$\langle x|k\rangle = \{1/\sqrt{V}\} e^{ik \cdot x} \quad \langle R|K\rangle = \{1/\sqrt{V}\} e^{iK \cdot R} \quad (97)$$
with box-normalization; we assume
$$\langle \lambda'[1_{p'}]|\lambda[1_p]\rangle = \delta_{\lambda'\lambda}\delta_{p',p}, \quad \langle \lambda'[1_{p'}]|0\rangle = 0 \quad (98)$$
for the photon states, and
$$\langle s|s'\rangle = \delta_{s,s'} \quad (99)$$
for the molecular internal states.

3.4 Polarization Fields

A generally useful form for the electric polarization field for a collection of charges $\{e_\alpha: \alpha = 1,....N\}$ with position coordinates $\{x_\alpha: \alpha = 1,.....N\}$ is conventionally written as in (55); if we now compute the divergence of its rhs we easily get,

$$\nabla \cdot \mathbf{P}(\mathbf{x}) = -\sum_\alpha e_\alpha \delta^3(\mathbf{x}_\alpha - \mathbf{x}) + \sum_\alpha e_\alpha \delta^3(\mathbf{R} - \mathbf{x}) \qquad (100)$$
$$= -\rho(\mathbf{x}) + eQ\,\delta^3(\mathbf{R} - \mathbf{x}) \qquad (101)$$

where $\rho(\mathbf{x})$ is the charge density, and $eQ = \sum_\alpha e_\alpha$ is the overall charge of the system. For a neutral system $Q = 0$, and the choice of $\mathbf{R}$ is irrelevant and $\mathbf{R}$ may be eliminated; for a charged system one needs an extra condition on $\mathbf{R}$ to ensure that $\mathbf{P}(\mathbf{x})$ satisfies its basic defining equation (37). Consider first a neutral two particle system, for example the hydrogen atom [20]; let the electron have coordinate $\mathbf{x}_1 = \mathbf{x}_e$ and charge $e_1 = -e$, while the proton has coordinate $\mathbf{x}_2 = \mathbf{x}_p$ and charge $e_2 = +e$. Then

$$\mathbf{P}(\mathbf{x}) = e_1 \int_{\mathbf{R}}^{\mathbf{x}_1} d\mathbf{z}\, \delta^3(\mathbf{z} - \mathbf{x}) + e_2 \int_{\mathbf{R}}^{\mathbf{x}_2} d\mathbf{z}\, \delta^3(\mathbf{z} - \mathbf{x})$$

$$= -e \int_{\mathbf{R}}^{\mathbf{x}_e} d\mathbf{z}\, \delta^3(\mathbf{z} - \mathbf{x}) + e \int_{\mathbf{R}}^{\mathbf{x}_p} d\mathbf{z}\, \delta^3(\mathbf{z} - \mathbf{x}) \qquad (102)$$

Since exchanging the limits of the second integral simply reverses its overall sign, the two terms may be combined into one integral in which $\mathbf{R}$ no longer appears,

$$\mathbf{P}(\mathbf{x}) \equiv -e \int_{\mathbf{x}_p}^{\mathbf{x}_e} d\mathbf{z}\, \delta^3(\mathbf{z} - \mathbf{x}) \qquad (103)$$

This manipulation generalizes at once [19]. Suppose there are altogether M nuclei with positive charges $\{eZ_a: a = 1,\ldots M\}$; then there must be MZ_a electrons with charge $-e$ to give electroneutrality. Let the nuclei have coordinates $\{\mathbf{x}_a: a = 1,\ldots M\}$, momenta $\{\mathbf{p}_a: a = 1,\ldots M\}$ and masses $\{m_a: a = 1,\ldots M\}$; similarly the electrons have coordinates $\{\mathbf{x}_{ai}: i = 1,\ldots Z_a\}$, momenta $\{\mathbf{p}_{ai}: i = 1,\ldots Z_a\}$ and mass m_e. Then (100) can be rewritten as

$$\mathbf{P}(\mathbf{x}) = -e \sum_{a,i} \int_{\mathbf{x}_a}^{\mathbf{x}_{ai}} d\mathbf{z}\, \delta^3(\mathbf{z} - \mathbf{x}) \qquad (104)$$

where the summations are over all nuclei (a) and all electrons (i), since (37) is seen to be satisfied by this choice. In this expression the typical integral is taken along a path C starting from a nucleus $\mathbf{x}_a$ and ending at an electron $\mathbf{x}_{ai}$. On such a curve C, $\mathbf{z}$ is a definite function $\mathbf{z}(\mathbf{x}_a, \mathbf{x}_{ai}, \lambda)$ of $(\mathbf{x}_a, \mathbf{x}_{ai})$ and a real parameter λ which varies between upper and lower limits corresponding to $\mathbf{x}_a$ and $\mathbf{x}_{ai}$ respectively. Provided $\partial \mathbf{z}/\partial \lambda$ is bounded and piecewise continuous on C, the integral is well-defined [25].

In many applications of the polarization field we are concerned with spatial integrals of its scalar product with electromagnetic field quantities that are expanded in plane waves; the spatial integration is then a fourier transform of the polarization field to wavevector space. In this case we have

$$P(k) = -e \sum_{a,i} \int_{x_a}^{x_{ai}} dz \, e^{ik.z} \qquad (105)$$

and k can often be identified with the wavevector of a photon; if the wavelength of the photon is much greater than the lengths $\{|z|\}$ of the paths that contribute significantly to (105) (e.g. optical photons, <u>bound</u> charges) only the limit $k \to 0$ is of interest. This reduces $P(k)$ to the dipole moment operator for the charge distribution (corresponding to negligible spatial variation in $P(x)$),

$$\lim_{k \to 0} P(k) = -e \sum_{a,i} \int_{x_a}^{x_{ai}} dz \qquad (106)$$
$$= -e \sum_{a,i} x_{ai} + e \sum_a Z_a x_a \equiv d.$$

One well-known and useful result that involves d is the *velocity-dipole formula* [26] that relates matrix elements of the momentum operators of the charges and d,

$$\langle i| \sum_\alpha (e_\alpha/m_\alpha) p_\alpha |f\rangle = (i/\hbar)(E_i - E_f)\langle i|d|f\rangle \qquad (107)$$

where the states $|i\rangle, |f\rangle$, are eigenstates of the Hamiltonian for the charged particles; this relation is proved by evaluating the commutator of H_{mol} and d. Another useful relation follows directly from the canonical commutation relations for x and p,

$$[d^s, \sum_\alpha (e_\alpha/m_\alpha) p_\alpha^t] = i\hbar \, \delta_{st} \sum_\alpha (e_\alpha^2/m_\alpha) \qquad (108)$$

where the sums, as in (107), are over all particles in the system. Note that both expressions (107),(108) contain the zero wavevector limit of the fourier transform of the current density for the particles (41). Since there are realistic physical situations where non-zero wavevectors are important, for example problems involving free charges, high-energy photons (X-rays, γ-rays), some birefringence phenomena, it is of interest to generalize (107),(108) for such cases; this requires the evaluation of the commutator of $P(k)$ with H_{mol} and $j(k')$ respectively, and these in turn are determined by suitable combinations of the commutator of the polarization field $P(x)$ with the particle momentum operators $\{p_a, p_{ai}\}$.

Straightforward calculations described elsewhere [19] show that

$$\langle i| \sum_\alpha (e_\alpha/m_\alpha) p_\alpha e^{ik.x_\alpha} |f\rangle = (i/\hbar)(E_i - E_f)\langle i|P(k)|f\rangle + i\,k \wedge \langle i|M(k)|f\rangle \qquad (109)$$

is the generalization of the velocity-dipole formula to arbitrary wavevector k (cf [27]); equation (109) is of course the fourier transform of equation (42) used earlier to define the magnetization field. As for the current density - electric polarization field commutator we easily find the generalization of (108) in the form

$$[P(k)^s, j(k')^t] = i\hbar \delta_{st} \sum_\alpha (e_\alpha^2/m_\alpha) e^{i(k+k').x_\alpha} - \{\hbar k \wedge Y(k,k')^t\}^s \qquad (110)$$

where the vector Y is given explicitly in [19].

The introduction of centre-of-mass and internal variables produces a useful

simplification of these polarization fields. We adapt the notation of (86) and set
$$\mathbf{p}_a = (m_a/M)\mathbf{P} + \boldsymbol{\pi}_a, \quad \mathbf{p}_{ai} = (m_i/M)\mathbf{P} + \boldsymbol{\pi}_{ai};$$ (111)
the corresponding centre-of-mass (**R**) and internal coordinates are given by
$$\mathbf{x}_a = \mathbf{R} + \mathbf{r}_a, \quad \mathbf{x}_{ai} = \mathbf{R} + \mathbf{r}_{ai}.$$ (112)
The electric polarization field (105) becomes
$$\mathbf{P}(\mathbf{k}) = e^{i\mathbf{k}\cdot\mathbf{R}}\,\mathbf{P}(\mathbf{k})^{int}$$ (113)
where

$$\mathbf{P}(\mathbf{k})^{int} = -e\sum_{a,i}\int_{\mathbf{r}_a}^{\mathbf{r}_{ai}}d\mathbf{z}\, e^{i\mathbf{k}\cdot\mathbf{z}}.$$ (114)

The magnetization field decomposes into two terms
$$\mathbf{M}(\mathbf{k}) = \{M^{-1}\mathbf{P}\wedge\mathbf{P}(\mathbf{k})^{int} + \mathbf{M}(\mathbf{k})^{int}\}e^{i\mathbf{k}\cdot\mathbf{R}}$$ (115)
where the first term is analogous to the *Röntgen current* in the classical electrodynamics of dielectric media [28] and $\mathbf{M}(\mathbf{k})^{int}$ involves the internal angular momentum and line integrals over the internal coordinates [19]. We note in passing that the discussion so far has only referred to the classical orbital angular momentum contributions to the magnetization field; the spin contribution can be included in an obvious fashion in $\mathbf{M}(\mathbf{k})^{int}$ since the magnetization is linear in the angular momentum operators.

3.5 Scattering at high energies

Having specified the Hamiltonian and a set of reference states, we now turn to the evaluation of the cross-section, equation (22), for electron and photon scattering by molecules. An exact evaluation is out of the question of course since it would require a full treatment of a many-body problem; accordingly we seek approximation techniques guided by physical considerations. In light scattering, one deals with photon wavelengths (thousands of Å) >> molecular dimensions (a few Å), so that it is usually valid to drop all dependence of the cross-section on the photon momentum (the electric dipole approximation); the essence of *diffraction* is that the probe beam particles have de Broglie wavelengths much more nearly comparable with the length scale over which the molecular charge density varies. This is of course just the condition that requires the retention of the plane-wave factors, $e^{i\mathbf{k}\cdot\mathbf{x}}$, that describe the momentum state of the probe beam. A 1 keV electron, for example, has a de Broglie wavelength of about 0.4Å and a hundred-fold increase in energy reduces the wavelength by a factor of ten to a value comparable with the electron Compton wavelength (higher energies correspond to 'relativistic' electrons which require relativistic electrodynamics). By way of comparison with the electron beam case we note that copper and molybdenum produce K_α radiation with wavelengths of 1.54Å and 0.71Å respectively, corresponding to photon energies of 8keV and 17 keV. These energies are very much greater than chemical binding energies (a few eV), and greater than most electron ionization energies. This can be put more formally as follows.

Let the energy required to remove particle α from the target molecule leaving it with a very small momentum P_α be denoted by R_α [2]. The average kinetic energy of

particle α is
$$K_\alpha = <s|p^2_\alpha/2m_\alpha|s> \qquad (116)$$
where the expectation value is taken with respect to the initial bound internal state of the molecule, and then
$$R_\alpha = K_\alpha + U_\alpha \qquad (117)$$
defines the average potential energy, U_α, of particle α. If the energy of a probe beam particle of momentum **p** is E_p, the *weak binding condition* [2] is the requirement that
$$K_\alpha \ll E_p ; \quad |U_\alpha| \ll E_p, \quad \alpha = 1,....N; \qquad (118)$$
these inequalities are generally satisfied for the probe beams described above for particles {α} in molecules. When the conditions (118) are valid the scattering takes place as though the particles in the target molecule are *free*, and as shown in [2] this greatly simplifies the calculation of the T-matrix elements. This method of calculation is also known as the 'impulse' approximation; if multiple scattering effects can be neglected it means that the complete wave scattered by the bound system is obtained from simple additivity for the waves scattered by the individual particles in the target molecule; each such scattering event is a simple two-body collision between a probe beam particle and a particle in the target considered to be free. This means that the T-matrix may be written in the form
$$T = \sum_\alpha t_\alpha \qquad (119)$$
where
$$t_\alpha = V_\alpha + V_\alpha \check{G}_o(E) t_\alpha \qquad (120)$$
in which $\check{G}_o(E)$ is now to be evaluated with the *neglect* of the interparticle interaction terms $\{U_{\alpha\beta}\}$ for the target particles (so for a molecule this means neglect of the interparticle Coulombic interactions). Unlike the Born approximation this does *not* require an assumption that the interaction between probe and target particles is *weak*. Of course the first Born approximation, $t_\alpha \approx V_\alpha$, becomes essentially exact whenever the inverse dependence on energy of the Green function $\check{G}$ becomes dominated by the high energy of the probe particles; trivially there is then no reference to the interparticle interactions in the target. Strictly speaking, in cases where the probe particles are identical to particles bound in the target (as in electron - molecule scattering) the Pauli principle should be invoked; however the exchange terms fall off much more rapidly than the direct terms as the energy increases and this fact leads to considerable simplification in the theory of scattering at high energies [2]. In the following discussion we shall take this result to justify the neglect of antisymmetrization from the outset. We begin with a conventional perturbative approach to the calculation of the cross-section, and subsequently look briefly at the generalization produced by the impulse approximation.

4 Electron - Molecule Scattering

For electron scattering there are no photons present in the initial and final states, so we can take these to be:
$$|i> = |k, K, s, 0>, \qquad E_i = E_k + E_K + E_s \qquad (121)$$
$$|f> = |k', K', s', 0>, \qquad E_f = E_{k'} + E_{K'} + E_{s'}. \qquad (122)$$
To order e^2 i.e. to first order in the dimensionless coupling constant α = 1/137, we require

only the first two terms of (16) for the T-matrix,

$$T_{fi} \approx \langle f|H_c|i\rangle + \sum_I \langle f|H_c|I\rangle (E_i - E_I)^{-1} \langle I|H_c|i\rangle \tag{123}$$

where H_c is given by equation (92); the instantaneous Coulomb interaction between the electron beam and the molecular charge density, equation (85), contributes to the first term in (123), while the field dependent terms, (82),(83) are needed for the second-order perturbation contribution. Each current operator contains a term linear in e and the interaction amounts to the exchange of a virtual photon. As we shall see shortly, the energy denominator in the electrodynamic contribution to (123) involves $(E_k - E_{k'})$ which can be small - it vanishes in the case of elastic scattering - rather than just E_k, so we are not justified in discarding this term at once on the basis of a large energy denominator argument; formally both types of contribution are required to maintain the gauge-invariance of the cross-section since the apparently clean separation between the instantaneous Coulombic interaction (first term in (123)) and the virtual photon exchange interaction (second term in (123)) is a consequence of choosing $p(x;x')^\perp = 0$ (Coulomb gauge condition). On the other hand, the conventional account of the scattering of fast electrons by atoms and molecules *is* based purely on the first-order perturbation treatment of the instantaneous Coulombic interaction, so we will need another argument to dispose of the extra contributions of an electrodynamic nature.

For the first-order perturbation term we have

$$\langle f|H_c|i\rangle = \langle k', K', s'|H_{e,Mol}|s, K, k\rangle$$
$$= V^{-2}\int d^3x \int d^3R\, e^{ix\cdot(k-k')}\, e^{iR\cdot(K-K')} \langle s'|H_{e,Mol}(\{x,R,r_\alpha\})|s\rangle. \tag{124}$$

Using equations (78),(84),(85) the double fourier integral can be evaluated at once to yield a δ-function guaranteeing overall conservation of momentum, and a matrix element of the internal state electric charge 'form-factor'

$$\Gamma(q) = \sum_\alpha Z_\alpha\, e^{iq\cdot r_\alpha} \tag{125}$$

where we have put $eZ_\alpha = e_\alpha$. If we define the electron momentum transfer vector by $\hbar q = \hbar(k' - k)$, we have

$$\langle f|H_c|i\rangle = (e^2/\varepsilon_0 q^2)(2\pi)^3 V^{-2}\, \Gamma(-q)_{s's}\, \delta^3(k'+ K' - k - K). \tag{126}$$

Notice that the internal states of the molecule are not involved in the momentum conservation condition, and that

$$\Gamma(0)_{s's} = Q\, \delta_{s's},\ eQ = \sum_\alpha e_\alpha \tag{127}$$

so that for small q the matrix element approximates to what would be expected for Rutherford scattering of an electron by a 'charge' eQ; for a neutral system the matrix element varies as q^2 for small q.

For the second-order perturbation theory contribution to order α there are two possible types of intermediate (virtual) states:

$$|a\rangle = |k, K', s', \lambda[1_Q]\rangle,\quad E_a = E_k + E_{K'} + E_{s'} + E_Q \tag{128}$$
$$|b\rangle = |k', K, s, \lambda[1_Q]\rangle,\quad E_b = E_{k'} + E_K + E_s + E_Q \tag{129}$$

corresponding to the two time-ordered Feynman diagrams shown in Figure 1.ii ($E_Q = \hbar Qc$ is the photon energy).

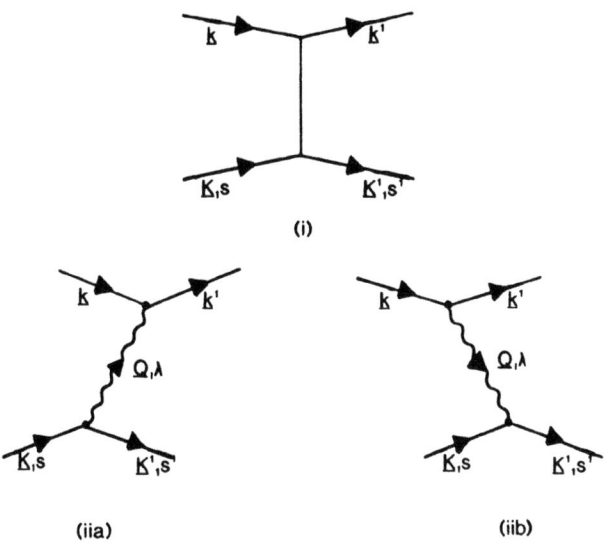

Figure 1. Time-ordered Feynman diagrams for electron-molecule scattering to order α in the Coulomb gauge; (i) the instantaneous Coulomb interaction, (ii) the two possible virtual photon exchange processes.

The summation over the intermediate states, {I}, amounts to a summation over the two polarization states of the photon,

$$\sum_\lambda \hat{e}_\lambda(\mathbf{Q}) \hat{e}_\lambda(\mathbf{Q}) = \mathcal{J}(\mathbf{Q}), \qquad (130)$$

where the tensor $\mathcal{J}$ is defined as $\mathcal{J}(\mathbf{G}) = (\mathbb{I} - \hat{\mathbf{G}}\hat{\mathbf{G}})$, and integration over all possible values of its momentum. Momentum is conserved at each vertex in a Feynman diagram, and so the virtual photon momentum $\mathbf{Q}$ must appear in δ-functions that make the $\mathbf{Q}$-integration immediate.

In diagram 1.iia, the electron absorbs a photon $(\mathbf{Q}, \lambda)$ and makes a transition $\mathbf{k} \to \mathbf{k}'$; this is possible because although momentum is conserved, energy is not in virtual processes. Diagram 1.iib describes the same transition occurring with emission of a photon. These processes are associated with the part of the matrix element

$$\langle \mathbf{k}' | -\int d^3r\, j_e(\mathbf{r}).\mathbf{A}(\mathbf{r}) | \mathbf{k} \rangle \qquad (131)$$

that is linear in the vector potential, where appropriate states for 1-photon absorption/emission involving the photon vacuum are to be understood. A straightforward calculation using (75),(79),(97) and (98) yields the matrix element as

$$-(e/m)\sqrt{\{\hbar^2/2V\varepsilon_0 E_Q\}}\, (\hbar \mathbf{k}).\hat{e}_\lambda(\mathbf{Q})\, (2\pi)^3\, V^{-1}\, \delta^3(\mathbf{k} - \mathbf{k}' \pm \mathbf{Q}) \qquad (132)$$

where the (+,-) signs are to be taken for absorption/emission respectively.

A similar calculation is required for the photon-molecule vertex where the interaction is mediated by the operator in equation (83); here the transition from a molecular state $|N\rangle \equiv |\mathbf{K},s\rangle$ to $|N'\rangle \equiv |\mathbf{K}',s'\rangle$ is accompanied by absorption/emission of a photon $(\mathbf{Q},\lambda)$. Integration over the molecular centre-of-mass states yields the customary momentum conserving δ-function, and we can write

$(e/m)<N'|-\sum_\alpha s_\alpha p_\alpha \cdot A(x_\alpha)|N>$

$= -(e/m)\sqrt{\{\hbar^2/2V\varepsilon_o E_Q\}} \Delta(K,\pm Q)_{s's} \cdot \hat{e}_\lambda(Q) (2\pi)^3 V^{-1} \delta^3(K - K' \pm Q)$ (133)

where $(e/m)s_\alpha = (e_\alpha/m_\alpha)$, and $(+Q, -Q)$ again refer to photon absorption/emission respectively. We defer for a moment discussion of the internal state matrix element $\Delta(K,Q)_{s's}$. For diagram (1.iia) the energy denominator is,

$$E_i - E_a = E_K - E_{K'} + E_s - E_{s'} - E_Q$$ (134)

and for diagram (1.iib)

$$E_i - E_b = E_k - E_{k'} - E_Q;$$ (135)

because of overall energy conservation in the scattering process $(E_i = E_f)$, we can rewrite the denominator (134) in the form,

$$E_i - E_a = -(E_k - E_{k'} + E_Q).$$ (136)

It is now a straightforward matter to put together the matrix elements and energy denominators, (132) - (136), for each diagram, and sum over the intermediate states. For diagram (1.iia), the integration over **Q** yields the result **Q** = **q**, whereas for diagram (1.iib), **-Q** = **q**. Consequently the matrix elements for both diagrams give the same net result involving $\Delta(K,-q)$. The energy denominators can then be combined using

$$\frac{1}{E_k - E_{k'} + E_q} - \frac{1}{E_k - E_{k'} - E_q} = \frac{-2E_q}{(E_k - E_{k'})^2 - E_q^2}$$ (137)

so that the final result for the second-order perturbation term is,

$(e^2/\varepsilon_o)(\hbar^3/m^2)(2\pi)^3 V^{-2} \delta^3(k' + K' - k - K) \Delta(K,-q)_{s's} k \cdot \mathfrak{Z}(q) \{(E_q^2 - (E_k - E_{k'})^2\}^{-1}.$ (138)

We now return to equation (133) to investigate further the matrix element $\Delta_{s's}$. Evaluation of the 1-photon absorption matrix element,

$$<K', s',[0]|-(e/m)\sum_\alpha s_\alpha p_\alpha \cdot A(x_\alpha)|\lambda[1_k], s, K>$$ (139)

shows that we can make the identification

$$<K', s'|\sum_\alpha s_\alpha p_\alpha e^{ik \cdot x_\alpha}|s, K> = (2\pi)^3 V^{-1} \delta^3(K - K' + k) \Delta(K, k)_{s's}.$$ (140)

Comparison of equations (109),(113),(115),(140) then shows that

$\Delta(K, k)_{s's} = (imE_{N'N}/\hbar) P(k)_{s's}^{int} + imk \wedge \{(\hbar K/M) \wedge P(k)_{s's}^{int} + M(k)_{s's}^{int}\}$ (141)

where

$$E_{N'N} = E_{K+k} + E_{s'} - E_K - E_s.$$ (142)

The gauge-invariant T-matrix element to order e^2 is the sum of the two contributions (126),(138),

$$T_{fi} = \lambda_c^2(e^2/\varepsilon_o)(2\pi)^3 V^{-2} \delta^3(q + K' - K) \mathcal{L}(k',k,K)_{s's}$$ (143)

with λ_c the electron Compton wavelength (= h/mc), and

$\mathcal{L}(k',k,K)_{s's} = \Gamma(-q)_{s's} x^{-2} + \Delta(K,-q)_{s's} p \cdot \mathfrak{Z}(q) \{m^2c^2(x^2 - \gamma_{kk'}^2)\}^{-1}$ (144)

where $p = \hbar k$ is the momentum of the incident electron, and x and $\gamma_{kk'}$ are the dimensionless parameters $\lambda_c q$ and $2\pi(E_k - E_{k'})/mc^2$ respectively. We can now apply equations (21) - (28) directly to obtain the differential cross-section for scattering an electron with momentum **p** into the interval [**p'**, **p'** + d^3p']. The transition rate is

$\tau(k',k)^{-1} = \lambda_c^4(e^4/\hbar\varepsilon_o^2) (2\pi)^4 V^{-3} \sum_{K'Ks's} w_{Ks} |\mathcal{L}(k',k,K)_{s's}|^2 \delta^3(q + K' - K)$
$* \delta(E_k + E_K + E_s - E_{k'} - E_{K'} - E_{s'}).$ (145)

The average over **K'** is immediate using (28), and serves simply to put **K'** = **K** - **q** in the energy conservation δ-function. If we define

$$\Im(\mathbf{k}',\mathbf{k}) = \sum_{K,s,s'} w_{Ks} |\mathcal{L}(\mathbf{k}',\mathbf{k},\mathbf{K})_{s's}|^2 \delta(E_\mathbf{k} + E_K + E_s - E_{\mathbf{k}'} - E_{K-q} - E_{s'}) \quad (146)$$

the cross-section takes the form

$$d\sigma = (2\alpha\lambda_c)^2 (p'/p) \Im(\mathbf{k}',\mathbf{k}) d\Omega_{p'}. \quad (147)$$

where α is the dimensionless fine structure constant; note that $\alpha\lambda_c = r_e$, the classical electron radius, so the cross-section is proportional to r_e^2.

The cross-section $d\sigma(\mathbf{k}',\mathbf{k})$ we have just calculated for electron scattering by molecules within the framework of non-relativistic quantum electrodynamics is exact to order α^2; it is also very general and encompasses a wide range of possible physical processes since, so far, there are no restrictions on the initial and final momenta $(\mathbf{k},\mathbf{k}')$. As far as 'structure' is concerned however we can confine attention to the cross-sections for *elastic scattering* defined by the condition $E_\mathbf{k} = E_{\mathbf{k}'}$; this leads to several simplifications in the cross-section formulae. Firstly, $E_\mathbf{k}$ and $E_{\mathbf{k}'}$ cancel in the energy conservation δ-function which therefore only refers to purely molecular quantities; thus the elastic scattering is restricted to an average over the molecular transitions $\mathbf{K},s \to \mathbf{K} - \mathbf{q},s'$. The restriction $E_\mathbf{k} = E_{\mathbf{k}'}$ implies the further condition $|\mathbf{p}| = |\mathbf{p}'|$, so the factor (p'/p) drops out of the cross-section. This condition on the magnitude of the momenta also leads to the relation $|\mathbf{q}| = 2k\sin(\phi/2)$ and writing $k = 2\pi/\lambda$ (where λ is the de Broglie wavelength for electrons, h/mv) leads to the familiar diffraction parameter $|\mathbf{q}| = s = (4\pi/\lambda)\sin(\theta)$ where, in accordance with the usual convention, the scattering angle ϕ has been replaced by 2θ.

We now attempt to simplify the elastic scattering cross-section for electrons by reducing it to an average over the initial density matrix for the internal states of the molecule. The elastic cross-section is

$$d\sigma^e = 4 r_e^2 \Im^e(\mathbf{k}',\mathbf{k}) d\Omega_{p'}. \quad (148)$$

where

$$\Im^e(\mathbf{k}',\mathbf{k}) = \sum_{K,s,s'} w_{Ks} |\mathcal{L}(\mathbf{k}',\mathbf{k})_{s's}|^2 \delta(E_K + E_s - E_{K-q} - E_{s'}) \quad (149)$$

and $\mathcal{L}(\mathbf{k}',\mathbf{k})$ is given by (144). As is well-known, an expression such as

$$A_s(\Delta E) = \sum_{s'} |\Omega_{s's}|^2 \delta(E_{s'} - E_s + \Delta E) \quad (150)$$

involving a sum over a complete set of intermediate states $\{|s'\rangle\}$, can be written in an equivalent form involving a time-integration over the *two-point fluctuation function* for the operator Ω in the state $|s\rangle$. This is done by introducing the fourier representation of the δ-function and using a unitary transformation to time-dependent operators,

$$\Omega(t) = e^{iHt/\hbar} \Omega e^{-iHt/\hbar}. \quad (151)$$

The result is

$$A_s(\Delta E) = \frac{1}{2\pi\hbar} \int_{-\infty}^{+\infty} dt \, e^{-i\Delta Et/\hbar} \langle s|\Omega^+(0)\Omega(t)|s\rangle. \quad (152)$$

If we choose $H = H_i$, the internal molecular Hamiltonian, we can apply this transformation to (149) which then reads

$$\mathfrak{F}^e(\mathbf{k}',\mathbf{k}) = \frac{1}{2\pi\hbar} \int_{-\infty}^{+\infty} dt \sum_\mathbf{K} \sum_s w_s <s|\mathcal{L}^+(0)\mathcal{L}(t)|s> w_\mathbf{K} e^{-i\Delta E_\mathbf{K} t/\hbar} \quad (153)$$

where
$$\Delta E_\mathbf{K} = E_\mathbf{K} - E_{\mathbf{K}-\mathbf{q}} = (\hbar^2/M)\mathbf{K}\cdot\mathbf{q} - E_R; \quad (154)$$
$E_R = \hbar^2 q^2/2M$ is the recoil energy of the molecular centre-of-mass when momentum $\hbar\mathbf{q}$ is transferred to it. Given a specific form for $w_\mathbf{K}$, the probability that the centre-of-mass has momentum $\hbar\mathbf{K}$ in the initial state, we can attempt to evaluate the sum over $\mathbf{K}$ in (153). In the case of an ideal gas, $w_\mathbf{K}$ is the usual Maxwell-Boltzmann velocity distribution expressed in terms of the wave-vector $\mathbf{K}$, and

$$\sum_\mathbf{K} w_\mathbf{K} \rightarrow \{\hbar^2/2\pi M k_B T\}^{3/2} \int d^3\mathbf{K} \exp(-\hbar^2 K^2/2M k_B T). \quad (155)$$

In the approximation that the fluctuation function is independent of $\mathbf{K}$, it follows easily that in this case we can use the result
$$\sum_\mathbf{K} w_\mathbf{K} \exp(-i\Delta E_\mathbf{K} t/\hbar) = \exp(iE_R t/\hbar - \mu t^2) \quad (156)$$
where $\mu = E_R k_B T/\hbar^2$. In the limit that $\mu \gg 1$ s^{-2} ($q \neq 0$, 'sufficiently high' T) this exponential function behaves like a δ-function in the remaining time integration,
$$\exp(iE_R t/\hbar - \mu t^2) \rightarrow \sqrt{(\pi/\mu)} \exp\{(iE_R t/\hbar)^2/4\mu\} \delta(t - t_o) \quad (157)$$
where t_o is the point of stationary phase
$$t_o = iE_R/2\mu\hbar = i\hbar/2k_B T \quad (158)$$
and putting these results back into $\mathfrak{F}^e(\mathbf{k}',\mathbf{k})$ we find

$$\mathfrak{F}^e(\mathbf{k}',\mathbf{k}) = \frac{e^{-E_R/4k_B T}}{\sqrt{4\pi E_R k_B T}} <<\mathcal{L}^+(0)\, \mathcal{L}(i\hbar/2k_B T)>> \quad (159)$$

where
$$<<\Omega^+(0)\,\Omega(t)>> = \sum_s w_s <s|\Omega^+(0)\,\Omega(t)|s>$$
$$= \mathrm{Tr}[\rho_o^i\, \Omega^+(0)\,\Omega(t)] \quad (160)$$
is a two-point fluctuation function for $\mathcal{L}$ evaluated as an average over the initial state ρ_o^i. Collecting together (148), (159) and (160) the cross-section can be displayed in the form

$$d\sigma^e = \frac{r_e^2\, e^{-E_R/k_B T}}{\sqrt{4\pi E_R k_B T}} \mathrm{Tr}[\rho_o^i\, \mathcal{L}^+(0)\,\mathcal{L}(i\hbar/2k_B T)]\, d\Omega_{p'} \quad (161)$$

in which all reference to individual eigenstates has been eliminated. Because of the thermal average the operators $\mathcal{L}^+$, $\mathcal{L}$ appear in a Matsubara type of representation defined by
$$\mathcal{L}(i\beta/2) = \exp(-\beta H_i/2)\, \mathcal{L}\, \exp(+\beta H_i/2) \quad (162)$$
with $\beta = \hbar/k_B T$. The same type of result is obtained if the full $\mathbf{K}$-dependence is retained in the cross-section. Let the terms in (149) be denoted as $f(\mathbf{K})$; then (156) is replaced by

$\sum_K w_K \exp(-i\Delta E_K t/\hbar) f(K) = \exp(iE_R t/\hbar - \mu t^2)\pi^{-3/2} \int d^3w \exp(-w^2) f\{w/\lambda_T + (t/2t_o)q\}$ (163)

(λ_T is the thermal de Broglie wavelength, $\hbar/\{2Mk_B T\}^{1/2}$ for the molecule). Only the spherical average of f survives the integration over $d\Omega(\hat{w})$, and there is then a modified form for $\mathcal{L}$ in the fluctuation function; the t-integration goes through as before, so the final result is again of the form (161). For forward scattering, $q = 0$, ΔE_K vanishes and so a different analysis is required which we do not pursue here.

In the case of elastic electron scattering the parameter $\gamma_{kk'} = 0$, so both terms in $\mathcal{L}$ contain the common factor x^{-2} (= $(q\lambda_c)^{-2}$). Reference to equation (144) shows that the electron-scattering cross-section depends on the components of $\Delta(K,-q)$, and hence on $P(-q)^{int}$ and $M(-q)^{int}$. Now $E_{N'N}$, equation (142) with $k = -q$, vanishes because of the energy conservation δ-function, and so the first term in (141) cannot contribute to the elastic cross-section. The first term in $\mathcal{L}$ can also be expressed in terms of the internal electric polarization field. We may write Γ in the form

$$\Gamma(k) = \sum_{a,i} \{e^{ik.r_a} - e^{ik.r_{ai}}\} \quad (164)$$

($Z_\alpha = -1$ for electrons), and then from (114) it follows at once that

$$\Gamma(k) = -ik.P(k)^{int} \quad (165)$$

which is essentially a Fourier transform version of (37); the amplitude $\mathcal{L}$ involves this fourier component evaluated at $k = -q$. If we define

$$u(p,q) = p.\mathfrak{Z}(q) \quad (166)$$

$\mathcal{L}$ can be put in the form

$$x^2 \mathcal{L} = -i\left[(1 + u.\hbar K/mMc^2)q + 1/mMc^2(q.\hbar K)u\right].P(-q)^{int}$$
$$- (i\hbar/mc^2)(k \wedge q).M(-q)^{int} \quad (167)$$
$$\approx -i q.P(-q)^{int} - (i\hbar/mc^2)(k \wedge q).M(-q)^{int} \quad (168)$$

since the terms involving the molecular mass M are of order V/c (relative to 1), where V is the average molecular speed. We see therefore that $\mathcal{L}$, equation (144), can be decomposed into two terms, one proportional to the -q fourier component of the internal electric polarization field, $P(-q)^{int}$, and the other proportional to the -q fourier component of the internal magnetization field, $M(-q)^{int}$, and these describe the electric and magnetic 'structure' of the molecule respectively. The final step required to bring (168) into line with the customary account is the neglect of the magnetization term; this requires that $(\hbar/mc^2)\|k \wedge M(-q)\| \ll \|P(-q)\|$ for non-relativistic electrons (the inequality requires that $\|k\|$ not be too big - the theory is not trustworthy for relativistic electron beams in any case). The resulting cross-section then depends only on the *momentum transfer* q and, from (165), is determined by the thermal fluctuation average of $\Gamma(-q)$, essentially the Fourier transform of the molecular charge density operator expressed in terms of particle coordinates relative to the centre-of-mass. Next we look at the corresponding calculation for X-ray photons.

5 Photon - Molecule Scattering; the Kramers - Heisenberg Formula

The diffraction experiment performed with electromagnetic radiation is idealized in the following way. The probe beam consists of n_k photons in the mode with wavevector k and polarization λ; as a result of the interaction with the molecule, photons in the mode k' with polarization λ' are found at the detector. We therefore take the initial and final states to be:

$$|i\rangle = |K, s, \lambda[n_k]\rangle, \qquad E_i = E_K + E_s + n_k E_k \qquad (169)$$

$$|f\rangle = |K', s', \lambda[n_k - 1], \lambda'[1_{k'}]\rangle, \qquad E_f = E_{K'} + E_{s'} + E_{k'} + (n_k - 1)E_k \qquad (170)$$

where $E_k = \hbar kc$ is the photon energy, and (K, s) refer to the molecular centre-of-mass and internal states respectively. To order e^2 the T-matrix element is again given by the two leading terms in the perturbation series, equation (123), where now the coupling is due to $H_{Mol,Rad}$, equation (83), which must be separated into its terms linear and quadratic in e. The linear term is essentially the operator in (133) and we put it in the same form,

$$V^{(1)} = -(e/m)\sum_\alpha s_\alpha p_\alpha \cdot A(x_\alpha); \qquad (171)$$

it contributes to the second-order perturbation term. The quadratic contribution may be written as

$$V^{(2)} = \int d^3r \, \gamma(r) \, A(r) \cdot A(r) \qquad (172)$$

where

$$\gamma(r) = (e^2/2m)\sum_\alpha t_\alpha \, \delta^3(r - x_\alpha) \qquad (173)$$

and $(e^2/2m)t_\alpha = e_\alpha^2/2m_\alpha$; this term contributes to the first-order perturbation theory matrix element. Once again, both contributions are required in order to maintain the formal gauge-invariance of the cross-section [25,27].

For the first-order perturbation term we have

$$\langle f|H_c|i\rangle = \int d^3r \, \langle \lambda'[1_{k'}], \lambda[n_k - 1]|A(r) \cdot A(r)|\lambda[n_k]\rangle \langle K', s'|\gamma(r)|s, K\rangle. \qquad (174)$$

Using (61), the photon contribution is easily reduced to

$$(\hbar/\varepsilon_o Vc)(n_k/kk')^{\frac{1}{2}} \, \hat{e}_\lambda(k) \cdot \hat{e}_{\lambda'}(k') \, e^{-iq \cdot r} \qquad (175)$$

where $\hbar q = \hbar(k - k')$ is the momentum transfer for the photon. The matrix element of $\gamma(r)$ can be evaluated using the plane waves (97) for the centre-of-mass motion,

$$\langle K', s'|\gamma(r)|s, K\rangle = (e^2/2m) \, V^{-1} \, e^{i(K - K') \cdot r} \, \Gamma(K - K')_{s's} \qquad (176)$$

and

$$\Gamma(Q) = \sum_\alpha t_\alpha \, e^{iQ \cdot r_\alpha} \qquad (177)$$

is a molecular form-factor involving only the internal variables that is similar to (125). Combining (175) and (176), and carrying out the integration over r we obtain the first-order matrix element (174) as

$$\langle f|H_c|i\rangle = (e^2/m)(\hbar/2\varepsilon_o c)(n_k/kk')^{\frac{1}{2}} (2\pi)^3 \, V^{-2} \, \delta^3(q + K' - K) \, \hat{e}_\lambda(k) \cdot \hat{e}_{\lambda'}(k') \, \Gamma(-q)_{s's}. \qquad (178)$$

corresponding to the Feynman diagram in figure 2.i.

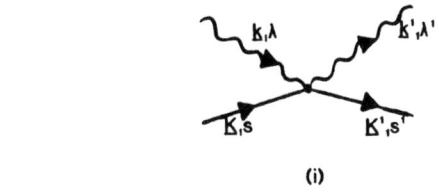

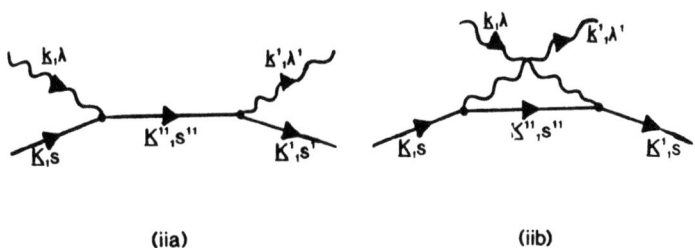

Figure 2. Time-ordered Feynman diagram for photon-molecule scattering to order α in the Coulomb gauge; (i) the 'A²' contact interaction, (ii) the two possible processes involving virtual excitation of the molecule.

The second-order scattering processes that involve virtual excitation of the molecule are shown in figure 2.ii(a,b). The two possible intermediate states are:

$$|a\rangle = |K'', s'', \lambda[n_k - 1]\rangle, \quad E_a = E_{K''} + E_{s''} + (n_k - 1)E_k \qquad (179)$$
$$|b\rangle = |K'', s'', \lambda[n_k], \lambda'[1_{k'}]\rangle, \quad E_b = E_{K''} + E_{s''} + E_{k'} + n_k E_k. \qquad (180)$$

In the notation of (133), diagram 2.iia corresponds to the steps $|N\rangle \to |N''\rangle$ accompanied by the absorption of a photon (k,λ), followed by $|N''\rangle \to |N'\rangle$ with emission of the photon (k',λ'). The numerator for this diagram is therefore the product of two such factors as displayed in (133), namely,

$$(e/m)^2 (\hbar/2\varepsilon_o c) (n_k/kk')^{1/2} (2\pi)^6 V^{-3} \delta^3(K - K'' + k) \delta^3(K'' - K' - k')$$
$$* \Delta(K'', -k')_{s's''} \Delta(K, k)_{s''s} : \hat{e}_\lambda(k)\hat{e}_{\lambda'}(k') \qquad (181)$$

In diagram 2.iib the sequence of photon absorption and emission is reversed, and so the numerator for this diagram is obtained from (181) by the interchange $k' \leftrightarrow -k$, i.e.

$$(e/m)^2 (\hbar/2\varepsilon_o c) (n_k/kk')^{1/2} (2\pi)^6 V^{-3} \delta^3(K - K'' - k') \delta^3(K'' - K' + k)$$
$$* \Delta(K'', k)_{s's''} \Delta(K, -k')_{s''s} : \hat{e}_\lambda(k)\hat{e}_{\lambda'}(k') \qquad (182)$$

The corresponding energy denominators are:

$$E_i - E_a = E_K + E_s - E_{K''} - E_{s''} + E_k \qquad (183)$$
$$E_i - E_b = E_K + E_s - E_{K''} - E_{s''} - E_{k'}. \qquad (184)$$

Equations (181) - (184) must then be combined, and summed over the intermediate states $\{|K'', s''\rangle\}$ to form the second-order perturbation term. Using (28), the sum over K'' for diagram 2.iia leads to the substitution of K'' by $K + k$, and for diagram 2.iib, $K'' \to K - k'$; the product of δ-functions is reduced to

$$V (2\pi)^{-3} \delta^3(q + K' - K) \qquad (185)$$

which expresses momentum conservation. The second-order term is then,

$$(e^2/m)(\hbar/2\varepsilon_o c)(n_k/kk')^{1/2}(2\pi)^3 V^{-2}\delta^3(q + K' - K) \hat{e}_\lambda(k)\hat{e}_{\lambda'}(k'):\Re(K,s,s',k,k') \quad (186)$$

where the dyadic $\Re$ is defined as

$$\Re(K,s',s,k',k) = m^{-1}\sum_{s''}\left\{\frac{\Delta(K+k,-k')_{s's''}\Delta(K,k)_{s''s}}{E_K + E_s - E_{K+k} - E_{s''} + E_k} + \frac{\Delta(K-k',k)_{s's''}\Delta(K,-k')_{s''s}}{E_K + E_s - E_{K-k'} - E_{s''} - E_{k'}}\right\} \quad (187)$$

Note that $\hat{e}_\lambda(k)$ is contracted with the Δ factor that contains k as its second argument (and similarly for $\hat{e}_{\lambda'}(k')$ with $\Delta(...,k')$). The overall gauge-invariant T-matrix element is the sum of (178) and (187)

$$T_{fi} = (e^2/m)(\hbar/2\varepsilon_o c)(n_k/kk')^{1/2}(2\pi)^3 V^{-2}\delta^3(q + K' - K)\,\hat{e}_{\lambda'}(k')^\alpha\,\hat{e}_\lambda(k)^\beta$$
$$* \{\Gamma(-q)_{s's}\delta_{\alpha\beta} + \Re(K,s',s,k',k)_{\alpha\beta}\} \quad (188)$$

and a straightforward repetition of the argument that lead from (143) to (147) provides the differential cross-section for scattering of a photon (k) to (k',λ'),

$$d\sigma(k',\lambda',k) = r_e^2(k'/k)\sum_{K,s,\lambda,s'} w_\lambda w_{K,s}\,|\mathfrak{C}(K,k',k,s',s,\lambda',\lambda)|^2$$
$$* \delta(E_k + E_K + E_s - E_{k'} - E_{K-q} - E_{s'})\,d\Omega_{k'} \quad (189)$$

where

$$\mathfrak{C} = \hat{e}_{\lambda'}(k')^\alpha \hat{e}_\lambda(k)^\beta \{\Gamma(-q)_{s's}\delta_{\alpha\beta} + \Re(K,s',s,k',k)_{\alpha\beta}\}. \quad (190)$$

In obtaining (189), the summation over the final molecular centre-of-mass momentum, K', has been carried out using the δ-function in (188) and, exactly as for electron scattering, it picks out the value $K' = K - q$ in the energy conservation δ-function.

For ease of comparison with the cross-section for electron scattering, equation (147), which was calculated for unpolarized electrons, we assume that the detector monitors all photons of wavevector k' irrespective of their polarization state. We define

$$\Xi(k',k)_{\alpha\beta\gamma\delta} = \sum_{\lambda,\lambda'} w_\lambda\,\hat{e}_{\lambda'}(k')^\alpha\,\hat{e}_\lambda(k)^\beta\,\hat{e}_{\lambda'}(k')^\gamma\,\hat{e}_\lambda(k)^\delta \quad (191)$$

so that the averaged cross-section for unpolarized photons is,

$$d\sigma(k',k) = r_e^2(k'/k)\,\Xi(k',k)_{\alpha\beta\gamma\delta}\sum_{K,s,s'} w_{K,s}\,\mathfrak{C}(K,k',k,s',s)_{\alpha\beta\gamma\delta}$$
$$* \delta(E_k + E_K + E_s - E_{k'} - E_{K-q} - E_{s'})\,d\Omega_{k'} \quad (192)$$

where the tensor $\mathfrak{C}$ is obtained by comparison with $|\mathfrak{C}|^2$ in (190). The discussion preceding the elastic electron scattering cross-section, equation (148), can be repeated here verbatim; (k'/k) becomes 1, and $E_k - E_{k'}$ drops out of the energy conservation delta function in (192). It then follows that $\mathfrak{C}$ in (190) plays the same role as $\mathcal{L}$ in equation (149), and analogous moves lead to a formula like equation (161) for the elastic X-ray diffraction cross-section with $\mathfrak{C}$ replacing $\mathcal{L}$. There is little simplification of the dyadic $\Re$ however; formally the sum over s" and the energy denominators in (187) can be replaced by an integral over time displaced Δ factors by changing to time-dependent operators using (151) and a Fourier representation of $(\Delta E + i0)^{-1}$, but this is only a start and much more analysis is required to make the cross-section intelligible. In the Discussion section, §6, we indicate briefly another approach. The sum over λ' in (191) can be carried out at once with the aid of equation (130); if by way of example we also assume that the incident beam of photons is unpolarized (commonly the case with X-rays) then (191) can be evaluated completely in terms of the tensor $\mathcal{3}$, equation (130) to yield

$$\Xi(\mathbf{k}',\mathbf{k})_{\alpha\beta\gamma\delta} = \tfrac{1}{2} [\mathcal{3}(\mathbf{k}')]_{\alpha\gamma}[\mathcal{3}(\mathbf{k})]_{\beta\delta}. \qquad (193)$$

Further reduction requires specific information about the tensor $\mathcal{C}$ which has to be averaged over $\mathbf{K}$ and s with the weight factor $w_{\mathbf{K}s}$. Its *scalar part* (which is all that would be required if only the first term in (190) is retained in the cross-section) reduces Ξ to

$$\Xi_{\alpha\beta\gamma\delta}\, \delta_{\alpha\beta}\, \delta_{\gamma\delta} = \tfrac{1}{2}(1 + \cos^2(\phi)) \qquad (194)$$

where ϕ is the scattering angle ($\mathbf{k}.\mathbf{k}' = \cos(\phi)$), and we recognize the classical Thomson cross-section [10]

$$\tfrac{1}{2}r_e^2(1 + \cos^2(\phi)) \qquad (195)$$

as a factor in the general formula (192).

6 Discussion

This chapter has presented a conventional quantum mechanical account of the diffraction experiment involving molecules in fluid phases as the targets, and electrons and photons as the projectiles. The calculations are based on the standard time-independent perturbation theory approach to scattering problems, in the context of non-relativistic quantum electrodynamics. They are not new, but perhaps are not as widely known in theoretical chemistry as they might be. The formulation of the cross-section that is directly related to the diffraction observables is reviewed in §2, and follows well-known lines. In §3, some rather general considerations about electrodynamics are presented, focusing particularly on the presence of arbitrary quantities in the Hamiltonian due to the existence of the gauge symmetry group for charged particles. In order to perform the calculations in §4, §5 it is convenient to specify a particular gauge (the Coulomb gauge), though as noted in §3, that leaves the issue of gauge invariance of calculated observables to be resolved.

After integrating out the plane-wave states for the probe beam particles and the molecular centre-of-mass motion (cf §4), the observables are expressed as matrix elements involving the molecular internal states $\{|s\rangle\}$. Perhaps it would be as well at this stage to alert the reader to an important issue that is customarily glossed over, namely, what do the states $\{|s\rangle\}$ *actually* refer to ? In the conventional treatment of molecular light scattering (see for example the comprehensive survey in [29]) these states are assumed to refer to the particular *molecular* species of interest, and the usual structure of electronic, vibrational and rotational terms is exposed via the notion of a potential energy surface and some version of the Born-Oppenheimer approximation ('clamped nucleus', 'adiabatic' etc.). There are no difficulties with such an approach for *diatomic* molecules, but for general polyatomic molecules (most of chemistry !) this argument fails for reasons that are explained in [23]; briefly, as noted in §1, the failure is that starting from the customary internal Hamiltonian, H_i, equation (89), there is no rational theory of chemical isomerism and the Born-Oppenheimer approximation does *not* remove this difficulty.

The essential technical point is that the states $\{|s\rangle, |s'\rangle, |s''\rangle...\}$ used in the perturbation formulae are supposed to form a complete set of states for the unperturbed Hamiltonian - here the molecular internal Hamiltonian. The stationary eigenstates of H_i are such a set. Now the complete set of states $\{|s\rangle\}$ associated with the Hamiltonian (89) written down for the particles that comprise, say, the species with formula, C_8H_8, describes *all* possible arrangements of these particles. Chemically, this set includes for example cyclooctatetraene, vinylbenzene, the species 3-vinyl hex-1,4-diyne which has a *chiral* centre (cf [30]) - *prima facie* unexpected since H_i commutes with the parity operator - and also all possible fragments and ions composed from C_8H_8 represented by states for bound particle clusters embedded in the continuum of H_i. This example may seem artificial; the point of course is that the argument could be given for *any* chemical formula sufficiently complicated to sustain isomers i.e. practically all. In a formal sense, this may not matter since the complete set of states is simply used as a device for articulating the perturbation theory based on (16) in terms of matrix elements and, as we saw in §4, the resulting amplitudes can be written in other ways that need not refer to the full set of states, e.g using the density matrix for the initial state ρ_0^i. However, in a practical sense, this is an issue, since physical considerations are often used as a guide as to which matrix elements/ energy denominators are important and which can be neglected.

One commonly made argument that seeks to justify the drastic step of throwing away completely the dyadic $\Re$ in (187) is this: since X-ray photons have high energies relative to atomic/molecular transition energies, the denominators in (187) are so large that the whole expression can be neglected [31]. The X-ray scattering is then determined purely by the form-factor $\Gamma(-q)$, equation (177), which could be put into the analogue of (161) for the thermally averaged X-ray scattering cross-section. Of course, such an approximation (as with the analogous move in the electron diffraction case) - it is just the first Born approximation - leads to a cross-section determined purely by an 'electric form-factor' evaluated at the wavevector component -q, and this is then well on the way towards a structural interpretation (after Fourier transformation of the cross-section). This may be the right thing to do, but a better justification seems required. The sum in $\Re$ is a sum over *virtual transitions* involving a complete set of states and, as just noted, these states describe not only the low lying excitations of the molecule but also *all* of its fragmentation products as well (all of the continuous spectrum). Processes such as ionization (the photoelectric effect) and unimolecular reactions yielding simpler molecules or atoms can all occur as real processes induced by X-rays, and all are brought about by field-molecule interactions mediated in part by the **p.A.** operator in (83). At first sight then it is not obvious that the energy denominators are always going to overwhelm all the matrix elements of the coupling operator(s) between all these different species; this is an open question. In any case the argument is not manifestly gauge-invariant; it amounts to saying that the X-ray diffraction cross-section can be obtained by first-order perturbation theory applied to the **A.A** term in the interaction operator *in the Coulomb gauge*. Obviously, a change of gauge according to (53) introduces the arbitrary field $\mathbf{p}(x;x')$ into the cross-section which then has any value. The second-order perturbation theory sum is required to cancel out such arbitrary terms to leave an overall invariant result; this can be seen as follows.

For the generalized Hamiltonian written explicitly in terms of $\mathbf{p}(x;x')$, the perturbation operator V in (71) takes the form,

$$V = -\sum_\alpha (e_\alpha/2m_\alpha)\{\mathbf{p}_\alpha.\mathbf{A}(\mathbf{x}_\alpha) + \mathbf{A}(\mathbf{x}_\alpha).\mathbf{p}_\alpha\} \quad (196)$$
$$+ \sum_\alpha (e_\alpha^2/2m_\alpha)\mathbf{A}(\mathbf{x}_\alpha).\mathbf{A}(\mathbf{x}_\alpha) \quad (197)$$

$$-\sum_\alpha e_\alpha \int d^3x \ E(x)^\perp \cdot p(x;x_\alpha) \quad (198)$$
$$+ \tfrac{1}{2}\varepsilon_0^{-1}\sum_{\alpha\beta} e_\alpha e_\beta \ \wp(x_\alpha;x_\beta)^\perp \quad (199)$$

where

$$A(x) = A(x)^\perp - \nabla_x \int d^3x' \ A(x')^\perp \cdot p(x';x) \quad (200)$$
$$\wp(x;x')^\perp = \int d^3x'' \ p(x'';x)^\perp \cdot p(x'';x')^\perp \quad (201)$$

and $p(x;x')$ is any solution of (50). The particle variables satisfy the usual canonical commutation relations, (59), and all the field terms can be expressed in terms of the annihilation and creation operators in H_{Rad}, so V is well-defined as an operator on the product Hilbert space of particles and field, but of course contains the arbitrary vector field $p(x;x')$. As noted in §3 one can choose it to be zero (Coulomb gauge theory), and one can just as well choose a line integral from for $p(x;x')$ which cancels the Coulomb energies in H_{Mol} almost everywhere in space (multipolar Hamiltonian) - there are infinitely many choices. What is really required is a demonstration that the choice of $p(x;x')^\perp$ is *irrelevant* as far as physical observables are concerned, so that the form of $p(x;x')$ used in any application is determined primarily by convenience. A heuristic demonstration of this result was sketched in §3.2. This leads us to consider the invariance of transition amplitudes; we can argue as follows.

The perturbation operator V, (196) - (199), can be divided into two types of terms

$$V[p] = V_1[p] + V_2[p] \quad (202)$$

where $V_1[p]$ is given by the sum of (196) and (198) and is linear in the charge parameter e, whereas $V_2[p]$ is the sum of (197) and (199), and is quadratic in e. Let the unperturbed Hamiltonian for the reference system of charges and electromagnetic field be H_0, equation (71). The results in [25] for the multipolar Hamiltonian can then be generalized to arbitrary $p(x;x')$ to give,

$$V_1[p] = V_1[0] + (i/\hbar)[H_0,A[p]] \quad (203)$$
$$V_2[p] = V_2[0] - (i/\hbar)[V_1[0],A[p]]$$
$$\quad -(1/2!)(i/\hbar)^2[A[P],[H_0,A[P]]] \quad (204)$$

where

$$A[P] = \int d^3x \int d^3x' \ A(x')^\perp \cdot p(x';x) \ \rho(x) \quad (205)$$

as in §3; in the commutators, $\rho(x)$ fails to commute with the particle momenta, while $A(x)^\perp$ fails to commute with H_{Rad}.

Let $|i\rangle$ and $|f\rangle$ be two distinct states of the reference system with energies E_i and E_f; a physical transition amplitude A_{if} between these two states contains the energy conservation factor $\delta(E_i - E_f)$. Absorption and emission of a photon can be described in first order perturbation theory, and from (203) one obviously has for real (energy-conserving) transitions

$$^{(1)}A_{if}[p] \propto \langle i|V_1[p]|f\rangle = \langle i|V_1[0]|f\rangle. \quad (206)$$

Second-order perturbation theory (linear response) yields the Kramers-Heisenberg formula, §5, for scattering a photon. For any $p(x;x')$ one can then show

$$^{(2)}A_{if}[p] - {}^{(2)}A_{if}[0] = (E_f - E_i)\Delta[p]_{if} \quad (207)$$

as in [25] for the multipolar Hamiltonian choice, and *on the energy shell*, the rhs vanishes thus confirming the invariance of the physical amplitude to second-order in the charge parameter e. These calculations refer to the complete amplitude with all contributions to (123) retained. It is a separate matter to determine whether particular terms are dominant in physically interesting cases. In the case of diffraction the first-order contribution suggests (this is difficult to see from the second-order term) that the cross-section may be proportional to a form-factor which, after Fourier transformation, can be interpreted in terms of something like a particle density. We now briefly describe another approach that confirms this result.

Although the perturbation theory approach described in §4, §5 is not 'wrong', it yields an intractable cross-section formula because the T-matrix elements contain interparticle correlations in each order of perturbation theory beyond the first, so that the resulting cross-section contains very complicated many-particle terms. For high-energy scattering however the arguments that lead to the *impulse approximation* (§3.5) suggest that these many-particle correlations are essentially irrelevant to the result. In the following we give a brief sketch of this development and refer the reader to the extensive literature cited in [2]. It is convenient to start again from the formalism presented in §2; it will transpire that the exact nature of the probe particle - target interaction is not really required. The cross-section is obtained directly from the transition rate τ^{-1} which can be put in the form,

$$\tau^{-1} = (2\pi/\hbar)\sum_n w_n \langle n|T^\dagger \delta(E_n - H_o) D_\phi T|n\rangle \qquad (208)$$

where D_ϕ is the operator that characterizes the detector. As in multiple-scattering theory, [2], the action of T can be expressed in terms of a sum of two-body scattering amplitudes for collisions between a single bound particle α in the target, and a probe beam particle,

$$T|n\rangle = \sum_\alpha T_\alpha |\psi_\alpha\rangle. \qquad (209)$$

The crucial step in the impulse approximation is to recognize that in the case of high-energy scattering (see §3.5) the T-matrix element involving bound particle α and a probe-beam particle that makes the transition $\mathbf{k} \to \mathbf{k}'$ can be expressed in the form,

$$\langle \mathbf{k}'|T_\alpha|\mathbf{k}\rangle = \langle \mathbf{k}'|t_\alpha|\mathbf{k}\rangle e^{-i\mathbf{q}\cdot\mathbf{x}_\alpha} \qquad (210)$$

where t_α is a t-matrix for the interaction between a probe-beam particle and a *free* target particle α, and $\mathbf{q} = \mathbf{k}' - \mathbf{k}$ defines the momentum transfer as usual. The exponential factor provides the appropriate phase relation between waves scattered from target particles α in various positions $\mathbf{x}_\alpha$ [2,32].

With this result the transition rate becomes

$$\tau^{-1} = (2\pi/\hbar)\sum_{\alpha,\beta}\sum_n w_n \langle n|t_\beta^\dagger e^{i\mathbf{q}\cdot\mathbf{x}_\beta} \delta(E_n - H_o) D_\phi t_\alpha e^{-i\mathbf{q}\cdot\mathbf{x}_\alpha}|n\rangle; \qquad (211)$$

let us consider the rate for transitions to all final states which is obtained by replacing D_ϕ by the unit operator. It is convenient to make a transformation to time-dependent operators, by introducing the Fourier representation of the δ-function, as in §4. We then have

$$\tau^{-1} = \frac{1}{\hbar^2}\int_{-\infty}^{+\infty} dt\, e^{i\omega t}\sum_{\alpha,\beta} \sum_n w_n \langle n|t_\beta^{+} e^{i\mathbf{q}\cdot\mathbf{x}_\beta} t_\alpha(t) e^{-i\mathbf{q}\cdot\mathbf{x}(t)_\alpha}|n\rangle \qquad (212)$$

where $\hbar\omega = (E_{k'} - E_k)$. The further reduction of this transition rate depends on particular circumstances, and we do not go into details except to note that the elastic scattering is determined by the $\omega = 0$ limit of equation (212). The important point is that there are physically realistic circumstances, including diffraction of electrons and X-rays by fluids, in which the transition rate (212) can be decomposed into factors; one of these factors is the two-point fluctuation function, equation (160), with the operator Ω given by something like

$$\Omega = \sum_\alpha e^{-i\mathbf{q}\cdot\mathbf{x}_\alpha} = \int d^3\mathbf{x}\, n(\mathbf{x})\, e^{-i\mathbf{q}\cdot\mathbf{x}} \qquad (213)$$

and $n(\mathbf{x}) = \sum_\alpha \delta^3(\mathbf{x} - \mathbf{x}_\alpha)$ is the particle density operator, and the other contains averages over the t-matrices $\{t_\alpha\}$ taken pairwise. Thus the impulse approximation shows that the form of the cross-section obtained from the first Born approximation can be obtained in a more general way that does not require the assumption of first-order perturbation theory.

In principle this generalized cross-section might be calculated theoretically; in practice the lack of a coherent theory for the states of the target referred to above prevents this, and instead the diffraction experiments are *interpreted* in terms of the particle density fluctuations related to Ω. An alternative way of stating the difficulty is that it lies in our ignorance of the probabilities, $\{w_s\}$, (19) that specify the initial internal state density matrix ρ_o^i. Suppose the fluid is in thermal equilibrium with its surroundings; then statistical physics might be taken to suggest that ρ_o^i is just the Gibbs state for the given temperature,

$$\rho_o^i = Z^{-1} \exp(-H_i/k_B T) \qquad (214)$$

where Z is the partition function for H_i. This leads to an obvious difficulty; equation (214) is a unique formula associated with the given Hamiltonian H_i and so will yield a unique value for a scattering cross-section like (161). But if H_i supports a family of chemical isomers, we would expect *each isomer* to have its own characteristic cross-sections for electron and photon scattering (that's how we can distinguish them), and each isomer should have its own density matrix ρ_o^i which cannot be the Gibbs state for the molecular Hamiltonian H_i. The difficulty for the quantum theory of molecular structure is that we lack a rational account of such density matrices [23].

7 References

[1] Paoloni,L. (1979) *Towards a Culture-based Approach to Chemical Education in Secondary Schools: The Role of Chemical Formulae in the Teaching of Chemistry*, Eur.J.Sci.Educ.,**1**, 365 - 377
Paoloni,L. (1982) *Reflections on Chemical Philosophy*, unpublished memoir

[2] Goldberger, M.L. & Watson, K.M. (1964) *Collision Theory*, John Wiley & Sons, Inc. New York

[3] Gottfried,K. (1966) *Quantum Mechanics.I:Fundamentals*, W.A.Benjamin, 1st Ed^n.

[4] Osborn,R.K. (1988) *Applied Quantum Mechanics*, World Scientific Publishing Co.Pte.Ltd

[5] Loudon,R. (1983) *Quantum Theory of Light*, 2nd Ed. O.U.P.

[6] Cohen-Tannoudji, C. Dupont-Roc,J. & Grynberg,G. (1989) *Photons and Atoms: Introduction to Quantum Electrodynamics*, John Wiley

[7] Dirac,P.A.M. (1952) *Les transformations de jauge en Électrodynamique*, Ann. Inst. Henri Poincaré, **13**, 1 - 42

[8] Woolley,R.G. (1975) *The Electrodynamics of Atoms and Molecules*, Adv.Chem.Phys., **33**,153 - 233

[9] Woolley,R.G. (1980) *Gauge invariant wave mechanics and the Power-Zienau-Woolley Transformation*, J.Phys., **A13**, 2795 - 2805

[10] Power,E.A. (1964) *Introductory Quantum Electrodynamics*, Longmans Green, London

[11] Woolley,R.G., (1971), *Molecular quantum electrodynamics*, Proc.Roy.Soc., **A321**, 557 - 572

[12] Babiker,M., Power,E.A., & Thirunamachandran,T., (1973), *Atomic field equations for Maxwell fields interacting with non-relativistic quantal sources*, Proc.Roy.Soc., **A332**, 187 - 197

[13] Power,E.A., & Zienau,S., (1959), *Coulomb gauge in non-relativistic electrodynamics and the shape of spectral lines*, Phil.Trans.Roy.Soc., **251**, 427 - 454

[14] Göppert-Mayer,M., (1931), *Über Elementarakte mit zwei Quantensprüngen*, Ann.Phys.(Leipzig), **9**, 273 - 294

[15] Dirac,P.A.M., (1934), *Discussion of the infinite distribution of electrons in the theory of the positron*, Proc.Camb.Phil.Soc., **30**, 150 - 163

[16] Fock,V., (1937), *Eigen-Time in Classical and Quantum Mechanics*, Phys.Zeits. d Sowjetunion, **12**, 404 - 425

[17] Peierls,R.E., (1934), *The vacuum in Dirac's theory of the Positive Electron*, Proc.Roy.Soc., **A146**, 420 - 441

[18] Woolley.R.G. (1974), *A reformulation of molecular quantum electrodynamics*, J.Phys., **B7**, 488 -499

[19] Woolley,R.G., (1996), *Gauge Invariance and the Electric Polarization Field*, Mol.Phys., **88**, 291 - 307

[20] Woolley,R.G. (1971), *On the hamiltonian theory of the molecule-electromagnetic field system*, Mol.Phys., **22**, 1013 - 1023

[21] Woolley,R.G. (1971), *On the theory of optical activity in quantum electrodynamics*, Mol.Phys., **22**, 555 -559

[22] Mandelstam, S. (1956) *Dynamical variables in the Bethe-Salpeter formalism*, Proc.Roy.Soc., **A233**, 248 - 266

[23] Sutcliffe,B.T. (1992) *The chemical bond and molecular structure*, J.Mol.Structure - Theochem, **259**, 29 - 58

[24] Woolley,R.G. (1991) *Quantum Chemistry beyond the Born-Oppenheimer approximation*, J.Mol. Structure - Theochem, **230**, 17 - 46

[25] Healy,W.P. & Woolley,R.G. (1978) *On the derivation of the Kramers-Heisenberg dispersion formula from non-relativistic quantum electrodynamics*, J.Phys., **B11**, 1131 - 1136

[26] Atkins,P.W., (1983), *Molecular Quantum Mechanics*, 2nd Ed^n, OUP, p.449

[27] Healy,W.P. (1977) *A generalization of the Kramers-Heisenberg dispersion formula*, Phys.Rev., **A16**, 1568 - 1574

[28] Power,E.A. & Thirunamachandran,T. (1971) *Three distribution identities and Maxwell's atomic field equations including the Röntgen current*, Mathematika, **18**, 240 - 246
[29] Barron,L.D. (1982) *Molecular Light Scattering and Optical Activity*, Cambridge University Press
[30] Sutcliffe,B.T. (1996) *The Development of the Idea of a Chemical Bond*, Int. J. Quant. Chem., **58**, 645 - 655
[31] James,R.W. (1965) *The Optical Principles of the Diffraction of X-rays*, Ch.3, Cornell University Press
[32] Fowler,T.K. (1958) *Quasi-elastic scattering of pions by nuclei*, Phys.Rev., **112**, 1325 - 1335

ASPECTS OF THE CHEMICAL BOND 1996

J. F. OGILVIE
Department of Chemistry
Oregon State University
Corvallis, OR 97331-4003 USA

1. Introduction

In a preceding essay [1] I outlined the historical, mathematical and physical bases of our present qualitative knowledge about the chemical bond, specifically about covalent binding in simple molecules. An expanded version [2] of that material afforded the opportunity to clarify and to extend the arguments somewhat, and to discuss the reaction to the previously printed version. In the present essay I consider important aspects of chemical binding previously ignored, namely ionic substances, the *transition state* purported to occur between reactants and products during the course of a simple chemical reaction, and a description of the mechanism of an intramolecular rearrangement. As for the covalent bond, these topics evolved largely during three quarters of a century since quantum mechanics was discovered and developed, but involve, more or less implicitly, many ideas based on the classical notion of molecular structure emanating from the nineteenth century. The final section consists of a discussion of the context of quantum chemistry in a contemporary scientific milieu.

2. Ionic materials

Whereas some chemical materials with simple stoichiometric descriptions such as H_2, CH_4, CO_2 and H_2O are either gaseous or volatile liquid substances under common ambient conditions, other materials with equally simple nominal descriptions such as Be, NaCl and SiO_2 are involatile and refractory solid substances. These properties are supposed to reflect the nature of chemical binding between the atomic centres. In isolated molecules of substances of the former group, there exist strong bonds between adjacent atomic centres, whereas when these substances are condensed into a liquid or solid state only weak binding between atomic centres in separate formula units is observed. In pure solid or liquid phases of substances in the latter group one has difficulty to distinguish atomic centres in explicit sets corresponding to simple formula units. The binding between atomic centres in such distinct units is less weak in the case of H_2O than in the other cases, consistent with the observation that solid or liquid H_2O near 300 K is less volatile than CH_4 or CO_2. Significant spatial attraction (intermolecular interaction) is observable in liquid and solid H_2O outside the simple formula unit that realistically represents the molecular entity H_2O in the dilute gaseous phase; likewise small but significant distortions of the geometric structure (defined according to internuclear distance and angle) that is characteristic of the gaseous phase are detectable in ice. Such distortions are much more pronounced for crystalline substances described as inorganic or mineralogical, for which in many cases no compound of the same stoichiometric ratio is characterised in the gaseous phase. The customary description of these substances has been known as the ionic model, but Hyde and O'Keeffe [3] have shown that this description is inadequate; the following discussion on this topic is based largely on a review published by Hyde [4], supplemented with material from subsequent authors. What is offered to replace the ionic model is still essentially a physical model, as quantal treatments of such extended structures remain largely impracticable.

The prototypical example of a binary ionic crystal is that of sodium chloride, described as a virtually infinite array of nominally spherical sodium and chloride ions in three-dimensional space. These ions are described as cations, Na^+, and anions, Cl^-; the latter or negative ions are supposed to pack closely in space, with positive ions inserted into appropriate interstices. For NaCl in its normal cubic close-packed form, the cations locate in octahedral interstices, so that each Na^+ has six equidistant adjacent anions; in other crystals cations are found in tetrahedral interstices, accordingly with four anions as immediate neighbours. The coordination at the cationic site is supposed to depend on a purely geometric criterion, the ratio of radii of cation and anion, being in the ranges [0.225, 0.414] for tetrahedral coordination, [0.414, 0.732] for octahedral coordination, and > 0.732 for cubic coordination with eight anionic neighbours.

The vital assumption underlying such a criterion is that an atomic ion of a particular type, classified according to element and to nominal state of oxidation, has a well defined radius that has little or no dependence on the environment, such as the nature or properties of counterions or adjacent atomic centres. Although this assumption, which predates quantum theories, is formally inconsistent with a quantal description of matter, its ramifications with respect to classical geometric and electromagnetic concepts have proved more important in recent practice. A radius of an isolated ion is clearly an arbitrary property, just as even an apparent or effective radius of an atom of a noble gas varies with the nature of the experiment for its measurement, apart from any further dependence on temperature or energy. In contrast the sum of atomic radii of a cation and an adjacent anion in a particular crystalline environment is empirically well defined, just like an equilibrium internuclear distance of a stable diatomic molecule (in an hypothetical state without residual energy). For conventional ions the radii are commonly taken from a list by Pauling [5], and comparison is also made with a more recent compilation [6]; in the (mostly, older) literature such ionic radii are found in several other sets.

Much evidence inconsistent with this ionic model exists. For instance, in metallic oxides MgO, CaO, SrO and BaO of group 2, the oxide anions can not be closely packed in every case because the distance between adjacent oxide centres varies through a large range, 298 pm for MgO to 391 pm for BaO; likewise the volume per O^{2-} varies in the range/10^{-29} m^3 from 1.38 for BeO to 4.21 for BaO [7]. Although all four oxides MgO - BaO possess a structure analogous to that of NaCl, ratios of conventional radii of ions in BaO, SrO and possibly also CaO deviate from the range nominally applicable to a close-packed octahedral structure.

The geometry of close packing is also the geometry of open packing [7, 8]. The latter arrangement allows distances between anions and between cations to be a maximum, not a minimum, consistent with the phenomenon of electrostatic repulsion between ions having like charges. The term eutactic, implying well arranged, is devised [8] to signify a spatial order as in close packing. Many structures contain irregular arrays of anions, which are hence not closely packed or eutactic; although these geometries commonly defy description in terms of arrays of anions, or of anionic polyhedra with cations at their centres, in many cases, such as Ca_2SiO_4 to be discussed, the arrangement of cations is regular, according to a familiar geometry.

For some substances, the coordination number is found to be unrelated to the ratio of the ionic radii. For instance Mg has octahedral coordination, denoted ^{VI}Mg, in MgO, but tetrahedral coordination, so ^{IV}Mg, in a mineral spinel, $MgAl_2O_4$, despite its being surrounded by oxygen centres in both cases. In such crystals of $^{IV}Mg^{VI}Al_2^{IV}O_4$, the larger cation Mg^{2+} ($R = 71|57$ pm) occupies the smaller, tetrahedral interstice and conversely the smaller cation Al^{3+} (67.5|53.5 pm) is located in the larger, octahedral cavities. (Of two values of ionic radius R specified in these cases, the former is listed by Shannon [6] and the latter by Pauling [5].) As an instance of multiple coordination numbers over a greater range, in ternary compounds of oxygen with metals in group 1 and in groups 4 – 12 sodium exhibits coordination to oxide ions numbering from two to twelve [9]; the ionic radius of Na^+ varies almost linearly from about 98 pm at coordination number four to nearly 140 pm at coordination number twelve [6]. The concept of an almost constant ionic radius invariant with coordination number is clearly tested severely under such conditions.

Although measurement of absolute electronic densities in crystals by means of xray diffraction (or other) experiments is problematical, accurate maps of electron density within a unit cell are claimed to be derived with this technique. For lithium fluoride in which the smallest distance between Li and F nuclei is 201 pm, the minimum electronic density along the line between the lithium and fluoride nuclei occurs at a distance 109 pm from the latter atomic centre, in contrast with a much larger nominal ionic radius of F^-, 133|136 pm. In crystalline sodium chloride, integration of the density of electronic charge within a sphere of radius corresponding to the minimum electronic density along the line between centres of a cation and of an adjacent anion indicates that the net deficiency of electronic charge within this volume, relative to Na as an electrically neutral atom, is 0.95 electrons, whereas the excess charge within the sphere about the chlorine nucleus is 0.70 electrons. The remaining negative charge, equivalent to 0.25 electrons, is then deduced to locate within the volumes between the spheres; there the electronic density is small, less than 2×10^{-7} electrons pm^{-3}, compared with 7×10^{-5} and 1.7×10^{-4} electrons pm^{-3} near sodium and chlorine nuclei respectively [10]. For MgO a minimum electronic density along the line between the atomic centres of Mg and O occurs 91 pm from Mg (compare with the ionic radii specified above); a sphere of this radius contains a net charge $+1.9$ electrons, but the corresponding sphere about O includes a charge only -0.9 electrons [11]. In this case 1.0 electron per couple of ions (Mg and O) is located in the space between such spheres. This result is consistent with the fact that the electron affinity of an oxygen atom in the gaseous phase is positive for only the first extra electron (*i.e.* $\Delta U^0 < 0$ for the process $O + e^- \rightarrow O^-$, but $\Delta U^0 > 0$ for the process $O^- + e^- \rightarrow O^{2-}$).

The size of the ion Si^{4+} with tetrahedral coordination is listed to be 40|26 pm. If the size of this ion is defined as the radius of a sphere containing ten electrons, quantal calculations [12] indicate a radius 61 pm. This value, similar to the value 58 pm for Si^{4+} in α-quartz [13], much exceeds the specified nominal values for coordination of this type. In any case, the idea that anions are necessarily much larger than cations is incorrect; old values of ionic radii are based on theoretical properties of free ions, not of these ions in crystals. In the latter environment anions are subject to a positive site potential (due to the collective electric field of surrounding cations) [14, 15]; this potential, typically $12z$ V with z being the net charge of the anion in units of the magnitude of the protonic charge, has the effect of contracting the distribution of charge. In this way O^{2-} appears to become stable in the crystalline field, whereas the free ion O^{2-} has indefinite extent. Conversely, cations are subject to a negative potential that has the effect of expanding the distribution of charge. All this evidence indicates that the customary ionic model is gravely deficient.

According to an alternative approach [4], an atomic centre is treated simply as an atom, not as an ion; the terms cation and anion are however retained to distinguish atomic centres of metallic elements from those of non-metallic elements. From a survey of numerous well determined structures of silicates and silica compounds consisting of SiO_4 tetrahedra connected by corners in framework, sheet and chain arrangements, O'Keeffe and Hyde showed that the distances between nearest Si centres lay within a notably narrow range, (306 ± 6) *pm*, with a few outliers at larger separations [16]. Moreover the variation of this range is no greater than for the distribution of lengths of Si-O bonds, (164 ± 6) *pm*; hence the distance between adjacent Si centres varies as little as that between adjacent Si and O centres. This result conforms to what Bragg and Claringbull [17] found earlier, that the distance 305 *pm* between nearest Si centres in a single chain is invariant in five chains existing in three pyroxene minerals. The distance between nearest Si centres not only in silicate glasses but also in gaseous molecular substances containing the angular moieties Si-O-Si, Si-NH-Si and Si-CH_2-Si lies also within the range/*pm* [300, 315], despite the variation of lengths of Si-O, Si-N and Si-C bonds and of interbond angles, 165 *pm* and 2.56 *rad*, 170 *pm* and 2.09 *rad*, and 188 *pm* and 1.90 *rad* in the same order [4]. These data convey the significance that, if silicon atoms in these structures be considered to be in contact, an almost constant atomic radius 153 *pm* might be ascribed to that interatomic interaction, which is deemed not a chemical bond. Analogous trends prevail in borates (containing the angular moiety B-O-B), phosphates (containing P-O-P) *et cetera*; the radii are found to be additive, for instance in borosilicates (containing B-O-Si). Such data yield two conclusions. Interactions between atomic centres, between which there is nominally no chemical bond in the traditional sense, that are second-nearest neighbours can play an important role in determining the geometric structure and thermodynamic stability of chemical compounds or materials, in both gaseous and condensed phases. Of these non-bonding interactions those between cations might be more important than those between anions because radii of cations generally exceed those of anions, particularly N, O and F.

Such an approach is proposed to be applicable to small molecules, both organic and inorganic, in the gaseous phase [18]. Extension of this approach to involatile solid compounds is consistent with results of quantal calculations on small isolated molecules in relation to likely structures of solid materials containing similar structural moieties [19, 20, 21]. In their development of conformational analysis, organic chemists recognise the importance of non-bonded interactions, and methods to predict and to analyse structures of molecular crystals depend on such phenomena [22]. The latter calculations commonly involve application of functions of potential energy in one dimension, typically of the form $V(d) = Ae^{-Bd} - C/d^6$ with d as internuclear distance; A, B and C are parameters characteristic of a particular couple of atomic centres. The concept of a hard sphere characterised with a constant radius for a particular atomic centre becomes accordingly replaced for quantitative applications by a distance variable in a range depending on the energy of intermolecular interaction. Although these functions are established for elemental constituents of organic molecular crystals and systems of biochemical importance, within computational approaches named molecular mechanics for instance, they are lacking for nominal cations such as Si^{4+} and O^{2-}. This approach involving hard spheres of specified radii nevertheless renders a qualitative understanding of structures of crystalline phases of simple inorganic compounds and minerals. Hence both α-quartz and cristobalite possess frameworks, topologically distinct but flexible, of SiO_4 tetrahedra connected at corners and collapsed from the most open possible geometries that would be expected if they were truly ionic. The collapse proceeds to the extent that the distance between nearest Si centres is decreased to about 306 pm. SiO_2 consists of tetrahedral SiO_4 moieties rather than octahedral SiO_6 moieties, not because six oxygens can not pack around Si (which occurs in $^{VI}Si^{IV}P_2^{II}O_7$ for instance), but because more than two silicons must then pack about an oxygen, denoted $^{>IV}Si^{>II}O_2$. The minimum condition would then be $^{VI}Si^{III}O_2$, which would lead to a distance between nearest silicon atomic centres significantly less than 306 pm, thus an extremely repulsive condition. For comparison, the maximum angle Si-O-Si would be $2\pi/3 = 2.09$ rad for oxygen coordinating three silicons, whereas this angle in quartz with oxygen coordinating two silicons is 2.56 rad. Great pressure serves to overcome such repulsions, transforming SiO_2 into a structure of the rutile (TiO_2) type with $^{VI}Si^{III}O_2$.

The structures of crystals and of even some isolated molecules thus represent a balance between strong attractions affecting nearest neighbours (as chemical bonds) and less strong repulsions between atoms that are second-nearest neighbours about a central atom (between cations about a central anion, and *vice versa*). The latter, non-bonding repulsions that operate over large distances and involve small electronic densities tend to cause extension (or stretching) of chemical bonds characterised by small distances and large electronic densities. These repulsions are not of primarily electrostatic origin because such conditions fail to explain the observed geometries of flexible structures such as quartz and cristobalite [23], or of structures of $CaCl_2$ type relative to the topologically identical rutile type

[24]. The importance of interactions between second-nearest neighbours, either between two cations or between two anions, depends on not only the relative sizes (pertinent to non-bonded interactions) of atomic centres but also their relative numbers or the stoichiometry of the substance. In a crystalline compound such as Li_3N repulsions between cations are more important than between anions because cations are preponderant; conversely in AlF_3 the reverse situation pertains. In less extreme cases the relative influences of size and stoichiometry are more subtle. Beside a table of atomic radii Hyde presented [4] a comparison of observed bond angles in cristobalites with those calculated from those radii, which indicates the success of this approach.

3. Application of an alternative approach to inorganic structures

Although diamond is an extremely hard and incompressible substance and a thermal conductor superior to metallic copper or silver, it is metastable under normal conditions with respect to graphite. These properties reflect a condition that the distance 252 pm between second-nearest neighbours is about twice the non-bonding radius 125 pm; hence non-bonding contacts are repulsive. If one associates 0.028 aJ with the potential energy of each contact between C atomic centres as second-nearest neighbours in diamond according to the pertinent function $V(d)$ [22], this energy 100 kJ per mole of carbon is comparable with many enthalpies of chemical reactions. Resistance to deformation or to compression of a crystal that necessarily decreases some interatomic distances, with a concomitant greatly increased repulsive energy, explains not only the large bulk modulus, stiffness and thermal conductivity of diamond but also its instability with respect to graphite; in the latter allotropic form atomic centres of carbon within layers, denoted ^{III}C to indicate the coordination number, experience a greatly decreased energy of repulsion, relative to ^{IV}C in diamond, because the number of second-nearest neighbours is decreased to a quarter that in diamond. About four fifths of the bulk modulus of diamond arises from these non-bonded repulsions [25], consistent with the concept of stretched bonds and compressed non-bonded contacts. To overcome these repulsions and hence to synthesise diamond, great pressure is required, likewise for boron nitride and silicon carbide in isostructural forms; these three substances are among the least compressible.

For many transformations under great pressure the coordination number increases, as from $^{IV}Zn^{IV}S$ in a tetrahedral structure to an octahedral structure of $^{VI}Na^{VI}Cl$ type. In such a case the greatest repulsions involve nearest neighbours, or bonded atomic centres; these repulsions are relieved with an increased coordination number as more numerous bonds imply weaker and longer bonds. As the volume of the unit cell decreases under an applied pressure, the linear dimensions also decrease, about 2.2 *per cent* for the edge of the unit cell in the case of this phase transformation of ZnS, but the bond length correspondingly increases about 5.6 *per cent* [4].

In other transformations under pressure, such as for the minerals *olivine* $\rightarrow$ *spinel* with formula unit $^{VI}Mg_2{}^{IV}Si^{IV}O_4$ or for *zircon* $\rightarrow$ *scheelite* with $^{VIII}Zr^{IV}Si^{III}O_4$ [4], no alteration of coordination number is involved. In the former case in which the volume of a unit cell (that contains formula units of the same number) decreases about ten *per cent*, the oxygen centres have approximately hexagonal eutaxy in *olivine* but cubic eutaxy in *spinel*; hence coordination numbers between anions remain invariant. Although the distances between nearest oxide anions decrease on transformation, so that the repulsive energy increases, and correspondingly the distances between magnesium cations decrease, the distances between nearest Mg and Si centres become increased. That this phase change relieves strong repulsions between Mg and Si at the expense of more numerous repulsions between Mg and Mg is consistent with the coordination number of Si by Mg increasing from nine in *olivine* to twelve in *spinel*.

That repulsions between cations are more important than those between anions is consistent with the fact that arrays of cations in many crystalline structures exhibit regularity whereas associated arrays of anions lack such regularity. A few notable instances include La_2O_3, $BaSO_4$, β-K_2SO_4 and Ca_2SiO_4 [4]. In the latter case, for which five phases exist at various temperatures, in three polymorphs only Ca and Si cations display a regular array, recognised to be of the Ni_2Si or $PbCl_2$ type, as is indeed the separate compound Ca_2Si; the molar volumes 51.8 $mL\ mol^{-1}$ of these silicates and 49.9 $mL\ mol^{-1}$ of the silicide are notably comparable [26].

The stoichiometry of binary fluorides, oxides and nitrides confirms the validity of this approach. For oxides of alkali metals, apart from peroxides with O-O bonds and suboxides with M-M bonds, the normal formula unit M_2O implies twice as many large cations M as small anions (O^{2-}). All these oxides have an *antifluorite* structure–tetrahedral coordination of the cations and eight cations about each anion, so $^{IV}M_2{}^{VIII}O$. Hence the co-ordination number of the large cation is small and that of the small anion is large, a condition inconsistent with traditional rules about the ratios of radii. These oxides are prepared only with great difficulty: the peroxides M_2O_2 and, with more massive alkali atoms, even superoxides MO_2 form readily instead. Even when a normal oxide forms it is reactive with atmospheric H_2O, CO_2 *et cetera*, indicating a modest enthalpy of formation and weak bonds. This poor stability is attributed [4] to large repulsions between cations in the normal binary compound. The effect is less severe with oxides of alkaline-earth metals (group 2) because a smaller stoichiometric ratio accompanies increased charge on cations. Conversely the stoichiometric ratio is decreased for alkali metals with fluorides as anions, in which binding is correspondingly stronger [27].

Analogously, if the anion carries a greater negative charge, the problem relative to normal alkali oxides is expected to be exacerbated, as there are then even more numerous cations per anion. Such a condition arises for nitrides of alkali metals, M_3N. For the more massive atoms of elements in group 1 these compounds have not been prepared: only Li_3N is known. Instead of nitrides, azides MN_3 are formed, of which NaN_3 has considerable stability. These polyanions N_3^- are obvious extensions of peroxides O_2^{2-}. For carbides no binary compound M_4C of group 1 is known, and only Be_2C of group 2, but Al_4C_3 and boron carbides of group 3. The stable compounds consist instead of ethynides, for instance CaC_2, containing C_2^{2-}, or polycations in Ca_2N and $Ca_{11}N_8$.

The traditional ionic model in chemistry, with its emphasis on relatively large anions, is thus obsolete: many properties and phenomena in the chemistry and physics of the solid state, and in mineralogy, are readily explained [4] as a natural consequence of repulsion between cations.

Measurements of density of electronic charge within the unit cell of crystals by means of xray diffraction can illuminate the nature of chemical binding. Although chemists might retain an impression that charge flows extensively in the formation of a covalent bond from two atoms, so as to accumulate in the region between closely adjacent atomic centres at the expense of electronic density in other directions, accurate measurements of charge density indicate that, for instance, approximately one tenth of an electron is displaced into the *overlap* region between two carbon atomic centres separated about 150 *pm* [28]. Formation of a covalent chemical bond nominally involving sharing of electrons remains consistent with a minimum of electronic density along the line of that bond between the nuclei. Such experimental results conform to those from calculations. The distribution of electronic charge associated with an isolated atomic centre hence alters little on formation of a covalent bond. Even though the extent is small, it is significant: nuclei are thereby held near their equilibrium separations. For instance, in the case of HCl at the equilibrium separation the internuclear repulsion is forty times as great as the binding energy or strength of the chemical bond. In preceding essays [1, 2] there was reported an interpretation [29] of electronic density within unit cells, by means of xray diffraction measurements, that for electronic charge to accumulate in the region between two nominally bonded atomic centres might not necessarily accompany formation of a bond [30]. That approach, combined with other work [31], depends on plots of the difference of electronic density between an experimental distribution and a quasi-classical density postulated for spherically averaged atoms, whether with or without allowance for atomic vibrations. Such an hypothetical distinction is fraught with risks of incorrect and unwarranted conclusions concerning covalent chemical bonds [32], but the choice of a reference state alternative to that of a spherical atomic centre for the comparison of experimental electronic densities is even more arbitrary [33].

A claim to detect orbitals (that lack physical existence) by means of experiments with scattered xrays would be astounding, but to compare experimental densities of electronic charge with those calculated accurately is intrinsically unobjectionable. For a complex of iron with tetraphenylporphyrin derivatives, a pattern of electronic density near the iron atomic centre was interpreted to indicate varied occupancy of d orbitals [34]; this pattern agrees with results of calculations [35] according to an *extended Huckel* method that involve an approximate one-electron hamiltonian and an inadequate Mulliken population analysis. As this method is inherently unreliable because of its lack of theoretical foundation, gross approximations, and neglect of important electronic interactions such as correlation [35], any agreement with experimental results is regarded as fortuitous. Improved calculations of electronic density according to superior procedures are required before such comparison with experimental data can be regarded as meaningful, but perhaps such crude calculations are required [36] to confer such alleged properties of nonexistent orbitals. In that crystallographic analysis [34], hydrogen atomic centres were eventually located directly, rather than being placed at idealised positions relative to adjacent atomic centres associated with greater electronic density on the basis of preconceived structural notions; the latter practice is common in experiments in which measurement of the diffraction pattern is less thorough. When crystallographic analyses based on xray diffraction enable detection of hydrogenic atomic centres at separations from more massive nuclei consistent with accurate distances inferred with other experimental techniques, a claim to detect electronic density associated with specific features of a basis set in a theoretical calculation might be more seriously entertained.

In an ionic material the minimum electronic density at some point along the line connecting the nuclei of a cation and an adjacent anion was found to be small [10]. In crystalline lithium hydride that has a structure analogous to that of NaCl, the internuclear distance between Li and adjacent H is 204 *pm*, compared with about 160 *pm* for the isolated diatomic molecule in the gaseous phase [37]. The measured electric dipolar moment of the latter free LiH molecules corresponds to nominal transfer of about 0.75 electron from lithium to the hydrogen atomic centre [37], indicating a strongly polar bond. For such an ionic (or, more properly, strongly polar) bond in a crystalline environment there might appear to occur a much greater transfer of electronic density from the region of one atomic centre to another than actually occurs in a covalent bond such as between carbon atoms mentioned above. Whether the electronic density in inorganic crystals is consistent with superposition of ions or of atoms becomes a pertinent question [33]. A careful analysis of crystalline Li_2BeF_4 according to both atomic and ionic models yielded almost identical indices of the goodness of fit of 6435 independent reflections of xrays, but supplemental measurements on only 22 weak reflections more sensitive to the electronic distribution indicate that charge density far from atomic centres is slightly better represented on a basis of neutral atoms than of ions [38]. As scattering of xrays in a crystal of Li_2BeF_4 is dominated by the electronic density present near the fluorine atomic centres, of which most electronic density is insensitive to the environment, a superior test to distinguish ions and atoms might be made with a crystal of LiH.

In conclusion of this discussion of chemical binding in inorganic species, the question whether crystals traditionally considered ionic consist of atomic ions or of nearly electrically neutral atomic centres remains generally unresolved. For the same reason the energy of an ionic bond appears not to differ greatly from that of a covalent bond at comparable internuclear separations typical of structures in these crystalline environments. Consistent with this condition, calculations of metallic lithium and sodium–and even methane–according to an ionic model provide moderate agreement with experimental measurements of cohesive energy [39], despite these substances being regarded as not ionic.

The problem of locating hydrogen atomic centres within unit cells of crystals by means of xray diffraction contrasts with the ease of calculating electronic structure and properties of H_2^+ and H_2. The latter systems have been favoured vehicles to test the virial theorem [40] in relation to relative contributions of kinetic and potential energies of electrons to the net binding energy, commonly as a function of internuclear distance. According to such calculations one can readily demonstrate that Bader's attribution [41] of the phenomenon, that accumulation of electronic charge in the internuclear region causes the potential energy of a system to increase rather than to decrease, only to a system that fails to obey the virial theorem is inaccurate. His definition of an atom, or functional group, within a molecule is not unique, despite his profound reflections on this subject [42]; other definitions would yield *atoms* of varied size and shape, as any partition of electronic density within a molecule is fundamentally arbitrary. If the electronic density about any isolated atomic centre were to show a radical alteration on formation of a chemical bond, in contravention of the argument above, the atomic centre most likely to exhibit such effects would be hydrogen. For such a reason, just as to extrapolate from results for an H atom (for which Schrodinger's equation is solved exactly as an atomic system with one electron–two bodies in total) to atoms with many electrons is invalid, so to infer much about another chemical bond from effects that one readily calculates for H_2 is unwise: the hydrogenic atomic centre is special in involving only a *valence* electron.

4. Mechanisms of reactions

A theory of absolute rates of chemical reactions based on purported thermodynamic properties of an energised species (formerly activated complex) is of long standing [43], but severe and convincing criticisms of that theory are of equal longevity [44]. The quintessential equation of this theory is expressed as

$$k' = \kappa(k_B T/h) e^{-\Delta H^\ddagger/(R_g T)} e^{\Delta S^\ddagger/R_g} \qquad (1)$$

in which k' is the coefficient of the rate of a bimolecular reaction between reactants A and B in solution at temperature T [45]:

$$A + B \to products \qquad (2)$$

a transmission coefficient κ measures the efficiency of passage of the energised species through a transition state to yield the products, although κ is almost invariably accorded a value unity; k_B is the Boltzmann constant (the ideal-gas constant R_g divided by the Avogadro constant N_A), h is the Planck constant, and $\Delta H^\ddagger$ and $\Delta S^\ddagger$ are respectively the standard molar changes of enthalpy and entropy for the conversion of reactants into the activated complex $AB^\ddagger$. The latter species, purported to be in equilibrium with the reactants, proceeds somehow to form the eventual products; confusion between a thermodynamic equilibrium and a steady state (if κ is unity) failed to daunt the originators of this approach. As k_B pertains to molecules rather than to moles, the dimensions of k' are the product of the reciprocal of a concentration of molecules per unit volume and a reciprocal of time [46]. The reaction of A and B is characterised empirically with an activation energy E_a according to an equation

$$k = A e^{-E_a/R_g T} \qquad (3)$$

that Arrhenius attributed to van't Hoff, in which the coefficient k pertains to an overall rate of reaction according to a defining rate law

$$-d[A]/dt = v = k[A][B] \qquad (4)$$

for the velocity v of the reaction at time t. The dimensions of k are therefore those of reciprocal of concentration and of reciprocal of time. To make the latter two equations consistent with the former one, one must include a volume in that equation. If one considers both equations in terms of molecules, instead of moles, per unit volume (m^3), one writes

$$k' = (k_B T/h) V^{\ddagger} e^{-\Delta H^{\ddagger}/(R_g T)} e^{\Delta S^{\ddagger}/R_g} \quad (5)$$
$$k' = (1/N_A) A e^{-E_a/(R_g T)} \quad (6)$$

Then if we assume, for the purpose of estimating an order of magnitude of the pertinent quantity, a spherical volume $V^{\ddagger} = \frac{4}{3}\pi R^3$, that volume is identified with $(1/N_A)$. Hence the radius R of the energised species is $\approx 7 \times 10^{-9}$ m, the same for all reactions [46]! From the combination of the latter two equations under these circumstances, $-\Delta S^{\ddagger}/J\ K^{-1}\ mol^{-1}$ lies in the range [0, 200] for selected reactions [46], but such values are meaningless as they depend on the invariance of $V^{\ddagger}$.

This theory of the *absolute rate* of reactions is fundamentally spurious [46]; its parameters $\Delta G^{\ddagger}, \Delta H^{\ddagger}, \Delta S^{\ddagger}, V^{\ddagger}$ et cetera have no meaning outside this context. Despite this severe failure, Gibbs energies, enthalpies and entropies of activation are commonly reported for reactions in solution, even though this theory takes no account of the influence of the solvent (or environment) on the mechanism; the research field called physical organic chemistry that had as its objective to investigate and to characterise reactions in solution according to such tenets is practically moribund. Within the classical concept of molecular structure, a reaction takes place on a hypersurface of potential energy: no hypersurface of Gibbs energy G is definable.

Moreover even the justification of the classical concept of molecular structure by means of the approach of Born and others is precariously applied to an energised species. The *transition state* of which some authors undertake unremittingly to calculate directly the geometry with computer programmes for molecular electronic structure is neither a quantum state nor a thermodynamic state: the result of such calculation can best be called a transition structure; as such a structure generally implies a point on a hypersurface of potential energy at which (adiabatic) electronic states might intersect or interact strongly, the approximation due to Born and Oppenheimer is least valid in these conditions. For some small stable molecules having nearly the minimum energy (relative to unexcited atoms) and with due account of perturbations (adiabatic and nonadiabatic effects that are artifacts of the approximate separation of electronic and nuclear motions) [47], the concept of molecular structure is demonstrated to be practicable; the corresponding experimental justification and detailed evaluation of a definable structure of any postulated *transition state* or of an effective structure within an actual quantum state of the energised species remains a worthy but formidable objective, to which experiments are currently directed. To the extent that such a species has necessarily only transient existence, the experimental evaluation of its intrinsic properties becomes correspondingly hampered, in accordance with Heisenberg's principle of indeterminacy. Improved theories of reaction kinetics that retain the notion of a transition state are based on a potential-energy surface [48, 49].

As a mechanism of a specific chemical reaction, we consider briefly a stereospecific electrocyclic process, namely the conversion of buta-1,3-diene to cyclobutene. Woodward and Hoffmann [50] postulated that the stereochemical course of electrocyclic closure to form a ring of carbon atoms is set according to the symmetry of a particular molecular orbital, that of greatest energy with which electrons are associated, in the acyclic precursor. Support for this generalisation came from calculations based on an *extended Huckel* theory (disreputable for reasons stated above), which also indicates application to reactions in which such a ring of carbon atoms was opened. Such an explanation is unsatisfactory because the structure of the product of the reaction of ring opening, rather than the structure of the reactant, appears to govern whether the reaction is conrotatory or disrotatory, apart from the problems of reliability associated with extended Huckel theory in supporting predictions of a rule based on the symmetry of a single hypothetical molecular orbital. As any orbital, atomic or molecular, is an artifact of a particular mathematical method, and as in a thorough calculation of electronic structure including electronic correlation no orbital remains at the end of the calculation, such an approach is inherently suspect.

Longuet-Higgins and Abrahamson developed an alternative and more systematic procedure by considering the symmetry of electronic states, hence generating correlations along the entire course between reactants and products, without engaging in numerical calculations [51]. In this way predictions eventuated for not only the interconversion of cyclobutene and buta-1,3-diene but also of cyclopropyl radical, cation and anion and each corresponding allyl species. Although the latter authors maintained reliance on atomic and molecular orbitals to produce the symmetries of electronic states, these symmetries can be in principle determined on analysis of rotational fine structure in molecular spectra; in that sense symmetries are observable properties. The so called conservation of orbital symmetry lacks physical foundation as it relies on constituents of a basis set that one can in principle select arbitrarily, without even regard for the accuracy of an ultimate energy that might result from an actual calculation employing them according to a conventional procedure.

5. Conclusion

In this essay I extend the discussion of chemical binding to include what are conventionally called ionic compounds and inorganic crystals, and summarise aspects of mechanisms of chemical reactions. If as a consequence of perusing this essay and its predecessor [2] the reader acquires a pessimistic view of the present status of understanding of fundamental chemical phenomena and properties, he might appreciate the paradox that, although "chemistry is demanding on the intellect, it is increasingly anti-intellectual" [52]. "Terms such as electronegativity, hybridisation and orbitals are used in meaningless explanations (in chemistry), devoid of intellectual content" [52]. Selinger's reflections in 1982 on the dichotomies between teaching and research in chemical science, between the lip-service to quantum mechanics as the basis of chemical science and the selective use of quantal terms in circular arguments to rationalise essentially classical observations, remain valid. As an imminent advance in quantum theory, or its total replacement due to obsolescence on the emergence of a successive theory at once more readily understandable and easily applicable, seems unlikely, what is needed to improve the internal consistency of chemical theory is a thorough reassessment of essential chemical knowledge worthy at each successive stage of the educational process.

Does the nature of the chemical bond matter [53]? "The most useless part of chemistry–theoretical–is widely taught, seldom understood or appreciated and its redeeming feature (its profound philosophical implications) ignored" [52]. In such grossly exaggerated claims as "great successes of quantum chemistry [actually quantum-chemical computations] that super high accuracy predictions can now be made" [54], the practitioners of calculations, so called *ab initio* (despite their calibrated basis sets) – or perhaps more accurately *ad nauseum*? – delude themselves as they seek to mislead their readers. The accuracy claimed for bond lengths is a few parts in 10^{-13} m [54]; as a length of a typical C-H or O-H bond is about 10^{-10} m, the ratio of this error is about one part in 400. In contrast the order of adiabatic effects in bond lengths is the ratio of an electronic to a nuclear mass [37], so a maximum about one part in 1800 in the case of the mass of a proton. Hence the errors of such computations of bond lengths are still typically at least a few times the magnitudes of the corresponding adiabatic effects, whereas nonadiabatic effects are entirely ignored despite having magnitudes comparable to those of adiabatic effects for small molecules [37], and these errors amount to a thousand times the meaningful uncertainties of equilibrium bond lengths and properties of small molecules [55]. If the nature of the chemical bond matters, we should expend a serious effort to ensure that during the teaching of chemistry as an "indoctrination of the student with current paradigms-the behavioural code of normal science" [52] chemical binding in gaseous and crystalline matter is provided a more firm foundation than the confused collection of empty rhetoric abounding in the present chemical literature.

References

1. Ogilvie, J. F. (1990) *J. Chem. Ed.* **67**, 280-289
2. Ogilvie, J. F. (1994) The nature of the chemical bond 1993, p. 171-198 in *Conceptual Trends in Quantum Chemistry*, editors Kryachko, E., and Calais, J.-L., **1**, Kluwer, Dordrecht, Netherlands
3. O'Keeffe, M. and Hyde, B. G. (1985) *Struct. Bonding*, **61**, 71-144
4. Hyde, B. G. (1986) *Proc. Roy. Soc. New South Wales*, **119**, 153-164
5. Pauling, L. C. (1960) *The Nature of the Chemical Bond*, third edition, Cornell University Press, Ithaca, NY USA
6. Shannon, R. D. (1976) *Acta Cryst.* **A32**, 751-767
7. O'Keeffe, M. (1977) *Acta Cryst.* **A33**, 924-927
8. Brunner, G. O. (1971) *Acta Cryst.* **A27**, 388-390
9. Hoppe, R. D. (1980) *Angew. Chem. Int. Ed. Eng.* **19**, 110-125
10. Harvey, K. B. and Porter, G. B. (1963) *Introduction to Physical and Inorganic Chemistry*, Addison-Wesley, Reading, MA USA
11. Bukowinski, M. S. T. (1982) *J. Geophys. Res.* **87**, 303-310
12. Clementi, E. and Roetti, C. (1974) *At. Data Nucl. Data Tables*, **14**, 177-478
13. Stewart, R. F. and Spackman, M. (1981) Charge Density Distributions, in *Structure and Bonding in Crystals*, eds. O'Keeffe, M. and Navrotsky, A., **1**, 279-298, Academic Press, New York, USA
14. Tosi, M. P. (1964) *Solid State Phys.* **16**, 1-120
15. O'Keeffe, M. J. (1990) *Solid State Chem.* **85**, 108-116
16. O'Keeffe, M. and Hyde, B. G. (1978) *Acta Cryst.* **B34**, 27-32
17. Bragg, L. and Claringbull, G. F. (1965) *The Crystal Structures of Minerals*, Bell, London, UK
18. Bartell, L. S. (1968) *J. Chem. Ed.* **45**, 754-767
19. Gibbs, G. V. (1982) *Amer. Mineral.* **67**, 421-450
20. O'Keeffe, M., Newton, M. D. and Gibbs, G. V. (1980) *Phys. Chem. Miner.* **6**, 305-312
21. Burdett, J. K. and Caneva, D. (1985) *Inorg. Chem.* **24**, 3866-3873
22. Kitaigorodsky, A. I. (1973) *Molecular Crystals and Molecules*, Academic Press, New York, USA
23. Fischer, R. and Zemann, J. (1975) *Tschermaks Min. Petr. Min.* **22**, 1-14
24. Barnighausen, H., Bossert, W. and Anselment, B. (1984) *Acta Cryst.* **A40**, supp. C96
25. O'Keeffe, M. and Hyde, B. G. (1981) The role of nonbonded forces in crystals, in *Structure and Bonding in Crystals*, editors O'Keeffe, M. and Navrotsky, A., **1**, 227-254, Academic Press, New York, USA
26. Hyde, B. G., Sellar, J. R. and Stenberg, L. (1986) *Acta Cryst.* **B42**, 423-429
27. O'Keeffe, M. and Hyde, B. G. (1985) *Nature*, **309**, 411-414
28. Dawson, B., unpublished results, quoted by Coulson, C. A. (1972) p. 103 in *Proc. R. A. Welch Conference on Chemical Research, XVI Theoretical Chemistry*, editor Milligan, W. O., Houston, USA
29. Dunitz, J. D. and Seiler, P. (1983) *J. Am. Chem. Soc.* **105**, 7056-7058 (1983)
30. Spackman, M. A. and Maslen, E. N. (1985) *Acta Cryst.* **A41**, 347-352
31. Savariault, J. M. and Lehmann, M. S. (1980) *J. Am. Chem. Soc.* **102**, 1298-1303
32. Schwarz, W. H. E., Valtazanos, P. and Ruedenberg, K. (1985) *Theoret. Chim. Acta*,

68, 471-506
33. Dunitz, J. D. (1988) *Bull. Chem. Soc. Jpn.* **61**, 1-11
34. Li, N., Coppens, P. and Landrum, J. (1988) *Inorg. Chem.* **27**, 482-488
35. Coppens, P. (1989) *J. Phys. Chem.* **93**, 7979-7984
36. Mezey, P. G., Daudel, R. and Czismadia, I. (1979) *Int. J. Quantum Chem.* **16**, 1009-1019
37. Ogilvie, J. F., Oddershede, J. and Sauer, S. P. A. (1994) *Chem. Phys. Lett.* **228**, 183-190
38. Seiler, P. and Dunitz, J. D. (1986) *Helv. Chim. Acta*, **69**, 1107-1112
39. Nieuwpoort, W. C. and Blasse, G. (1968) *J. Inorg. Nucl. Chem.* **30**, 1635-1637
40. Kutzelnigg, W. (1990) The Physical Origin of the Chemical Bond, p. 1-43, in *The Concept of the Chemical Bond*, editor Maksic, Z. B., Springer-Verlag, Berlin, Germany
41. Bader, R. F. W. (1990) *Atoms in Molecules - a Quantum Theory*, p. 324, Oxford University Press, London, UK
42. Bader, R. F. W., Popelier, P. L. A. and Keith, T. A. (1994) *Angew. Chem. Int. Ed. Engl.* **33**, 620-631
43. Eyring, H. (1935) *J. Chem. Phys.* **3**, 107-115
44. Guggenheim, E. A. and Weiss, J. (1938) *Trans. Faraday Soc.* **34**, 57-70 and adjacent papers and discussion.
45. Wynne-Jones, W. F. K. and Eyring, H. (1935) *J. Chem. Phys.* **3**, 492-502
46. Moelwyn-Hughes, E. A. (1936) *J. Chem. Phys.* **4**, 292
47. Ogilvie, J. F. (1996) New tests of models in chemical binding- extra-mechanical effects and molecular properties, p. 41-53 in *Fundamental Principles of Molecular Modeling*, editors Gans, W., Amman, A., and Boeyens, J. C. A., Plenum Press, New York, USA
48. Quack, M. and Troe, J. (1981) Statistical methods in scattering, p. 199-276, in *Theoretical Chemistry: Advances and Perspectives*, **6B**, Academic Press, New York, USA
49. Troe, J. (1986) *J. Phys. Chem.* **90**, 357-365
50. Woodward, R. B. and Hoffmann, R. (1965) *J. Am. Chem. Soc.* **87**, 395-396
51. Longuet-Higgins, H. C. and Abrahamson, E. W. (1965) *J. Am. Chem. Soc.* **87**, 2045-2046
52. Selinger, B. (1982) *Chem. Austr.* **49**, 448-455
53. Ogilvie, J. F. (1982) *S. Afr. J. Sci.* **92**, 57-59
54. Handy, N. C. and Lee, A. M. (1996) *Chem. Phys. Lett.* **252**, 425-430
55. Ogilvie, J. F., Uehara, H. and Horiai, K. (1995) *J. Chem. Soc. Faraday Trans.* **91**, 3007-3013

LIE SYMMETRIES IN QUANTUM MECHANICS

D. RODNEY TRUAX
Department of Chemistry
The University of Calgary
2500 University Dr. NW
Calgary, Alberta
Canada T2N 1N4

1. Introduction

The evolution of a N-dimensional quantum mechanical system where $N = 1$ or 2 is described by the solutions to a time-dependent Schrödinger equation [1, 2, 3],

$$S_N \Psi(\mathbf{x}, t) = 0, \qquad (1)$$

where the Schrödinger operator is given by

$$S_N = \sum_{\sigma=1}^{N} \partial_{x_\sigma x_\sigma} + 2i\partial_\tau - 2V(\mathbf{x}, \tau). \qquad (2)$$

and $V(\mathbf{x}, \tau)$ is, for the moment, an arbitrary time-dependent potential. The symbol $\mathbf{x} = (x_1, x_2)$ refers to the 2-tuple of coordinates in the 2-dimensional configuration space [4]. If $N = 1$, then we shall write $\mathbf{x} = x_1 = x$. The symbol ∂_σ represents the partial derivative with respect to x_σ. The analytical solution of Eq. (1) is possible for certain choices of the τ-dependent potential, $V(\mathbf{x}, \tau)$, because in those cases, Eq. (1) admits Lie symmetries. We shall work with space-time or kinematical symmetries of Eq. (1).

Generators of space-time symmetries of Eq. (1) have the form [5, 6, 7]

$$\mathcal{L} = A(\mathbf{x}, \tau)\partial_\tau + \sum_{\sigma=1}^{N} B^\sigma(\mathbf{x}, \tau)\partial_{x_\sigma} + C(\mathbf{x}, \tau), \qquad (3)$$

where the coefficients of the partial differentials depend on both the spatial variables and time. For the generator $\mathcal{L}$ to be a symmetry of (1), we require that it transform solutions of (1) into solutions [5]. This means that $\mathcal{L}$ must satisfy the operator equation

$$[\mathcal{S}_N, \mathcal{L}] = \lambda(\mathbf{x}, \tau)\mathcal{S}_N, \tag{4}$$

where $\lambda(\mathbf{x}, \tau)$ is an, as yet, unspecified function of time and the coordinates.

In earlier work [6, 7, 8], we showed that the generators of space-time symmetries for 1- and 2-dimensional Schrödinger equations are constants of the motion and form Lie algebras. We were able to utilize that algebraic structure and its representation theory to construct spaces of R-separable solutions [5]. The existence of Lie symmetries was also shown to be directly relevant to the development of displacement operator coherent states and squeezed states [9, 10, 11, 12]. Also, we have found this approach to be quite useful in the investigation of supersymmetry of time-dependent systems [7].

Other authors have used the technique of time-dependent invariants to study the behaviour of time-dependent systems. For example, see the selected works cited in Ref. [13].

In this chapter, we shall compute the Lie algebra of generators of kinematical symmetries for 1- and 2-dimensional, time-dependent Schrödinger equations (1), when the potential has the form

$$V(\mathbf{x}, \tau) = g^{(2)}(\tau)\mathbf{x} \cdot \mathbf{x} + \sum_{\sigma=1}^{N} g_\sigma^{(1)}(\tau)x_\sigma + g^{(0)}(\tau). \tag{5}$$

We shall call this potential the time- or τ-dependent oscillator. Each of the τ-dependent coefficents $g^{(2)}$, $g_1^{(1)}$, $g_2^{(1)}$, and $g^{(0)}$ is a real, differentiable, and piecewise continuous function. The maximal kinematical algebra is obtained when $V(\mathbf{x}, \tau)$ is of this form. We refer to this Lie algebra as the Schrödinger algebra and denote it by $(\mathcal{SA})_N$. In Section 2, we outline our calculations and note that the Schrödinger algebra, $(\mathcal{SA})_N$, is a real Lie algebra. The real Lie algebra is useful in the characterization of R-separable solutions [5]) to Eq. (1). Next, we determine the generators of the complexification of $(\mathcal{SA})_N$, which we denote by $(\mathcal{SA})_N^c$. We have found that operators in $(\mathcal{SA})_N^c$ are most useful when computing solutions to the Schrödinger equation in a specific R-separable coordinate system. The basis for the complexified algebra can be expressed as linear combinations of the basis for the real Schrödinger algebra.

In Section 3, we outline a method to obtain solutions for both the 1-dimensional and 2-dimensional Schrödinger equations in one particular separable coodinate system is. In each case, it is unnecessary to work with the entire algebra $(\mathcal{SA})_N^c$. For one dimension, the oscillator subalgebra, $os(1)$, of $(\mathcal{SA})_1^c$ is sufficient. For two dimensions, we use the subalgebra $os(2) \times so(2)$. The R-separable solutions span an irreducible representation space of the appropriate Lie algebra. The solutions are obtained by selecting commuting constants of the motion and finding a set of common eigenfunctions. It is important to recognize that these eigenfunctions are not necessarily energy eigenfunctions.

Coherent states are quantum analogues of classical states and satisfy the minimum uncertainty relation. Briefly, we describe how to compute discplacement operator coherent states (DOCS) in Section 4. By using the kinematical symmetries and their algebraic structure, the analysis is straightforward. Expectation values for position and momentum are presented and uncertainty relations derived.

In Section 5, a four examples are worked out: the harmonic oscillator in one and two dimensions and the driven oscillator in one and two dimensions.

In the final section, Section 6, we outline how the techniques of Sections 2 and 3 can be extended to Schrödinger equations that include velocity dependence or field dependence. The former case is an application of the 1-dimensional driven oscillator, while the latter makes use of results for the 2-dimensional driven oscillator.

2. Lie Symmetries

2.1. REAL SCHRÖDINGER ALGEBRAS, $(\mathcal{SA})_N$

When we substitute Eq. (2) for $\mathcal{S}_2$ and Eq. (2) for $\mathcal{L}$ in Eq. (4), we obtain a set of coupled partial differential equations for the coefficients A, B^1, B^2, and C. A full discussion can be found in Ref. [14]; here, we present only the results of their solution for the case of one and two dimensions when $V(\mathbf{x}, \tau)$ is restricted to the form (5).

We adopt the following convention for indices: $\sigma, \nu = 1, 2$ label coordinates; $j, k = 1, 2, 3$, and $\alpha = 1, 2$.

The coefficient A depends only on the time variable, τ, and it takes the form

$$A(\tau) = \sum_{j=1}^{3} \beta_j \hat{\phi}_j(\tau), \tag{6}$$

where each β_j is a constant and each τ-dependent function $\hat{\phi}_j$ is a real solution of the third-order, ordinary differential equation

$$\dddot{A} + 8g^{(2)}\dot{A} + 4\dot{g}^{(2)}A = 0. \tag{7}$$

The 'dot' over a symbol indicates ordinary differentiation by τ.

From the theory of ordinary differential equations [15], we are guaranteed the existence of three, linearly independent solutions, $\hat{\phi}_j$. The general solution (6) is a linear combination of these three solutions.

The coefficients B^1 and B^2 depend on both spatial variables and time. We have

$$B^1(\mathbf{x},\tau) = \sum_{j=1}^{3} \beta_j \left\{ \tfrac{1}{2}\dot{\phi}_j(\tau)x_1 + \hat{\mathcal{E}}_{j,1}(\tau) \right\} + \sum_{\alpha=1}^{2} \beta_{\alpha;1} \{\chi_\alpha(\tau)\}$$
$$+ \beta_2^1 \left(x_2 - \hat{\mathcal{E}}_2(\tau)\right), \tag{8}$$

and

$$B^2(\mathbf{x},\tau) = \sum_{j=1}^{3} \beta_j \left\{ \tfrac{1}{2}\dot{\phi}_j(\tau)x_2 + \hat{\mathcal{E}}_{j,2}(\tau) \right\} + \sum_{\alpha=1}^{2} \beta_{\alpha;2} \{\chi_\alpha(\tau)\}$$
$$+ \beta_2^1 \left(-(x_1 - \hat{\mathcal{E}}_1(\tau))\right), \tag{9}$$

where $\beta_{\alpha;\sigma}$ and β_2^1 are constants. For a 1-dimensional system, we have only the coefficient B^1 in Eq. (8) with $\beta_2^1 = 0$.

In Eqs. (8) and (9), thes function, χ_1 and χ_2, form a set of two linearly independent, real solutions to the second-order ordinary differential equation

$$\ddot{a} + 2g^{(2)}(\tau)a = 0. \tag{10}$$

For χ_1 and χ_2 to be linearly independent, their Wronskian, $W(\chi_1,\chi_2)$, must be nonzero [15] over the domain of definition of τ. In fact, the Wronskian must be a constant [6, 15], and so we can always pick χ_1 and χ_2 such that

$$W(\chi_1,\chi_2) = \chi_1\dot{\chi}_2 - \dot{\chi}_1\chi_2 = 1. \tag{11}$$

The Wronskian plays an important role in calculating relationships among the different τ-dependent functions and in proving closure of the Lie algebra of generators of kinematical symmetries.

The real solutions, χ_1 and χ_2, to Eq. (10) are fundamental to our analysis. Indeed, the reader may show that χ_1^2, χ_2^2, and $\chi_1\chi_2$ are three real solutions of Eq. (7). Therefore, we choose [6, 7, 14]

$$\hat{\phi}_1(\tau) = \chi_1^2(\tau), \quad \hat{\phi}_2(\tau) = \chi_2^2(\tau), \quad \hat{\phi}_3(\tau) = 2\chi_1(\tau)\chi_2(\tau). \tag{12}$$

The τ-dependent functions $\hat{\mathcal{E}}_{j,\sigma}$ are

$$\hat{\mathcal{E}}_{1,\sigma}(\tau) = -\chi_1(\tau)\mathcal{C}_{1,\sigma}(\tau), \quad \hat{\mathcal{E}}_{2,\sigma}(\tau) = -\chi_2(\tau)\mathcal{C}_{2,\sigma}(\tau),$$
$$\hat{\mathcal{E}}_{3,\sigma}(\tau) = -\chi_1(\tau)\mathcal{C}_{2,\sigma}(\tau) - \chi_2(\tau)\mathcal{C}_{1,\sigma}, \tag{13}$$

and $\hat{\mathcal{E}}_\sigma$ are

$$\hat{\mathcal{E}}_\sigma(\tau) = \chi_1(\tau)\mathcal{C}_{2,\sigma}(\tau) - \chi_2(\tau)\mathcal{C}_{1,\sigma}(\tau). \tag{14}$$

Each of these functions depends on the following functions

$$C_{\alpha,\sigma}(\tau) = \int^{\tau} ds\, g_\sigma^{(1)}(s)\chi_\alpha(s) = c_{\alpha,\sigma}(\tau) + C_{\alpha,\sigma}^o, \qquad (15)$$

where $C_{\alpha,\sigma}^o$ is an integration constant and

$$c_{\alpha,\sigma}(\tau) = \int_{\tau_o}^{\tau} ds\, g_\sigma^{(1)}(s)\chi_\alpha(s). \qquad (16)$$

The lower limit, τ_o is an initial time and we shall set $\tau_o = 0$.

The final coefficient, C, can be written as follows

$$\begin{aligned}C(\mathbf{x},\tau) = & \sum_{j=1}^{3} \beta_j \left\{ -\tfrac{i}{4}\dddot{\phi}_j(\tau)\mathbf{x}\cdot\mathbf{x} - i\sum_{\sigma=1}^{N} \hat{\mathcal{E}}_{j,\sigma}(\tau)x_\sigma + \tfrac{N}{4}\dot{\phi}_j(\tau) \right. \\ & \left. + \sum_{\sigma=1}^{N} i\hat{\mathcal{D}}_{j,\sigma}(\tau) + ig^{(0)}(\tau) \right\} \\ & + \sum_{\alpha=1}^{2}\sum_{\sigma=1}^{2} \beta_{\alpha;\sigma}\left\{-ix_\sigma\dot{\chi}_\alpha(\tau) + i\mathcal{C}_{\alpha,\sigma}(\tau)\right\} \\ & + \beta_2^1\left\{i\left(x_1\dot{\hat{\mathcal{E}}}_2(\tau) - x_2\dot{\hat{\mathcal{E}}}_1(\tau)\right) + i\mathcal{F}_{1,2}(\tau)\right\} + \beta\{i\}, \qquad (17)\end{aligned}$$

where β is a constant. Again note that when $N = 1$, $\beta_2^1 = 0$. The function $\hat{\mathcal{D}}_{j,\sigma}$ is defined by

$$\hat{\mathcal{D}}_{j,\sigma}(\tau) = \int^{\tau} ds\, \hat{\mathcal{E}}_{j,\sigma}(s)g_\sigma^{(1)}(s), \qquad (18)$$

where $\hat{\mathcal{E}}_{j,\sigma}$ is given by Eq. (13). In particular, we find that [14]

$$\begin{aligned}\hat{\mathcal{D}}_{1,\sigma}(\tau) &= -\tfrac{1}{2}\mathcal{C}_{1,\sigma}^2(\tau), \\ \hat{\mathcal{D}}_{2,\sigma}(\tau) &= -\tfrac{1}{2}\mathcal{C}_{2,\sigma}^2(\tau), \\ \hat{\mathcal{D}}_{3,\sigma}(\tau) &= -\mathcal{C}_{1,\sigma}(\tau)\mathcal{C}_{2,\sigma}^2(\tau), \qquad (19)\end{aligned}$$

where $\mathcal{C}_{\alpha,\sigma}$ is given by Eq. (15). The remaining function is

$$\hat{\mathcal{F}}_{\sigma,\nu}(\tau) = \int^{\tau} ds\, \hat{\mathcal{E}}_\sigma(s)g_\nu^{(1)}(s) - \int^{\tau} ds\, \hat{\mathcal{E}}_\nu(s)g_\sigma^{(1)}(s), \qquad (20)$$

where $\hat{\mathcal{E}}_\sigma$ is given by Eq. (14). This definition reduces to [14]

$$\hat{\mathcal{F}}_{\sigma,\nu}(\tau) = \mathcal{C}_{2,\sigma}(\tau)\mathcal{C}_{1,\nu}(\tau) - \mathcal{C}_{1,\sigma}(\tau)\mathcal{C}_{2,\nu}(\tau). \qquad (21)$$

Combining Eqs. (6), (8), (9), and (17), for the coefficients with expression (3) for the generator $\mathcal{L}$, we obtain

$$\mathcal{L} = \sum_{j=1}^{3} \beta_j \{\hat{L}_j\} + \beta_2^1 \{\hat{L}_{1,2}\}$$
$$+ \sum_{\sigma=1}^{2} \beta_{1,\sigma} \{P_\sigma\} + \sum_{\sigma=1}^{2} \beta_{2,\sigma} \{B_\sigma\} + \beta \{E\}. \quad (22)$$

The coefficients of the constants are a set of differential operators which are generators of space-time transformations. These operators also form a basis for a $\frac{1}{2}(N^2 + 3N + 8)$-dimensional Lie algebra, called the Schrödinger algebra, $(\mathcal{SA})_N$, where $N = 1, 2$. The differential operators are

$$\hat{L}_j = \hat{\phi}_j \partial_\tau + \sum_{\sigma=1}^{N} (\tfrac{1}{2}\dot{\hat{\phi}}_j x_\sigma + \hat{\mathcal{E}}_{j,\sigma}) \partial_\sigma - \tfrac{i}{4}\ddot{\hat{\phi}}_j \mathbf{x} \cdot \mathbf{x}$$
$$- i \sum_{\sigma=1}^{N} x_\sigma \dot{\hat{\mathcal{E}}}_{j,\sigma} + \tfrac{N}{4}\dot{\hat{\phi}}_j + \sum_{\sigma=1}^{N} i\hat{\mathcal{D}}_{j,\sigma} + ig^{(0)}\hat{\phi}_j, \quad (23)$$

where $N = 1$ or $N = 2$ and $j = 1, 2, 3$. These three generators satisfy the commutation relations

$$[\hat{L}_1, \hat{L}_2] = \hat{L}_3, \quad [\hat{L}_3, \hat{L}_1] = -2\hat{L}_1, \quad [\hat{L}_3, \hat{L}_2] = +2\hat{L}_2, \quad (24)$$

of an $sl(2, \mathbf{R})$ Lie algebra and we say that the three operators $\hat{L}_j$ form a basis for $sl(2, \mathbf{R})$. The next generator in Eq. (22) occurs only for the 2-dimensional case and it is a generalized angular momentum operator with the form

$$L_{1,2} = (x_2 - \hat{\mathcal{E}}_2)\partial_1 - (x_1 - \hat{\mathcal{E}}_1)\partial_2 + i(x_1\dot{\hat{\mathcal{E}}}_2 - x_2\dot{\hat{\mathcal{E}}}_1) + i\hat{\mathcal{F}}_{1,2},$$
$$L_{2,1} = -L_{1,2}. \quad (25)$$

This single generator forms a basis for an $so(2)$ Lie algebra. The remaining operators, of which there are three if $N = 1$ and five if $N = 2$, form a basis for the Heisenberg-Weyl algebras w_1 and w_2, respectively. These operators are

$$P_\sigma = \chi_1 \partial_\sigma - ix_\sigma \dot\chi_1 + i\mathcal{C}_{1,\sigma},$$
$$B_\sigma = \chi_2 \partial_\sigma - ix_\sigma \dot\chi_2 + i\mathcal{C}_{2,\sigma}, \quad 1 \leq \sigma \leq 2,$$
$$E = i \quad (26)$$

and have the nonzero commutation relations

$$[B_\sigma, P_\sigma] = E, \quad 1 \le \sigma \le 2. \tag{27}$$

The remaining commutation relations are

$$[\hat{L}_j, \hat{L}_{1,2}] = 0, \tag{28}$$

for $j = 1, 2, 3$;

$$[\hat{L}_1, P_\nu] = 0, \quad [\hat{L}_2, P_\nu] = -B_\nu, \quad [\hat{L}_3, P_\nu] = -P_\nu,$$
$$[\hat{L}_1, B_\nu] = P_\nu, \quad [\hat{L}_2, B_\nu] = 0, \quad [\hat{L}_3, B_\nu] = B_\nu, \tag{29}$$

for $1 \le \nu \le 2$, and when $N = 2$

$$[\hat{L}_{1,2}, P_\nu] = -\delta_{\nu,2} P_1 + \delta_{\nu,1} P_2, \quad [\hat{L}_{1,2}, B_\nu] = -\delta_{\nu,2} B_1 + \delta_{\nu,1} B_2, \tag{30}$$

for $1 \le \nu \le 2$. Now, with these results, we can say that the Schrödinger algebra, $(\mathcal{SA})_N$ has the following structure in one dimension

$$(\mathcal{SA})_1 = sl(2, \mathbf{R}) \diamond w_1, \tag{31}$$

and in two dimensions

$$(\mathcal{SA})_2 = (sl(2, \mathbf{R}) \oplus so(2)) \diamond w_2. \tag{32}$$

2.2. COMPLEXIFICATION OF $(\mathcal{SA})_N$

In Ref. [7, 8, 14], we showed that it was advantageous to work with the complexified Lie algebra $(\mathcal{SA})_N^c$. To obtain the operator basis for $(\mathcal{SA})_N^c$, we use complex solutions of Eq. (10). Complex solutions can be written as linear combinations of the real solutions χ_1 and χ_2 in the following way:

$$\xi(\tau) = \sqrt{\tfrac{1}{2}}(\chi_1(\tau) + i\chi_2(\tau)), \quad \bar{\xi}(\tau) = \sqrt{\tfrac{1}{2}}(\chi_1(\tau) - i\chi_2(\tau)), \tag{33}$$

where the 'bar' indicates the complex conjugate. The Wronskian of the complex solutions is

$$W(\xi, \bar{\xi}) = \xi \dot{\bar{\xi}} - \dot{\xi} \bar{\xi} = -i, \tag{34}$$

which follows from Eq. (11) and definition (33). Now, we define the complex functions

$$C_\sigma(\tau) = \int^\tau ds\, g_\sigma^{(1)}(s) \xi(s) = c_\sigma(\tau) + C_\sigma^o, \tag{35}$$

where $\mathcal{C}_\sigma^o$ is a complex integration constant and

$$c_\sigma(\tau) = \int_0^\tau ds\, g_\sigma^{(1)}(s)\xi(s), \qquad (36)$$

and their complex conjugates. These functions are related to their real counterparts by the linear transformation

$$\begin{aligned}
\mathcal{C}_\sigma &= \sqrt{\tfrac{1}{2}}(\mathcal{C}_{1,\sigma} + i\mathcal{C}_{2,\sigma}), \\
c_\sigma &= \sqrt{\tfrac{1}{2}}(c_{1,\sigma} + ic_{2,\sigma}), \\
\mathcal{C}_\sigma^o &= \sqrt{\tfrac{1}{2}}\left(\mathcal{C}_{1,\sigma}^o + i\mathcal{C}_{2,\sigma}^o\right).
\end{aligned} \qquad (37)$$

There are a corresponding set of equations for their complex conjugates. With these definitions, we can write the complex operators $\mathcal{J}_{\sigma,\pm}$ in terms of the real generators P_σ and B_σ. We have

$$I = 1, \qquad (38)$$

$$\mathcal{J}_{\sigma-} = \tfrac{1}{\sqrt{2}}(P_\sigma + iB_\sigma) = \xi\partial_\sigma - ix_\sigma\dot{\xi} + i\mathcal{C}_\sigma, \qquad (39)$$

$$\mathcal{J}_{\sigma+} = \tfrac{1}{\sqrt{2}}(-P_\sigma + iB_\sigma) = -\bar{\xi}\partial_\sigma + ix_\sigma\dot{\bar{\xi}} - i\bar{\mathcal{C}}_\sigma, \qquad (40)$$

Next, we give the complex versions (without 'hats') of the real operators (with 'hats'). First, we deal with the operator for the generalized angular momentum. The complex form of $\hat{\mathcal{E}}_\sigma$ is

$$\mathcal{E}_\sigma(\tau) = i\left(\xi(\tau)\bar{\mathcal{C}}_\sigma(\tau) - \bar{\xi}(\tau)\mathcal{C}_\sigma(\tau)\right), \qquad (41)$$

and the complex function $\mathcal{F}_{1,2}$ is

$$\mathcal{F}_{1,2}(\tau) = i\left(\bar{\mathcal{C}}_1(\tau)\mathcal{C}_2(\tau) - \mathcal{C}_1(\tau)\bar{\mathcal{C}}_2(\tau)\right). \qquad (42)$$

Substitution of the first equation in (37) and its complex conjugate into (41) and (42) yields the result

$$\mathcal{E}_\sigma(\tau) = \hat{\mathcal{E}}_\sigma(\tau), \qquad \mathcal{F}_{1,2}(\tau) = \hat{\mathcal{F}}_{1,2}(\tau). \qquad (43)$$

Hence, we have

$$\begin{aligned}
\mathcal{L}_{1,2} &= i\hat{L}_{1,2}, \\
&= i\{(x_2 - \mathcal{E}_2)\partial_1 - (x_1 - \mathcal{E}_1)\partial_2 \\
&\quad + i\left(x_1\dot{\mathcal{E}}_2 - x_2\dot{\mathcal{E}}_1\right) + i\mathcal{F}_{1,2}\},
\end{aligned} \qquad (44)$$

For the complex operators corresponding to the real generators $\hat{L}_j$, we need ϕ_j, where

$$\phi_1(\tau) = \xi^2(\tau), \quad \phi_2(\tau) = \bar{\xi}^2(\tau), \quad \phi_3(\tau) = 2\xi(\tau)\bar{\xi}(\tau). \quad (45)$$

The function, ϕ_3, plays an important role in future calculations and is always real and is normally positive. In addition, the $\mathcal{E}_{j,\sigma}$ functions have the form

$$\mathcal{E}_{1,\sigma}(\tau) = -\xi(\tau)C_\sigma(\tau), \quad \mathcal{E}_{2,\sigma}(\tau) = -\bar{\xi}(\tau)\bar{C}_\sigma(\tau),$$
$$\mathcal{E}_{3,\sigma}(\tau) = -\xi(\tau)\bar{C}_\sigma - \bar{\xi}(\tau)C_\sigma(\tau). \quad (46)$$

Finally, the $\mathcal{D}_{j,\sigma}$ functions are

$$\mathcal{D}_{1,\sigma}(\tau) = -\tfrac{1}{2}C_\sigma^2(\tau), \quad \mathcal{D}_{2,\sigma}(\tau) = -\tfrac{1}{2}\bar{C}_\sigma^2(\tau),$$
$$\mathcal{D}_{3,\sigma}(\tau) = -C_\sigma \bar{C}_\sigma. \quad (47)$$

Each set of the above complex functions can be expressed as a linear combination of the corresponding real functions [14]. This means that the complex operators, which we denote by $\mathcal{M}_\pm$ and $\mathcal{M}_3$, can be expressed as linear combinations of the the real operators $\hat{L}_j$. We have

$$\begin{aligned}
\mathcal{M}_- &= \tfrac{i}{2}\left(\hat{L}_1 - \hat{L}_2 + i\hat{L}_3\right), \\
&= i\Big\{\phi_1 \partial_\tau + \sum_{\sigma=1}^N \left(\tfrac{1}{2}\dot{\phi}_1 x_\sigma + \mathcal{E}_{1,\sigma}\right)\partial_\sigma - \tfrac{i}{4}\ddot{\phi}_1 \mathbf{x}\cdot\mathbf{x} \\
&\quad -i\sum_{\sigma=1}^N x_\sigma \dot{\mathcal{E}}_{1,\sigma} + \tfrac{N}{4}\dot{\phi}_1 + i\sum_{\sigma=1}^N \mathcal{D}_{1,\sigma} + ig^{(0)}\phi_1\Big\}, \quad (48) \\
\mathcal{M}_+ &= \tfrac{i}{2}\left(\hat{L}_1 - \hat{L}_2 - i\hat{L}_3\right), \\
&= i\Big\{\phi_2 \partial_\tau + \sum_{\sigma=1}^N \left(\tfrac{1}{2}\dot{\phi}_2 x_\sigma + \mathcal{E}_{2,\sigma}\right)\partial_\sigma - \tfrac{i}{4}\ddot{\phi}_2 \mathbf{x}\cdot\mathbf{x} \\
&\quad -i\sum_{\sigma=1}^N x_\sigma \dot{\mathcal{E}}_{2,\sigma} + \tfrac{N}{4}\dot{\phi}_2 + i\sum_{\sigma=1}^N \mathcal{D}_{2,\sigma} + ig^{(0)}\phi_2\Big\}, \quad (49) \\
\mathcal{M}_3 &= i\left(\hat{L}_1 + \hat{L}_2\right), \\
&= i\Big\{\phi_3 \partial_\tau + \sum_{\sigma=1}^N \left(\tfrac{1}{2}\dot{\phi}_3 x_\sigma + \mathcal{E}_{3,\sigma}\right)\partial_\sigma - \tfrac{i}{4}\ddot{\phi}_3 \mathbf{x}\cdot\mathbf{x} \\
&\quad -i\sum_{\sigma=1}^N x_\sigma \dot{\mathcal{E}}_{3,\sigma} + \tfrac{N}{4}\dot{\phi}_3 + i\sum_{\sigma=1}^N \mathcal{D}_{3,\sigma} + ig^{(0)}\phi_3\Big\}. \quad (50)
\end{aligned}$$

Next, we compute the commutation relations for these operators. This is most easily done by using the expressions for the complex operators in terms of the real generators and the commutation relations for $(\mathcal{SA})_N$ in Eqs. (24), (27), and (28) through (30). We find that the three operators $\mathcal{M}_\pm$ and $\mathcal{M}_3$ form a basis for a $su(1,1)$ algebra with commutation relations

$$[\mathcal{M}_+, \mathcal{M}_-] = \mathcal{M}_3, \quad [\mathcal{M}_3, \mathcal{M}_\pm] = \pm 2\mathcal{M}_\pm. \tag{51}$$

Denoting the complexification of the Heisenberg-Weyl subalgebra by w_N^c, its basis, $\{\mathcal{J}_{\sigma\pm}, I\}$ satisfy the nonzero commutation relations

$$[\mathcal{J}_{\sigma-}, \mathcal{J}_{\sigma+}] = I. \tag{52}$$

The remaining commutation relations are as follows:

$$[\mathcal{M}_3, \mathcal{L}_{1,2}] = [\mathcal{M}_\pm, \mathcal{L}_{1,2}] = 0; \tag{53}$$

$$[\mathcal{M}_-, \mathcal{J}_{\sigma-}] = 0, \quad [\mathcal{M}_+, \mathcal{J}_{\sigma-}] = \mathcal{J}_{\sigma+}, \quad [\mathcal{M}_3, \mathcal{J}_{\sigma-}] = -\mathcal{J}_{\sigma-},$$
$$[\mathcal{M}_-, \mathcal{J}_{\sigma+}] = -\mathcal{J}_{\sigma-}, \quad [\mathcal{M}_+, \mathcal{J}_{\sigma-}] = 0, \quad [\mathcal{M}_3, \mathcal{J}_{\sigma+}] = +\mathcal{J}_{\sigma+}, \tag{54}$$

and

$$[\mathcal{L}_{1,2}, \mathcal{J}_{\nu-}] = -i\delta_{\nu,2}\mathcal{J}_{1-} + i\delta_{\nu,1}\mathcal{J}_{2-}, \quad [\mathcal{L}_{1,2}, \mathcal{J}_{\nu+}] = -i\delta_{\nu,2}\mathcal{J}_{1+} + i\delta_{\nu,1}\mathcal{J}_{2+}. \tag{55}$$

Therefore, the algebraic structure of the complexification of the Schrödinger algebra in one dimension is

$$(\mathcal{SA})_1^c = su(1,1) \diamond w_1^c, \tag{56}$$

and in two dimensions

$$(\mathcal{SA})_2^c = (su(1,1) \oplus so(2)) \diamond w_2^c, \tag{57}$$

where $(\mathcal{SA})_N^c$ denotes the complexified Schrödinger algebra.

Frequently, a subalgebra of $(\mathcal{SA})_N^c$ is sufficient to compute the solution space [7, 7, 16]. For a 1-dimensional, time-dependent oscillator, the subalgebra consisting of operators $\{\mathcal{M}_3, \mathcal{J}_\pm, I\}$ and its representation theory is sufficient to obtain solutions to the Schrödinger equation (1). We denote this subalgebra by $os(1)$. For a 2-dimensional, time-dependent oscillator, the subalgebra with basis $\{\mathcal{M}_3, \mathcal{L}_{1,2}, \mathcal{J}_\pm, I\}$ and its representation theory play an instrumental role in the determination of sets of solutions of the Schrödinger equation. The next section will be devoted to this subject.

3. Solutions to the Time-dependent Schrödinger Equation

3.1. SYSTEMS IN ONE DIMENSION

The oscillator algebra, $os(1)$ is spanned by the operators $\mathcal{M}_3$, $\mathcal{J}_\pm$, and I which satisfy the nonzero commutation relations

$$[\mathcal{M}_3, \mathcal{J}_\pm] = \pm \mathcal{J}_\pm, \quad [\mathcal{J}_-, \mathcal{J}_+] = I. \tag{58}$$

Notice that we have dropped the σ subscript on x and the generators of w_1^c to simplify notation. We have called $os(1)$ an oscillator algebra because of the isomorphism between it and the algebra normally constructed from the time-independent harmonic oscillator Hamiltonian and the raising and lowering operators $a^\dagger$ and a that factor the Hamiltonian [1, 17]. However, it is important to recognize that $\mathcal{M}_3$ is not a Hamiltonian, but we have

$$\mathcal{J}_+ \mathcal{J}_- + \tfrac{1}{2} = \mathcal{M}_3 - \tfrac{1}{2}\phi_3 S_1. \tag{59}$$

Therefore, on the solution space of the Schrödinger equation (1), the operator $\mathcal{M}_3$ is acting like a Hamiltonian.

The Casimir operator for $os(1)$ can be written

$$\mathbf{C} = \mathcal{J}_+ \mathcal{J}_- - \mathcal{M}_3 = -\tfrac{1}{2}(\phi_3 S_1 + 1), \tag{60}$$

where the second equality follows from Eq. (59). The Casimir operator is an invariant operator, commuting with all operators in $os(1)$, and in particular,

$$[\mathbf{C}, \mathcal{M}_3] = [\mathbf{C}, \mathcal{J}_\pm] = 0. \tag{61}$$

For the three commuting constants of the motion, $\mathbf{C}$, $\mathcal{M}_3$, and I, we can find a set of common eigenfunctions. These eigenfunctions, denoted by $\{\Psi_n : n \in \mathbf{Z}^+\}$, span a representation space of $os(1)$. The symbol $\mathbf{Z}^+$ means the set of nonnegative integers. In addition, we require that each Ψ_n satisfy Eq. (1). Then, according to Eq. (60), we obtain the specific representation space [7, 8, 18], labeled by $\uparrow_{-1/2}$, in which

$$\mathbf{C}\Psi_n = -\tfrac{1}{2}\Psi_n, \quad \mathcal{M}_3 \Psi_n = \left(n + \tfrac{1}{2}\right)\Psi_n, \tag{62}$$

$$\mathcal{J}_-\Psi_n = \sqrt{n}\,\Psi_{n-1}, \quad \mathcal{J}_+\Psi_n = \sqrt{n+1}\,\Psi_{n+1}, \quad I\Psi_n = \Psi_n, \tag{63}$$

for each $n \in \mathbf{Z}^+$. The spectrum of $\mathcal{M}_3$, denoted $\text{Sp}(\mathcal{M}_3)$, is $\text{Sp}(\mathcal{M}_3) = \{n + \tfrac{1}{2} : n \in \mathbf{Z}^+\}$, that is, $\text{Sp}(\mathcal{M}_3)$ is bounded below. The fact that the spectra of $\mathbf{C}$ and I are constant implies that the representation $\uparrow_{-1/2}$ of $os(1)$ under is irreducible. The function Ψ_0 represents an extremal state corresponding to the lower bound of $Sp(\mathcal{M}_3)$. The extremal state, Ψ_0 is annihilated by the operator $\mathcal{J}_-$, that is,

$$\mathcal{J}_-\Psi_0 = 0. \tag{64}$$

As we have mentioned, the algebraic treatment of the time-independent harmonic oscillator provides an example of this approach. In that case, the set of states, ψ_n, that we obtain are energy eigenstates. The condition, $a\psi_0 = 0$, defines the ground state. We restrict the use of the terms ground and excited states to refer to energy eigenstates only. Thus, a ground state is an example of an extremal state. However, not all extremal states are ground states. For example, Ψ_0 is not generally a ground state since the set of states $\{\Psi_n\}$ are not necessarily energy eigenstates.

When we combine Eq. (50) with the second equation in (62), we obtain a first-order partial differential equation for Ψ_n, namely,

$$\phi_3 \Psi_\tau + \left(\tfrac{1}{2}\dot{\phi}_3 x + \mathcal{E}_{3,1}\right) \Psi_x$$
$$= \left\{\tfrac{1}{4}\ddot{\phi}_3 x^2 + ix\dot{\mathcal{E}}_{3,1} - \tfrac{1}{4}\dot{\phi}_3 - iD_3 - ig^{(0)}\phi_3 - i\left(n+\tfrac{1}{2}\right)\right\}\Psi, \tag{65}$$

where the label n on the wave function has been suppressed for notational convenience. It is remarkable that this equation can be integrated without direct knowledge of the τ-dependent coefficients, $g^{(r)}$, $r = 0, 1, 2$, in the potential. We use the method of characteristics [7, 8, 16, 19]. The R-separable solution has the general form

$$\Psi_n(x,\tau) = \exp(i\mathcal{R})\psi_n(\zeta)\Xi_n(\eta), \tag{66}$$

where the R-factor [20] is

$$\mathcal{R}(x,\tau) = \tfrac{1}{4}\frac{x^2}{\phi_3}\left(\dot{\phi}_3 - \dot{\phi}_3^o\right) + \left(\frac{x}{\phi_3^{1/2}}\right)\left(\frac{\mathcal{E}_{3,1}}{\phi_3^{1/2}} - \frac{\mathcal{E}_{3,1}^o}{(\phi_3^o)^{1/2}} + \tfrac{1}{2}B_{3,1}\dot{\phi}_3^o\right), \tag{67}$$

and the similarity variables, ζ and η, are

$$\zeta = \frac{x}{\phi_3^{1/2}} - B_{3,1}(\tau), \quad \eta = \tau. \tag{68}$$

The η-dependent function in Eq. (66) is

$$\Xi_n(\eta) = \left(\frac{\phi_3^o}{\phi_3}\right)^{\frac{1}{4}} \left(\frac{\xi^o\bar{\xi}(\eta)}{\bar{\xi}^o\xi(\eta)}\right)^{\frac{1}{2}\left(n+\frac{1}{2}\right)} \exp\left[-i\left(\Lambda_3(\eta) + G^{(0)}(\eta)\right)\right], \tag{69}$$

where

$$\xi^o = \xi(0), \quad \phi_3^o = \phi_3(0), \quad \mathcal{E}_{3,1}^o = \mathcal{E}_{3,1}(0), \tag{70}$$

are constants. Note that ϕ_3^o and $\mathcal{E}_{3,1}^o$ are real constants.

The remaining τ- or η-dependent functions are

$$B_{3,\sigma}(\tau) = \int_0^\tau ds \, \frac{\mathcal{E}_{3,\sigma}(s)}{\phi_3^{3/2}(s)} = b_{3,\sigma}(\tau) - b_{3,\sigma}^o, \tag{71}$$

where the real function $b_{3,\sigma}$ is

$$b_{3,\sigma}(\tau) = \frac{i}{\phi_3^{1/2}} \left[\xi(\tau)\bar{C}_\sigma(\tau) - \bar{\xi}(\tau)C_\sigma(\tau)\right], \tag{72}$$

with $b_{3,\sigma}^o = b_{3,\sigma}(0)$, $\sigma = 1$, and

$$G^{(0)}(\eta) = \int_0^\eta ds \, g^{(0)}(s), \tag{73}$$

$$\Lambda_{3,\sigma}(\eta) = \int_0^\eta ds \left(\frac{\mathcal{E}_{3,\sigma}^2(s)}{\phi_3^2(s)} + \frac{\mathcal{D}_{3,\sigma}(s)}{\phi_3(s)}\right) - B_{3,\sigma}(\eta)\left(\frac{\mathcal{E}_{3,\sigma}^o}{(\phi_3^o)^{1/2}}\right)$$

$$+ \tfrac{1}{4}B_{3,\sigma}^2(\tau)\dot\phi_3^o, \tag{74}$$

with $\sigma = 1$.

The function $\psi_n(\zeta)$ satisfies a second-order, ordinary differential equation in ζ

$$\frac{d^2\psi}{d\zeta^2} - 2i(\alpha^o\zeta + \beta^o)\frac{d\psi}{d\zeta} + (2n + 1 - A^o\zeta^2 + B^o\zeta + C^o)\psi = 0. \tag{75}$$

The constants α^o and β^o used in this equation are

$$\alpha^o = \tfrac{1}{2}\dot\phi_3^o, \quad \beta_\sigma^o = \frac{\mathcal{E}_{3,\sigma}^o}{(\phi_3^o)^{1/2}} = \phi_3^o b_{3,\sigma}^o, \tag{76}$$

and are real constants, since ϕ_3 and $\mathcal{E}_3$ are real. The constants A^o, B^o, and C^o depend upon α^o, β^o, and $b_{3,1}^o$ in the following way

$$A^o = 1 + (\alpha^o)^2, \quad B_1^o = 2\left(b_{3,1}^o - \alpha^o\beta_1^o\right), \quad C^o = \left(b_{3,1}^o\right)^2 - (\beta_1^o)^2 - i\alpha^o. \tag{77}$$

The quantities, A^o and B_1^o, are real numbers.

Since $Sp(\mathcal{M}_3)$ is bounded below, the condition (64) yields the following first-order differential equation for ψ_0

$$\frac{d\psi_0}{d\zeta} + \left[(1 - i\alpha^o)\zeta - b_{3,1}^o - i\beta^o\right]\psi_0 = 0. \tag{78}$$

This equation has the solution

$$\psi_0(\zeta) = \mathcal{N} \exp\left[-\tfrac{1}{2}(1 - i\alpha^\circ)\zeta^2 + (b^\circ_{3,1} + i\beta^\circ)\zeta\right], \quad (79)$$

where $\mathcal{N}$ is an integration constant. Therefore, the extremal state is

$$\begin{aligned}\Psi_0(x,\tau) &= \left(\frac{1}{\pi\phi_3}\right)^{\frac{1}{4}} \exp\left(-(b^\circ_{3,1})^2/2\right) \exp(i\mathcal{R}) \\ &\quad \times \exp\left[-(1 - i\alpha^\circ)\zeta^2/2 + (b^\circ_{3,1} + i\beta^\circ_1)\zeta\right] \\ &\quad \times \left(\frac{\xi^\circ \bar{\xi}(\eta)}{\bar{\xi}^\circ \xi(\eta)}\right)^{\frac{1}{4}} \exp\left[-i\left(\Lambda_3(\eta) + G^{(0)}(\eta)\right)\right]. \quad (80)\end{aligned}$$

The integration constant $\mathcal{N} = (1/\pi\phi^\circ_3)^{1/4} \exp\left(-(b^\circ_3)^2/2\right)$ was determined from the normalization condition

$$\int_{-\infty}^{+\infty} dx\, \bar{\Psi}_n(x,\tau)\Psi_n(x,\tau) = 1, \quad (81)$$

with $n = 0$.

The wave functions for states with higher n can be reached by repeated application of the raising operator $\mathcal{J}_+$. The normalized state Ψ_n can be written

$$\Psi_n = \frac{1}{n!}(\mathcal{J}_+)^n \Psi_0. \quad (82)$$

Frequently, all we need is Eq. (82) to compute expectation values, however sometimes it is convenient to have a coordinate realization of Ψ_n. We can obtain this algebraically by a method outlined in Ref. [8], and the result is

$$\begin{aligned}\Psi_n(x,\tau) &= \left(\frac{1}{n!}\right)^{\frac{1}{2}} \left(\frac{1}{2}\right)^{\frac{n}{2}} \left(\frac{\bar{\xi}^\circ}{\xi^\circ}\right)^{\frac{1}{4}} \exp(i\mathcal{R})\psi_n(\zeta) \\ &\quad \times \left(\frac{\phi^\circ_3}{\phi_3}\right)^{\frac{1}{4}} \left(\frac{\bar{\xi}(\eta)}{\xi(\eta)}\right)^{\frac{1}{2}\left(n+\frac{1}{2}\right)} \\ &\quad \times \exp\left[-i\left(\Lambda_{3,1}(\eta) + G^{(0)}(\eta)\right)\right], \quad (83)\end{aligned}$$

where

$$\psi_n(\zeta) = H_n\left(\zeta - b^\circ_{3,1}\right) \left(\frac{1}{\pi\phi^\circ_3}\right)^{\frac{1}{4}} \exp\left[-(1 - i\alpha^\circ)\zeta^2/2 + (b^\circ_{3,1} + i\beta^\circ_1)\zeta\right], \quad (84)$$

and $H_n\left(\zeta - b_{3,1}^o\right)$ is a Hermite polynomial with Rodrigues formula [21]

$$H_n\left(\zeta - b_{3,1}^o\right) = (-)^n \exp\left[\zeta^2 - 2b_{3,1}^o\zeta\right] \partial_\zeta^n \exp\left[-\zeta^2 + 2b_{3,1}^o\zeta\right]. \quad (85)$$

Because of Eqs. (59) and (62), the states Ψ_n are also eigenstates of the operator $\mathcal{J}_+\mathcal{J}_-$ with eigenvalue n, a nonnegative integer. Therefore, it is natural to refer to the functions Ψ_n as eigenstates of the number operator, $\mathcal{J}_+\mathcal{J}_-$ or more simply as number operators eigenstates, with $\mathcal{J}_+\mathcal{J}_-$ being implied. Also, these functions form a complete set of discrete states and often the calculation of expectation values is made easier in this basis. As an example, kinematical symmetries and the states Ψ_n, in particular the extremal state Ψ_0, play an instrumental role in the construction of displacement operator coherent states and the calculation of their properties.

3.2. SYSTEMS IN TWO DIMENSIONS

For the 2-dimensional, time-dependent oscillator, we shall consider the subalgebra $os(2) \times \{\mathcal{L}_{1,2}\}$ which is spanned by the basis $\{\mathcal{M}_3, \mathcal{L}_{1,2}, \mathcal{J}_{1\pm}, \mathcal{J}_{2\pm}, I\}$. From Eqs. (52), (54), and (55) we see that these operators satisfy the nonzero commutation relations

$$[\mathcal{M}_3, \mathcal{J}_{1\pm}] = \pm\mathcal{J}_{1\pm}, \quad [\mathcal{M}_3, \mathcal{J}_{2\pm}] = \pm\mathcal{J}_{2\pm},$$
$$[\mathcal{L}_{1,2}, \mathcal{J}_{1+}] = +i\mathcal{J}_{2+}, \quad [\mathcal{L}_{1,2}, \mathcal{J}_{1-}] = +i\mathcal{J}_{2-},$$
$$[\mathcal{L}_{1,2}, \mathcal{J}_{2+}] = -i\mathcal{J}_{1+}, \quad [\mathcal{L}_{1,2}, \mathcal{J}_{2-}] = -i\mathcal{J}_{1-},$$
$$[\mathcal{J}_{\sigma-}, \mathcal{J}_{\sigma+}] = I. \quad (86)$$

Now, we point out two important relationships

$$\sum_{\sigma=1}^{2} \mathcal{J}_{\sigma+}\mathcal{J}_{\sigma-} = -\tfrac{1}{2}\phi_3 S_2 + \mathcal{M}_3 - 1, \quad (87)$$

which is the 2-dimensional analogue of Eq. (59), and

$$i(\mathcal{J}_{1-}\mathcal{J}_{2+} - \mathcal{J}_{1+}\mathcal{J}_{2-}) = \mathcal{L}_{1,2}, \quad (88)$$

which will be needed later.

A more convenient algebra [7] to work with can be obtained by defining the two commuting operators, $\mathcal{D}_0$ and $\mathcal{F}_0$,

$$\mathcal{D}_0 = \tfrac{1}{2}(\mathcal{M}_3 + \mathcal{L}_{1,2}), \quad \mathcal{F}_0 = \tfrac{1}{2}(\mathcal{M}_3 - \mathcal{L}_{1,2}), \quad (89)$$

and a new basis for w_2^c

$$
\begin{aligned}
\mathcal{A}_- &= \sqrt{\tfrac{1}{2}}(\mathcal{J}_{1-} + i\mathcal{J}_{2-}), \\
&= \sqrt{\tfrac{1}{2}}\left[\xi(\partial_{x_1} + i\partial_{x_2}) - i\dot{\xi}(x_1 + ix_2) + i(C_1 + iC_2)\right], \quad (90)
\end{aligned}
$$

$$
\begin{aligned}
\mathcal{A}_+ &= \sqrt{\tfrac{1}{2}}(\mathcal{J}_{1+} - i\mathcal{J}_{2+}), \\
&= \sqrt{\tfrac{1}{2}}\left[-\bar{\xi}(\partial_{x_1} - i\partial_{x_2}) + i\dot{\bar{\xi}}(x_1 - ix_2) - i(\bar{C}_1 - i\bar{C}_2)\right], \quad (91)
\end{aligned}
$$

$$
\begin{aligned}
\mathcal{C}_- &= \sqrt{\tfrac{1}{2}}(\mathcal{J}_{1-} - i\mathcal{J}_{2-}) \\
&= \sqrt{\tfrac{1}{2}}\left[\xi(\partial_{x_1} - i\partial_{x_2}) - i\dot{\xi}(x_1 - ix_2) + i(C_1 - iC_2)\right], \quad (92)
\end{aligned}
$$

$$
\begin{aligned}
\mathcal{C}_+ &= \sqrt{\tfrac{1}{2}}(\mathcal{J}_{1+} + i\mathcal{J}_{2+}), \\
&= \sqrt{\tfrac{1}{2}}\left[-\bar{\xi}(\partial_{x_1} + i\partial_{x_2}) + i\dot{\bar{\xi}}(x_1 + ix_2) - i(\bar{C}_1 + i\bar{C}_2)\right]. \quad (93)
\end{aligned}
$$

The new basis satisfies a set of commutation relations that reveals the algebraic structure more completely. We have

$$[\mathcal{D}_0, \mathcal{F}_0] = 0, \quad (94)$$

$$[\mathcal{D}_0, \mathcal{A}_-] = [\mathcal{D}_0, \mathcal{A}_+] = 0, \quad [\mathcal{D}_0, \mathcal{C}_-] = -\mathcal{C}_-, \quad [\mathcal{D}_0, \mathcal{C}_+] = +\mathcal{C}_+, \quad (95)$$

$$[\mathcal{F}_0, \mathcal{A}_-] = -\mathcal{A}_-, \quad [\mathcal{F}_0, \mathcal{A}_+] = +\mathcal{A}_+, \quad [\mathcal{F}_0, \mathcal{C}_-] = [\mathcal{F}_0, \mathcal{C}_+] = 0, \quad (96)$$

and for the subalgebra $\{\mathcal{A}_\pm, \mathcal{A}_\pm, I\}$, we see that

$$
\begin{aligned}
[\mathcal{A}_-, \mathcal{A}_+] &= [\mathcal{C}_-, \mathcal{C}_+] = I, \\
[\mathcal{A}_\pm, \mathcal{C}_\pm] &= [\mathcal{A}_\pm, \mathcal{C}_\mp] = 0,
\end{aligned} \quad (97)
$$

which still has the Heisenberg-Weyl structure. There are two Casimir operators, $\mathbf{C}_1$ and $\mathbf{C}_2$, having the form

$$\mathbf{C}_1 = \mathcal{A}_+ \mathcal{A}_- - \mathcal{F}_0 I = -\tfrac{1}{4} S_2 - \tfrac{1}{2}, \quad (98)$$

$$\mathbf{C}_2 = \mathcal{C}_+ \mathcal{C}_- - \mathcal{D}_0 I = -\tfrac{1}{4} S_2 - \tfrac{1}{2}, \quad (99)$$

where we have used Eqs. (87) and (88) to obtain the second equality in each case.

In the new basis, the algebra almost has the structure of a direct sum of two 1-dimensional oscillator subalgebras, $os(1)$, where the two subalgebras are spanned by $\{\mathcal{F}_0, \mathcal{A}_\pm, I\}$ and $\{\mathcal{D}_0, \mathcal{C}_\pm, I\}$. The only element they have in common is the central element I. This means that we can look at the product of two representations, one for each $os(1)$ algebra. As the set of commuting constants of the motion, we select C_1, C_2, $\mathcal{D}_0$, $\mathcal{F}_0$, and I. We designate the eigenfunctions $\Psi_{n,m}$ and require that they satisfy the Schrödinger equation (1). Then, we have

$$C_1 \Psi_{n,m} = -\tfrac{1}{2}\Psi_{n,m}, \qquad C_2 \Psi_{n,m} = -\tfrac{1}{2}\Psi_{n,m}, \tag{100}$$

$$\mathcal{F}_0 \Psi_{n,m} = \left(n+\tfrac{1}{2}\right)\Psi_{n,m}, \qquad \mathcal{D}_0 \Psi_{n,m} = \left(m+\tfrac{1}{2}\right)\Psi_{n,m}, \tag{101}$$

$$\mathcal{A}_- \Psi_{n,m} = \sqrt{n}\,\Psi_{n-1,m}, \qquad \mathcal{C}_- \Psi_{n,m} = \sqrt{m}\,\Psi_{n,m-1}, \tag{102}$$

$$\mathcal{A}_+ \Psi_{n,m} = \sqrt{n+1}\,\Psi_{n+1,m}, \qquad \mathcal{C}_+ \Psi_{n,m} = \sqrt{m+1}\,\Psi_{n,m+1}, \tag{103}$$

$$I \Psi_{n,m} = \Psi_{n,m}, \tag{104}$$

where $n, m \in \mathbf{Z}^+$. The solutions $\Psi_{n,m}$ span the representation space [22] which we denote by $\uparrow_{-1/2} \times \uparrow_{-1/2}$. Since $\mathrm{Sp}(\mathcal{D}_0) = \{m + 1/2 : m \in \mathbf{Z}^+\}$ and $\mathrm{Sp}(\mathcal{F}_0) = \{n + 1/2 : n \in \mathbf{Z}^+\}$ are bounded below, we have

$$\mathcal{A}_- \Psi_{0,0} = \mathcal{C}_- \Psi_{0,0} = 0. \tag{105}$$

This condition defines the extremal state, $\Psi_{0,0}$. Also, since $\mathcal{A}_-$ commutes with $\mathcal{C}_+$ and $\mathcal{C}_-$ with $\mathcal{A}_+$, we have

$$\mathcal{A}_- \Psi_{0,m} = \mathcal{C}_- \Psi_{n,0} = 0, \tag{106}$$

for all n, m in $\mathbf{Z}^+$. The fact that $\mathcal{A}_-$ and $\mathcal{A}_+$ commute with $\mathcal{C}_-$ and $\mathcal{C}_+$ implies that each 'level' in $\mathrm{Sp}(\mathcal{F}_0)$ has an infinite-fold degeneracy, that is, there are an infinite number of values of the quantum number n for each value of m. The converse is true too; for each value of m, there are an infinite number of values of n. Thus, $\mathrm{Sp}(\mathcal{D}_0)$ has an infinite-fold degeneracy too.

Since $\Psi_{n,m}$ is a common eigenfunction of $\mathcal{D}_0$ and $\mathcal{F}_0$, it is also an eigenfunction of both $\mathcal{M}_3$ and $\mathcal{L}_{1,2}$. Since $\mathcal{M}_3 = \mathcal{D}_0 + \mathcal{F}_0$ and $\mathcal{L}_{1,2} = \mathcal{D}_0 - \mathcal{F}_0$, we observe that

$$\mathcal{M}_3 \Psi_{n,m} = (n+m+1)\Psi_{n,m}, \qquad \mathcal{L}_{1,2}\Psi_{n,m} = (m-n)\Psi_{n,m}, \tag{107}$$

according to Eq. (101). Note that the spectrum of $\mathcal{L}_{1,2}$,

$$\mathrm{Sp}(\mathcal{L}_{1,2}) = \{m - n : n, m \in \mathbf{Z}^+\}, \tag{108}$$

is the set of positive and negative integers, which is the spectrum normally associated with generators of $so(2)$ symmetry. The spectrum of $\mathcal{M}_3$ is the set of positive integers.

We shall work with the two equations in Eq. (107). The first equation yields the first order, partial differential equation

$$\phi_3 \Psi_\tau + \left(\tfrac{1}{2}\dot\phi_3 x_1 + \mathcal{E}_{3,1}\right) \Psi_{x_1} + \left(\tfrac{1}{2}\dot\phi_3 x_2 + \mathcal{E}_{3,2}\right) \Psi_{x_2} = \left\{\tfrac{i}{4}\ddot\phi_3 \mathbf{x}\cdot\mathbf{x} + i x_1 \dot{\mathcal{E}}_{3,1} + i x_2 \dot{\mathcal{E}}_{3,2} - \tfrac{1}{2}\dot\phi_3 \right.$$
$$\left. - i\mathcal{D}_{3,1} - i\mathcal{D}_{3,2} + i g^{(0)}\phi_3 - i(n+m+1)\right\}\Psi. \quad (109)$$

where for notational convenience we have suppressed the n and m labels on the wave function. We can solve for Ψ by using the method of characteristics [7, 8, 16, 19]. The result is the $\mathcal{R}$-separable solution

$$\Psi_{n,m}(x_1, x_2, \tau) = \exp(i\mathcal{R}_\mathcal{M})\psi_{n,m}(\zeta_1,\zeta_2)\Xi_{n,m}(\eta), \quad (110)$$

where the similarity variables are

$$\zeta_\sigma = \frac{x_\sigma}{\phi_3^{1/2}} - \mathcal{B}_{3,\sigma}(\tau), \quad \eta = \tau. \quad (111)$$

These variables are the 2-dimensional analogues of those defined for the 1-dimensional case in Eq. (68). The τ-dependent function, $\mathcal{B}_{3,\sigma}$, is defined in Eq. (71). The $\mathcal{R}$-factor is

$$\mathcal{R}_\mathcal{M} = \sum_{\sigma=1}^{2}\left\{\tfrac{1}{4}\frac{x_\sigma^2}{\phi_3}\left(\dot\phi_3 - \dot\phi_3^o\right) + \frac{x_\sigma}{\phi_3^{1/2}}\left(\frac{\mathcal{E}_{3,\sigma}}{\phi_3^{1/2}} + \tfrac{1}{2}\dot\phi_3^o \mathcal{B}_{3,\sigma} - \beta_\sigma^o\right)\right\}, \quad (112)$$

where β_σ^o is given in Eq. (76). The η-dependent function, $\Xi_{n,m}$, is defined by

$$\Xi_{n,m}(\eta) = \left(\frac{\phi_3^o}{\phi_3}\right)^{\frac{1}{2}} \left(\frac{\xi^o\bar\xi(\eta)}{\bar{\xi^o}\xi(\eta)}\right)^{\frac{1}{2}(n+m+1)}$$
$$\times \exp\left[-i\left(\sum_{\sigma=1}^{2}\Lambda_{3,\sigma}(\eta) + G^{(0)}(\eta)\right)\right], \quad (113)$$

where the η-dependent functions, $G^{(0)}$ and $\Lambda_{3,\sigma}$ are given by Eqs. (73) and (74), respectively.

Next, we require that $\Psi_{n,m}$ also satisfy the second eigenvalue equation in (107). Substituting Eq. (110) into the second equation, we find after some careful algebraic manipulation the unknown function, $\psi_{n,m}$, must satisfy the first order partial differential equation

$$\left(\zeta_2 - b_{3,2}^o\right)\psi_{\zeta_1} - \left(\zeta_1 - b_{3,1}^o\right)\psi_{\zeta_2} =$$
$$i\left\{-\zeta_1 a_2^o + \zeta_2 a_1^o - \phi_3^o\left(b_{3,1}^o \dot b_{3,2}^o - \dot b_{3,1}^o b_{3,2}^o\right) + (m-n)\right\}\psi, \quad (114)$$

where, according to Eq. (76)

$$a_\sigma^o = \tfrac{1}{2}\dot\phi_3^o b_{3,\sigma}^o + \beta_\sigma^o,$$
$$= \tfrac{1}{2}\dot\phi_3 b_{3,\sigma}^o + \phi_3^o \dot b_{3,\sigma}^o. \tag{115}$$

Again, for notational convenience, we have dropped the n and m subscripts on the wave function.

Solving this equation by the method of characteristics, we find that

$$\psi_{n,m}(\zeta_1, \zeta_2) = \exp(i\mathcal{R}_\mathcal{L}) R_{n,m}(\rho) \exp[i(m-n)\theta], \tag{116}$$

where the new variables of separation are

$$\rho^2 = \left(\zeta_1 - b_{3,1}^o\right)^2 + \left(\zeta_2 - b_{3,2}^o\right)^2, \tag{117}$$

and

$$\tan\theta = \frac{\zeta_2 - b_{3,2}^o}{\zeta_1 - b_{3,1}^o}. \tag{118}$$

The variables (ρ, θ) form a polar coordinate system and we can write

$$\zeta_1 - b_{3,1}^o = \rho\cos\theta, \quad \zeta_2 - b_{3,2}^o = \rho\sin\theta. \tag{119}$$

The $\mathcal{R}$-factor has the form

$$\mathcal{R}_\mathcal{L} = \sum_{\sigma=1}^2 a_\sigma^o \left(\zeta_\sigma - b_{3,\sigma}^o\right). \tag{120}$$

Now, the total wave function can be expressed as

$$\Psi_{n,m}(x_1, x_2, \tau) = \exp(i\mathcal{R}_\mathcal{M})\exp(i\mathcal{R}_\mathcal{L})R_{n,m}(\rho)\exp[i(m-n)\theta]. \tag{121}$$

Next, we use the condition in Eq. (105) to determine the extremal state, $\Psi_{0,0}$. The equation implies that

$$A_-\Psi_{0,0} = e^{i\mathcal{R}_\mathcal{M}} e^{i\mathcal{R}_\mathcal{L}} R_{n,m}(\rho) \Xi_{0,0}(\eta) = 0, \tag{122}$$

that

$$\left(e^{-i\mathcal{R}_\mathcal{M}} e^{-i\mathcal{R}_\mathcal{L}} a_- e^{i\mathcal{R}_\mathcal{M}} e^{i\mathcal{R}_\mathcal{L}}\right) R_{n,m}(\rho) = 0, \tag{123}$$

since $\mathcal{A}_-$ has no dependence on τ-derivatives. Then, we complete the transformation to ρ and θ variables. Define the following operators

$$\begin{aligned}A_- &= e^{-i\mathcal{R}_\mathcal{M}}e^{-i\mathcal{R}_\mathcal{L}}\mathcal{A}_- e^{i\mathcal{R}_\mathcal{M}}e^{i\mathcal{R}_\mathcal{L}},\\ &= \tfrac{1}{2}\left(\frac{\xi}{\bar{\xi}}\right)^{\frac{1}{2}} e^{i\theta}\left[\partial_\rho + \frac{i}{\rho}\partial_\theta + (1-i\alpha^o)\rho\right],\end{aligned}\qquad(124)$$

$$\begin{aligned}A_+ &= e^{-i\mathcal{R}_\mathcal{M}}e^{-i\mathcal{R}_\mathcal{L}}\mathcal{A}_+ e^{i\mathcal{R}_\mathcal{M}}e^{i\mathcal{R}_\mathcal{L}},\\ &= \tfrac{1}{2}\left(\frac{\bar{\xi}}{\xi}\right)^{\frac{1}{2}} e^{-i\theta}\left[-\partial_\rho + \frac{i}{\rho}\partial_\theta + (1+i\alpha^o)\rho\right],\end{aligned}\qquad(125)$$

$$\begin{aligned}C_- &= e^{-i\mathcal{R}_\mathcal{M}}e^{-i\mathcal{R}_\mathcal{L}}\mathcal{C}_- e^{i\mathcal{R}_\mathcal{M}}e^{i\mathcal{R}_\mathcal{L}},\\ &= \tfrac{1}{2}\left(\frac{\xi}{\bar{\xi}}\right)^{\frac{1}{2}} e^{-i\theta}\left[\partial_\rho - \frac{i}{\rho}\partial_\theta + (1-i\alpha^o)\rho\right],\end{aligned}\qquad(126)$$

$$\begin{aligned}C_+ &= e^{-i\mathcal{R}_\mathcal{M}}e^{-i\mathcal{R}_\mathcal{L}}\mathcal{C}_+ e^{i\mathcal{R}_\mathcal{M}}e^{i\mathcal{R}_\mathcal{L}},\\ &= \tfrac{1}{2}\left(\frac{\bar{\xi}}{\xi}\right)^{\frac{1}{2}} e^{i\theta}\left[-\partial_\rho - \frac{i}{\rho}\partial_\theta + (1+i\alpha^o)\rho\right],\end{aligned}\qquad(127)$$

where α^o can be found in Eq. (76). The five operators, $A_\pm$, $C_\pm$, and I, satisfy the same commutation relations as do $\mathcal{A}_\pm$, $\mathcal{C}_\pm$, I.

Now, Eq. (123) is equivalent to

$$A_- R_{0,0}(\rho) = 0,\qquad(128)$$

and so by substituting Eq. (124) for A_-, we obtain the ordinary differential equation

$$\frac{dR_{0,0}}{d\rho} + (1-i\alpha^o)\rho R_{0,0} = 0.\qquad(129)$$

The solution to this equation is

$$R_{0,0}(\rho) = \mathcal{N}\exp\left[-(1-i\alpha^o)\rho^2\right],\qquad(130)$$

where $\mathcal{N}$ is an integration constant. Therefore, the complete extremal state wave function is

$$\begin{aligned}\Psi_{0,0}(x_1,x_2,\tau) &= \exp(i\mathcal{R}_\mathcal{M})\exp(i\mathcal{R}_\mathcal{L})\exp\left[-(1-i\alpha^o)\rho^2/2\right]\left(\frac{1}{\pi\phi_3}\right)^{\frac{1}{2}}\\ &\quad\times\left(\frac{\xi^o\bar{\xi}(\eta)}{\bar{\xi}^o\xi(\eta)}\right)^{\frac{1}{2}}\exp\left[-i\left(\sum_{\sigma=1}^2 \Lambda_{3,\sigma} + G^{(0)}\right)\right],\end{aligned}\qquad(131)$$

where $\Xi_{0,0}$ is given by expression (113) and the integration constant $\mathcal{N}$ has been fixed by normalization [23]

$$\int_{-\infty}^{+\infty}\int_{-\infty}^{+\infty} dx_1 dx_2\, \bar{\Psi}_{n,m}(x_1,x_2,\tau)\Psi_{n,m}(x_1,x_2,\tau) = 1. \qquad (132)$$

The reader can check that the condition $c_-\Psi_{0,0} = 0$ is also satisfied. Notice that the extremal state has no θ dependence.

The higher states can be computed by repeated application of the raising operators $\mathcal{A}_+$ and $\mathcal{C}_+$ and normalizing. We express this by writing

$$\Psi_{n,m} = \left(\frac{1}{n!m!}\right)^{\frac{1}{2}} (\mathcal{A}_+)^n (\mathcal{C}_+)^m \Psi_{0,0}. \qquad (133)$$

This is often sufficient for computational purposes, however we find it interesting that we can derive algebraically an expression for $R_{n,m}(\rho)$. We quote the result here

$$\Psi_{n,m}(x_1,x_2,\tau) = \exp(i\mathcal{R}_\mathcal{M})\exp(i\mathcal{R}_\mathcal{L})\exp[i(m-n)\theta]$$

$$\times \left(\frac{1}{\pi\phi_3}\right)^{\frac{1}{2}} (1-i\alpha^o)^{k'} (-)^k \frac{k!}{\sqrt{n!m!}} \left(\frac{\bar{\xi}^o}{\xi^o}\right)^{\frac{1}{2}}$$

$$\times \rho^{|n-m|} \exp\left[-(1-i\alpha^o)\rho^2/2\right] L_k^{(|n-m|)}\left((1-i\alpha^o)\rho^2\right)$$

$$\times \left(\frac{\bar{\xi}}{\xi}\right)^{\frac{1}{2}(n+m+1)} \exp\left[-i\left(\sum_{\sigma=1}^{2}\Lambda_{3,\sigma}+G^{(0)}\right)\right], \qquad (134)$$

where α^o is given in Eq. (76) and k and k' are defined as

$$k = \tfrac{1}{2}(n+m-|n-m|), \qquad k' = \tfrac{1}{2}(n+m+|n-m|). \qquad (135)$$

The function $L_k^{(|n-m|)}(((1-i\alpha^o)\rho^2))$ is a generalized Laguerre polynomial [21]. We outline the derivation of Eq. (134) in Appendix I.

3.3. COMMENTS

For both the 1-dimensional and 2-dimensional Schrödinger equations with potential (5), it is surprising that they admit separable solutions. At first sight, the space and time variables seem to be hopelessly entangled. The fact that these equations have a large number of Lie space-time symmetries in fact guarantees the existence of separable solutions [5]. Indeed, these equations separate into other coordinate systems. By selecting other sets of commuting constants of the motion and appropriate subalgebras of $(\mathcal{SA})_N$

containing them, we can construct different solution spaces. The analysis is too lengthy to go into here. For the 1-dimensional Schrödinger equation, we chose the operator $\hat{L}_1 + \hat{L}_2$ (or $\mathcal{M}_3$ in $(\mathcal{SA})_1^c$ and the subalgebra $os(1)$ containing $\mathcal{M}_3$) to diagonalize. In the case of the 2-dimensional Schrödinger equation, we selected two constants of the motion, $\hat{L}_1 + \hat{L}_2$ and $\hat{L}_{1,2}$ (or $\mathcal{M}_3$ and $\mathcal{L}_{1,2}$ in $(\mathcal{SA})_2^c$ and the subalgebra $os(2) \times so(2)$ containing $\mathcal{M}_3$ and $\mathcal{L}_{1,2}$). In each case, the structure of $(\mathcal{SA})_N^c$ and the appropriate subalgebras faciliates the calculation.

The solutions, Ψ_n in Eq. (84) and $\Psi_{n,m}$ in Eq. (134), form a complete set of discrete states. Also, they are eigenstates of number operators. In the 1-dimensional case, the number operator is $\mathcal{J}_+\mathcal{J}_-$, which has the eigenvalue spectrum, $\text{Sp}(\mathcal{J}_+\mathcal{J}_-) = \{n : n \in \mathbf{Z}^+\}$, while for two dimensions, there are two number operators, $\mathcal{A}_+\mathcal{A}_-$ and $\mathcal{C}_+\mathcal{C}_-$, involved and they have eigenvalue spectra $\text{Sp}(\mathcal{A}_+\mathcal{A}_-) = \{n : n \in \mathbf{Z}^+\}$ and $\text{Sp}(\mathcal{C}_+\mathcal{C}_-) = \{m : m \in \mathbf{Z}^+\}$, respectively. Frequently, we shall refer to these states as number operator eigenstates.

It has already been mentioned that these states are not necessarily energy eigenstates. It is important to remember this. Nevertheless, these states form a convenient set of states with which to perform calculations. In the next section, we shall see how space-time symmetries and number operator eigenstates can be useful in the construction of coherent states and examination of their dymanics.

4. Coherent States

Coherent states are quantum analogues of classical states for a system. Schrödinger investigated them initially [24] and theoretical investigation into them and their properties have intensified with the publication of Glauber's paper [25] in 1963. Now, there is an extensive literature on their theory and their applications. For a representative selection of work on this subject, we refer the reader to the compendium by Klauder and Skagerstam [26].

It is a well-known fact [27, 28] that for the harmonic oscillator there are three types of equivalent coherent states, discplacement operator coherent states (DOCS), annihilation operator coherent states (AOCS), and minimum uncertainty coherent states (MUCS). For the 1-dimensional oscillator, this arises from the algebraic structure of $os(1)$. Indeed, we expect that for any system which admits $os(1)$ symmetry, this equivalence will be maintained. In Ref. [9], we explored this fact by studying coherent states of well-known systems which admit isomorphic kinematical algebras: the harmonic and repulsive oscillators, a free particle, the driven oscillator, and a linear potential.

Next, we define in general terms the DOCS for the 1- and 2-dimensional Schrödinger equations discussed previously. Then, we outline how to obtain expectation values for position and momentum in each case. Application to specific examples will be made in Section 5.

For the 1-dimensional time-dependent harmonic oscillator, the DOCS wave function, $|\alpha\rangle$, is given formally by the definition

$$|\alpha\rangle = D(\alpha|0\rangle, \tag{136}$$

where α is a complex parameter

$$\alpha = |\alpha|e^{i\delta}. \tag{137}$$

We use Dirac notation for convenience and $|0\rangle$ corresponds to the extremal state (80). The displacement operator, $D(\alpha)$, is

$$D(\alpha) = \exp\left(\alpha \mathcal{J}_+ - \bar{\alpha}\mathcal{J}_-\right), \tag{138}$$

where the operators $\mathcal{J}_\pm$ are given by Eqs. (39) and (40). Since $\mathcal{J}_-$ and $\mathcal{J}_+$ are Hermitian conjugates and α complex, the operator $D(\alpha)$ is unitary. A more convenient form for $D(\alpha)$ is

$$D(\alpha) = e^{-\frac{1}{2}|\alpha|^2} \exp\left(\alpha \mathcal{J}_+\right) \exp\left(-\bar{\alpha}\mathcal{J}_-\right). \tag{139}$$

This relationship is frequently refered to as a Baker-Campbell-Hausdorff or BCH relation.

Although an explicit coordinate realization of the coherent state can be obtained [9], it is not necessary for the calculations we wish to do. Expectation values for position and momentum can be calculated from the following operator relationships

$$x = \bar{\xi}\mathcal{J}_- + \xi\mathcal{J}_+ + i\left(\xi\bar{C}_1 - \bar{\xi}C_1\right), \tag{140}$$

$$p_x = \bar{\dot{\xi}}\mathcal{J}_- + \dot{\xi}\mathcal{J}_+ + i\left(\dot{\xi}\bar{C}_1 - \bar{\dot{\xi}}C_1\right). \tag{141}$$

Letting $\mathcal{O}$ denote an operator, we compute its expectation value for the state $|\alpha\rangle$ according to

$$\langle\mathcal{O}\rangle = \langle\alpha|\mathcal{O}|\alpha\rangle = \langle 0|D^{-1}(\alpha)\mathcal{O}D(\alpha)|0\rangle. \tag{142}$$

Since we have [29]

$$D^{-1}(\alpha)\mathcal{J}_-D(\alpha) = \mathcal{J}_- + \alpha I, \quad D^{-1}(\alpha)\mathcal{J}_+D(\alpha) = \mathcal{J}_+ + \bar{\alpha}I, \tag{143}$$

the expectation values of x and p_x are

$$\langle x(\tau) \rangle = \alpha\bar{\xi} + \bar{\alpha}\xi + i\left(\xi\bar{C}_1 - \bar{\xi}C_1\right), \qquad (144)$$

$$\langle p_x(\tau) \rangle = \alpha\dot{\bar{\xi}} + \bar{\alpha}\dot{\xi} + i\left(\dot{\xi}\bar{C}_1 - \dot{\bar{\xi}}C_1\right), \qquad (145)$$

where we have made use of Eq. (64). If $x^o = \langle x(0) \rangle$ and $p_x^o = \langle p_x(0) \rangle$ are initial position and momentum, then using the Wronskian (34), we get [9, 16, 10]

$$\alpha = i\left(p_x^o \xi^o - x^o \dot{\xi}^o\right) + iC_1^o, \qquad (146)$$

and its complex conjugate. These equations lead to

$$\langle x(\tau) \rangle = i\left\{ \bar{\xi}(\tau)\left[\xi^o p_x^o - \dot{\xi}^o x^o - c_1(\tau)\right] \right.$$
$$\left. - \xi(\tau)\left[\bar{\xi}^o p_x^o - \dot{\bar{\xi}}^o x^o - \bar{c}_1(\tau)\right]\right\}, \qquad (147)$$

$$\langle p_x(\tau) \rangle = i\left\{ \dot{\bar{\xi}}(\tau)\left[\xi^o p_x^o - \dot{\xi}^o x^o - c_1(\tau)\right] \right.$$
$$\left. - \dot{\xi}(\tau)\left[\bar{\xi}^o p_x^o - \dot{\bar{\xi}}^o x^o - \bar{c}_1(\tau)\right]\right\}, \qquad (148)$$

where c_1 is given by Eq. (36). The uncertainty in position and momentum can be computed in a similar fashion and is found to be

$$(\Delta x)^2 = \xi(\tau)\bar{\xi}(\tau), \quad (\Delta p_x)^2 = \dot{\xi}(\tau)\dot{\bar{\xi}}(\tau), \qquad (149)$$

which are real and positive. The uncertainty product is

$$(\Delta x)^2(\Delta p_x)^2 = \xi(\tau)\bar{\xi}(\tau)\dot{\xi}(\tau)\dot{\bar{\xi}}(\tau). \qquad (150)$$

Generally speaking, the uncertainty product is dependent on τ. Initially, the minimum uncertainty is satisfied and then increases with 'time' τ. Only when $g^{(2)}$ is a positive constant is the minimum uncertainty satisfied for all time [9, 16].

The 2-dimensional case is an extension of the 1-dimensional picture described above. The coherent state is now specified by two complex parameters, α and γ, where

$$\alpha = |\alpha|e^{i\delta}, \quad \gamma = |\gamma|e^{i\epsilon}. \qquad (151)$$

The states are defined by

$$|\alpha, \gamma\rangle = D(\alpha, \gamma)|0\rangle. \qquad (152)$$

where the displacement operator is

$$D(\alpha, \gamma) = \exp\left(\alpha A_+ - \bar{\alpha}A_- + \gamma C_+ - \bar{\gamma}C_-\right). \qquad (153)$$

LIE SYMMETRIES IN QUANTUM MECHANICS

Since the operators $\mathcal{A}_\pm$ and $\mathcal{C}_\pm$ commute with each other it is convenient to write $D(\alpha, \gamma)$ as a product

$$D(\alpha,\gamma) = D(\alpha)D(\gamma) = D(\gamma)D(\alpha), \qquad (154)$$

where $D(\alpha)$ and $D(\gamma)$ is an operator of the form (138) with the operators $\mathcal{A}_\pm$ replacing the operators $\mathcal{J}_\pm$ in the former and γ and $\mathcal{C}_\pm$ replacing α and $\mathcal{J}_\pm$ in the latter. The BCH relation (139) applies in each case.

To calculate expectation values for position and momentum, we write the the operators x_σ and p_{x_σ} in terms of the raising and lowering operators, $\mathcal{A}_\pm$ and $\mathcal{C}_\pm$, as follows: for the two components of position

$$x_1 = \sqrt{\tfrac{1}{2}}\left(\xi(\mathcal{A}_+ + \mathcal{C}_+) + \bar{\xi}(\mathcal{A}_- + \mathcal{C}_-)\right) + i\left(\xi\bar{\mathcal{C}}_1 - \bar{\xi}\mathcal{C}_1\right), \qquad (155)$$

$$x_2 = i\sqrt{\tfrac{1}{2}}\left(\xi(\mathcal{A}_+ - \mathcal{C}_+) - \bar{\xi}(\mathcal{A}_- - \mathcal{C}_-)\right) + i\left(\xi\bar{\mathcal{C}}_2 - \bar{\xi}\mathcal{C}_2\right), \qquad (156)$$

and for the two components of momentum

$$p_{x_1} = \sqrt{\tfrac{1}{2}}\left(\dot{\xi}(\mathcal{A}_+ + \mathcal{C}_+) + \dot{\bar{\xi}}(\mathcal{A}_- + \mathcal{C}_-)\right) + i\left(\dot{\xi}\bar{\mathcal{C}}_1 - \dot{\bar{\xi}}\mathcal{C}_1\right), \qquad (157)$$

$$p_{x_2} = i\sqrt{\tfrac{1}{2}}\left(\dot{\xi}(\mathcal{A}_+ - \mathcal{C}_+) - \dot{\bar{\xi}}(\mathcal{A}_- - \mathcal{C}_-)\right) + i\left(\dot{\xi}\bar{\mathcal{C}}_2 - \dot{\bar{\xi}}\mathcal{C}_2\right). \qquad (158)$$

The calculation of expection values proceeds as in the 1-dimensional case. For position, we have

$$\langle x_1(\tau)\rangle = \sqrt{\tfrac{1}{2}}\left(\xi(\bar{\alpha} + \bar{\gamma}) + \bar{\xi}(\alpha + \gamma)\right) + i\left(\xi\bar{\mathcal{C}}_1 - \bar{\xi}\mathcal{C}_1\right), \qquad (159)$$

$$\langle x_2(\tau)\rangle = i\sqrt{\tfrac{1}{2}}\left(\xi(\bar{\alpha} - \bar{\gamma}) - \bar{\xi}(\alpha - \gamma)\right) + i\left(\xi\bar{\mathcal{C}}_2 - \bar{\xi}\mathcal{C}_2\right), \qquad (160)$$

and for momentum, we find that

$$\langle p_{x_1}(\tau)\rangle = \sqrt{\tfrac{1}{2}}\left(\dot{\xi}(\bar{\alpha} + \bar{\gamma}) + \dot{\bar{\xi}}(\alpha + \gamma)\right) + i\left(\dot{\xi}\bar{\mathcal{C}}_1 - \dot{\bar{\xi}}\mathcal{C}_1\right), \qquad (161)$$

$$\langle p_{x_2}(\tau)\rangle = i\sqrt{\tfrac{1}{2}}\left(\dot{\xi}(\bar{\alpha} - \bar{\gamma}) - \dot{\bar{\xi}}(\alpha - \gamma)\right) + i\left(\dot{\xi}\bar{\mathcal{C}}_2 - \dot{\bar{\xi}}\mathcal{C}_2\right). \qquad (162)$$

The analogues of expression (146) and its complex conjugate are

$$\alpha = i\sqrt{\tfrac{1}{2}}\left(\xi^\circ(p^\circ_{x_1} + ip^\circ_{x_2}) - \dot{\xi}^\circ(x^\circ_1 + ix^\circ_2) + \mathcal{C}^\circ_1 + i\mathcal{C}^\circ_2\right), \qquad (163)$$

$$\gamma = i\sqrt{\tfrac{1}{2}}\left(\xi^\circ(p^\circ_{x_1} - ip^\circ_{x_2}) - \dot{\xi}^\circ(x^\circ_1 - ix^\circ_2) + \mathcal{C}^\circ_1 - i\mathcal{C}^\circ_2\right), \qquad (164)$$

and their complex conjugates. Combining Eqs. (159) through (164), we obtain the analogues of Eqs. (147) and (148) in two dimensions,

$$\langle x_\sigma(\tau)\rangle = i\left\{\bar{\xi}(\tau)\left[\xi^\circ_\sigma p^\circ_{x_\sigma} - \dot{\xi}^\circ x^\circ_\sigma - c_\sigma(\tau)\right]\right.$$

$$-\xi(\tau)\left[\bar{\xi}^{\circ}p_{x_{\sigma}}^{\circ} - \bar{\dot{\xi}}^{\circ}x_{\sigma}^{\circ} - \bar{c}_{\sigma}(\tau)\right]\Big\}, \tag{165}$$

$$\langle p_{x_{\sigma}}(\tau)\rangle = i\Big\{\bar{\dot{\xi}}(\tau)\left[\xi^{\circ}p_{x_{\sigma}}^{\circ} - \dot{\xi}^{\circ}x_{\sigma}^{\circ} - c_{\sigma}(\tau)\right]$$
$$-\dot{\xi}(\tau)\left[\bar{\xi}^{\circ}p_{x_{\sigma}}^{\circ} - \bar{\dot{\xi}}^{\circ}x_{\sigma}^{\circ} - \bar{c}_{\sigma}(\tau)\right]\Big\}. \tag{166}$$

The uncertainty for position and momentum are identical to Eq. (150) for each coordinate momentum pair

$$(\Delta x_{\sigma})^2 (\Delta p_{x_{\sigma}})^2 = \xi(\tau)\bar{\xi}(\tau)\dot{\xi}(\tau)\bar{\dot{\xi}}(\tau), \tag{167}$$

for $\sigma = 1, 2$.

The operator $\mathcal{L}_{1,2}$ can be expressed as

$$\mathcal{L}_{1,2} = C_+ C_- - A_+ A_-, \tag{168}$$

and its expectation value is

$$\langle \mathcal{L}_{1,2}(\tau) \rangle = |\gamma|^2 - |\alpha|^2, \tag{169}$$

a constant, which makes sense since it is a constant of motion. The orbital angular momentum operator, L_z, can be written

$$L_z = i(x_2 \partial_{x_1} - x_1 \partial_{x_2}),$$
$$= C_+ C_- - A_+ A_- + i\sqrt{\tfrac{1}{2}}\left[(\mathcal{C}_1 + i\mathcal{C}_2)A_+ - (\bar{\mathcal{C}}_1 - i\bar{\mathcal{C}}_2)A_- \right.$$
$$\left. -(\mathcal{C}_1 - i\mathcal{C}_2)C_+ + (\bar{\mathcal{C}}_1 + i\bar{\mathcal{C}}_2)C_-\right] - i(\bar{\mathcal{C}}_1\mathcal{C}_2 - \mathcal{C}_1\bar{\mathcal{C}}_2). \tag{170}$$

In terms of initial positions and momenta, its expectation value is

$$\langle L_z(\tau) \rangle = (x_1^{\circ} p_{x_2}^{\circ} - x_2^{\circ} p_{x_1}^{\circ}) - i\left(\xi^{\circ}\bar{c}_2 - \bar{\xi}^{\circ}c_2\right)p_{x_1}^{\circ} + i\left(\xi^{\circ}\bar{c}_1 - \bar{\xi}^{\circ}c_1\right)p_{x_2}^{\circ}$$
$$+ i\left(\dot{\xi}^{\circ}\bar{c}_2 - \bar{\dot{\xi}}^{\circ}c_2\right)x_1^{\circ} - i\left(\dot{\xi}^{\circ}\bar{c}_1 - \bar{\dot{\xi}}^{\circ}c_1\right)x_2^{\circ} - (\bar{c}_1 c_2 - c_1 \bar{c}_2). \tag{171}$$

As expected, the initial angular momentum is modified with time as the system evolves.

5. Examples

A large number of systems are subsumed under the potential (5). Several examples are the harmonic oscillator, the driven harmonic oscillator, a free particle, the linear potential, and repulsive oscillator, to name a few of the more common ones. Many of them have been investigated using the above techniques [8, 9, 16, 12]. Here, we will focus only on the harmonic oscillator

LIE SYMMETRIES IN QUANTUM MECHANICS

and the driven harmonic oscillator in one and two dimensions. The former example will illustrate our analysis with a well-known example and the latter will be needed in Section 6.

5.1. HARMONIC OSCILLATOR

5.1.1. *One Dimension*

For the 1-dimensional harmonic oscillator, the potential is given by

$$V(x) = \tfrac{1}{2}\omega^2 x^2, \tag{172}$$

where ω^2 is a positive constant and

$$g^{(2)}(\tau) = \tfrac{1}{2}\omega^2, \quad g^{(1)}(\tau) = g^{(0)}(\tau) = 0. \tag{173}$$

We can write the τ-independent Hamiltonian for this system as follows:

$$H = -\tfrac{1}{2}\partial_{xx} + \tfrac{1}{2}\omega^2 x^2. \tag{174}$$

The differential equation (10) has the form

$$\ddot{a} + \omega^2 a = 0, \tag{175}$$

and it has two real solutions which we write as

$$\chi_1 = \tfrac{1}{\sqrt{\omega}}\cos\omega\tau, \quad \chi_2 = \tfrac{1}{\sqrt{\omega}}\sin\omega\tau. \tag{176}$$

Using Eq. (33), the two complex solutions are

$$\xi = \sqrt{\tfrac{1}{2\omega}}e^{i\omega\tau}, \quad \bar{\xi} = \sqrt{\tfrac{1}{2\omega}}e^{-i\omega\tau}, \tag{177}$$

from which we obtain the three functions ϕ_1, ϕ_2, and ϕ_3:

$$\phi_1 = \tfrac{1}{2\omega}e^{2i\omega\tau}, \quad \phi_2 = \tfrac{1}{2\omega}e^{-2i\omega\tau}, \quad \phi_3 = \tfrac{1}{\omega}. \tag{178}$$

Therefore, the generators have the form

$$\mathcal{J}_- = \sqrt{\tfrac{1}{2\omega}}e^{i\omega\tau}(\partial_x + \omega x), \quad \mathcal{J}_+ = \sqrt{\tfrac{1}{2\omega}}e^{-i\omega\tau}(-\partial_x + \omega x), \tag{179}$$

$$\mathcal{M}_- = \tfrac{1}{2\omega}e^{2i\omega\tau}(i\partial_\tau - \omega x \partial_x - \omega^2 x^2 - \tfrac{1}{2}\omega), \tag{180}$$

$$\mathcal{M}_+ = \tfrac{1}{2\omega}e^{-2i\omega\tau}(i\partial_\tau + \omega x \partial_x - \omega^2 x^2 + \tfrac{1}{2}\omega), \tag{181}$$

$$\mathcal{M}_3 = \tfrac{i}{\omega}\partial_\tau. \tag{182}$$

These operators satisfy the commutation relations (51), (52), and (54) for the $(\mathcal{SA})_n^c$ Lie algebra. Although we shall not need $\mathcal{M}_\pm$, we include them here for completion. For the remaining examples, we shall not give them but they can be constructed from the appropriate τ-dependent functions given in Section 2.2. To obtain the wave function, Ψ_n, from Eq. (83), we need to determine the form of several τ-dependent functions. Since many of these rely on $g_1^{(1)}$ which is zero here, we can write immediately that

$$C_1(\tau) = \mathcal{E}_{3,1}(\tau) = \mathcal{D}_{3,1}(\tau) = \Lambda_{3,1}(\tau) = b_{3,1}(\tau) = 0. \tag{183}$$

This also means that the constants

$$C_1^o = \mathcal{E}_{3,1}^o = \beta_1^o = b_{3,1}^o = 0. \tag{184}$$

Therefore, we have

$$B_{3,1}(\tau) = 0, \tag{185}$$

and because $\dot{\phi}_3 = 0$, we see that the constants

$$\dot{\phi}_3^o = \alpha^o = 0. \tag{186}$$

Finally, since $g^{(0)}(\tau) = 0$, we have

$$G^{(0)}(\tau) = 0. \tag{187}$$

The $\mathcal{R}$-separable coordinates are

$$\zeta = \omega^{1/2} x, \qquad \eta = \tau, \tag{188}$$

and the $\mathcal{R}$-factor

$$\mathcal{R}(x,\tau) = 0. \tag{189}$$

Therefore, the wave function, Ψ_n, takes the form

$$\Psi_n(x,\tau) = \left(\frac{1}{n!}\right)^{\frac{1}{2}} \left(\frac{1}{2}\right)^{\frac{n}{2}} \left(\frac{\omega}{\pi}\right)^{\frac{1}{4}} H_n(\omega^{1/2} x)$$
$$\times \exp\left(-\omega x^2/2\right) \exp\left[-i\left(n+\tfrac{1}{2}\right)\omega\tau\right]. \tag{190}$$

In this example, the operator $\mathcal{M}_3$ is proportional to the energy operator on the solution space $\mathcal{F}_{S_1} = \{\Psi_n : n \in \mathbf{Z}^+\}$. Since each ψ_n is an eigenfunction of $\mathcal{M}_3$ by construction, then each function is also an eigenfunction of the energy operator. Thus, each Ψ_n is an energy eigenstate. The extremal state,

$$\Psi_0(x,\tau) = \left(\frac{\omega}{\pi}\right)^{\frac{1}{4}} \exp\left(-\omega x^2/2\right) \exp\left[-i\omega\tau/2\right], \tag{191}$$

is the ground state wave function. These functions correspond to the well-known solutions of the 1-dimensional harmonic oscillator [1, 3].

Let us now turn to the coherent state $|\alpha\rangle$. From Eqs. (144), (145), (147), and (148) in Section 4, the expectation values for position and momentum in the coherent state $|\alpha\rangle$ are

$$\langle x(\tau)\rangle = \sqrt{\tfrac{2}{\omega}}|\alpha|\cos(\omega\tau - \delta), \quad \langle p_x(\tau)\rangle = -\sqrt{2\omega}|\alpha|\sin(\omega\tau - \delta), \quad (192)$$

or

$$\langle x(\tau)\rangle = x^o \cos\omega\tau + \tfrac{1}{\omega}p_x^o \sin\omega\tau, \quad \langle p_x(\tau)\rangle = p_x^o \cos\omega\tau - \omega x^o \sin\omega\tau. \quad (193)$$

These two pairs of equations are connected by the formulae

$$|\alpha|^2 = \tfrac{1}{2\omega}\left[(p_x^o)^2 + \omega^2(x^o)^2\right], \quad \tan\delta = \frac{p_x^o}{\omega x^o}. \quad (194)$$

The trajectory in phase space is an ellipse and the expectation values satisfy the classical equations of motion. Also, they are minimum uncertainty states for all time since

$$(\Delta x)^2(\Delta p_x)^2 = \tfrac{1}{4}. \quad (195)$$

It is of interest to compare these results to predictions made by computing expectation values in the number operator basis set in Eq. (190). In this basis, the expectation values of position and momentum are both zero. This would correspond to the trivial solution to Newton's equation of motion. The uncertainty product

$$(\Delta x)^2(\Delta p_x)^2 = \tfrac{1}{4}(2n+1)^2, \quad (196)$$

depend upon the quantum number n. However, the ground state, $n = 0$, does satisfy the minimum uncertainty condition and is a coherent state.

5.1.2. Two Dimensions

In our next example, we examine the harmonic oscillator in two dimensions. The τ-dependent functions, $g^{(2)}$, $g_1^{(1)}$, $g_2^{(1)}$, and $g^{(0)}$ are

$$g^{(2)} = \tfrac{1}{2}\omega^2, \quad g_1^{(1)} = g_2^{(1)} = g^{(0)} = 0, \quad (197)$$

and the τ-independent Hamiltonian is

$$H = -\tfrac{1}{2}\{\partial_{x_1 x_1} + \partial_{x_2 x_2}\} + \tfrac{1}{2}\omega^2 \mathbf{x}\cdot\mathbf{x}. \quad (198)$$

The appropriate differential equation to solve to obtain the Lie symmetries is Eq. (175). Its real and complex solutions are still given by Eqs. (176) and (177), respectively. The function, ϕ_3 can be found in Eq. (178).

The 2-dimensional analogues of Eqs. (183), (184), and (185) are

$$C_\sigma(\tau) = \mathcal{E}_{3,\sigma}(\tau) = \mathcal{D}_{3,\sigma}(\tau) = \Lambda_{3,\sigma}(\tau) = b_{3,\sigma}(\tau) = 0, \tag{199}$$

which means that the constants

$$C_\sigma^o = \mathcal{E}_{3,\sigma}^o = \beta_\sigma^o = b_{3,\sigma}^o = 0, \tag{200}$$

on account of Eq. (197). Also, we have

$$\mathcal{B}_{3,\sigma}(\tau) = 0, \tag{201}$$

and Eqs. (186) and (187).

The generators that form a basis for $os(2) \times so(2)$ are

$$I = 1,$$
$$\mathcal{A}_- = \tfrac{1}{4\omega} e^{i\omega\tau}\left[(\partial_{x_1} + i\partial_{x_2}) + \omega(x_1 + ix_2)\right],$$
$$\mathcal{A}_+ = \tfrac{1}{4\omega} e^{-i\omega\tau}\left[-(\partial_{x_1} - i\partial_{x_2}) + \omega(x_1 - ix_2)\right],$$
$$\mathcal{C}_- = \tfrac{1}{4\omega} e^{i\omega\tau}\left[(\partial_{x_1} - i\partial_{x_2}) + \omega(x_1 - ix_2)\right],$$
$$\mathcal{C}_+ = \tfrac{1}{4\omega} e^{i\omega\tau}\left[-(\partial_{x_1} + i\partial_{x_2}) + \omega(x_1 + ix_2)\right], \tag{202}$$

and the two symmetry operators

$$\mathcal{D}_0 = \tfrac{i}{2}\left(\tfrac{1}{\omega}\partial_\tau + x_2\partial_{x_1} - x_1\partial_{x_2}\right),$$
$$\mathcal{F}_0 = \tfrac{i}{2}\left(\tfrac{1}{\omega}\partial_\tau - x_2\partial_{x_1} + x_1\partial_{x_2}\right). \tag{203}$$

These operators satisfy the commutation relations (94) through (97). Recall from Eqs. (89), that the operators $\mathcal{D}_0$ and $\mathcal{F}_0$ are linear combinations of

$$\mathcal{M}_3 = \tfrac{1}{\omega}\partial_\tau, \quad \mathcal{L}_{1,2} = i(x_2\partial_{x_1} - x_1\partial_{x_2}). \tag{204}$$

In formulating an expression for the wave function $\Psi_{n,m}$, we note that the two $\mathcal{R}$-factors are

$$\mathcal{R}_\mathcal{M} = \mathcal{R}_\mathcal{L} = 0, \tag{205}$$

because of Eqs. (199) through (201). The coordinates of separation are

$$\zeta_1 = \omega^{1/2} x_1, \quad \zeta_2 = \omega^{1/2} x_2, \quad \eta = \tau, \tag{206}$$

or in terms of polar coordinates (119)

$$\omega^{1/2} x_1 = \rho\cos\theta, \quad \omega^{1/2} x_2 = \rho\sin\theta, \quad \eta = \tau. \tag{207}$$

Thus, we have

$$\Psi_{n,m}(x_1, x_2, \tau) = \left(\frac{\omega}{\pi}\right)^{\frac{1}{4}} \frac{k!}{\sqrt{n!m!}} (-)^k \exp\left[i(m-n)\theta\right]$$
$$\times \rho^{|n-m|} \exp\left(-\rho^2/2\right) L_k^{(|n-m|)}(\rho^2)$$
$$\times \exp\left[-i(n+m+1)\omega\tau\right]. \qquad (208)$$

where k ia given by Eq. (135). The extremal state is simply

$$\Psi_{0,0}(x_1, x_2, \tau) = \left(\frac{\omega}{\pi}\right)^{\frac{1}{4}} \exp\left(-\rho^2/2\right) \exp\left(-i\omega\tau\right). \qquad (209)$$

These wave functions are energy eigenstates since they are eigenfunctions of the operator $\mathcal{M}_3$, which in this case is proportional to the energy operator on the solution space.

For the coherent state $|\alpha, \gamma\rangle$, the expectation values for position and momentum can be calculated from Eqs. (165) and (166),

$$\langle x_\sigma(\tau)\rangle = x_\sigma^o \cos\omega\tau + \tfrac{1}{\omega} p_{x_\sigma}^o \sin\omega\tau,$$
$$\langle p_{x_\sigma}(\tau)\rangle = p_{x_\sigma}^o \cos\omega\tau - \omega x_\sigma^o \sin\omega\tau. \qquad (210)$$

Also, we can write

$$\langle x_\sigma(\tau)\rangle = G_\sigma \cos(\omega\tau - \vartheta_\sigma), \quad \langle p_{x_\sigma}(\tau)\rangle = -\omega G_\sigma \sin(\omega\tau - \vartheta_\sigma), \qquad (211)$$

where $\sigma = 1, 2$. Using Eqs. (151), (163), and (164), we find that

$$G_1^2 = (x_1^o)^2 + \tfrac{1}{\omega^2}(p_{x_1})^2 = \tfrac{1}{\omega}\left[|\alpha|^2 + |\gamma|^2 + 2|\alpha||\gamma|\cos(\delta-\epsilon)\right],$$
$$G_2^2 = (x_2^o)^2 + \tfrac{1}{\omega^2}(p_{x_2})^2 = \tfrac{1}{\omega}\left[|\alpha|^2 + |\gamma|^2 - 2|\alpha||\gamma|\cos(\delta-\epsilon)\right], \qquad (212)$$

and

$$\tan\vartheta_1 = \frac{p_{x_1}^o}{\omega x_1^o} = \frac{|\alpha|\sin\delta + |\gamma|\sin\epsilon}{|\alpha|\cos\delta + |\gamma|\cos\epsilon},$$

$$\tan\vartheta_2 = \frac{p_{x_2}^o}{\omega x_2^o} = -\frac{|\alpha|\cos\delta - |\gamma|\cos\epsilon}{|\alpha|\sin\delta - |\gamma|\sin\epsilon}. \qquad (213)$$

Either of Eqs. (210) or (211) correspond to the classical solutions to the classical equations of motion. For each degree of freedom, the minimum uncertainty relation applies

$$(\Delta x_\sigma)^2 (\Delta p_{x_\sigma})^2 = \tfrac{1}{4}. \qquad (214)$$

Orbital angular momentum is conserved since $L_z = \mathcal{L}_{1,2}$ and $\mathcal{L}_{1,2}$ is a constant of the motion. We have

$$\langle L_z \rangle = \langle \mathcal{L}_{1,2} \rangle = |\gamma|^2 - |\alpha|^2 = x_1^o p_{x_2}^o - x_2^o p_{x_1}^o, \tag{215}$$

the initial orbital angular momentum.

5.2. DRIVEN HARMONIC OSCILLATOR

5.2.1. *One Dimension*
For this 1-dimensional system, we have

$$g^{(2)}(\tau) = \tfrac{1}{2}\omega^2, \tag{216}$$

and we assume that $g_1^{(1)}(\tau)$ and $g^{(0)}(\tau)$ are real and nonzero. Furthermore, we set $g_1^{(1)}(0) = g_1^o$. With these constraints, we can write the τ-dependent Hamiltonian as

$$H(\tau) = -\tfrac{1}{2}\partial_{xx} + \tfrac{1}{2}\omega^2 x^2 + g_1^{(1)}(\tau)x + g^{(0)}(\tau). \tag{217}$$

The appropriate differential equation to solve is (175) and its real solutions are given by (176) and the complex solutions by (177). The function, C_1, can be written as in Eq. (35) where $c_1(\tau)$, given by Eq. (36), is

$$c_1(\tau) = \sqrt{\tfrac{1}{2\omega}} \int_0^\tau ds\, g_1^{(1)}(s) e^{i\omega s}. \tag{218}$$

From Eq. (218) and its complex conjugate, all remaining τ-dependent functions can be found.

The two operators $\mathcal{J}_\pm$ are

$$\mathcal{J}_- = \sqrt{\tfrac{1}{2\omega}} e^{i\omega\tau}\left\{\partial_x + \omega x + i\sqrt{2\omega}e^{-i\omega\tau}C_1\right\}, \tag{219}$$

$$\mathcal{J}_+ = -\sqrt{\tfrac{1}{2\omega}} e^{-i\omega\tau}\left\{\partial_x - \omega x + i\sqrt{2\omega}e^{+i\omega\tau}\bar{C}_1\right\}. \tag{220}$$

Now, we obtain a value for the integration constant in Eq. (35). By computing the operator $\mathcal{J}_+\mathcal{J}_- + \tfrac{1}{2}$, which has the form of a Hamiltonian, and taking the limit $\tau \to 0$, and identifying it with $H(0)$, we find the integration constant

$$C^o = -\tfrac{i}{\omega\sqrt{2\omega}} g_1^o. \tag{221}$$

This is the method we have adopted to evaluate C_1^o and its complex conjugate.

From Eqs. (35), (218), and (221) and their complex conjugate, all remaining τ-dependent functions can be found. For example, according to Eq. (46),

$$\mathcal{E}_{3,1}(\tau) = -\sqrt{\tfrac{1}{2\omega}}\left(e^{i\omega\tau}\bar{C}_1 + e^{-i\omega\tau}C_1\right), \quad \mathcal{E}_{3,1}^o = 0, \tag{222}$$

where the second equality is the initial value. Hence, β_1^o, which is given by Eq. (76), is

$$\beta_1^o = 0. \tag{223}$$

We shall leave the function $\mathcal{D}_{3,1}$ in the form defined in Eq. (55).

From Eq. (50), the operator $\mathcal{M}_3$ can be expressed as

$$\mathcal{M}_3 = i\left\{\tfrac{1}{\omega}\partial_\tau + \mathcal{E}_{3,1}\partial_x - ix\dot{\mathcal{E}}_{3,1} + i\mathcal{D}_{3,1} + \tfrac{i}{\omega}g^{(0)}\right\}, \tag{224}$$

where ϕ_3 is given by Eq. (178). The operators (219), (220), and (222) satisfy the commutation relations (58) for an $os(1)$ Lie algebra.

To obtain an expression for the wave function, we shall need several other functions of τ and their initial values. First, from Eq. (72), we find that

$$b_{3,1}(\tau) = i\sqrt{\tfrac{1}{2}}\left[e^{i\omega\tau}\bar{C}_1 - e^{-i\omega\tau}C_1\right], \quad b_{3,1}^o = -\tfrac{1}{\omega^{3/2}}g_1^o, \tag{225}$$

where the second equality is the initial value. The function $\mathcal{B}_{3,1}$ is given by Eq. (71). Eqs. (222) and (74) together with Eq. (186) imply that

$$\Lambda_{3,1}(\eta) = \int_0^\tau ds\,\left(\omega^2\mathcal{E}_{3,1}^2(s) + \omega\mathcal{D}_{3,1}(s)\right), \tag{226}$$

The function $G^{(0)}(\eta)$ is given by Eq. (73).

The $\mathcal{R}$-separable coordinates (68) are

$$\zeta = \omega^{1/2}x - \mathcal{B}_{3,1}(\tau), \quad \eta = \tau, \tag{227}$$

with $\mathcal{R}$-factor

$$\mathcal{R}(x,\tau) = \omega x \mathcal{E}_{3,1}. \tag{228}$$

Therefore, the wave function is

$$\begin{aligned}\Psi_n(x,\tau) &= \left(\tfrac{1}{n!}\right)^{\frac{1}{2}}\left(\tfrac{1}{2}\right)^{\frac{n}{2}}\left(\tfrac{\omega}{\pi}\right)^{\frac{1}{4}}\exp\left[i\omega x \mathcal{E}_{3,1}\right]\\ &\quad H_n(\zeta - b_{3,1}^o)\exp\left[-\zeta^2/2 + b_{3,1}^o\zeta\right]\exp\left[-i\left(n+\tfrac{1}{2}\right)\omega\eta\right]\\ &\quad \times \exp\left[-i\left(\Lambda_{3,1}(\eta) + G^{(0)}(\eta)\right)\right],\end{aligned} \tag{229}$$

and the extremal state is

$$\Psi_0(x,\tau) = \exp\left[i\omega x \mathcal{E}_{3,1}\right] \left(\frac{\omega}{\pi}\right)^{\frac{1}{4}} \exp\left[-\zeta^2/2 + b^o_{3,1}\zeta\right]$$
$$\times \exp\left[-i\omega\eta/2\right] \exp\left[-i\left(\Lambda_{3,1}(\eta) + G^{(0)}(\eta)\right)\right]. \quad (230)$$

Neither of these states are energy eigenstates which is, of course, to be expected.

We shall look briefly at some properties of the coherent state $|\alpha\rangle$. Combining the values for the functions χ_1, χ_2, ξ, and $\bar{\xi}$ in Eqs. (176) and (177) with expressions (147) and (148) for the expectation values, we get

$$\langle x(\tau)\rangle = (x^o + \tfrac{1}{\omega})\cos\omega\tau + \tfrac{1}{\omega}(p^o_x - \sqrt{\omega})\sin\omega\tau$$
$$\langle p_x(\tau)\rangle = (p^o_x - \sqrt{\omega}c_{1,1})\cos\omega\tau - (\omega x^o + \sqrt{\omega}c_{2,1})\sin\omega\tau, \quad (231)$$

where $c_{1,1}$ and $c_{2,1}$ are real functions defined by Eq. (16). It is interesting that the uncertainty product is independent of the interaction $g^{(1)}(\tau)x$, and Eq. (195) still holds.

5.2.2. *Two Dimensions*

For the driven oscillator in two dimensions, the τ-dependent Hamiltonian has the form

$$H = -\tfrac{1}{2}\{\partial_{x_1 x_1} + \partial_{x_2 x_2}\} + \tfrac{1}{2}\omega^2 \mathbf{x}\cdot\mathbf{x} + g_1^{(1)}(\tau)x_1 + g_2^{(1)}(\tau)x_2 + g^{(0)}(\tau), \quad (232)$$

where $g^{(2)}$ is given by Eq. (216) while $g_1^{(1)}$, $g_2^{(1)}$, and $g^{(0)}$ are real functions of τ, with $g_\sigma^{(1)}(0) = g^o_\sigma$, $\sigma = 1, 2$.

The differential equation (175) and its real and complex solutions (176) and (177) still apply. In addition, we shall need ϕ_3 given by Eq. (178). The function C_σ can be found in Eq. (35) with

$$c_\sigma(\tau) = \sqrt{\tfrac{1}{2\omega}} \int_0^\tau ds\, g_\sigma^{(1)}(s) e^{i\omega s}, \quad (233)$$

and

$$C^o_\sigma = -\tfrac{i}{\omega\sqrt{2\omega}} g^o_\sigma. \quad (234)$$

Therefore, we can write $\mathcal{A}_\pm$ and $\mathcal{C}_\pm$ as follows:

$$\mathcal{A}_- = \sqrt{\tfrac{1}{4\omega}} e^{i\omega\tau} \left[(\partial_{x_1} + i\partial_{x_2}) + \omega(x_1 + ix_2)\right] + i\sqrt{\tfrac{1}{2}}(C_1 + iC_2),$$

$$\mathcal{A}_+ = \sqrt{\tfrac{1}{4\omega}} e^{-i\omega\tau} \left[-(\partial_{x_1} - i\partial_{x_2}) + \omega(x_1 - ix_2)\right] - i\sqrt{\tfrac{1}{2}}(\bar{C}_1 - i\bar{C}_2),$$

$$\mathcal{C}_- = \sqrt{\tfrac{1}{4\omega}} e^{i\omega\tau} \left[(\partial_{x_1} - i\partial_{x_2}) + \omega(x_1 - ix_2)\right] + i\sqrt{\tfrac{1}{2}}(C_1 - iC_2),$$

$$\mathcal{C}_+ = \sqrt{\tfrac{1}{4\omega}} e^{-i\omega\tau} \left[-(\partial_{x_1} + i\partial_{x_2}) + \omega(x_1 + ix_2)\right] - i\sqrt{\tfrac{1}{2}}(\bar{C}_1 + i\bar{C}_2).$$

(235)

The only new functions that we need are $\mathcal{E}_\sigma$, which is defined in Eq. (41),

$$\mathcal{E}_\sigma(\tau) = i\sqrt{\tfrac{1}{2\omega}}\left(e^{i\omega\tau}\bar{C}_\sigma - e^{-i\omega\tau}C_\sigma\right), \tag{236}$$

and $\mathcal{F}_{1,2}$ given by (42). Therefore, we can write the generator $\mathcal{L}_{1,2}$ as in Eq. (44).

For multidimensional analogues of Eqs. (222), (223), (225), and (226), we replace in each of these equations the subscript "1" with a "σ". Thus, the operator, $\mathcal{M}_3$, is

$$\mathcal{M}_3 = i\left\{\tfrac{1}{\omega}\partial_\tau + \sum_{\sigma=1}^{2}\mathcal{E}_{3,\sigma}\partial_{x_\sigma}\right.$$
$$\left. -i\sum_{\sigma=1}^{2} x_\sigma \dot{\mathcal{E}}_{3,\sigma} + i\sum_{\sigma=1}^{2} \mathcal{D}_{3,\sigma} + \tfrac{i}{\omega}g^{(0)}\right\}. \tag{237}$$

The operators $\mathcal{D}_0$ and $\mathcal{F}_0$ can then be constructed from Eqs. (236) and (237), according to Eq. (89).

The $\mathcal{R}_\mathcal{M}$-separable coordinates are

$$\zeta_1 = \omega^{1/2}x_1 - \mathcal{B}_{3,1}, \quad \zeta_2 = \omega^{1/2}x_2 - \mathcal{B}_{3,2}, \quad \eta = \tau, \tag{238}$$

with $\mathcal{R}$-factor

$$\mathcal{R}_\mathcal{M} = \sum_{\sigma=1}^{2}\omega x_\sigma \mathcal{E}_{3,\sigma}. \tag{239}$$

The $\mathcal{R}$-factor, $\mathcal{R}_\mathcal{L} = 0$ and the polar coordinates are defined

$$\zeta_1 - b^o_{3,1} = \rho\cos\theta, \quad \zeta_2 - b^o_{3,2} = \rho\sin\theta. \tag{240}$$

The wave functions are then

$$\Psi_{n,m}(x_1, x_2, \tau) = \left(\frac{\omega}{\pi}\right)^{\frac{1}{2}}\frac{k!}{\sqrt{n!m!}}\exp\left[i\sum_{\sigma=1}^{2}\omega x_\sigma \mathcal{E}_{3,\sigma}\right]$$
$$\times \exp\left[i(m-n)\theta\right]\exp\left(-\rho^2/2\right)$$
$$\times (-)^k \rho^{|n-m|} L_k^{(|n-m|)}(\rho^2)$$
$$\times \exp\left[-i(n+m+1)\omega\eta\right]$$
$$\times \exp\left[-i\left(\sum_{\sigma=1}^{2}\Lambda_{3,\sigma}(\eta) + G^{(0)}(\eta)\right)\right], \tag{241}$$

where k is given by Eq. (135), and the extremal state is

$$\Psi_{0,0}(x_1,x_2,\tau) = \exp\left[i\sum_{\sigma=1}^{2}\omega x_\sigma \mathcal{E}_{3,\sigma}\right]\left(\frac{\omega}{\pi}\right)^{\frac{1}{2}}\exp\left(-\rho^2/2\right)$$
$$\times \exp\left(-i\omega\eta\right)\exp\left[-i\left(\Lambda_{3,1}(\eta)+G^{(0)}(\eta)\right)\right]. \quad (242)$$

Neither of these states are energy eigenstates.

As in the 1-dimesional driven oscillator, the trajectories predicted for the coherent state $|\alpha,\gamma\rangle$ will depend upon the τ-dependent functions $g_\sigma^{(1)}$. Expressions for position and momentum are

$$\langle x_\sigma(\tau)\rangle = \left(x_\sigma^o + \sqrt{\tfrac{1}{\omega}}c_{2,\sigma}\right)\cos\omega\tau + \tfrac{1}{\omega}\left(p_{x_\sigma}^o - \sqrt{\omega}c_{1,\sigma}\right)\sin\omega\tau,$$
$$\langle p_{x_\sigma}(\tau)\rangle = \left(p_{x_\sigma}^o - \sqrt{\omega}\right)\cos\omega\tau - \left(\omega x_\sigma^o + \sqrt{\omega}c_{2,\sigma}\right)\sin\omega\tau, \quad (243)$$

where $c_{1,\sigma}$ and $c_{2,\sigma}$ are given by Eq. (16).

The uncertainty relation is given by Eq. (214) and is independent of the linear contributions to the potential. This was also the situation in the 1-dimensional driven oscillator.

The orbital angular momentum, L_z, is not a constant of the motion in the coherent state and so its expectation value will vary with time according to Eq. (171),

$$\langle L_z(\tau)\rangle = x_1^o p_{x_2}^o - x_2^o p_{x_1}^o + \sqrt{\tfrac{1}{\omega}}\left(c_{2,1}p_{x_2}^o - c_{2,2}p_{x_1}^o\right)$$
$$+\sqrt{\omega}\left(c_{1,1}x_2^o - c_{1,2}x_1^o\right) + c_{1,1}c_{2,2} - c_{2,1}c_{1,2}, \quad (244)$$

in the coherent state $|\alpha\rangle$. The real functions, $c_{\alpha,\sigma}(\tau)$, are defined in Eq. (16).

6. Transformation of the Schrödinger Equation

In this section, we shall show how to incorporate new terms into the Schrödinger operator. In our treatment, we will work with specific equations rather than investigate the general problem. In our first example, we introduce a velocity-dependent potential into a 1-dimensional Schrödinger equation with a suitable time-dependent transformation. In the second example, we shall work with a Schrödinger equation for a spinless particle of charge e and mass M interacting with uniform, axially symmetric magnetic field. This is a modification of the problem treated in Ref. [7]. In our example, here, we include a nonzero scalar field.

LIE SYMMETRIES IN QUANTUM MECHANICS

Basically, the general idea is a simple one. Suppose that we have a time-dependent Schrödinger equation of the type

$$S_N \Phi(x, \tau) = 0, \tag{245}$$

where we leave the specific nature of the Schrödinger operator undefined at this point. Is there a operator T that will transform Eq. (245) into Eq.(1) according to

$$T S_N T^{-1} T \Phi = 0, \tag{246}$$

where we identify

$$\mathcal{S}_N = T S_N T^{-1}, \quad \Psi = T\Phi. \tag{247}$$

We are particularly interested in those Schrödinger operators $\mathcal{S}_N$ that can be related to S_N by a transformation of the form

$$T = \exp(i\varphi(\tau)K), \tag{248}$$

where K is a first order differential operator of the form

$$K = i \sum_{\sigma=1}^{N} Q_\sigma(\mathbf{x}, \tau) \partial_\sigma + Q(\mathbf{x}, \tau), \tag{249}$$

and $\varphi(\tau)$ is some τ-dependent function. The operator K is assumed to be Hermitian. If that is the case, then T is unitary. Now, if such a transformation T exists, then the solution to Eq. (245) can be obtained since the solution to Eq. (1) is known when V is given by Eq. (5). Naturally, this places restrictions on the form of the interaction in the Eq. (245).

One important mathematical equation that is central to our analysis is

$$\begin{aligned} e^{(i\varphi(\tau)K)} S_N e^{(-i\varphi(\tau)K)} &= S_N + [i\varphi(\tau)K, S_N] \\ &\quad + \tfrac{1}{2!}[i\varphi(\tau)K, [i\varphi(\tau)K, S_N]] + \cdots, \\ &= \mathcal{S}_N. \end{aligned} \tag{250}$$

For the first equality in Eq. (250), see the comment in Ref. [29].

From Eq. (4), performing the inverse transformation, we see that

$$[T^{-1}\mathcal{S}_N T, T^{-1}\mathcal{L}T] = [S_n, L] = T^{-1}\lambda T T^{-1}\mathcal{S}_N T = \Lambda S_N, \tag{251}$$

where we define the generators of space-time symmetries of Eq. (245) in terms of the generators, $\mathcal{L}$ of Eq. (1) by

$$L = T^{-1}\mathcal{L}T. \tag{252}$$

Since the transformation T is unitary, the commutation relations for the generators $\mathcal{L}$ are preserved. This means that the Lie algebra, $(SA)_N^c$, formed by the generators L, is isomorphic to the Lie algebra, $(\mathcal{SA})_N^c$, formed by the generators $\mathcal{L}$. The same comment is true for the oscillator subalgebras. This is quite important since, in one dimension, the set of states $\Phi_n = T^{-1}\Psi_n$, for n a nonnegative integer, span the representation space of the irreducible representation, $\uparrow_{-1/2}$, of the oscillator subalgebra of $(SA)_1^c$ as do the states Ψ_n for the oscillator subalgebra of $(\mathcal{SA})_1^c$. A similar comment applies to the case of two dimensions.

For 1-dimensional systems, the set of states $\{\Phi_n : n \in \mathbf{Z}^+\}$ do form a discrete, complete set of orthonormal states. They are not energy eigenstates, but they are very useful for computational purposes. For example, we can construct coherent and squeezed states from the extremal state, Φ_0, and the displacement operators formed from the generators of space-time symmetries. The same will be true for 2-dimensional systems. An investigation of the properties of coherent and squeezed states of each model systems discussed in this section are in progress and will be published elsewhere [30].

We calculate the symmetries and wave functions for the two systems mentioned in the first paragraph of this section.

6.1. A VELOCITY DEPENDENT POTENTIAL

We take as our starting point the equation [31]

$$S_1 \Phi(x,\tau) = \left\{\partial_{xx} - ig\partial_x + 2i\partial_\tau - \omega^2 x^2\right\} \Phi(x,\tau), \qquad (253)$$

where g and ω^2 are constants. To transform this equation to Eq. (1) with S_1 given by Eq.(2), we choose the operator K to be

$$K = -i\partial_x. \qquad (254)$$

Setting up Eq. (250) and computing the result we obtain

$$\begin{aligned} S_1 &= \exp(i\varphi(\tau)K) S_1 \exp(-i\varphi(\tau)K) \\ &= \partial_{xx} - i(2\dot\varphi + g)\partial_x + 2i\partial_\tau \\ &\quad -\omega^2 x^2 - 2\omega^2 \varphi x - \omega^2 \varphi^2. \end{aligned} \qquad (255)$$

We choose φ such that

$$2\dot\varphi + g = 0. \qquad (256)$$

Integrating this equation, we obtain

$$\varphi(\tau) = -\tfrac{1}{2}g\tau, \qquad (257)$$

where we have set the integration constant to be zero. The operator S_1 becomes

$$S_1 = \partial_{xx} + 2i\partial_\tau - \omega^2 x^2 + \omega^2 g\tau x - \tfrac{1}{4}\omega^2 g^2 \tau^2, \qquad (258)$$

and the corresponding Schrödinger equation is Eq. (1).

Now, identify

$$g^{(2)}(\tau) = \tfrac{1}{2}\omega^2, \quad g_1^{(1)}(\tau) = -\tfrac{1}{2}\omega^2 g\tau, \quad g^{(0)}(\tau) = \tfrac{1}{8}\omega^2 g^2 \tau^2, \qquad (259)$$

and refering to Section 5.2.1, compute the τ-dependent functions,

$$c_1(\tau) = \tfrac{1}{2}g\sqrt{\tfrac{1}{2\omega}}\left[1 - (1 - i\omega\tau)e^{i\omega\tau}\right], \quad C_1^o = -\tfrac{g}{2}\sqrt{\tfrac{1}{2\omega}}, \qquad (260)$$

$$\mathcal{E}_{3,1}(\tau) = \mathcal{E}_{3,1}^o = \tfrac{g}{2\omega}, \quad \beta_1^o = \tfrac{g}{2}\sqrt{\tfrac{1}{\omega}}, \qquad (261)$$

$$\mathcal{D}_{3,1} = -\tfrac{g^2}{8\omega}\left(1 + \omega^2\tau^2\right). \qquad (262)$$

With these values, the symmetries associated with Eq. (258) are then given by the symmetries for the driven oscillator, Eqs. (219), (220), and (224).

Solutions, Ψ_n, to the Schrödinger equation (1) with Schrödinger operator (258) can be obtained directly from Eq. (229) with the substitution of Eqs. (260) through (262) and the following functions:

$$\mathcal{B}_{3,1}(\tau) = b_{3,1}(\tau) = \sqrt{\tfrac{\omega}{4}}g\tau, \quad b_{3,1}^o = 0, \qquad (263)$$

$$G^{(0)}(\eta) = \tfrac{1}{24}\omega^2 g^2 \eta^3, \qquad (264)$$

$$\Lambda_{3,1}(\eta) = -\tfrac{1}{24}g^2\omega^2 - \tfrac{1}{8}g^2\eta, \qquad (265)$$

and Eq. (186). Eq. (230) gives the corresponding extremal state, Ψ_0.

The solutions, Φ_n, to Eq. (253) can be obtained from

$$\Phi_n = \exp\left(-\tfrac{i}{2}g\tau K\right)\Psi_n, \qquad (266)$$

for $n \in \mathbf{Z}^+$. Each Φ_n is normalized since the transformation, $\mathcal{T}$, is unitary. The symmetry generators associated with Eq. (253) are

$$M_3 = \mathcal{T}^{-1}M_3\mathcal{T}, \quad J_- = \mathcal{T}^{-1}J_-\mathcal{T}, \quad J_+ = \mathcal{T}^{-1}J_+\mathcal{T}, \qquad (267)$$

and these operators along with the unit operator, I, form an $os(1)$ Lie algebra. The wave functions, Φ_n, satisfy the equations

$$\mathbf{C}\Phi_n = -\tfrac{1}{2}\Phi_n, \quad M_3\Phi_n = \left(n + \tfrac{1}{2}\right)\Phi_n, \qquad (268)$$

$$J_-\Phi_n = \sqrt{n}\Phi_{n-1}, \quad J_+\Phi_n = \sqrt{n+1}\Phi_{n+1}, \quad I\Phi_n = \Phi_n, \qquad (269)$$

where $\mathbf{C} = J_+J_- - M_3 I$, simply because of transformation (267) and Eqs. (62) and (63). Therefore, the eigenstates of M_3 span the irreducible representation space, $\uparrow_{-1/2}$, of $os(1)$. Also, we have

$$J_-\Phi_0 = 0. \qquad (270)$$

We do not need the specific from of the wave functions, Φ_n, nor do we require the operators M_3; these can be left for the interested reader to compute. However, we will give the operators $J_\pm$, as examples. We have

$$\begin{aligned}
J_- &= T^{-1} J_- T, \\
&= \sqrt{\tfrac{1}{2w}} e^{iw\tau} \left[\partial_x + w \left(x + \tfrac{1}{2} g\tau \right) + i\sqrt{2w} e^{-iw\tau} C_1 \right], \\
J_+ &= T^{-1} J_+ T, \\
&= -\sqrt{\tfrac{1}{2w}} e^{-iw\tau} \left[\partial_x - w \left(x + \tfrac{1}{2} g\tau \right) + i\sqrt{2w} e^{iw\tau} \bar{C}_1 \right],
\end{aligned} \tag{271}$$

6.2. SPINLESS PARTICLE IN AN ELECTROMAGNETIC FIELD

The time-dependent Schrödinger equation for a particle of charge e and mass m in an electromagnetic field is given by [1, 7, 32]

$$\left\{ \tfrac{1}{2m} [\mathbf{p} - e\mathbf{A}(\mathbf{x}, t)]^2 + \phi'(\mathbf{x}, t) \right\} \Phi = i\hbar \partial_t \Phi. \tag{272}$$

Leaving the scalar potential, $\phi'(\mathbf{x}, t)$, for the moment, suppose that the scalar potential, $\mathbf{A}(\mathbf{x}, t)$, describes a uniform, time-independent magnetic induction, $\mathbf{B}$. Choosing the cylindrical gauge

$$A_1 = -\tfrac{1}{2} B x_2, \quad A_2 = \tfrac{1}{2} B x_1, \tag{273}$$

where B is the magnitude of the induction $\mathbf{B}$, and supposing that the momentum in the x_3 direction is zero [33] and ϕ' is independent of x_3, we obtain the Schrödinger equation

$$S_2 \Phi = 0, \tag{274}$$

where

$$\begin{aligned}
S_2 &= \sum_{\sigma=1}^{2} \partial_{x_\sigma x_\sigma} + iw(x_2 \partial_{x_1} - x_1 \partial_{x_2}) + 2i\partial_\tau \\
&\quad - \tfrac{1}{4} w^2 \mathbf{x} \cdot \mathbf{x} - 2\phi(\mathbf{x}, t).
\end{aligned} \tag{275}$$

In Eq. (275), we have set

$$\hbar = 1, \quad \tau = t/m, \quad w = eB, \quad \phi = m\phi'. \tag{276}$$

Furthermore, let us suppose that the scalar potential has the form

$$\phi = f^{(2)}(\tau) \mathbf{x} \cdot \mathbf{x} + \sum_{\sigma=1}^{2} f^{(1)}_\sigma x_\sigma + f^{(0)}(\tau). \tag{277}$$

Then, the Schrödinger operator in Eq. (275) becomes

$$S_2 = \sum_{\sigma=1}^{2} \partial_{x_\sigma x_\sigma} + iw(x_2 \partial_{x_1} - x_1 \partial_{x_2}) + 2i\partial_\tau$$
$$-2g^{(2)}(\tau)\mathbf{x}\cdot\mathbf{x} - 2\sum_{\sigma=1}^{2} f_\sigma^{(1)} x_\sigma - 2g^{(0)}(\tau), \qquad (278)$$

where
$$2g^{(2)}(\tau) = \tfrac{1}{4}w^2 + 2f^{(2)}(\tau), \quad g^{(0)} = f^{(0)}(\tau). \qquad (279)$$

To define the transformation (248) for this situation, we take the operator K to be
$$K = L_z = i(x_2 \partial_{x_1} - x_1 \partial_{x_2}). \qquad (280)$$

Performing the transformation acoording to Eq. (250) yields

$$e^{i\varphi K} S_2 e^{-i\varphi K} = \sum_{\sigma=1}^{2} \partial_{x_\sigma s_\sigma} + (w + 2\dot\phi)K + 2i\partial_\tau$$
$$-2g^{(2)}(\tau)\mathbf{x}\cdot\mathbf{x} - 2\sum_{\sigma=1}^{2} g_\sigma^{(1)}(\tau)x_\sigma - 2g^{(0)}(\tau), \quad (281)$$

where $g_\sigma^{(1)}$ is defined to be

$$g_1^{(1)}(\tau) = f_1^{(1)}(\tau)\cos\varphi + f_2^{(1)}(\tau)\sin\varphi,$$
$$g_2^{(1)}(\tau) = -f_1^{(1)}(\tau)\sin\varphi + f_2^{(1)}(\tau)\cos\varphi. \qquad (282)$$

If we set the coefficient of K to zero in Eq. (281), that is, if we take

$$\phi(\tau) = -\tfrac{1}{2}w\tau, \qquad (283)$$

then, we obtain the Schrödinger operator

$$S_2 = \sum_{\sigma=1}^{2} \partial_{x_\sigma s_\sigma} + 2i\partial_\tau - 2g^{(2)}(\tau)\mathbf{x}\cdot\mathbf{x} - 2\sum_{\sigma=1}^{2} g_\sigma^{(1)}(\tau)x_\sigma - 2g^{(0)}(\tau), \quad (284)$$

where now

$$g_1^{(1)}(\tau) = f_1^{(1)}(\tau)\cos\tfrac{1}{2}w\tau - f_2^{(1)}(\tau)\sin\tfrac{1}{2}w\tau,$$
$$g_2^{(1)}(\tau) = f_1^{(1)}(\tau)\sin\tfrac{1}{2}w\tau + f_2^{(1)}(\tau)\cos\tfrac{1}{2}w\tau. \qquad (285)$$

with $g_\sigma^o = f_\sigma^o$ and the Schrödinger equation is now identical to Eq. (1).

We can solve this equation by the methods outlined in Section 3.2 without specific knowledge of the τ-dependent functions, $g^{(2)}$, $g_\sigma^{(1)}$, or $g^{(0)}$. However, we shall develop results for the more specific case in which $g^{(2)}$ is a constant

$$g^{(2)}(\tau) = \tfrac{1}{2}\omega^2. \tag{286}$$

This means, of course, that $f^{(2)}$ is a constant and $f^{(2)} = \omega^2 - (1/4)w^2$. Now, the results of Section 5.2.2 for the 2-dimensional driven oscillator apply. The symmetries are given by Eqs. (235), (44), and (237). Eq. (241) is the wave function for an arbitrary state, while Eq. (242) is the extremal state wave function for the Schrödinger equation (1) with Schrödinger operator (284).

The solution, $\Phi_{n,m}$, to Eq. (274) can be obtained from Eq. (241) by the transformation

$$\Phi_{n,m} = e^{-i\varphi L_z} \Psi_{n,m}, \tag{287}$$

with φ given by Eq. (283). Because of Eqs. (100) through (104) and (287), the functions $\Phi_{n,m}$ satisfy

$$\mathbf{C}_1 \Phi_{n,m} = -\tfrac{1}{2}\Phi_{n,m}, \quad \mathbf{C}_2 \Phi_{n,m} = -\tfrac{1}{2}\Phi_{n,m}, \tag{288}$$

$$f_0 \Phi_{n,m} = \left(n + \tfrac{1}{2}\right)\Phi_{n,m}, \quad d_0 \Phi_{n,m} = \left(m + \tfrac{1}{2}\right)\Phi_{n,m}, \tag{289}$$

$$a_- \Phi_{n,m} = \sqrt{n}\,\Phi_{n-1,m}, \quad c_- \Phi_{n,m} = \sqrt{m}\,\Phi_{n,m-1}, \tag{290}$$

$$a_+ \Phi_{n,m} = \sqrt{n+1}\,\Phi_{n+1,m}, \quad c_+ \Phi_{n,m} = \sqrt{m+1}\,\Phi_{n,m+1}, \tag{291}$$

$$I \Phi_{n,m} = \Phi_{n,m}, \tag{292}$$

where $n, m \in \mathbf{Z}^+$. The operators in this set of equations are the transformed operators which are the symmetries of Eq. 274. We have

$$M_3 = e^{-i\varphi K} \mathcal{M}_3 e^{i\varphi K}, \quad L_{1,2} = e^{-i\varphi K} \mathcal{L}_{1,2} e^{i\varphi K}, \tag{293}$$

and we define

$$d_0 = \tfrac{1}{2}(M_3 + L_{1,2}), \quad f_0 = \tfrac{1}{2}(M_3 - L_{1,2}). \tag{294}$$

We give an explicit construction for the operators $a_\pm$ and $c_\pm$. They are

$$\begin{aligned}
a_- &= e^{-i\varphi K} \mathcal{A}_- e^{i\varphi K}, \\
&= \sqrt{\tfrac{1}{4\omega}}\, e^{i\omega\tau} e^{-i\varphi} \left[(\partial_{x_1} + i\partial_{x_2}) + \omega(x_1 + ix_2)\right] + i\sqrt{\tfrac{1}{2}}(C_1 + iC_2), \\
a_+ &= e^{-i\varphi K} \mathcal{A}_+ e^{i\varphi K}, \\
&= \sqrt{\tfrac{1}{4\omega}}\, e^{-i\omega\tau} e^{i\varphi} \left[-(\partial_{x_1} - i\partial_{x_2}) + \omega(x_1 - ix_2)\right] - i\sqrt{\tfrac{1}{2}}(\bar{C}_1 - i\bar{C}_2),
\end{aligned} \tag{295}$$

and

$$\begin{aligned}
c_- &= e^{-i\varphi K} C_- e^{i\varphi K}, \\
&= \sqrt{\tfrac{1}{4\omega}} e^{i\omega\tau} e^{i\varphi} \left[(\partial_{x_1} - i\partial_{x_2}) + \omega(x_1 - ix_2)\right] + i\sqrt{\tfrac{1}{2}}(C_1 - iC_2), \\
c_+ &= e^{-i\varphi K} C_+ e^{i\varphi K}, \\
&= \sqrt{\tfrac{1}{4\omega}} e^{-i\omega\tau} e^{-i\varphi} \left[-(\partial_{x_1} + i\partial_{x_2}) + \omega(x_1 + ix_2)\right] - i\sqrt{\tfrac{1}{2}}(\bar{C}_1 + i\bar{C}_2).
\end{aligned} \quad (296)$$

The above operators satisfy the commutation relations in Eqs. (94) through (97) for an $os(2) \times so(2)$ Lie algebra. The Casimir operators are

$$\begin{aligned}
C_1 &= a_+ a_- - f_0 I = -\tfrac{1}{4} S_2 - \tfrac{1}{2}, \\
C_2 &= c_+ c_- - d_0 I = -\tfrac{1}{4} S_2 - \tfrac{1}{2}.
\end{aligned} \quad (297)$$

Also, for the extremal state, we have the conditions

$$a_- \Phi_{0,0} = c_- \Phi_{0,0} = 0. \quad (298)$$

Acknowledgement

The author wishes to gratefully acknowledge the financial support of the Natural Sciences and Engineering Research Council of Canada.

Appendix I

In Section 3.2, Eq. (134), we gave the form of the wave functionfor a 2-dimensional τ-dependent oscillator. To derive the radial part of the wave function, we start by combining Eqs. (116), (131) and (133). In Eq. (133), substitute Eq. (116) for $\Psi_{n,m}$ and Eq. (131) for $\Psi_{0,0}$ and solve for $R_{n,m}(\rho)$ and noting that the operators A_+ and C_+ have no time derivatives

$$R_{n,m}(\rho) = \left(\frac{1}{\pi \phi_3^o}\right)^{\frac{1}{2}} \left(\frac{1}{n!m!}\right)^{\frac{1}{2}} \left(\frac{\bar{\xi}^o \xi}{\xi^o \bar{\xi}}\right)^{\frac{1}{2}(m+n)}$$
$$\times \exp\left[-i(m-n)\theta\right] A_+^n C_+^m \exp(-\kappa \rho^2 / 2), \quad (299)$$

where $\kappa = 1 - i\alpha^o$ and A_+ and C_+ are given in Eqs. (125) and (127), respectively. Recognizing that the operators, A_+ and C_+, have no time derivatives and defining,

$$a^\dagger = \tfrac{1}{2} e^{-i\theta} \left[-\partial_\rho + \frac{i}{\rho} \partial_\theta + (1 + i\alpha^o)\rho\right], \quad (300)$$

$$c^\dagger = \tfrac{1}{2} e^{i\theta} \left[-\partial_\rho - \frac{i}{\rho} \partial_\theta + (1 + i\alpha^o)\rho\right], \quad (301)$$

we can write

$$R_{n,m}(\rho) = Q e^{-i(m-n)\theta} \left(a^\dagger\right)^n \left(c^\dagger\right)^m e^{-\kappa\rho^2/2}. \tag{302}$$

The symbol Q is a constant and

$$Q = \left(\frac{1}{\pi\phi_3^o}\right)^{\frac{1}{2}} \left(\frac{1}{n!m!}\right)^{\frac{1}{2}} \left(\frac{\bar{\xi}^o}{\xi^o}\right)^{\frac{1}{2}(m+n)}. \tag{303}$$

Now, we shall make a few definitions and cite a few results which simplify calculations. First, it is straightforward to show that

$$[a^\dagger, c^\dagger] = 0. \tag{304}$$

Next, define the three operators

$$X_k = -\partial_\rho + \kappa\rho + \frac{k}{\rho}, \tag{305}$$

and

$$\mathcal{X}_k = X_k + \frac{i}{\rho}\partial_\theta, \quad \mathcal{Y}_k = X_k - \frac{i}{\rho}\partial_\theta. \tag{306}$$

Then, we have

$$a^\dagger = \tfrac{1}{2}e^{-i\theta}\mathcal{X}_0, \quad c^\dagger = \tfrac{1}{2}e^{-i\theta}\mathcal{Y}_0. \tag{307}$$

We leave it to the reader to prove the following two useful relationships

$$\mathcal{X}_l e^{-ik\theta} = e^{-ik\theta}\mathcal{X}_{k+l}, \quad \mathcal{Y}_l e^{ik\theta} = e^{ik\theta}\mathcal{Y}_{k+l}. \tag{308}$$

Using Eq. (308), we can quickly show that

$$e^{-ik\theta}\mathcal{X}_l e^{ik\theta} = \mathcal{X}_{l-k}, \quad e^{ik\theta}\mathcal{Y}_l e^{-ik\theta} = \mathcal{Y}_{l-k}. \tag{309}$$

Starting with Eq. (307) and repeatedly applying Eq. (308), we obtain the two results

$$\left(a^\dagger\right)^k = \tfrac{1}{2^k}e^{-ik\theta}\mathcal{X}_{k-1}\mathcal{X}_{k-2}\cdots\mathcal{X}_1\mathcal{X}_0, \tag{310}$$

and

$$\left(c^\dagger\right)^k = \tfrac{1}{2^k}e^{ik\theta}\mathcal{Y}_{k-1}\mathcal{Y}_{k-2}\cdots\mathcal{Y}_1\mathcal{Y}_0. \tag{311}$$

Furthermore, since

$$\begin{aligned}
\mathcal{X}_l \rho^k e^{-\kappa\rho^2/2} &= \mathcal{Y}_l \rho^k e^{-\kappa\rho^2/2} \\
&= X_l e^{-\kappa\rho^2/2} \\
&= \left[2\kappa\rho^{k+1} + (l-k)\rho^{k-1}\right]e^{-\kappa\rho^2/2},
\end{aligned} \tag{312}$$

we have

$$\left(a^\dagger\right)^k e^{-\kappa\rho^2/2} = \tfrac{1}{2^k} e^{-ik\theta} \mathcal{X}_{k-1}\mathcal{X}_{k-2}\cdots\mathcal{X}_1\mathcal{X}_0 e^{-\kappa\rho^2/2},$$
$$= \kappa^k e^{-ik\theta} \rho^k e^{-\kappa\rho^2/2}, \quad (313)$$

and

$$\left(c^\dagger\right)^k e^{-\kappa\rho^2/2} = \tfrac{1}{2^k} e^{-ik\theta} \mathcal{Y}_{k-1}\mathcal{Y}_{k-2}\cdots\mathcal{Y}_1\mathcal{Y}_0 e^{-\kappa\rho^2/2},$$
$$= \kappa^k e^{ik\theta} \rho^k e^{-\kappa\rho^2/2}. \quad (314)$$

One final relationship that we need is

$$-\partial_\rho + \kappa\rho - \frac{\beta}{\rho} = -\rho^{-\beta} e^{\kappa\rho^2/2} \partial_\rho \rho^\beta e^{-\kappa\rho^2/2}, \quad (315)$$

where β is an integer.

Suppose that $m > n$. Then, with the help of Eq. (314), Eq. (302) becomes

$$R_{n,m}(\rho) = Q\kappa^m e^{-i(m-n)\theta} \left(a^\dagger\right)^n e^{im\theta} \rho^m e^{-\kappa\rho^2/2}, \quad (316)$$

which can be easily rearranged to give

$$R_{n,m}(\rho) = Q\kappa^m e^{in\theta} \left(e^{-im\theta} a^\dagger e^{im\theta}\right)^n \rho^m e^{-\kappa\rho^2/2}, \quad (317)$$

Combining Eqs. (307) and (317), we see that

$$R_{n,m}(\rho) = Q\kappa^m \tfrac{1}{2^n} e^{in\theta} \left(e^{-i\theta} e^{-im\theta} \mathcal{X}_0 e^{im\theta}\right)^n \rho^m e^{-\kappa\rho^2/2},$$
$$= Q\kappa^m \tfrac{1}{2^n} e^{in\theta} \left(e^{-i\theta} \mathcal{X}_{-m}\right)^n \rho^m e^{-\kappa\rho^2/2}, \quad (318)$$

where we have used Eq. (309) to obtain the second line of Eq. (318). According to Eq. (308), Eq. (318) becomes

$$R_{n,m}(\rho) = Q\kappa^m \tfrac{1}{2^n} e^{in\theta} \left(e^{-in\theta} \mathcal{X}_{n-m-1}\mathcal{X}_{n-m-2}\cdots\mathcal{X}_{1-m}\mathcal{X}_{-m}\right) \rho^m e^{-\kappa\rho^2/2},$$
$$= Q\kappa^m \tfrac{1}{2^n} \mathcal{X}_{n-m-1}\mathcal{X}_{n-m-2}\cdots\mathcal{X}_{1-m}\mathcal{X}_{-m} \rho^m e^{-\kappa\rho^2/2}, \quad (319)$$

Since the argument upon which $\mathcal{X}_{-m}$ operates is independent of the angle θ, Eq. (319) reduces to

$$R_{n,m}(\rho) = Q\kappa^m \tfrac{1}{2^n} X_{n-m-1} X_{n-m-2}\cdots X_{1-m} X_{-m} \rho^m e^{-\kappa\rho^2/2},$$
$$= Q\kappa^m \tfrac{1}{2^n} \left(-\partial_\rho + \kappa\rho - \frac{m-n+1}{\rho}\right) \left(-\partial_\rho + \kappa\rho - \frac{m-n+2}{\rho}\right)$$
$$\cdots \left(-\partial_\rho + \kappa\rho - \frac{m-1}{\rho}\right) \left(-\partial_\rho + \kappa\rho - \frac{m}{\rho}\right) \rho^m e^{-\kappa\rho^2/2}, \quad (320)$$

where we have used the definition (305) to obtain the second equality in Eq. (320). Noting the operator identity in Eq. (315), we can write Eq. (320) as

$$
\begin{aligned}
R_{n,m}(\rho) = {} & Q\kappa^m \tfrac{1}{2^n}(-)^n \left(\rho^{-(m-n+1)} e^{\kappa\rho^2/2} \partial_\rho \rho^{(m-n+1)} e^{-\kappa\rho^2/2}\right) \\
& \times \left(\rho^{-(m-n+2)} e^{\kappa\rho^2/2} \partial_\rho \rho^{(m-n+2)} e^{-\kappa\rho^2/2}\right) \\
& \times \quad \cdots\cdots\cdots\cdots\cdots \\
& \times \left(\rho^{-(m-1)} e^{\kappa\rho^2/2} \partial_\rho \rho^{(m-1)} e^{-\kappa\rho^2/2}\right) \\
& \times \left(\rho^{-m} e^{\kappa\rho^2/2} \partial_\rho \rho^m e^{-\kappa\rho^2/2}\right) \rho^m e^{-\kappa\rho^2/2},
\end{aligned}
\qquad (321)
$$

which becomes

$$
\begin{aligned}
R_{n,m}(\rho) &= Q\kappa^m \tfrac{1}{2^n}(-)^n \rho^{-(m-n)} e^{\kappa\rho^2/2} \left(\frac{1}{\rho}\frac{d}{d\rho}\right)^n \left[\rho^{2m} e^{-\kappa\rho^2}\right], \\
&= Q\kappa^m (-)^n n! \rho^{(m-n)} e^{-\kappa\rho^2/2} \\
&\quad \times \left\{\frac{1}{2^n n!}\rho^{-2(m-n)} e^{\kappa\rho^2} \left(\frac{1}{\rho}\frac{d}{d\rho}\right)^n \left[\rho^{2m} e^{-\kappa\rho^2}\right]\right\}.
\end{aligned}
\qquad (322)
$$

The Rodrigues formula for generalized Laguerre polynomials is [21]

$$
L_p^{(\alpha)}(z) = \tfrac{1}{p!} z^{-\alpha} e^z \frac{d^n}{dz^n}\left[e^{-z} z^{n+\alpha}\right]. \qquad (323)
$$

If we set $z = \kappa\rho^2$, then we get

$$
L_p^{(\alpha)}(\kappa\rho^2) = \frac{1}{2^p p!}\rho^{-2\alpha} e^{\kappa\rho^2} \left(\frac{1}{\rho}\frac{d}{d\rho}\right)^p \left[\rho^{2(n+\alpha)} e^{-\kappa\rho^2}\right]. \qquad (324)
$$

If we set $\alpha = m - n$ and $p = n$, and substitute the resulting equation into Eq. (322), we find the result

$$
R_{n,m}(\rho) = Q\kappa^m (-)^n n! \rho^{(m-n)} e^{-\kappa\rho^2/2} L_n^{(m-n)}(\kappa\rho^2). \qquad (325)
$$

Now, suppose $n > m$. Then, Eq. (302) becomes

$$
R_{n,m}(\rho) = Q e^{-i(m-n)\theta} \left(c^\dagger\right)^n \left(a^\dagger\right)^m e^{-\kappa\rho^2/2}, \qquad (326)
$$

where we have taken Eq. (304) into account. Now the analysis follows the above procedure closely. Eq. (314) and Eq. (326) together imply that

$$
R_{n,m}(\rho) = Q\kappa^n e^{-im\theta} \left(e^{in\theta} c^\dagger e^{-in\theta}\right)^m \rho^n e^{-\kappa\rho^2/2}, \qquad (327)
$$

which becomes upon substitution of Eq. (307),

$$R_{n,m}(\rho) = Q\kappa^n \tfrac{1}{2m} e^{-im\theta} \left(e^{i\theta} e^{in\theta} \mathcal{Y}_0 e^{-in\theta}\right)^m \rho^n e^{-\kappa\rho^2/2}. \qquad (328)$$

Substituting Eq. (309) into Eq. (328) yields

$$R_{n,m}(\rho) = Q\kappa^n \tfrac{1}{2m} e^{-im\theta} \left(e^{i\theta} \mathcal{Y}_{-n}\right)^m \rho^n e^{-\kappa\rho^2/2}. \qquad (329)$$

Making use of Eq. (308), we find that

$$R_{n,m}(\rho) = Q\kappa^n \tfrac{1}{2m} \mathcal{Y}_{m-n-1} \mathcal{Y}_{m-n-2} \cdots \mathcal{Y}_{1-n} \mathcal{Y}_{-n} \rho^n e^{-\kappa\rho^2/2}, \qquad (330)$$

which leads to the result

$$R_{n,m}(\rho) = Q\kappa^n \tfrac{1}{2m} X_{m-n-1} X_{m-n-2} \cdots X_{1-n} X_{-n} \rho^n e^{-\kappa\rho^2/2}. \qquad (331)$$

Substituting Eq. (305) for X_k and noting Eq. (315), we obtain the following expression for $R_{n,m}$:

$$\begin{aligned} R_{n,m}(\rho) &= Q\kappa^n (-)^m m! \rho^{(n-m)} e^{-\kappa\rho^2/2} \\ &\times \left\{ \frac{1}{2^m m!} \rho^{-2(n-m)} e^{\kappa\rho^2} \left(\frac{1}{\rho}\frac{d}{d\rho}\right)^m \left[\rho^{2n} e^{-\kappa\rho^2}\right] \right\}. \end{aligned} \qquad (332)$$

Combining Eqs. (324) and (332), we see that

$$R_{n,m}(\rho) = Q\kappa^n (-)^m m! \rho^{(n-m)} e^{-\kappa\rho^2/2} L_m^{(n-m)}(\kappa\rho^2). \qquad (333)$$

Both expressions (325) and (333) can be represented by Eq. (134).

References

1. A. Messiah, *Quantum Mechanics*, 3rd edition (Wiley, New York, 1961).
2. P. A. M. Dirac, *The Principles of Quantum Mechanics*, 4th edition (Oxford U. P., London, 1958).
3. C. Cohen-Tannoudji, B. Diu, and F. LaLoë, *Quantum Mechanics*, Vol. I and II. (Wiley, New York, 1977).
4. The usual time-dependent Schrödinger equation is [1]

$$\left\{ \sum_{\sigma=1}^N \frac{\hbar^2}{m_\sigma} \partial_{y_\sigma y_\sigma} - 2U(\mathbf{y},t) \right\} \Phi = -2i\hbar \partial_t \Phi.$$

Let $x_\sigma = y_\sigma \sqrt{m_\sigma}/\hbar$ and $\tau = t/\hbar$. Then, we obtain Eqs. (1) and (2). An alternative definition is given in Section 6.2.
5. W. Miller, Jr., *Symmetry and Separation of Variables*. (Addison-Wesley, Reading, MA, 1977).
6. D. R. Truax, J. Math. Phys. **22**, (1981).

7. V. A. Kostelecký, V. I. Man'ko, M. M. Nieto, and D. R. Truax, Phys. Rev. A **48**, 951 (1993).
8. D. R. Truax, J. Math. Phys. **23**, 43 (1982).
9. S. Gee and D. R. Truax, Phys. Rev. A **29**, 1627 (1984).
10. M. M. Nieto and D. R. Truax, Forschritte der , in press.
11. M. M. Nieto and D. R. Truax, "Displacement-operator squeezed states. I. Time-dependent systems having isomorphic symmetry algebras" J. Math. Phys., submitted.
12. M. M. Nieto and D. R. Truax, "Displacement-operator squeezed states. II. Time-dependent systems having isomorphic symmetry algebras" J. Math. Phys., submitted.
13. There is a rapidly growing literature on the subject of time-dependent invariants. We ask for the indulgence of those authors whose work is not mentioned here. Selected references are: H. R. Lewis, Jr., J. Math. Phys. **9**, 1976 (1968); **25**, 1139 (1984); H. R. Lewis, Jr. and W. B. Riesenfeld, J. Math. Phys. **10**, 1458 (1969). We would refer the reader to the work of V. I. Man'ko and V. V. Dodonov in the book *Invariants and the Evolution of Nonstationary Quantum Systems*, edited by M. A. Markov, (Nova Science, Commack, New York, 1989). There are many references cited in this volume. G. Schrade, V. I. Man'ko, W. P. Schleich, and R. J. Glauber, Quantum Semiclass. Opt. **7**, 307-325 (1995). J. G. Hartley and J. R. Ray, Phys. Rev. A **25**, 2388 (1982); J. G. Hartley and J. R. Ray, Phys. Rev. D **25**, 382 (1982); J. R. Ray, Phys. Rev. A **28**, 2603-2605 (1983). U. Niederer, Helv. Phys. Acta **47**, 167-172 (1974). C. P. Boyer, Helv. Phys. Acta, **47**, 589 (1974). C. P. Boyer, R. T. Sharp, and P. Winternitz, J. Math. Phys. **17**, 1439 (1976). S. Kais and R. D. Levine, Phys. Rev. A **34**, 4615 (1986). R. L. Anderson, S. Kumei, and C. E. Wulfman, Rev. Mex. Fis. bf 21, 1, 35 (1972).
14. A. Kalivoda and D. R. Truax, "Symmetry of the N-dimensional Schrödinger equation," to be published.
15. E. A. Coddington, *An Introduction to Ordinary Differential Equations.* (Prentice-Hall, Englewood Cliffs, N. J., 1961).
16. A. Kalivoda and D. R. Truax, "Lie symmetry and analytically solvable, time-dependent Schrödinger equations in dimension," to be published.
17. W. Miller, Jr., *Symmetry Groups and their Applications.* (Academic, New York, 1972).
18. W. Miller, Jr., *Lie Theory and Special Functions.* (Academic, New York, 1968).
19. E. C. Zachmanoglou and D. W. Thoe, *Introduction to Partial Differential Equations with Applications* (Williams and Wilkens, Baltimore, Md., 1976).
20. In the theory of separation of variables [5], an R-separable solution is written as

$$\Psi(x,\tau) = \exp(i\mathcal{R})\psi(\zeta)\Xi(\eta),$$

where ζ and η are called R-separable coordinates. The function $\mathcal{R}$ is called an R-factor. If $\mathcal{R} = 0$, then we say that we have ordinary separation of variables. Note that $\mathcal{R}$ cannot be written as $f(\zeta) + g(\eta)$, otherwise we would have ordinary separation.
21. W. Magnus, F. Oberhettinger, and R. P. Soni, *Formulas and Theorems for the Special Functions of Mathematical Physics* (Springer, Berlin, 1966).
22. Each $os(1)$ subalgebra has three classes of of irreducible representations [18]. Therefore, there are nine possible classes of irreducible representations for $os(1) \times os(1)$. We have selected only one, $\uparrow_{-1/2} \times uparrow_{-1/2}$ to work with.
23. W. Kaplan, *Advanced Calculus*, 2nd edition (Addison-Wesley, Reading, MA, 1973).
24. E. Schrödinger, Naturwissenschaften **14**, 664 (1926).
25. R. J. Glauber, Phys. Rev. **131**, 2766 (1963).
26. J. R. Klauder and B.-S. Skagerstam, *Coherent States - Applications in Physics and Mathematical Physics* (World Scientific, Singapore, 1985).
27. M. M. Nieto, "Coherent States with Classical Motion; From an Analytical Method Complementary to Group Theory" in *Group Theoretical Methods in Physics*, Pro-

ceedings, Vol. II, of International Seminar at Zvenigorod, 1982, ed. by M. A. Markov (Nauka, Moscow, 1983), p. 174. A copy of this article can also be found in [26].
28. M. M. Nieto and L. M. Simmons, Jr., Phys. Rev. D **20**, 1321, 1332 (1979). There are six papers in this interesting series. The others are M. M. Nieto and L. M. Simmons, Jr., Phys. Rev. D **20**, 1342 (1979); M. M. Nieto, Phys. Rev. D **22**, 391 (1980); V. P. Gutschick and M. M. Nieto, Phys. Rev. D **22**, 403 (1980); M. M. Nieto, L. M. Simmons, Jr., and V. P. Gutschick, Phys. Rev. D **23**, 927 (1981).
29. The theorem that we refer to can be stated as

$$e^A B e^{-A} = B + [A, B] + \tfrac{1}{2!}[A, [A, B]] + \cdots$$

See, for example, Ref. [17], Chapter 5 for a proof. This equation is quite useful and we use it in other places in this chapter.
30. M. M. Nieto and D. R. Truax, to be published.
31. We make the following comment on the Lagrangian $T - V$ where $V = V(x, \dot{x}, \tau)$. If $V = \tfrac{1}{2}\omega^2 x^2 - g'\dot{x}$, then replace $\dot{x}$ with the momentum and so the classical expression becomes $V = \tfrac{1}{2}\omega^2 x^2 + \tfrac{1}{2}g p_x$, where p_x is the classical momentum.
32. See V. I. Man'ko and V. V. Dodonov in *Invariants and the Evolution of Nonstationary Quantum Systems*, edited by M. A. Markov, (Nova Science, Commack, New York, 1989), and references therein.
33. This is a simplification and not a necessary requirement. See Ref. [7].

THE INTERPLAY BETWEEN QUANTUM CHEMISTRY AND MOLECULAR DYNAMICS SIMULATIONS

S. M. KAST, J. BRICKMANN
*Institut für Physikalische Chemie
und Darmstädter Zentrum für wissenschaftliches Rechnen
Technische Hochschule Darmstadt
Petersenstraße 20
D-64287 Darmstadt, Germany*

R. S. BERRY
*Department of Chemistry
and The James Franck Institute
The University of Chicago
5735 South Ellis Avenue
Chicago, IL 60637, U.S.A.*

1. Introduction

The ultimate goal of physical theories is to find the most concise and general set of descriptors and their formal relations in order to explain and predict one or more aspects of physical reality. In chemistry the basic constituting descriptors are nuclei and electrons. They together build up higher orders of formalization: ions, atoms, functional groups, molecules, and so on. Explaining measurable observables in chemistry now involves taking into account the behavior of a huge number of the order 10^{23} of these basic elements over a time scale of the experiment. Statistical mechanics [1,2] is capable of this task in principle. We are then faced with the problems (a) to describe the interaction energies of our descriptors depending on their spatial positions and orientations, and (b) to solve complicated many-dimensional integrals.

Concerning the latter point, since only simplest systems can be solved analytically, simulation techniques come into play, all of which have the common goal of generating a statistically significant sample of configurations of a finite system. Averaging over these samples ideally approaches experimental observables. Purely configurational Monte Carlo (MC) or time-resolved molecular dynamics (MD) simulation methods can be employed to solve this computationally expensive task numerically [3-8]. No less demanding is the first issue: The interaction energy and forces of atomic and molecular systems are determined by the solution of, e.g., the electronic Schrödinger equation in the adiabatic approximation, which is usually not analytically possible. Quantum mechanical electronic structure theory, also termed quantum chemistry, therefore developed a variety of numerical techiques [9,10].

Ideally, results of both parts can be combined directly: Quantum chemistry yields instantaneous potential and forces for a given state which are needed to propagate the system to the next state, and so forth. This is the basis for the computationally demanding *ab initio* MD simulation techniques [11,12]. For systems containing many thousands of atoms, such as those relevant in biology, these first principle methods are not (yet) applicable. One therefore has to separate the task to some extent: Again ideally and by using the adiabatic approximation, the solution of the electronic Schrödinger equation as a function of the nuclear coordinates yields the potential energy surfaces (PES) of a chemical system and can be used directly for MC simulations. This expression can furthermore be differentiated or the Hellmann-Feynman theorem [13,14] is used to yield the forces that enter the molecular dynamics propagation.

In fact, we gain no acceleration of the total process at all if the PES is calculated beforehand for *every* part of configuration space that might possibly be accessible because exactly this is done 'on the fly' in *ab initio* dynamical simulations. A great increase of performance is obtained if the basic and transferable descriptors of the PES are known. This is the area of molecular mechanics (MM) or so-called empirical force fields [15,16]. Over their long history it turned out that the total PES of a chemical system can, to a certain accuracy, be described as a sum over parameterized analytical terms connecting only a fraction of atoms of the whole sample, involving, e.g., bond stretching, angle bending, dihedral angle, improper torsion, and nonbonded distance-dependent terms. These coordinates constitute one example of a set of PES descriptors. If we assume that the set is complete then we can put all our knowledge about certain points of the quantum mechanical PES into a parametrization process and expect that the PES and the forces are sufficiently described even in regions of configurational space which were not originally used for the parametrization. It is this *predictive* character that is responsible for greatly enhanced capabilities of molecular simulations. In this sense, force fields, not just as a speed-up tool for the ultimate simulation of physical reality, have a right in their own because they represent a highly *compressed* representation of the total PES, or, put in different terms, they demonstrate what science is all about: the art of data compression [17]. Compression of data (the total *ab initio* PES) in the form of short relations yields an increase in predictive power and deeper insight into the physics of interaction. Information is wasted, though gained during a simulation, if the system passes regions of configuration space that were sampled earlier. Studying the PES itself for instance of cluster systems reveals the connection between its topography, i.e. the location of minima and connecting saddles, and the dynamics [18,19]. Certain multidimensional surfaces focus systems to glassy, some to crystalline states. Particularly if an analytical form is available one is able to investigate the influence of a parameter variation or different functional forms on distortions and changing topographical features of the PES.

However, force field parametrization is often termed to be rather art than science. This conclusion stems from the apparent difficulties involved in the parametrization process: Are there enough and well chosen reference observables, is the functional form adequate, is the level of quantum chemical theory sufficient, are the results compatible with experiment, which parameters are transferable - these questions demonstrate that a high degree of physical intuition and trial-and-error procedures are needed together with common fitting techniques to accomplish the task [20]. Intuition itself can be viewed as the output of a well trained human neural network: Information about chemically similar structures is gathered, processed, compressed, and finally put into suggestions for transferable sets of functional terms and parameters. Great efforts have been and are made towards developing more reliable and more general force fields which combine quantum chemical and/or experimental data in the parametrization process, and which are supposed to be applicable to a wide range of compounds and complex chemical systems [21-25].

This method is tedious but works reasonably well for nonreactive processes, whereas the whole concept of molecular mechanical force fields seems to break down for reactions since we have to transfer functional terms and/or parameters in a certain manner from the reactant to the product state. Some attempts have been made towards extending common force fields to reacting systems [26,27], but a general procedure is yet unclear. It is therefore reasonable to leave unchanged those parts of a force field which belong to groups not involved in the reaction process directly, and to treat the reactive center "quantum mechanically", i.e. in an MD simulation to calculate the instantaneous energy and forces of the reaction region using quantum chemical methods, and classical force fields for the rest of the system. This is the area of "hybrid quantum mechanical/molecular mechanical" (QM/MM) simulations [28,29].

From this introduction it seems desirable to have at hand a straightforward and automated method for deriving potential energy functionals and parameters. We shall therefore consider some aspects about *efficiently using results from* quantum chemistry and molecular dynamics rather than quantum chemistry itself. After a short summary of basic classical MD simulation technique, we will treat in some more detail a recently developed constant temperature MD simulation procedure [30,31]. Constant pressure, *ab initio*, and QM/MM MD methods will be considered briefly. The next major part of this article is concerned with parametrization problems. We will derive an efficient fitting procedure based on the results of constant temperature MD algorithms, and illustrate its value with some examples. We do not intend to give a comprehensive overview of simulations and parameter estimation based on quantum chemical results in general, but we hope to elucidate some novel ideas.

2. Molecular Dynamics Simulations

2.1. PRINCIPLES

Classical MD simulations are based on the numerical solution of the coupled set of Newtonian equations of motion of a (conservative) many-body system

$$m_i \ddot{r}_i(t) = F_i(t) = -\nabla_{r_i} V[r_i(t)] \;, \tag{1}$$

$$p_i(t) = m_i \dot{r}_i(t) = m_i v_i(t) \;. \tag{2}$$

Here, r denotes the cartesian position vector of particle i, F the force acting on it, m its mass, V the potential, v the velocity, and p the momentum. The solution yields the trajectory of a single phase space point representing one subsystem of an ensemble. If the forces can be derived from a potential [as in Eq. (1)] which is not explicitly time dependent, the total energy is a constant of motion. Numerical integration strategies are based on a *discretization* of the continuous time t in time steps of length Δt. For large systems (up to some 10^5 particles) the algorithms have to be highly optimized in order to keep the numerical effort as small as possible. A very special type of differential equation has to be integrated. Consequently, general methods like Runge-Kutta integrators are not widely used in the field of MD simulations.

There are essentially two different integration schemes commonly in use, the (Gear) predictor-corrector method [32] and, most importantly, the Verlet algorithm [33] and variants that are algebraically related to it [34], e.g. "velocity Verlet" [35]. The standard Verlet algorithm reads

$$r_{i,n+1} = 2r_{i,n} - r_{i,n-1} + \frac{\Delta t^2}{m_i} F_{i,n} + O(\Delta t^4) \;, \tag{3}$$

$$v_{i,n} = \frac{r_{i,n+1} - r_{i,n-1}}{2\Delta t} + O(\Delta t^2) \;, \tag{4}$$

n is the step number. It is important to note that the velocity appears *implicitly*. It can be computed only *after* a configuration space propagation step is finished.

With the above algorithm approximately the microcanoncial ensemble (*NVE*, constant number of particles N, volume V, and energy E) is generated, more precisely, the *NVE* ensemble with additional constraints due to momentum conservation. To mimic real systems, it is important to incorporate coupling to externally imposed conditions, mainly to maintain constant temperature T

[canonical (*NVT*) ensemble] and possibly constant pressure p [isothermal-isobaric (*NpT*) ensemble]. The total system Hamiltonian H therefore has to be extended.

2.2. CONSTANT TEMPERATURE SIMULATIONS

2.2.1. *General Outline*

Simulations in the canonical ensemble require coupling the equations of motion of the simulated system to an external heat reservoir which allows the system to pass among phase space hypersurfaces of different total energy. The canonical phase space density reads

$$\rho_{NVT}(P,Q) \propto \exp[-\beta H(P,Q)] \qquad (5)$$

where P, Q denote all momenta and coordinates of a system and $\beta = 1/k_B T$. During the course of an MD simulation typically the kinetic (velocity space) temperature T_k according to the equipartition theorem

$$\left\langle \sum_{i=1}^{N} \frac{p_i^2}{2m_i} \right\rangle = \frac{3}{2} N k_B T_k \qquad (6)$$

can be monitored easily. In contrast, the configurational or position space temperature, which should be equal to the kinetic temperature if the canonical density is achieved, must be derived by inverting simulated distributions. If these inversions are numerically integrated, the different truncation errors of configuration and velocity space generally lead to deviations of kinetic and configurational temperature different from the desired one unless special (non-MD) techniques, for instance hybrid Monte Carlo [36,37], are applied to generate the canonical ensemble.

In the past two decades many different methods for conducting MD simulations at constant temperature have been proposed. They vary in the way the system and the heat bath are coupled and can be roughly divided into deterministic and stochastic methods. We shall review the most important ones.

Deterministic Methods. Some authors constrained the kinetic energy to a fixed value which, if taken as a mean value, would correspond to the desired kinetic temperature. The natural fluctuations of the kinetic energy are ignored. The canonical distribution in configuration space is obtained [38-41]. Berendsen *et al.* [42] developed a so-called "weak coupling" velocity scaling algorithm to adjust the kinetic temperature rapidly to the desired value. However, the method does not generate states in the canonical ensemble. Furthermore, local temperature differences may be preserved over rather long times resulting in an inefficient equilibration procedure.

The most widely used deterministic method was derived by Nosé [43] and Hoover [44], in which the simulated system is extended by an additional degree of freedom, s. The potential associated with s (f denotes the system's number of degrees of freedom) is given as

$$V_s = (f+1) k_B T \ln s \tag{7}$$

while the kinetic energy reads

$$K_s = p_s^2/(2Q) \tag{8}$$

The "mass" Q is an adjustable parameter. The extra degree of freedom acts on the particle velocities via

$$v_i = p_i/(m_i s) \; . \tag{9}$$

It can be shown that the total system including the extra degree of freedom, characterized by the Hamiltonian $H = V + V_s + K + K_s$, is microcanonical, whereas the simulated system alone obeys the canonical phase space density. While in older applications of the Nosé-Hoover scheme problems occurred with regard to ergodicity and strong dependencies on the coupling parameter [45], these problems have been resolved in recent work [46]: The extra degree of freedom is itself thermalized by additional Nosé terms, thus damping down eventual energy oscillations between system and reservoir. For related work see also [47,48]. Deterministic methods were reviewed by Nosé [49].

Stochastic Methods. Stochastic dynamics provide an alternative route to generating canonical MD ensembles. Andersen [50] proposed a procedure where the velocities of randomly selected system particles are chosen at distinct times from a Maxwellian distribution at the desired temperature. This technique and related approaches [51-53] all generate trajectories which sample the canonical ensemble, but the coupling parameters are not easy to estimate for a given application. Other strategies use Brownian dynamics algorithms to integrate the strict Langevin equations of a many-body system [54,55]. The canonical ensemble is sampled, but the computational effort is very high.

2.2.2. Finite-mass Stochastic Collision Dynamics

General Model. Recently, a new stochastic approach to constant temperature dynamics has been developed by two of us [30,31]. The strategy is based on central impulsive collisions between system particles and imaginary heat bath particles of finite mass, therefore offering a rather intuitive approach to the system-reservoir coupling problem. The method is an adaptation of the Rayleigh model [56] of an ensemble of frictionless pistons subject to one-dimensional collisions with heat bath particles of Maxwellian velocity distribution. The technique is computationally efficient, easy to implement into existing MD code, and does not suffer the difficulties observed with the deterministic thermalization of small systems. The algorithm can be characterized as follows.

The force vector acting on atom i at time step n

$$F_{i,n} = F_{i,n}^{(s)} + F_{i,n}^{(c)} \qquad (10)$$

is the sum of the systematic part

$$F_{i,n}^{(s)} = -\nabla_{r_{i,n}} V(r_1, ..., r_N) \qquad (11)$$

and the collision part

$$F_{i,n}^{(c)} = m_i \frac{2\alpha}{\Delta t}(u_{i,n} - v_{i,n-1}) = \frac{2m_i}{\Delta t} \frac{m_r}{1+m_r}(u_{i,n} - v_{i,n-1}) \qquad (12)$$

where $\alpha = m_{b,i}/(m_i + m_{b,i}) = m_r/(1+m_r)$ with $m_r = m_{b,i}/m_i$, $m_{b,i}$ being the mass of the heat bath particles. Only the velocity of step $n-1$ is available when the forces at step n are calculated. The velocity vector of the bath particles is a random number vector denoted by u_i. The velocity distribution $f_{b,i}(u_i)$ is independently the same for each spatial direction and is taken to be Maxwellian at the bath temperature T_b:

$$f_{b,i}(u_i) = \left(\frac{m_{b,i}}{2\pi k_B T_b}\right)^{1/2} \exp\left(\frac{-m_{b,i} u^2}{2 k_B T_b}\right). \qquad (13)$$

The bath particles do not interact with each other. The random variables u_i are taken to be *independent*, i.e. the bath is assumed to be infinitely large.

Inserting Eq. (4) in Eqs. (10)-(12) and the result in Eq. (3) we obtain a set of inhomogeneous finite difference equations that can be solved analytically if the systematic force is a linear function of the coordinates, i.e. for ensembles of free particles or one-dimensional harmonic oscillators. For these systems we can then look into the statistical and dynamical properties of the resulting stochastic processes without actually simulating numerically and gain some insight into algorithmic stability and reliability. Therefore the detailed analytic study of stochastic finite difference propagators has recently received some more attention in the literature [57,58]. The stochastic process is non-Markovian in nature and the theoretical analysis is therefore not possible by means of a master equation approach but by direct solution of the stochastic difference equation [59].

Outline of the Analytic Methodology. We shall briefly describe the methodology with a simpler Markovian model that could have easily been solved by induction, but the more complicated formalism is necessary for the full non-Markovian model. If configuration space is not considered and no potential-derived forces act on the system, we can define a particle colliding with heat bath particles as *Rayleigh* particle. Let v be the velocity vector of the Rayleigh particle with mass m_1 in Cartesian coordinates, u be the velocity vector of a heat bath particle with mass m_2. We consider all particles to be point masses. From energy and momentum conservation the velocity change of the Rayleigh particle from state n to state $n+1$ due to a collision reads

$$v_{n+1} - v_n = 2\alpha(u_n - v_n) \tag{14}$$

where $\alpha = m_2/(m_1+m_2)$. This represents a *discrete-time, continuous-state, real stochastic process* [60]. Rewriting Eq. (14) in one dimension gives

$$v_{n+1} = (1-\eta)v_n + \eta u_n \tag{15}$$

with $\eta = 2\alpha$. This is an inhomogeneous first order linear difference equation with constant coefficients of the type

$$f(n+k) + b_1 f(n+k-1) + \ldots + b_k f(n) = Q(n), \tag{16}$$

and we ask for the general expression f_n depending on k initial conditions. The general solution to the inhomogeneous equation is the general solution to the homogeneous equation where $Q(n)$ in Eq. (16) is set to zero, plus any special solution to the inhomogenous one. With the ansatz

$$v_n = \lambda^n \tag{17}$$

a special solution to the homogeneous equation is

$$\lambda = 1-\eta. \tag{18}$$

The general solution $f_{gh}(n)$ to the homogeneous equation is a linear combination of all linearly independent solutions $g_i(n)$ to the homogeneous expression:

$$f_{gh}(n) = \sum_i c_i g_i(n) = \sum_i c_i \lambda_i^n, \tag{19}$$

and for the Markovian model one obtains

$$v_{n,gh} = c_1(1-\eta)^n. \tag{20}$$

A special solution $f_{si}(n)$ to the inhomogeneous equation is found by variation of constants. In our case, we achieve

$$v_{n,si} = \eta \sum_{l=0}^{n-1} \frac{\lambda^n}{\lambda^{l+1}} u_l = \eta \sum_{l=0}^{n-1} \lambda^{n-1-l} u_l = \eta \sum_{l'=0}^{n-1} \lambda^{l'} u_{l'}, \tag{21}$$

because the sequence of the random variables u_l is exchangeable. Constructing the general solution to the inhomogenous equation $f_{gi}(n) = f_{gh}(n) + f_{si}(n)$, we obtain with $v_n = v_0$ for $n = 0$

$$v_n = v_0(1-\eta)^n + \eta \sum_{l=0}^{n-1}(1-\eta)^l u_l .\qquad(22)$$

With the general solution at hand we now proceed to derive stationary statistical and dynamical properties. Since $1-\eta \in]-1;1[$, v_n must become stationary in the limit $n \to \infty$. Furthermore, in this limit the velocity becomes independent of the initial conditions. Since if $n \to \infty$ the random variable v_n can be expressed as a sum of Gaussian distributed random variables, whose probability density is also Gaussian. Hence the stochastic process is completely determined in the long time limit by the first and second moments of its probability density. For $n \to \infty$ in Eq. (22) the first moment $\langle v_n \rangle$ tends to zero, because $\langle u_l \rangle = 0$. The angular brackets denote ensemble averaging. The second moment $\langle v_n^2 \rangle$ is

$$\langle v_n^2 \rangle = (1-\eta)^{2n} \langle v_0^2 \rangle + 2(1-\eta)^n \eta \sum_{l=0}^{n-1}(1-\eta)^l \langle v_0 u_l \rangle$$
$$+ \eta^2 \sum_{l=0}^{n-1}\sum_{k=0}^{n-1}(1-\eta)^{l+k}\langle u_l u_k \rangle .\qquad(23)$$

Due to the properties of the random variable u_l we have to consider the conditions $\langle v_0 u_l \rangle = 0$ and $\langle u_l u_k \rangle = \delta_{lk} k_B T / m_2$ where δ_{lk} is the Kronecker symbol. Thus in the long time limit one gets

$$\langle v^2 \rangle = \lim_{n \to \infty} \langle v_n^2 \rangle = \eta^2 \frac{k_B T}{m_2} \sum_{l=0}^{\infty}(1-\eta)^{2l} = \frac{\eta^2}{1-(1-\eta)^2} \frac{k_B T}{m_2} .\qquad(24)$$

With $\eta = 2m_2/(m_1+m_2)$ Eq. (24) simplifies to

$$\langle v^2 \rangle = \frac{k_B T}{m_1} .\qquad(25)$$

In the stationary case the second moment equals the second central moment. Thermal equilibrium between the heat bath and an ensemble of Rayleigh particles is reached. The unnormalized velocity autocorrelation function in the long time limit $C_v(m)$ is defined as

$$C_v(m) = \lim_{n \to \infty} \langle v_n v_{n+m} \rangle = \lim_{n \to \infty}\Big[(1-\eta)^{2n+m} \langle v_0^2 \rangle$$
$$+ (1-\eta)^m \eta \sum_{l=0}^{n+m-1}(1-\eta)^l \langle v_0 u_l \rangle + (1-\eta)^{n+m} \eta \sum_{l=0}^{n-1}(1-\eta)^l \langle v_0 u_l \rangle\qquad(26)$$
$$+ \eta^2 \sum_{l=0}^{n-1}\sum_{k=0}^{n-1}(1-\eta)^{l+k+m}\langle u_l u_k \rangle \Big],$$

which becomes simply

$$C_v(m) = \langle v^2 \rangle (1-\eta)^m . \tag{27}$$

Due to $(1-\eta) = (m_1-m_2)/(m_1+m_2)$ the correlation function decays monotonically if $m_2 < m_1$ and shows a damped oscillatory behavior if $m_2 > m_1$ since, for larger mass ratios, the particle's momentum changes its sign on average and the energy transfer becomes less effective. The case $m_2 = m_1$ stands for complete randomization of the velocities of the Rayleigh particles after a single collision. This corresponds to a thermalization process as proposed by Andersen [50]. In the stationary limit the stochastic process (15) is characterized by constant zero mean and velocity autocorrelation function which depends only on the time difference m.

Non-Markovian Model: a) Free Particle. With the "velocity Verlet" [35] formulation of a velocity propagation step we obtain for a one-dimensional potential-free particle

$$v_n = \beta v_{n-1} - \alpha v_{n-2} + \alpha(u_n + u_{n-1}) \tag{28}$$

with $\beta = 1-\alpha$, $\alpha = m_2/(m_1+m_2)$. The predicted v_n depends on the velocities of the last *two* time steps. The solution according to the last section yields quite complicated n-dependent expressions. With $n \to \infty$, regardless of the heat bath particle mass, the result becomes stationary and independent of the initial conditions, the limiting distribution of v being Gaussian and hence canonical. The second moment of the velocity distribution turns out to be $k_B T/m_1$ and thus exact thermal equilibrium between system and heat bath is achieved. There is no integration error and hence no temperature deviation from the bath temperature.

From the velocity time correlation function two limiting cases to characterize the dynamics can be derived: Firstly, if the mass ratio between system particles and heat bath particles and the time step both tend to zero, the discrete correlation function approaches the continuous correlation function from Langevin dynamics of a free particle. Secondly, for a critical mass ratio $m_r = (3-8^{1/2})/(8^{1/2}-2) \sim 0.2071$ (cf. Markovian model: $m_r = 1/2$) the dynamics switches from monotonically decreasing correlation functions to oscillatory behavior, In the region of mass ratios around the critical value the dynamics is similar to that induced by the Andersen thermostat. This means one is able to switch between Langevin-type and Andersen-type thermalization by means of only one control parameter, the mass ratio, without using different integration schemes. We can further prove that the dynamics is strictly ergodic, an important characteristic for small systems.

Non-Markovian Model: b) Harmonic Oscillator. For a one-dimensional oscillator subject to the heat bath we obtain in configuration space

$$r_{n+1} = (2-\Delta t^2\omega^2-\alpha)r_n - r_{n-1} + \alpha r_{n-2} + 2\alpha\Delta t\, u_n \quad . \tag{29}$$

We then define

$$\varepsilon = \Delta t^2 \omega^2 = 4\pi^2 (\Delta t/T)^2 \tag{30}$$

with T here being the vibration period (not to be confused with the temperature, the meaning is obvious from the context). ε has the meaning of a resolution or graining parameter since it is inversely proportional to the square of the number of time steps per vibration period $T/\Delta t$. We now have to deal with a linear third-order difference equation having complicated solutions. Applying the formalism described above again leads to the proof of the canonical density and ergodicity for configuration and velocity space, but only under certain circumstances: One can prove that the inequality

$$\varepsilon < 2 - 2\alpha \tag{31}$$

must be satisfied in order to get a stationary limiting distribution and therefore algorithmic stability. The period must be resolved into at least $T/\Delta t \approx 4.44$ steps for small values of α.

Interesting now is the fact that configurational and kinetic temperatures generally differ and deviate from the heat bath temperature. With the short notation T_R as the ratio of actual and desired (bath) temperature the results are

$$T_R^{(r)} = \frac{4(1-\alpha)(2-2\alpha+\alpha\varepsilon)}{8-8\alpha-6\varepsilon+2\alpha\varepsilon+\varepsilon^2} \tag{32}$$

for configuration space and

$$T_R^{(v)} = \frac{2(1-\alpha)}{2(1-\alpha)-\varepsilon} \tag{33}$$

for velocity space. Both temperatures coincide if $\varepsilon = 4\alpha/(1+2\alpha)$. Again, Langevin dynamics is recovered if both the inverse period resolution and the heat bath coupling parameter approach zero, and all temperatures coincide with the bath temperature. Thus, this deviation can quantitatively be attributed to the truncation error of the Taylor series expansion of Verlet integration.

From the analytic time correlation functions $c(m;\alpha,\varepsilon)$ we can compute correlation times according to

$$\tau_c(\alpha,\varepsilon) = \Delta t \sum_{m=0}^{\infty} |c(m;\alpha,\varepsilon)| - \frac{1}{2}\Delta t \quad . \tag{34}$$

Because it is important in the following, this relation is depicted in Figure 1, illustrating the behavior of τ_c in configuration space. Similar results are obtained for velocity space. We identify a minimum as a function of period resolution and coupling parameter. In this parameter region the equilibration performance reaches a maximum. An ensemble is thermalized after only a few time steps.

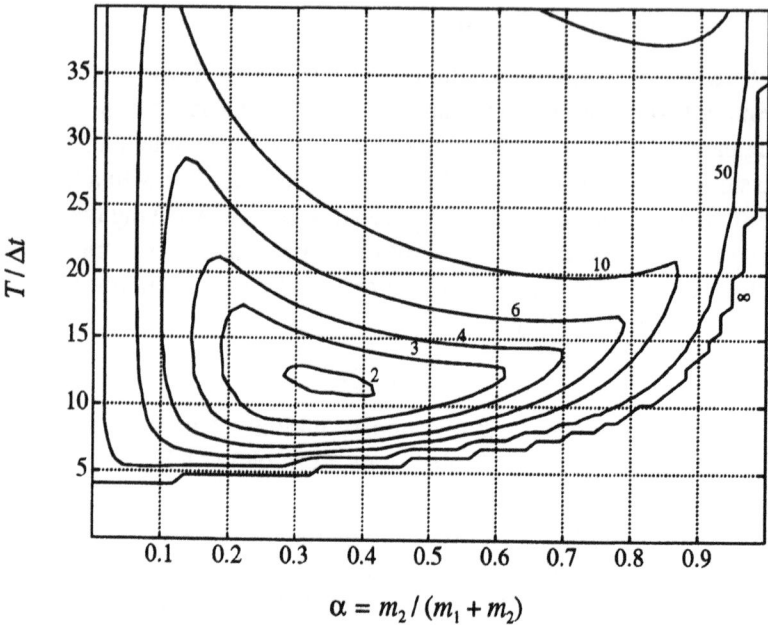

$$\alpha = m_2 / (m_1 + m_2)$$

Figure 1. Contour plot of the position correlation time of the harmonic oscillator, measured as multiples of the integration time step Δt, as a function of the coupling parameter between oscillator and heat bath α and the resolution of the vibration period $T/\Delta t$.

Conclusions. The detailed analytical analysis of a finite difference propagation algorithm not only increases our insight into algorithmic stability, artifacts, and possible pitfalls, but allows us to tailor the simulation parameters with respect to certain desired features as will be shown later. The model is fast and robust and has been used for several MD simulations of realistic systems [61-64].

2.3. CONSTANT PRESSURE SIMULATIONS

MD simulations at a constant pressure are not only neccessary for the equilibration phase of a simulation of large systems when the initial configuration is usually far from equilibrium, but are useful e.g. for the simulation of phase transitions in solids and clusters or the calculation of some special thermodynamic quantities, for example the adiabatic compressibility. We only review the most popular methods here. The instantaneous system pressure at time step n is

$$P_n = \frac{N}{V} k_B T_n + \frac{1}{3V} \sum_{i=1}^{N} r_{i,n} \cdot F_{i,n} \qquad (35)$$

where V is the container volume, T_n is the instantaneous temperature. Ensemble averaging of Eq. (35) gives the thermodynamic pressure. The most popular method for allowing an MD system to evolute at a constant pressure is due to Berendsen et al. [42]. In this approach, an extra term is added to the equations of motion to couple the system to an external "pressure bath":

$$\mu = \underline{1} - \frac{\beta \Delta t}{3\tau_p}(P_0 \underline{1} - P) \tag{36}$$

where τ_p is the compressibility, β is a coupling constant, μ is a scaling tensor, and P_0 is the target pressure. At every time step, the box volume is scaled in order to induce a pressure change. After calculation of the pressure tensor P and the scaling tensor μ, the new coordinates of the particles are

$$r'_i = \mu r_i \tag{37}$$

and the new simulation cell volume

$$V' = |\mu| V . \tag{38}$$

This scaling procedure proved to be fast and stable although the ensemble properties are not entirely clear.

2.4. FORCES BY ELECTRONIC STRUCTURE THEORY

2.4.1. *Ab Initio Molecular Dynamics*

We confine ourselves to electronically adiabatic, classical nuclear motion, i.e. the nuclei are constrained to move on a Born-Oppenheimer hypersurface. If we have available the exact stationary instantaneous electronic wavefunction $\Psi(r)$ from the Hamiltonian H, the forces acting on the nuclei and entering Eq. (3) are determined by the Hellmann-Feynman theorem [13,14]

$$F = -\frac{\partial E}{\partial r} = -\langle \Psi(r) | \frac{\partial H}{\partial r} | \Psi(r) \rangle . \tag{39}$$

Of course, only approximate wavefunctions can be computed. Direct application of a straightforward scheme, namely performing a full electronic structure calculation at each time step, propagation in time, and so forth, will therefore not yield trajectories conserving the total energy. Furthermore, for large and highly disordered systems the conventional wavefunction calculation at every time step would be computationally prohibitive. About a decade ago Car and Parrinello [65] therefore invented an ingenious method for coping with both drawbacks based on some old ideas: Firstly, the variational minimization of the electronic energy is replaced by a dynamic annealing procedure; secondly, electronic energy annealing and propagation of the nuclear degrees of freedom are done in parallel. We only treat those issues being important in the following. There are extensive reviews in the literature, e.g. [11,12].

Minimization. Conventional numerical quantum chemistry involves the diagonalization of huge matrices in order to reach self-consistency and a stationary minimum energy as a function of basis function coefficients. Alternatively, the minimum energy configuration in an arbitrary hyperspace can be treated analogously to PES minima of statistical mechanical systems: These are visited as the system is cooled down, kinetic energy is removed, and phase space trajectories get trapped. If the cooling is done sufficiently slowly the system will eventually find the global energy minimum. Algorithms based upon this analogy are termed *simulated annealing* (SA) [66-68]. They can be shown to be special cases of the general class of *genetic algorithms* [69]. Applications of SA in chemistry are numerous and have been summarized recently [70]. Generally, a penalty function depending on variables to be optimized is defined whose minimum is searched by cooling down trajectories of the variables generated by MC or MD techniques.

Let us as usual represent the electronic state in terms of a set of occupied orbitals, $\psi_1, ..., \psi_n$. Each orbital is expanded in some finite basis $\phi_1, ..., \phi_m$ with coefficients c such that

$$\psi_i = \sum_k c_k^i \phi_k . \qquad (40)$$

Basis and coefficients are assumed to be all real for simplicity. The electronic configuration space is then completely defined by the set $C = \{c_k^i\}$, and we search for the minimum of the electronic energy $E_{elec}(C)$ with additional orthonormality constraints

$$\sigma_{ij} = \sum_k c_k^i c_k^j - \delta_{ij} = 0 . \qquad (41)$$

Since this kind of holonomic constraint can more easily be realized with MD based simulation protocols, e.g. by using SHAKE [71], than with MC methods, a classical Lagrangian for the coefficients is defined:

$$L_{elec} = K_{elec} - V_{elec} = \frac{1}{2}\mu \sum_i \sum_k (\dot{c}_k^i)^2 - E_{elec}(C) \qquad (42)$$

with a "fictitious" mass μ associated with the coefficients. The equations of motion combined with the constraints, Eq. (41), are then derived to take the general form

$$\mu \ddot{c}_k^i = -\frac{\partial E_{elec}}{\partial c_k^i} - \sum_j \lambda_{ij} \frac{\partial \sigma_{ij}}{\partial c_k^i} \qquad (43)$$

where λ is a Lagrange multiplier. The equations of motion can then be coupled to an external thermostat (cf. Sec. 2.2.) and successively cooled down or the fictitious electronic kinetic energy is periodically quenched whereby eventually a minimum and therefore the optimal wavefunction is reached.

Parallel Dynamics. Since we already formulated the electronic annealing problem in a dynamical way, it is straightforward to write down a solution to the ultimate problem, namely the integration of the *nuclear* equations of motion *in parallel*. In the Car-Parrinello (CP) scheme a total, extended Lagrangian is defined

$$L_{total} = K_{elec} + K_{nucl} - V_{elec} - V_{nucl} \qquad (44)$$

where the index "nucl" refers to nuclear degrees of freedom which can be some core ions when pseudopotentials are used or, in the simplest case, the nuclei themselves with an internuclear Coulomb potential. Two sets of classical equations of motion are then derived, one for the coefficient motion, Eq. (43), and one for the nuclei using Hellmann-Feynman forces. The extended Lagrangian will then conserve the total energy

$$E_{total} = K_{elec} + K_{nucl} + V_{elec} + V_{nucl} \qquad (45)$$

within the accuracy of the integration procedure and the Hellmann-Feynman forces. However, for conservative nuclear dynamics,

$$E_{nucl} = K_{nucl} + V_{elec} + V_{nucl} \qquad (46)$$

must be conserved, thereby putting some bound on the fictitious coefficient masses μ which must be small. Although seemingly arbitrary, these masses have a well defined meaning if the CP scheme is derived from first principles, the quantum variational principle [72].

The dynamics is started with a well minimized electronic subsystem, i.e. $K_{elec} = \partial V_{elec}/\partial C = 0$. The success of the CP method seems qualitatively to be related to the fact that, when the nuclei start moving, the coefficients are left behind and accelerated. Eventually, they will catch up with the nuclei and even overshoot them after which they will be pulled back. The fictitious kinetic energy will therefore oscillate instead of becoming fully quenched. Small fictitious masses and therefore high frequency of the electronic compared to the nuclear degrees of freedom will then maintain adiabatic coefficient motion and keep the nuclear system always close to the Born-Oppenheimer ground state.

2.4.2. Hybrid QM/MM Methods

The CP method as described above is still computationally extremely demanding and time steps several orders of magnitude smaller than in conventional

MD are required. If we want to use direct quantum chemistry for calculating forces we could alternatively treat "quantum mechanically" a certain core of atoms only, and "classically", i.e. molecular mechanically with force fields, the rest of the system. This represents a different solution circumventing the last section's two drawbacks: Only a small subset of QM atoms is treated with expensive numerical quantum chemistry, and energy conservation is usually not an issue since the quantum subsystem is thermalized by the surrounding MM particles which themselves are usually thermalized with an external heat bath.

Several methods are conceivable to subdivide the total system Hamiltonian, the two most popular are due to Warshel and Weiss [73] and Karplus and co-workers [74]. We will focus on the latter approach. The total effective Hamiltonian is written as

$$H_{eff} = H_{QM} + H_{MM} + H_{QM/MM} \tag{47}$$

where H_{QM} is the usual nonrelativistic electronic Hamiltonian in atomic units

$$H_{QM} = -\frac{1}{2}\sum_i \nabla_i^2 + \sum_{i,j} \frac{1}{r_{ij}} - \sum_{i,\alpha} \frac{Z_\alpha}{r_{i\alpha}} + \sum_{\alpha,\beta} \frac{Z_\alpha Z_\beta}{R_{\alpha\beta}} \tag{48}$$

(i,j and α,β indicate electronic and nuclear coordinates, respectively, R,r denote distances, and Z nuclear charges) of the quantum mechanical subsystem, H_{MM} is just the sum of kinetic energy K_{MM} and some molecular mechanical potential function V_{MM}, e.g. the CHARMM function [75,76]

$$V_{MM} = \sum_{bonds} k_b(r-r_0)^2 + \sum_{angles} k_\alpha(\alpha-\alpha_0)^2 + \sum_{impropers} k_\phi(\phi-\phi_0)^2$$

$$+ \sum_{j>i}\left(\frac{A_{ij}}{r_{ij}^{12}} - \frac{B_{ij}}{r_{ij}^6} + \frac{q_i q_j}{4\pi\varepsilon_0 r_{ij}}\right) + \sum_{torsions} k_\tau[1+\cos(n\tau-\delta)] \tag{49}$$

(r: bond length, r_0: equilibrium bond length, k_b: force constant (bond), α: bending angle, α_0: equilibrium bending angle, k_α: force constant (bending), τ: torsional angle, n: multiplicity, δ: phase angle, k_τ: force constant (torsion), ϕ: improper torsional angle, k_ϕ: force constant (improper torsion), ϕ_0: equilibrium improper torsional angle, r_{ij}: distance, A_{ij}, B_{ij}: Lennard-Jones parameters, q_i, q_j: charges), and $H_{QM/MM}$ denotes the interaction between the two subsystems in atomic units

$$H_{QM/MM} = -\sum_{i,M} \frac{q_M}{r_{iM}} + \sum_{\alpha,M} \frac{Z_\alpha q_M}{R_{\alpha M}} + \sum_{\alpha,M}\left(\frac{A_{\alpha M}}{R_{\alpha M}^{12}} - \frac{B_{\alpha M}}{R_{\alpha M}^6}\right) \tag{50}$$

(subscripts i and α refer to the QM electrons and nuclei, respectively, and M to the MM atoms). The first term in Eq. (50) describes the electrostatic interaction between electrons of QM atoms and MM partial charges q_M, the second the electrostatics between MM partial charges and QM nuclei, the third some Lennard-Jones interaction between QM and MM atoms. The total energy is then

$$E = \langle \Psi | H_{QM} + H_{QM/MM} | \Psi \rangle + E_{MM} \tag{51}$$

since the wavefunction Ψ depends only parameterically on the position of the MM atoms. Forces are calculated preferably by analytical derivatives of the energy and enter the usual propagation scheme.

3. Aspects of Parametrization

3.1. GENERAL REMARKS

When we look at an expression like Eq. (49) we find that, in order to define an empirical potential energy function, we have to a) decompose the system into independent contributions from fractions of atoms, b) formulate some functional terms, and c) adjust the parameters. One of the ultimate goals of theoretical chemistry would be to *derive* all of the above issues from an *ab initio* PES, thereby learning about chemical similarities, possibly *derive* such things like functional groups from first principles. Unfortunately, it is unclear whether these exist and what is a unique and systematic way to do this, either step by step or all at once. Some attempts have been made toward this goal by Dinur and Hagler [21,22]. They try to use as much information as can be obtained from stationary and nonstationary points on an *ab initio* PES and try to separate contributions. However, intuition is definitely needed to a) define some suitable partitioning and b) find appropriate functional forms by trial and error. The huge amount of reference data seems to require enormously complicated model function containing even third order oupling terms.

Simulating dynamics itself does not generate new information once we have found the perfect decomposition and functional form, because what we actually do by simulations is *uncompressing data*. Some unique functional form or model, together with the equations of motion themselves and some initial conditions constitutes a maximally compressed data set. Together with the information about how to solve the equations of motion, i.e. with the "number of bits" of a simulation program or an analytic solution, our compressed data set contains all the information we could ever learn about a particular system, if and only if the model is capable to predict the energetics correctly for *any* configuration and the propagation method is free of errors.

The optimal functional form of a chemical force field is yet unclear. We shall therefore assume that we have found one suitable model function and discuss the last issue, the determination of parameters. Traditionally, least squares fitting is employed [77]. We define a penalty function

$$S(\phi) = \sum_i w_i [O_i(\phi) - O_{i,\text{ref}}]^2 \qquad (52)$$

that depends on a set of n parameters $\phi = \{\phi_1,...,\phi_n\}$ and which connects observables calculated from the parameterized model $O_i(\phi)$ with reference observables $O_{i,\text{ref}}$. The statistical weights are denoted w_i. The parametrization process now involves the minimization of S with respect to ϕ, i.e.

$$\frac{\partial S(\phi)}{\partial \phi_k} = 2 \sum_i w_i [O_i(\phi) - O_{i,\text{ref}}] \frac{\partial O_i(\phi)}{\partial \phi_k} = 0 \qquad (53)$$

for all ϕ_k. Typical reference observables derived from an *ab initio* PES in use for molecular mechanics parametrization are, e.g., interaction energy, location of minima, gradients, or normal frequencies by diagonalization of the mass-weighted Hessian. The penalty function is usually minimized using iterative linearized methods [78] or general nonlinear methods [77]. However, problems are often reported due to either linear dependencies among parameters whenever matrix inversions are needed, or the penalty function represents a rugged landscape and convergence to an acceptable result is bad. Particularly the latter problem, local minima nearby the initial parameter values, is important if parameters are fitted with respect to transferability from one molecule to another. The optimal fitting strategy therefore should be inversion-free and be able to find the global minimum of S, as is done by simulated annealing.

3.2. OPTIMAL SIMULATED ANNEALING

We present only a brief conceptual outline of a novel SA variant. Details will be reported elsewhere [79]. In the spirit of the Car-Parrinello scheme we treat the parameters ϕ_i as dynamical variables and define a set of classical equations of motion

$$m_{\phi,k} \ddot{\phi}_k = -\frac{\partial S(\phi)}{\partial \phi_k} + R \qquad (54)$$

where the forces are the negative of the middle part of Eq. (53) and again "fictitious" parameter masses $m_{\phi,k}$ have to be introduced. R represents a heat bath coupling term. Given the model observables $O_i(\phi)$ we typically need to compute derivatives of, e.g., model PES gradients and Hessians with respect to potential function parameters which can be done automatically and analytically for virtually any conceivable model function [79].

By coupling the equations of motion to an external heat bath that generates the canonical distribution the system is allowed to sample not only nearby minima but to cross some barriers. Cooling down then guides the sample ideally down to the global minimum of S. This global property can only be guaranteed for infinitesimally slow cooling. We use the Verlet integration scheme together with the constant temperature model described in Sec. 2.2.2.

Again, we are left with the question of properly choosing the parameter masses. In fact the mass determines the local stiffness of the parameter "potential" S. If we choose one uniform mass, a fixed time step, and the local curvature of S is different for each ϕ_k, then the parameters will evolve on different time scales. This represents an inefficient propagation procedure since we need many more time steps for slow modes than for the fastest mode. This is in some sense the inverse situation to real physical molecular dynamics: The masses are given their physical value, and one unique time step for all degrees of freedom is clearly a waste of computer time. For this reason, efficient multiple time step methods have been derived [80]. Since we deal with unphysical dynamics we could equally justify choosing one unique time step and adjust the masses with respect to motion on the same time scale. Additionally we keep in mind that, according to Figure 1, there is a point in α, ε space of the thermalized harmonic oscillator for which the correlation time is minimal, i.e. fastest generation of the canonical density. Accordingly we need a prescription for the mass, given a certain heat bath coupling parameter, period resolution, and time step.

The simplest way to accomplish this task is to approximate some local environment $S(\phi_k)$ quadratically, i.e.

$$F_k = -m_{\phi,k} \omega_k (\phi_k - \phi_{k,0}) \; . \tag{55}$$

Let Δ denote a difference operator between time steps $n+1$ and n. We then obtain for a local average

$$\frac{1}{m_{\phi,k}^2} \frac{\langle \Delta F_k^2 \rangle}{\langle \Delta \phi_k^2 \rangle} = \omega_k^4 \tag{56}$$

and therefore

$$m_{\phi,k} = \left(\frac{\langle \Delta F_k^2 \rangle}{\langle \Delta \phi_k^2 \rangle} \right)^{1/2} \Delta t^2 \frac{1}{\varepsilon} \tag{57}$$

if we recall Eq. (30). This means, one measures changes of both force and parameter value from step to step, averages over some interval, prescribes a certain value for the time step and the resolution parameter, and calculates optimal fictitious masses. This process has to be repeated every several time steps. The constant temperature algorithm of Sec. 2.2.2. is robust enough to cope with this crude but efficient approximation.

Alternatively, we can derive a similar equation like Eq. (57) if instead of the systematic force the total force including the heat bath contribution is taken into account. Recall that

$$F_k = F_k^{(c)} + F_k^{(s)} = m_{\phi,k} \frac{2\alpha}{\Delta t} \left(u_n - \frac{\phi_{k,n} - \phi_{k,n-2}}{2\Delta t} \right) - m_{\phi,k} \omega_k (\phi_k - \phi_{k,0}) \quad (58)$$

and accordingly

$$\frac{\Delta t^4}{m_{\phi,k}^2} \langle \Delta F_k^2 \rangle = \langle [2\alpha\Delta t (u_n - u_{n-1}) - \alpha(r_{n-2} - r_{n-3}) + (\alpha+\epsilon)(r_{n-1} - r_n)]^2 \rangle . \quad (59)$$

Due to the correlation properties $\langle u_l u_k \rangle = \delta_{lk} k_B T/m_2$, $\langle u_l \phi_{k,l+1} \rangle = 2\alpha\Delta t^2 k_B T/m_2$ as is easily derived by multiplicating Eq. (29) (r replaced by ϕ) by u_l and ensemble averaging, $\langle u_l \phi_{k,l<n+1} \rangle = 0$, and $\langle \phi_{k,n} \phi_{k,m} \rangle = \langle \phi_k^2 \rangle c(\alpha, \epsilon; n-m)$ where c is the normalized autocorrelation function, we get

$$\frac{\Delta t^4}{m_{\phi,k}^2} \langle \Delta F_k^2 \rangle = f_1(\alpha) \Delta t^2 \frac{k_B T}{m_2} + f_2(\alpha, \epsilon) \frac{k_B T}{m_{\phi,k} \omega_k^2} \quad (60)$$

and do not further specify f_1 and f_2. Dividing by $\langle \phi_k^2 \rangle = f_3(\alpha, \epsilon) k_B T/(m_{\phi,k} \omega_k^2)$ finally yields

$$m_{\phi,k} = \left(\frac{\langle \Delta F_k^2 \rangle}{\langle \Delta \phi_k^2 \rangle} \right)^{1/2} \Delta t^2 f(\alpha, \epsilon) . \quad (61)$$

Again, prescription of α, ϵ, and Δt is sufficient to determine masses from measurable quantities.

The crucial point of any SA application is the choice of a suitable cooling schedule. Cooling too fast will most likely trap the system in a nearby local minimum, cooling too slowly will make the procedure inefficient. Several schedules are conceivable [68], and we opted for a modified Huang schedule [81] that reads

$$\beta_{k+1} = \beta_k + \frac{\lambda}{\sigma_k} . \quad (62)$$

This scheme relates the temperature of cycle $k+1$ to that of cycle k, where λ has to be chosen between 0 and 1, and σ_k is the fluctuation of the total "energy" in S space. This schedule cools fast in the early phase and more slowly with decreasing temperature, thereby allowing the system to thoroughly sample the minimum region.

3.3. EXTENDED SYSTEMS REVISITED

Suppose we have available not only quantum chemical results but accurate experimental data as well. If we assume to know the suitable functional form of our model PES, we have to ask for a method to include thermally averaged observables into the fitting procedure. Ideally, this sort of parametrization should be done "on the fly", i.e. *during* the simulation to obtain exactly those thermal observables. An attempt in this direction has recently been made by Njo et al. [82]. In the spirit of the Berendsen thermostat and manostat [42] these authors define a dynamic coupling term which relates the deviation of an *instantaneous* observable q from the thermal reference value q_0 to a change of a parameter p:

$$\frac{dp}{dt} = \pm \frac{C}{\tau_p}(q - q_0) \tag{63}$$

where τ_p is a relaxation time governing the response of the system, C is a unit conversion constant. This relation only holds if there is a monotonic relationship between the change of the parameter and the change of the observable.

Alternatively, one can use the SA formalism described above. The reference observable now is a thermally averaged magnitude, and we are faced with the problem of calculating the derivative of a model ensemble average with respect to parameters. In the isothermal-isobaric ensemble an averaged model observable is given as

$$\langle O_i \rangle = \frac{1}{Z} \int O_i(\Omega) \, e^{-\beta(H+pV)} \, d\Omega \tag{64}$$

with pressure p, volume V, partition function Z, and Ω denotes all coordinates, momenta, and volume. With the short notation $\langle O_i \rangle = G/Z$ one gets

$$\frac{\partial \langle O_i \rangle}{\partial \phi_k} = \frac{1}{Z^2}(G'Z - Z'G) = \frac{G'}{Z} - \langle O_i \rangle \frac{Z'}{Z} . \tag{65}$$

On the other hand we have

$$Z' = \frac{\partial Z}{\partial \phi_k} = \frac{\partial}{\partial \phi_k} \int e^{-\beta(H+pV)} d\Omega = -\beta \int e^{-\beta(H+pV)} \frac{\partial U}{\partial \phi_k} d\Omega \tag{66}$$

with the potential energy U. Multiplying by $1/Z$ yields

$$\frac{Z'}{Z} = -\beta \left\langle \frac{\partial U}{\partial \phi_k} \right\rangle . \tag{67}$$

Analogously we have

$$G' = \frac{\partial G}{\partial \phi_k} = \frac{\partial}{\partial \phi_k} \int O_i e^{-\beta(H+pV)} d\Omega = -\beta \int O_i e^{-\beta(U+pV)} \frac{\partial U}{\partial \phi_k} d\Omega \tag{68}$$

and

$$\frac{G'}{Z} = -\beta \left\langle O_i \frac{\partial U}{\partial \phi_k} \right\rangle . \tag{69}$$

Inserting Eqs. (67) and (69) into Eq. (65) finally gives the expression

$$\frac{\partial \langle O_i \rangle}{\partial \phi_k} = -\beta \left(\left\langle O_i \frac{\partial U}{\partial \phi_k} \right\rangle - \langle O_i \rangle \left\langle \frac{\partial U}{\partial \phi_k} \right\rangle \right) . \tag{70}$$

All the ensemble averages on the rhs of Eq. (70) can be computed from short finite trajectories if the relaxation time is short, e.g. for the solid state and small parameter changes. By analogy with extended-system Car-Parrinello procedures it is then obvious how to combine the simulation of the real system with the simultaneous SA dynamic optimization of the parameter space. In this case, the real system simulation is the rate limiting part. The major advantage of this approach is that there is no need for a monotonic relation between parameter change and observable response.

3.4. EXAMPLES

The following two sections briefly illustrate the results obtained above with some examples of realistic systems. Details will be published elsewhere.

3.4.1. *Lennard-Jones Parameters from Quantum Chemical Data*

The interaction between N,N,N-trimethylamine-N-oxide (TMAO) and water has been studied [83]. *Ab initio* geometry optimization were performed, first, for the isolated N-oxide on the HF/6-31G** level of theory. Second, partial charges were computed by fitting to the electrostatic potential [84]. Third, two molecules of water according to the TIPS3P model [85] were added and the intermolecular orientation was optimized with MP2/6-31G** while keeping the individual molecules rigid. The total model potential function consists of Coulomb interaction between the atomic partial charges, plus the Lennard-Jones (LJ) model for the van der Waals interaction between atoms i,j separated by r_{ij}

$$V_{LJ}(r_{ij}) = 4\varepsilon_{ij}[(\sigma_{ij}/r_{ij})^{12} - (\sigma_{ij}/r_{ij})^6] , \tag{71}$$

where the well depth ε_{ij} and the contact distance σ_{ij} are calculated as geometric and arithmetic means respectively. Each atom type of TMAO is assigned an LJ parameter tuple; thus we have to determine a total of eight parameters, ε_O, ε_N, ε_C, ε_H, and σ_i accordingly. Intermolecular distances of the optimized *ab initio* dihydrate complex were taken as reference observables. We then need the derivative of, e.g., some distance in the optimized model r_0 with respect to force field parameters, approximately [86]

$$\frac{\partial r_0}{\partial \phi_k} = -F^{-1}(q_0,\phi)\frac{\partial \nabla V(q_0,\phi)}{\partial \phi_k}\frac{\partial r_0}{\partial q_0} \tag{72}$$

where V is the potential, F the Hessian, and q_0 denotes the set of Cartesian coordinates representing the minimized model PES.

The parameter optimization was performed with an initially very low bath temperature. The constant temperature method of Sec. 2.2.2. will under these conditions drive the system rapidly down in the direction of steepest descent, but, in contrast to direct quenching, there is still enough kinetic energy to cross some low-lying barriers. The resulting evolution of the model parameters as a function of the number of annealing cycles is depicted in Figure 2. Parameter mass scaling was done after each cycle comprising only ten time steps.

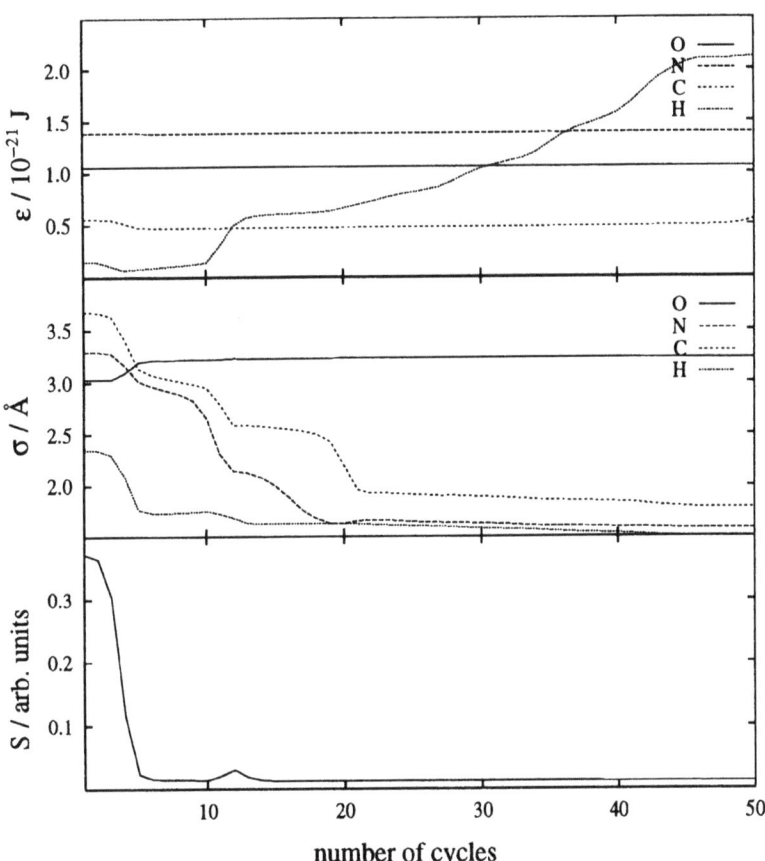

Figure 2. Evolution of the annealing run for the Lennard-Jones parameters of TMAO in the presence of two water molecules as function of the number of annealing cycles. Top: Well depth parameters ε; middle: Contact distance parameters σ; bottom: Penalty function. Values represent averages over each cycle.

After steep initial descent, the parameters cross a major barrier of the penalty function which then further decreases on average. This system is clearly statistically underdetermined. However, the annealing run reveals some topographical features of the penalty function "PES". Every visible plateau in the parameter evolution represent potential test sets that must be approved or discarded against independent observables. In this case, the parameters corresponding to the first major well in S turn out, though not globally optimal, to very well represent the experimental crystal structure in an MD simulation.

3.4.2. Ab Initio Data and Thermal Averages

A model function for solid fluoroapatite $Ca_{10}(PO_4)_6F_2$ has been parameterized [87] for use in MD simulations. The intermolecular contributions consist of Coulomb terms and Born-Mayer [88,89] nonbonded interactions. The intramolecular phosphate interaction is described by harmonic bond distance and angle bending terms and additional 1,3 distance dependent harmonic Urey-Bradley terms.

Starting from the experimental crystal structure, partial charges were determined for rigid phosphate by a modified fit to the quantum chemical electrostatic potential which includes crystal effects [90]. The LANL1DZ basis set [91] was used. Phosphate was treated quantum mechanically in the gas phase again with the LANL1DZ basis set to determine the optimal geometry and normal frequencies. This data was used for a rapid annealing fit similar to Sec. 3.4.1. Here, we additionally need the derivative of eigenvalues λ_i with respect to model function parameters [78]

$$\frac{\partial \lambda_i}{\partial \phi_k} = u_i^T \frac{\partial F(q_0, \phi)}{\partial \phi_k} u_i \tag{73}$$

with F now being the mass-weighted Hessian and u_i the i-th eigenvector. The annealing procedure leads to agreement between *ab initio* and model optimal geometry within 0.01 Å and for the frequencies within 1 cm^{-1}.

The parametrization was then completed by fitting the Born-Mayer parameters with respect to agreement between experimental and thermally averaged theoretical crystal structure as obtained from MD simulations. To this end the Cartesian experimental atomic sites have been used as reference observables. Mainly due to insufficient sampling, the penalty function turns out to be extremely rugged and therefore a very small cooling parameter $\lambda = 0.01$ in Eq. (62) had to be used. The time step in the real system was 1 fs and the forces in parameter space were periodically updated after sampling 50 time steps in real space. In order to eliminate the possibility of rotation, the axes of the simulation cell were matched to the experimental structure at each real space time step [92]. Parameter masses were adjusted every ten steps in parameter space.

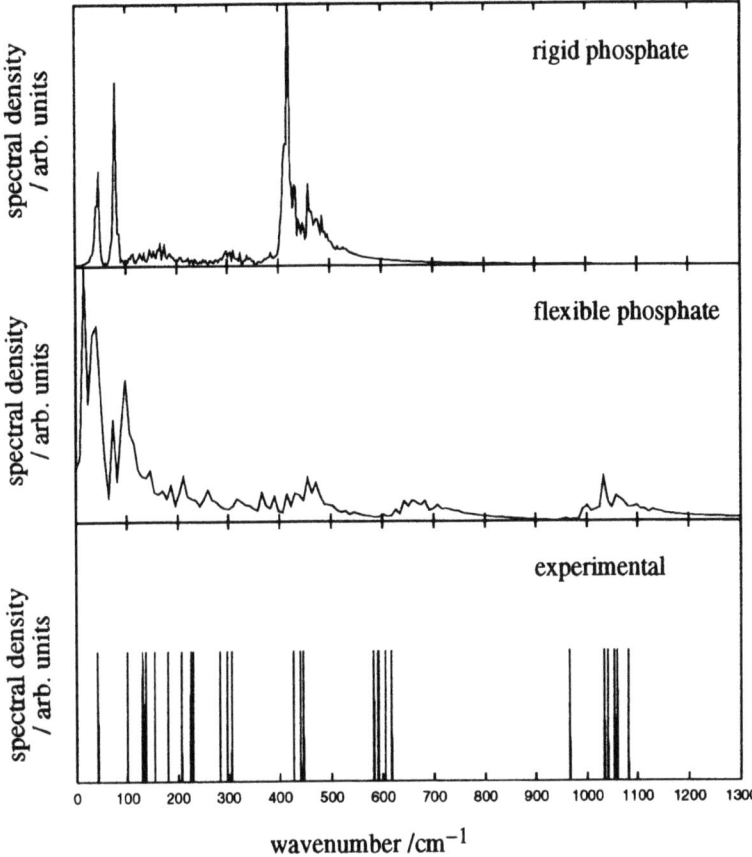

Figure 3. Theoretical and experimental IR spectra of fluorapatite. Top: Theoretical spectrum with rigid phosphate; middle: Theoretical spectrum with flexible phosphate; bottom: Experiment.

For both the rigid phosphate model for which parameters could be determined by a trial-and-error grid search, and for the converged parameters of the flexible phosphate model, infrared spectra were computed and depicted in Figure 3. The agreement between flexible model and experiment is excellent. Although the method definitely needs some further refinement and study, the results are very promising.

4. Conclusions

Which lessons have we learned? First of all, without definite information about the "correct" functional form for the description of an *ab initio* PES, there cannot be a unique way to determine parameters. The quality of a set that fits data well can be obscured by choosing an inappropriate functional model. Topographies depend on both functionals and parameters whose contributions cannot be decoupled. It is even unclear whether it is reasonable to search for the global minimum of some penalty function as long as we have no idea whether we have used a sufficient amount of reference data.

On the other hand, simulation techniques which do not need model functions have the potential to generate all the data we could possibly need for deriving both functionals and parameters. This is where the most fruitful interplay between quantum chemistry and molecular dynamics can be identified: Since it is *a priori* difficult to decide what are the important configurational regions we should look at in detail with quantum chemical methods, it is better to let the system decide for itself - by means of CP or QM/MM methods - and use the generated information *afterwards* or *simultaneously* in the sense of data compression and thereby learning.

The question of generally valid, suitable functional forms remains unanswered. But for the efficient extraction of parameters from raw data some aspects and methods have been discussed in this article. Simulated annealing has the ability to overcome optimization difficulties that usually require human interference. Automated and fast data handling will be more and more important in the future as computing power further increases.

5. Acknowledgements

S. M. K. thanks the Fond der Chemischen Industrie for financial support, and the Alexander von Humboldt-Stiftung for a postdoctoral fellowship at The University of Chicago.

6. References

1. McQuarrie, D. A. (1976) *Statistical Mechanics*, Harper and Row, New York.
2. Chandler, D. (1987) *Introduction to Modern Statistical Mechanics*, Oxford University Press, New York.
3. Van Gunsteren, W. F. and Berendsen, H. J. C. (1990) *Angew. Chem.* **102**, 1020.
4. Allen, M. P. and Tildesley, D. J. (1990) *Computer Simulation of Liquids*, Clarendon Press, Oxford.
5. Heermann, D. W. (1986) *Computer Simulation Methods in Theoretical Physics*, Springer-Verlag, Berlin.
6. Ciccotti, G. and Hoover, W. G. (1986) *Molecular Dynamical Simulation of Statistical-Mechanical Systems*, North-Holland, Amsterdam.
7. Hoover, W.G. (1986) *Molecular Dynamics*, Springer-Verlag, Berlin.
8. Brickmann, J., Kast, S. M., Vollhardt, H., and Reiling, S. (1995) in E. Yurtsever (ed.), *Proceedings of the NATO Advanced Study Institute, Series C*, 470, pp. 217-253.
9. Christoffersen, R. E. (1989) *Basic Principles and Techniques of Molecular Quantum Mechanics*, Springer-Verlag, New York.
10. Szabo, A. and Ostlund, N. S. (1989) *Modern Quantum Chemistry*, McGraw-Hill, New York.
11. Remler, D. K. and Madden, P. A. (1990) *Molec. Phys.* **70**, 921.

12. Tuckerman, M. E., Ungar, P. J., von Rosenvinge, T., and Klein, M. L. (1996) *J. Phys. Chem.* **100**, 12878.
13. Feynman, R. P. (1979) *Phys. Rev.* **56**, 340.
14. Hellmann, H. (1937) *Einführung in die Quantenchemie*, Deutick.
15. Burkert, U. and Allinger, N. L. (1982) *Molecular Mechanics*, ACS Monograph 177, American Chemical Society, Washington, D.C.
16. Bowen, J. P. and Allinger, N. L. (1991) in Lipkowitz, K. B. and Boyd, D. B. (eds.), *Reviews in Computational Chemistry, Vol. 2*, VCH Publishers, New York, pp. 81-97.
17. Li, M. and Vitányi, P. (1993) *An Introduction to Kolmogorov Complexity and its Applications*, Springer-Verlag, New York.
18. Berry, R. S. (1993) *Chem. Rev.* **93**, 2379.
19. Berry, R. S. (1994) *J. Phys. Chem.* **98**, 6910.
20. Liang, G., Fox, P. C., and Bowen, J. P. (1996) *J. Comp. Chem.* **17**, 940.
21. Dinur, U. and Hagler, A. T. (1991) in Lipkowitz, K. B. and Boyd, D. B. (eds.), *Reviews in Computational Chemistry, Vol. 2*, VCH Publishers, New York, pp. 99-164.
22. Maple, J. R., Hwang, M.-J., Stockfisch, T. P., Dinur, U., Waldman, M., Ewig, C. S., and Hagler, A. T. (1994) *J. Comp. Chem.* **15**, 162.
23. Cornell, W. D., Cieplak, P., Bayly, C. I., Gould, I. R., Merz, K. M. Jr., Ferguson, D. M., Spellmeyer, D. C., Fox, T., Caldwell, J. W., and Kollman, P. A. (1995) *J. Am. Chem. Soc.* **117**, 5179.
24. Halgren, T. A. (1996) *J. Comp. Chem.* **17**, 490; and subsequent articles.
25. Allinger, N. L., Chen, K., and Lii, J.-H. (1996) *J. Comp. Chem.* **17**, 642; and subsequent articles.
26. Eksterowics, J. E. and Houk, K. N. (1993) *Chem. Rev.* **93**, 2439.
27. Jensen, F. (1994) *J. Comp. Chem.* **15**, 1199.
28. Gao, J. (1995) in Lipkowitz, K. B. and Boyd, D. B. (eds.), *Reviews in Computational Chemistry, Vol. 7*, VCH Publishers, New York, pp. 119-185.
29. Gao, J. (1996) *Acct. Chem. Res.* **29**, 298.
30. Kast, S. M., Nicklas, K., Bär, H.-J., and Brickmann, J. (1994) *J. Chem. Phys.* **100**, 566.
31. Kast, S. M. and Brickmann, J. (1996) *J. Chem. Phys.* **104**, 3732.
32. Gear, C. W. (1971) *Numerical initial value problems in ordinary differential equations*, Prentice-Hall, Englewood Cliffs, NJ.
33. Verlet, L. (1967) *Phys. Rev.* **159**, 98.
34. Berendsen, H. J. C. and Van Gunsteren, W. F. (1985) *Proceedings of the Enrico Fermi Summer School, Molecular dynamics simulation of statistical mechanical systems*, Soc. Italiana di Fisica, Bologna.
35. Swope, W. C., Andersen, H. C., Berens, P. H, and Wilson, K. R. (1982) *J. Chem. Phys.* **76**, 637.
36. Duane, S., Kennedy, A. D., Pendleton, B. J., and Roweth, D. (1987) *Phys. Lett.* **B 195**, 216.
37. Forrest, B. M. and Suter, U. W. (1994) *J. Chem. Phys.* **101**, 2616.
38. Woodcock, L. V. (19971) *Chem. Phys. Lett.* **10**, 257.
39. Evans, D. J., Hoover, W. G., Failor, B. H., Moran, B., and Ladd, A. J. C. (1983) *Phys. Rev. A* **28**, 1016.
40. Evans, D. J. and Morriss, G. P. (1984) *Comput. Phys. Rep.* **1**, 297.
41. Esparza, C. H. and Kronmüller, H. (1989) *Molec. Phys.* **68**, 1341.
42. Berendsen, H. J. C., Postma, J. P. M., Van Gunsteren, W. F., DiNola, A., and Haak, J. R. (1984) *J. Chem. Phys.* **81**, 3684.
43. Nosé, S. (1984) *Molec. Phys.* **52**, 255.
44. Hoover, W. G. (1985) *Phys. Rev. A* **31**, 1695.
45. Toxvaerd, S. and Olsen, O. H. (1990) *Ber. Bunsenges. Phys. Chem.* **94**, 274.

46. Martyna, G. J., Klein, M. L., and Tuckerman, M. (1992) *J. Chem. Phys.* **97**, 2635.
47. Holian, B. L., Voter, A. F., and Ravelo, R. (1995) *Phys. Rev. E* **52**, 2338.
48. Hoover, W. G. and Holian, B. L. (1996) *Phys. Lett. A* **211**, 253.
49. Nosé, S. (1991) *Prog. Theor. Phys. Suppl.* **103**, 1.
50. Andersen, H. C. (1980) *J. Chem. Phys.* **72**, 2384.
51. Tanaka, H., Nakanishi, K., and Watanabe, N. (1983) *J. Chem. Phys.* **78**, 2626.
52. Ciccotti, G. and Tenenbaum, A. (1980) *J. Stat. Phys.* **23**, 767.
53. Bonomi, E. (1985) *J. Stat. Phys.* **39**, 167.
54. Schneider, T. and Stoll, E. (1978) *Phys. Rev. B* **17**, 1302.
55. Van Gunsteren, W. F., Berendsen, H. J. C., and Rullmann, J. A. C. (1981) *Molec. Phys.* **44**, 69.
56. Strutt, J. W. (Baron Rayleigh) (1891) *Phil. Mag.* **32**, 424; (1902) *Scientific Papers*, Vol. 3, Cambridge University Press, London.
57. Barkai, E. and Fleurov, V. (1995) *Phys. Rev. E* **52**, 137.
58. Mishra, B. and Schlick, T. (1996) *J. Chem. Phys.* **105**, 299.
59. Gelfond, A. O. (1958) *Differenzenrechnung*, VEB Verlag der Wissenschaften, Berlin.
60. Papoulis, A. (1991) *Probability, Random Variables, and Stochastic Processes*, 3rd ed., McGraw-Hill, New York.
61. Reiling, S. and Brickmann, J. (1995) *Macromol. Theory Sim.* **4**, 725.
62. Hauptmann, S., Mosell, T., Reiling, S., and Brickmann, J. (1996) *Chem. Phys.* **208**, 57.
63. Mosell, T., Schrimpf, G., Hahn, C., and Brickmann, J. (1996) *J. Phys. Chem.* **100**, 4571.
64. Mosell, T., Schrimpf, G., and Brickmann, J. (1996) *J. Phys. Chem.* **100**, 4582.
65. Car, R. and Parrinello, M. (1985) *Phys. Rev. Lett.* **55**, 2471.
66. Kirkpatrick, S., Gelatt, G. D., and Vechi, M. P. (1983) *Science* **220**, 671.
67. Vanderbilt, D. and Louie, S. G. (1984) *J. Comp. Phys.* **56**, 259.
68. Van Laarhoven, P. J. M. and Aarts, E. H. L. (1987) *Simulated Annealing: Theory and Applications*, Kluwer, Dordrecht.
69. Heistermann, J. (1994) *Genetische Algorithmen*, Teubner, Stuttgart.
70. Kalivas, J. H., ed. (1995) *Adaption of Simulated Annealing to Chemical Optimization Problems, Data Handling in Science and Technology*, Vol. 15, Elsevier, Amsterdam.
71. Ryckaert, J. P., Ciccotti, G., and Berendsen, H. J. C. (1977) *J. Comp. Phys.* **23**, 327.
72. Kryachko, E. S. (1993) *Int. J. Quantum Chem.* **49**, 109.
73. Warshel, A. and Weiss, R. M. (1980) *J. Am. Chem. Soc.* **102**, 6218.
74. Field, M. J., Bash, P. A., and Karplus, M. (1990) *J. Comp. Chem.* **11**, 700.
75. Brooks, B. R., Bruccoleri, R. E., Olafson, B. D., States, D. J., Swaminathan, S., and Karplus, M. (1983) *J. Comp. Chem.* **4**, 187.
76. Nilsson, L. and Karplus, M. (1986) *J. Comp. Chem.* **7**, 591.
77. Press, W. H., Teukolsky, S. A., Vetterling, W. T., and Flannery, B. P. (1992) *Numerical Recipes in Fortran*, 2nd ed., Cambridge University Press, Cambridge.
78. Ganda-Kesuma, F. S. and Miller, K. J. (1994) *J. Comp. Chem.* **15**, 1291.
79. Kast, S. M. and Berry, R. S., manuscript in preparation.
80. Tuckerman, M., Berne, B. J., and Martyna, G. J. (1992) *J. Chem. Phys.* **97**, 1990.
81. Fischer, T. H., Petersen, W. P., and Lüthi, H. P. (1995) *J. Comp. Chem.* **16**, 923.
82. Njo, S. L., van Gunsteren, W. F., and Müller-Plathe, F. (1995) *J. Chem. Phys.* **102**, 6199.
83. Kast, K. M., Brickmann, J., Kast, S. M., and Berry, R. S., manuscript in preparation.
84. Williams, D. E. (1991) in Lipkowitz, K. B. and Boyd, D. B. (eds.), *Reviews in Computational Chemistry*, Vol. 2, VCH Publishers, New York, pp. 219-271.
85. Jorgensen, W. L., Chandrasekharm, J., Madura, J. D., Impey, R. W., and Klein, M. L. (1983) *J. Chem. Phys.* **79**, 926.
86. Lifson, S. and Warshel, A. (1968) *J. Chem. Phys.* **49**, 5116.

87. Hauptmann, S., Kast, S. M., Dufner, H., and Brickmann, J., manuscript in preparation.
88. Huggins, M. L. and Mayer, J. E. (1933) *J. Chem. Phys.* **1**, 643.
89. Fumi, F. G. and Tosi, M. P. (1964) *J. Phys. Chem. Solids* **25**, 31.
90. Dufner, H. (1995) Thesis, Technische Hochschule Darmstadt, D17.
91. Hay, P. J. and Wadt, W. R. (1985) *J. Chem. Phys.* **82**, 270.
92. Ferro, D. R. and Hermans, J. (1977) *Acta Cryst. A* **33**, 345.

THE PERMUTATION GROUP IN MANY-ELECTRON THEORY

STEN RETTRUP
Department of Chemistry
Copenhagen University
Universitetsparken 5
DK-2100 Copenhagen Ø
Denmark

1. Introduction

Molecular properties can today be calculated by *ab initio* quantum chemistry methods to a high degree of accuracy. To a large extend it is due to the tremendous progress in the development of efficient numerical algorithms for solving quantum chemistry problems and the simultaneously increased computational capabilities on modern computers.

One of these algorithms is based on the theory of the symmetric group algebra which has been continuously developed during the last century. Particularly, the long series of papers by Alfred Young from the beginning of the present century introduced many of the concepts which are presently used in the application of symmetric group methods to the many-electron problem. A complete collection of the papers is available in the book edited by Coxeter *et al.* (1977). The results in the articles which particularly consider the representation theory of the permutation group have been extensively treated in the monograph by Rutherford (1948).

The introduction of the symmetric group methods in quantum mechanics is to a large extend due to the classical book from 1931 by Herman Weyl (1950). As Robinson formulates it in the foreword to Young's collected papers (Coxeter *et al.*, 1977) : "Young's work had attracted Weyl's attention, and suddenly he was famous". Since the permutation group is fundamental in the theory of many-particle systems the subject has been treated extensively in the celebrated monographs by Wigner (1959), Robinson (1961), Boerner (1963) and Hamermesh (1964). The various concepts in connection with the symmetric group was elegantly summarized by Coleman (1968) who provided in a few pages the basic terminology and many useful relations.

The application of the symmetric group as a computational procedure for studying many-electron systems goes back to 1955 where Kotani *et al.* described the close relationship between many-electron spin eigenfunctions and the irreducible representations of the permutation group. It was shown how the symmetric group could be used in general for evaluating Hamiltonian matrix elements between configuration state functions with well-defined spin quantum numbers. The method was later applied by Gerratt (1971) in his presentation of spin-coupled wave functions with particular emphasis on valence bond theory.

The concept of spin-free quantum chemistry was introduced by Matsen (1964) who demonstrated how the Pauli exclusion principle could be formulated without explicit use of spin variables in functions describing the electronic states. During the seventies a large number of different spin-free techniques were developed. Some of the procedures were directly based on the symmetric group. Other approaches utilized the theory of unitary groups. Primarily, the different schemes were developed in order to provide efficient tools for evaluating electronic Hamiltonian matrix elements between configuration state functions with particular emphasis on the application in configuration interaction (CI) calculations. The proceedings of a workshop organized by Hinze (1981) provides an overview and many references to the original work. A few years later, particularly, the methods based on the symmetric group has been presented by Duch and Karwowski (1985) and very recently, an extensive review of the many different approaches and methods is given in the monograph by Pauncz (1995). The book also provides a very comprehensive bibliography containing 404 references to the original publications.

As indicated above the literature on symmetric group methods is today very large and contains many different strategies and algorithms for the implementation in computer programmes to be used for studying the electronic structure of molecules. Since the various methods and techniques have recently been reviewed extensively by Pauncz (1995) it is the scope of the present article to provide a fundamental introduction to the symmetric group approach in quantum chemistry. Particularly, the author wishes to emphasize the generality of the method in many-electron theory.

2. The Electronic Schrödinger Equation

The majority of electronic structure calculations are based on the non-relativistic electronic Schrödinger equation where the nuclear coordinates,

$\mathbf{R} \equiv (\mathbf{R}_1, \mathbf{R}_2, \ldots, \mathbf{R}_K)$, for a collection of K nuclei are kept fixed as parameters. As such $\mathbf{R}$ just represents a given molecular geometry which has been chosen in advance. In this model description of molecules, usually called the Born-Oppenheimer approximation, only the electronic coordinates are treated as dynamical variables. If the vector $\mathbf{r} \equiv (\mathbf{r}_1, \mathbf{r}_2, \ldots, \mathbf{r}_N)$ denotes the set of Cartesian coordinates for the N electrons, $\mathbf{r}_i \equiv (x_i, y_i, z_i)$, and $\sigma \equiv (\sigma_1, \sigma_2, \ldots, \sigma_N)$ indicates the corresponding set of spin variables the electronic Schrödinger equation may be written

$$\hat{H}\,\Psi(\mathbf{r}, \sigma; \mathbf{R}) = E(\mathbf{R})\,\Psi(\mathbf{r}, \sigma; \mathbf{R}). \qquad (2-1)$$

For convenience the explicit dependence of $\mathbf{R}$ will be omitted in the following. I.e. $\Psi(\mathbf{r}, \sigma) \equiv \Psi(\mathbf{r}, \sigma; \mathbf{R})$. The Hamiltonian is given by

$$\hat{H} = \sum_{i=1}^{N} \hat{h}(i) + \sum_{i<j}^{N} \frac{1}{r_{ij}} + \sum_{a<b}^{K} \frac{Q_a Q_b}{R_{ab}} \qquad (2-2)$$

where the individual electrons are denoted by $i, j = 1, 2, \ldots, N$, and the one-electron operators $\hat{h}(i)$ are defined as

$$\hat{h}(i) = -\frac{1}{2}\nabla_i^2 - \sum_{a=1}^{K} \frac{Q_a}{r_{ia}}. \qquad (2-3)$$

The distances in the operators are defined as $R_{ab} = |\mathbf{R}_a - \mathbf{R}_b|$, $r_{ia} = |\mathbf{r}_i - \mathbf{R}_a|$, $r_{ij} = |\mathbf{r}_i - \mathbf{r}_j|$ and Q_a is the charge of nucleus a.

2.1. PERMUTATION SYMMETRY

We first note that the Hamiltonian we are considering does not contain operators which act on the spin variables. It means that if we let

$$P = \begin{pmatrix} 1 & 2 & 3 & \ldots & N \\ p_1 & p_2 & p_3 & \ldots & p_N \end{pmatrix} \qquad (2-4)$$

represent an abstract permutation and $\hat{P}^\sigma$ the corresponding permutation operator which acts separately on the spin variables

$$\hat{P}^\sigma \Psi(\mathbf{r}_1, \ldots, \mathbf{r}_N, \sigma_1, \ldots, \sigma_N) = \Psi(\mathbf{r}_1, \ldots, \mathbf{r}_N, \sigma_{p_1}, \ldots, \sigma_{p_N}) \qquad (2-5)$$

then the Hamiltonian commutes with with $\hat{P}^\sigma$

$$[\hat{P}^\sigma, \hat{H}] = 0 \quad \text{for all} \quad \hat{P}^\sigma \in \mathcal{S}_N^\sigma. \qquad (2-6)$$

Similarly, we have for the permutation operator, $\hat{P}^r$,

$$\hat{P}^r \Psi(\mathbf{r}_1, \ldots, \mathbf{r}_N, \sigma_1, \ldots, \sigma_N) = \Psi(\mathbf{r}_{p_1}, \ldots, \mathbf{r}_{p_N}, \sigma_1, \ldots, \sigma_N) \qquad (2-7)$$

which only permutes spatial variables

$$[\hat{P}^r, \hat{H}] = 0 \quad \text{for all} \quad \hat{P}^r \in \mathcal{S}_N^r \qquad (2-8)$$

since the Hamiltonian is symmetric in all the electronic coordinates. From the definition of $\hat{P}^\sigma$ and $\hat{P}^r$ it is obvious that all the elements of the two different realizations of the abstract permutation group $\mathcal{S}_N$ commute. I.e.

$$[\hat{Q}^r, \hat{P}^\sigma] = 0 \quad \text{for all} \quad \hat{Q}^r \in \mathcal{S}_N^r \quad \text{and} \quad \hat{P}^\sigma \in \mathcal{S}_N^\sigma. \qquad (2-9)$$

As a consequence the symmetry group of the electronic Hamiltonian is the direct product group, $\mathcal{S}_N^r \otimes \mathcal{S}_N^\sigma$, of the two groups we have indicated by $\mathcal{S}_N^r$ and $\mathcal{S}_N^\sigma$. It means that the exact solutions, $\Psi(\mathbf{r}, \boldsymbol{\sigma})$, to the electronic Schrödinger equation, must belong to a specific irreducible representation of the group in the case of non-degenerate states. Otherwise, they may be chosen to do so if the states are degenerate.

2.2. THE PAULI PRINCIPLE

The electrons are fermions which obey the Pauli principle. It means that the electronic wave function must belong to the totally antisymmetric representation of the permutation group, $\mathcal{S}_N$, where the elements are the operators which simultaneously permute Cartesian and spin coordinates

$$\hat{P} = \hat{P}^r \hat{P}^\sigma. \qquad (2-10)$$

In the most simple form the condition may be stated by requiring, that if $\hat{P}_{ij}$ represents an operator which interchanges simultaneously the Cartesian and the spin coordinates of any two electrons i and j

$$\hat{P}_{ij} \Psi(\ldots \mathbf{r}_i \ldots \mathbf{r}_j \ldots \sigma_i \ldots \sigma_j \ldots) = \Psi(\ldots \mathbf{r}_j \ldots \mathbf{r}_i \ldots \sigma_j \ldots \sigma_i \ldots)$$

then the function changes sign under the action of the permutation operator

$$\hat{P}_{ij} \Psi(\mathbf{r}, \boldsymbol{\sigma}) = -\Psi(\mathbf{r}, \boldsymbol{\sigma}). \qquad (2-11)$$

More generally, for arbitrary permutations it implies

$$\hat{P} \Psi(\mathbf{r}, \boldsymbol{\sigma}) = \epsilon_P \Psi(\mathbf{r}, \boldsymbol{\sigma}) \qquad (2-12)$$

where ϵ_P represents the parity (± 1) of the permutation P.

In order to study separately the transformation properties under permutation of Cartesian coordinates and spin coordinates it is convenient to define the usual antisymmetric projection operator $\hat{A}$ as

$$\hat{A} = \frac{1}{N!} \sum_{P \in \mathcal{S}_N} \epsilon_P \hat{P} = \frac{1}{N!} \sum_{P \in \mathcal{S}_N} \epsilon_P \hat{P}^r \hat{P}^\sigma \qquad (2-13)$$

which is normalized according to the relation

$$\hat{A}\,\hat{A} = \hat{A}. \qquad (2-14)$$

Alternatively, we can now state the above antisymmetry restriction by requiring the wave function to be an eigenfunction of the antisymmetric projection operator

$$\hat{A}\,\Psi(\mathbf{r},\sigma) = \Psi(\mathbf{r},\sigma). \qquad (2-15)$$

In other words the last equation states that the allowed wave functions must belong to the antisymmetric representation of $\mathcal{S}_N$.

2.3. N-ELECTRON SPIN EIGENFUNCTIONS

The electronic Hamiltonian in Eq. (2-2) does not contain any spin-dependent terms explicitly. It implies that the wave function can be chosen to be simultaneously eigenfunction of the total spin operators $\hat{S}^2$ and $\hat{S}_z$ with the corresponding eigenvalues $S(S+1)$ and M (in atomic units). In order to utilize this property in the next section we consider briefly the basic connection between electronic spin eigenfunctions and the permutation group.

Let a spin eigenfunction for a system consisting of N electrons be denoted

$$\Theta_{SM;\rho} \equiv \Theta_{S,M;\rho}(\sigma_1, \sigma_2, \ldots, \sigma_N). \qquad (2-16)$$

It satisfies the eigenvalue equations

$$\hat{S}_z\,\Theta_{S,M;\rho} = M\,\Theta_{S,M;\rho} \qquad (2-17)$$

$$\hat{S}^2\,\Theta_{S,M;\rho} = S(S+1)\,\Theta_{S,M;\rho}. \qquad (2-18)$$

If we denote the complete set of linearly independent spin eigenfunctions by

$$\{\Theta_{S,M;f^{N,S}}\} \equiv \{\Theta_{S,M;1}, \Theta_{S,M;2}, \Theta_{S,M;3}, \ldots, \Theta_{S,M;f^{N,S}}\} \qquad (2-19)$$

it is well-known (Pauncz, 1979) that the total number of linearly independent spin eigenfunctions is simply given by

$$f^{N,S} = \begin{pmatrix} N \\ \frac{1}{2}N - S \end{pmatrix} - \begin{pmatrix} N \\ \frac{1}{2}N - S - 1 \end{pmatrix}. \qquad (2-20)$$

Since the spin operators $\hat{S}_z$ and $\hat{S}^2$ are symmetric in all the spin coordinates

$$[\hat{S}_z, \hat{P}^\sigma] = [\hat{S}^2, \hat{P}^\sigma] = 0 \qquad (2-21)$$

we get from Eq. (2-17) and Eq. (2-18) that the set of spin eigenfunctions $\{\Theta_{S,M;f^{N,S}}\}$ forms a basis for an $f^{N,S}$-dimensional irreducible representation of the permutation group S_N^σ

$$\hat{P}^\sigma \{\Theta_{S,M;f^{N,S}}\} = \{\Theta_{S,M;f^{N,S}}\} \, \mathbf{U}^{N,S}(P) \qquad (2-22)$$

for all $P^\sigma \in S_N^\sigma$. The actual form of the representation matrices is not unique but it depends on the choice of spin eigenfunctions. It is noted that the representation matrices $\mathbf{U}^{N,S}(P)$ are independent of the M-value.

2.4. SPIN-FREE QUANTUM CHEMISTRY

From the last section we know that we can classify the electronic wave function with respect to the spin quantum numbers S and M. Consequently, we can always express the wave function as a simple linear combination of the form

$$\Psi_{SM}(\mathbf{r}, \sigma) = \sum_{\rho=1}^{f^{N,S}} \Phi_\rho(\mathbf{r}) \, \Theta_{S,M;\rho}(\sigma) \qquad (2-23)$$

where the Cartesian coordinates and the spin coordinates are separated into product terms consisting of a spatial function $\Phi_\rho(\mathbf{r})$ and a spin eigenfunction $\Theta_{S,M;\rho}(\sigma)$. Naturally, the function does not in general satisfy the antisymmetry condition required by the Pauli principle but the set of spatial functions

$$\{\Phi_{f^{N,S}}\} = \{\Phi_1(\mathbf{r}), \Phi_2(\mathbf{r}), \ldots, \Phi_{f^{N,S}}(\mathbf{r})\} \qquad (2-24)$$

is forced to obey symmetry transformation properties under permutation operators which are induced by the choice of spin basis.

In order to determine the explicit form of the transformation properties to be satisfied by the spatial functions, it is convenient to write the electronic wave function as a simple scalar (matrix) product of two vectors. Let

$\{\Theta_{S,M;f^{N,S}}\}^T$ represent the transpose of the row vector $\{\Theta_{S,M;f^{N,S}}\}$. With this definition, we can in a compact way write the electronic wave function as

$$\Psi_{SM}(\mathbf{r}, \boldsymbol{\sigma}) = \{\Phi_{f^{N,S}}\} \{\Theta_{S,M;f^{N,S}}\}^T. \qquad (2-25)$$

If we let $\mathbf{W}^{N,S}(P)$ denote the transformation matrix belonging to the set of Cartesian functions

$$\hat{P}^r \{\Phi_{f^{N,S}}\} = \{\Phi_{f^{N,S}}\} \mathbf{W}^{N,S}(P) \qquad (2-26)$$

we obtain from the Pauli principle

$$\hat{P} \Psi_{SM}(\mathbf{r}, \boldsymbol{\sigma}) = \epsilon_P \{\Phi_{f^{N,S}}\} \{\Theta_{S,M;f^{N,S}}\}^T \qquad (2-27)$$

$$= \{\Phi_{f^{N,S}}\} \mathbf{W}^{N,S}(P) \mathbf{U}^{N,S}(P)^T \{\Theta_{S,M;f^{N,S}}\}^T.$$

In the last equation $\mathbf{U}^{N,S}(P)^T$ denotes the transpose of the spin representation matrix $\mathbf{U}^{N,S}(P)$ defined in Eq. (2-22). It implies

$$\mathbf{W}^{N,S}(P)\mathbf{U}^{N,S}(P)^T = \epsilon_P \mathbf{1} \qquad (2-28)$$

where $\mathbf{1}$ represents an $f^{N,S}$-dimensional unit matrix. From the last equation we finally get

$$\mathbf{W}^{N,S}(P) = \epsilon_P \mathbf{U}^{N,S}(P^{-1})^T. \qquad (2-29)$$

It is easy to verify that the matrices $\mathbf{W}^{N,S}(P)$ for $P \in \mathcal{S}_N$ also constitute an irreducible matrix representation of the permutation group. Usually the representation is called the adjoint representation or the dual representation to $\mathbf{U}^{N,S}(P)$.

In the case where $\mathbf{U}^{N,S}(P)$ is a real orthogonal representation corresponding to a real orthogonal spin basis the matrices are obtained by the simple relation

$$\mathbf{W}^{N,S}(P) = \epsilon_P \mathbf{U}^{N,S}(P) \qquad (2-30)$$

which is most frequently used.

Consider now the Schrödinger equation where we have inserted Eq. (2-23)

$$\hat{H} \sum_{\rho=1}^{f^{N,S}} \Phi_\rho(\mathbf{r})\Theta_{S,M;\rho}(\boldsymbol{\sigma}) = E \sum_{\rho=1}^{f^{N,S}} \Phi_\rho(\mathbf{r})\Theta_{S,M;\rho}(\boldsymbol{\sigma}). \qquad (2-31)$$

Since the spin eigenfunctions are linearly independent we obtain that each of the spatial functions must satisfy the following spin-free Schrödinger equation

$$\hat{H}\,\Phi_\rho(\mathbf{r}) = E\,\Phi_\rho(\mathbf{r}) \quad \text{for} \quad \rho = 1, 2, \ldots, f^{N,S} \qquad (2-32)$$

with the additional condition that the allowed spin-free solutions $\Phi_\rho(\mathbf{r})$ must transform according to an irreducible representation of the permutation group which is dual to the desired spin representation. Essentially, the conventional antisymmetry requirement for many-electron wave functions has been replaced by a condition on the spin-free wave functions.

The $f^{N,S}$ linearly independent solutions to Eq. (2-32), however, do not represent different physical situations but they describe all the same electronic state. It is a consequence of the fact that the solutions are simply related to each other by permutational symmetry of the electronic coordinates. It means that in order to study electronic properties of a considered state we can always choose any one of the components, as for instance $\Phi_1(\mathbf{r})$.

Eq. (2-32) is a very fundamental equation and represents the spin-free counterpart of conventional many-electron theory. As such it may provide a starting point for both symmetric group and unitary group approaches to electronic structure calculations.

3. Approximate N-Electron Functions

In the case of approximate solutions to the Schrödinger equation the antisymmetry requirement is easily imposed on a given trial function by using that the antisymmetrizer $\hat{A}$ for any permutation P satisfy the relation

$$\hat{P}\,\hat{A} = \epsilon_P\,\hat{A}. \qquad (3-33)$$

It means that if $\Xi(\mathbf{r}, \boldsymbol{\sigma})$ is a primitive approximate solution then we can obtain a properly antisymmetrized trial function as

$$\Psi^o(\mathbf{r}, \boldsymbol{\sigma}) = \hat{A}\,\Xi(\mathbf{r}, \boldsymbol{\sigma}). \qquad (3-34)$$

It was very early realized by Dirac (1926) that starting from a simple single particle description consisting of spin orbitals the trial functions in the last equation leads to the concept of determinantal functions. Today such functions are usually called Slater determinants due to the developments by Slater (1929; 1931).

It is well-known that a single determinant does not in general form an eigenfunction of the total spin operators $\hat{S}^2$ and $\hat{S}_z$. Consequently, determinantal wave functions do not in general represent pure multiplets but are mixtures of for instance singlets, triplets, etc. In order to satisfy the spin symmetry constraint automatically in addition to the Pauli principle the unitary and symmetric group approaches have been very successful.

In the conventional unitary group approach the starting point is a set of orthonormal spin-free one-electron functions (orbitals) from where symmetry adapted tensor products are constructed to describe N-electron states. Usually these states are chosen as orthonormal Gel'fand states. Alternatively, the N-electron states may be described as Weyl states, but they have the disadvantage of being non-orthonormal. A detailed treatment is found in the pioneering article by Paldus (1976).

The symmetric group approach is not restricted to handle spin-free N-electron functions which are constructed from a one-electron basis. The form of trial function is in that description completely arbitrary. It is only for computational and conceptual reasons that the spin-free N-electron functions usually are generated from a one-electron basis.

Basically, there are two ways of presenting the symmetric group approach. One does not give any explicit reference to spin functions at all, but constructs the spin-free N-electron functions by using standard Wigner (or Young tableau) projection operators corresponding to an irreducible representation which is dual to the spin representation considered. In the other approach the spin-coupling is introduced from the beginning and it is the one which has been chosen in the following presentation since it automatically establishes the connection to the applied spin-coupling.

3.1. CONFIGURATION STATE FUNCTIONS

A configuration state function is defined as

$$\Psi^o_{S,M;\mu}(\mathbf{r}, \boldsymbol{\sigma}) \equiv \hat{A} \{\Phi^o(\mathbf{r}) \, \Theta_{S,M;\mu}(\boldsymbol{\sigma})\} \qquad (3-35)$$

where we for convenience have used the subscript μ to denote the applied spin-coupling. In the most primitive case the spatial trial function, $\Phi^o(\mathbf{r})$, is chosen as a simple product of spatial orbitals

$$\Phi^o(\mathbf{r}) = \varphi_1(\mathbf{r}_1) \, \varphi_2(\mathbf{r}_2) \, \varphi_3(\mathbf{r}_3) \, \ldots \, \varphi_N(\mathbf{r}_N). \qquad (3-36)$$

In this case the configuration state function can be written as a linear combination of Slater determinants. The explicit form of the linear combination, however, depends on the choice of spin eigenfunction, $\Theta_{S,M;\mu}(\boldsymbol{\sigma})$.

The connection between the exact wave function in Eq. (2-23) and the form of the configuration state function in Eq. (3-35) is not obvious. But if we expand the configuration state function as

$$\Psi^o_{S,M;\mu}(\mathbf{r},\sigma) = \frac{1}{N!} \sum_{P \in S_N} \epsilon_P \hat{P}^r \Phi^o(\mathbf{r}) \, \hat{P}^\sigma \Theta_{S,M;\mu}(\sigma)$$

$$= \frac{1}{N!} \sum_{\rho=1}^{f^{N,S}} \left(\sum_{P \in S_N} \epsilon_P U^{N,S}_{\rho\mu}(P) \hat{P}^r \Phi^o(\mathbf{r}) \right) \Theta_{S,M;\rho}(\sigma)$$

and define the matric basis units of the group algebra (Matsen, 1964) or Wigner operators as

$$\hat{\omega}^{N,S}_{\rho\mu} \equiv \frac{f^{N,S}}{N!} \sum_{P \in S_N} W^{N,S}_{\mu\rho}(P^{-1}) \hat{P}^r \qquad (3-37)$$

we obtain

$$\Psi^o_{S,M;\mu}(\mathbf{r},\sigma) = \frac{1}{f^{N,S}} \sum_{\rho=1}^{f^{N,S}} \left(\hat{\omega}^{N,S}_{\rho\mu} \Phi^o(\mathbf{r}) \right) \Theta_{S,M;\rho}(\sigma). \qquad (3-38)$$

Clearly, the last equation has the same form as the exact wave function in Eq. (2-23). But is also shows how the matric basis units guarantees that the projected set of spatial trial functions transform correctly since they satisfy the operator identity

$$\hat{P}^r \hat{\omega}^{N,S}_{\rho\mu} = \sum_{\nu=1}^{f^{N,S}} \hat{\omega}^{N,S}_{\nu\mu} W^{N,S}_{\nu\rho}(P). \qquad (3-39)$$

Totally there are $f^{N,S} \times f^{N,S}$ linearly independent matric basis units. The choice of spin-coupling μ in Eq. (3-35) is hence to select the μ'th set of projections. I.e.

$$\{\hat{\omega}^{N,S}_{1\mu}\Phi^o(\mathbf{r}), \hat{\omega}^{N,S}_{2\mu}\Phi^o(\mathbf{r}), \ldots, \hat{\omega}^{N,S}_{f^{N,S}\mu}\Phi^o(\mathbf{r})\}.$$

Since there are $f^{N,S}$ possible choices for the value of μ, we can, in general, from a given spatial trial function, $\Phi^o(\mathbf{r})$, construct $f^{N,S}$ linearly independent configuration state functions. It should be noted, however, that in cases where the spatial trial function is constructed in such a way that it has intrinsic permutational symmetry with respect to interchange of electronic coordinates then some of the projections become linearly dependent. One example is the case where some of the orbitals in a product of one-electron functions are doubly occupied. Another example is the case where the spatial N-electron function is constructed using symmetric and antisymmetric geminals.

3.2. MATRIX ELEMENTS

Let $\hat{\Omega}$ be a symmetric N-electron operator which does not contain any spin dependent terms

$$[\hat{P}^r, \hat{\Omega}] = [\hat{P}^\sigma, \hat{\Omega}] = 0 \quad \text{for all} \quad \hat{P} \in \mathcal{S}_N. \qquad (3-40)$$

Consider now a general matrix elements between two configuration state functions, say $\hat{A}\Phi\Theta_\rho$ and $\hat{A}\Phi'\Theta_{\rho'}$, where we to simplify notation have omitted the superscript $^\circ$ on the spatial trial functions and the explicit dependence on electronic coordinates and the spin quantum numbers S and M

$$\langle \hat{A}\Phi'\Theta_{\rho'} | \hat{\Omega} | \hat{A}\Phi\Theta_\rho \rangle = \langle \Phi'\Theta_{\rho'} | \hat{\Omega} | \hat{A}\Phi\Theta_\rho \rangle$$

$$= \sum_{P \in \mathcal{S}_N} \epsilon_P \langle \Phi' | \hat{\Omega} | \hat{P}^r \Phi \rangle \langle \Theta_{\rho'} | \hat{P}^\sigma | \Theta_\rho \rangle.$$

If the matrix representation of the operator $\hat{P}^\sigma$ in the given spin basis is represented by the matrix $\mathbf{P}$ with the elements

$$P_{\rho'\rho} = \langle \Theta_{\rho'} | \hat{P}^\sigma | \Theta_\rho \rangle \qquad (3-41)$$

we may write the last equations as

$$\langle \hat{A}\Phi'\Theta_{\rho'} | \hat{\Omega} | \hat{A}\Phi\Theta_\rho \rangle = \sum_{P \in \mathcal{S}_N} \epsilon_P \langle \Phi' | \hat{\Omega} | \hat{P}^r \Phi \rangle P_{\rho'\rho}. \qquad (3-42)$$

In this form the evaluation of matrix elements involves explicit knowledge of the form and structure of the spin eigenfunctions used in order to determine the required spin matrix elements, $P_{\rho'\rho}$, over the permutation group operators.

Alternatively, if $\mathbf{S}$ denotes the overlap matrix between a set of spin eigenfunctions, $\Theta_\mu, \mu = 1, 2, \ldots, f^{N,S}$, with the elements

$$S_{\rho'\rho} = \langle \Theta_{\rho'} | \Theta_\rho \rangle \qquad (3-43)$$

we have by using Eq. (2-22)

$$\mathbf{P} = \mathbf{S}\,\mathbf{U}(P) \qquad (3-44)$$

with the simplified notation $\mathbf{U}(P) = \mathbf{U}^{N,S}(P)$.

By inserting the last equation in Eq. (3-42) we obtain

$$\langle \hat{A}\Phi'\Theta_{\rho'}| \hat{\Omega} |\hat{A}\Phi\Theta_{\rho}\rangle = \sum_{P\in\mathcal{S}_N} \epsilon_P \langle \Phi'| \hat{\Omega} |\hat{P}^r\Phi\rangle [\mathbf{SU}(P)]_{\rho'\rho} \qquad (3-45)$$

where $[\mathbf{SU}(P)]_{\rho'\rho}$ is an element of the matrix product $\mathbf{SU}(P)$.

The last expression is conceptually different from the previous formula (Eq. (3-42)) since it emphasizes the central role of the group theoretical representation matrices, $\mathbf{U}(P)$. Obviously, if the spin basis is orthonormal the two expressions become identical. It means, that in this case the formulas may be considered as a purely group theoretical problem. *Any* unitary (or real orthogonal) irreducible representation of $\mathcal{S}_N$ may be used as long as it is equivalent to a spin representation without knowledge of the explicit form of the underlying spin basis. Accordingly, in this sense, the formulation becomes a genuine spin-free approach. Fortunately, however, it does not mean that the exact information of the spin-coupling is not available. It is a consequence of the fact that the one-electron spin basis is only two dimensional which implies that there exists only *one* set of spin eigenfunctions which transform correctly after a given irreducible representation of the permutation group. Accordingly, we can always uniquely project out the explicit form of the spin eigenfunctions corresponding to a given irreducible representation of $\mathcal{S}_N$ as long as the representation is consistent with an allowed spin representation.

4. Concluding Remarks

There is a practical drawback with both the two general expressions (Eq. (3-42) and Eq. (3-45)) for Hamiltonian matrix elements between configuration state functions if they are to be used directly as they stand. It is due to the summation over all the elements of the permutation group which leads to the so-called *N!-problem*. In methods based on fully non-orthogonal spatial orbitals as it might be the case in valence bond theory the full summation has to be carried out (Gerratt, 1971). Accordingly, the number of electrons which can be studied by means of the formulas are rather limited and is of the order 10-12.

Fortunately, in configuration interaction calculations based on orthonormal molecular orbitals, the expressions simplify considerably and only very few terms become non-zero as shown by Karwowski (1973) and Sarma and Rettrup (1977) who provided explicit formulas which were suited for computer implementation. As a consequence, the formulas have formed the basis for the development of two very efficient CI programmes based on the symmetric group graphical approach (Duch and Karwowski, 1985; Rettrup et al., 1987).

As indicated in the introduction an extensive monograph on the application of the symmetric group has very recently been published by Pauncz (1995). Accordingly, the author refer the reader to this book in order to get an more complete overview of other symmetric group formulations and a more complete set of references. The book also provides a very useful description of various ways of constructing irreducible representation matrices of S_N.

5. Acknowledgment

The author is very grateful to Ph.D. student Britt Friis-Jensen for valuable discussions and for carefully reading the manuscript. The project has received financial support from The Danish Natural Science Research Council.

References

H. Boerner. *Representations of Groups*. North Holland Publ., Amsterdam, 1963. (German edition, Springer-Verlag, Berlin, 1955).

A.J. Coleman. The Symmetric Group Made Easy. In P.-O. Löwdin, editor, *Advances in Quantum Chemistry*, volume 4, pages 83–108. Academic Press, London, 1968.

H.S.M Coxeter, G.F.D. Duff, D.A.S. Fraser, G. de B. Robinson, and P.G. Rooney, editors. *The Collected Papers of Alfred Young, 1873-1940*, volume 21 of *Matematical Expositions*. University of Toronto Press, Toronto, 1977.

P.A.M. Dirac. On the Theory of Quantum Mechanics. *Proc. Roy Soc. (London)*, A112:661, 1926.

W. Duch and J. Karwowski. Symmetric Group Approach to Configuration Interaction Methods. In G.H.F. Diercksen, editor, *Computer Physics Reports*, volume 2, pages 93–170. North-Holland Physics Publishing Division, Amsterdam, 1985.

J. Gerratt. General theory of spin-coupled wave functions for atoms and molecules. In D.R. Bates, editor, *Advances in Atomic and Molecular Physics*, volume 7, pages 141–221. Academic Press, London, 1971.

M. Hamermesh. *Group Theory and Its Applications to Physical Problems*. Addison-Wesley Publ., London, 1964.

J. Hinze, editor. *The Unitary Group for the Evaluation of Electronic Energy Matrix Elements*, volume 22 of *Lecture Notes in Chemistry*. Springer-Verlag, Berlin, 1981.

J. Karwowski. Matrix Elements of One- and Two-Electron Operators. *Theoret. Chim. Acta (Berl.)*, 29:151–166, 1973.

M. Kotani, A. Amemiya, I. Ishiguro, and T. Kimura. *Tables of Molecular Integrals*. Maruzen, Tokyo, 1955.

F.A. Matsen. Spin-Free Quantum Chemistry. In P.-O. Löwdin, editor, *Advances in Quantum Chemistry*, volume 1, pages 59–114. Academic Press, London, 1964.

J. Paldus. Many-Electron Correlation Problem. A Group Theoretical Approach. In H. Eyring and D. Henderson, editors, *Theoretical Chemistry: Advances and Perspectives*, volume 2, pages 131–290. Academic Press, New York, 1976.

R. Pauncz. *Spin Eigenfunctions: Construction and Use.* Plenum, New York, 1979.

R. Pauncz. *The Symmetric Group in Quantum Chemistry.* CRC Press, Boca Raton, 1995.

S. Rettrup, G.L. Bendazzoli, S. Evangelisti, and P. Palmieri. A Symmetric Group Approach to the Calculation of Electronic Correlation Effects in Molecules. In J.S. Avery, J.P. Dahl, and Aa.E. Hansen, editors, *Understanding Molecular Properties*, pages 533–546. Reidel Publ. Co., Dordrecht, Holland, 1987.

G. de B. Robinson. *Representation Theory of Symmetric Groups.* University of Toronto Press, Toronto, 1961.

D.E. Rutherford. *Substitutional Analysis.* Edingburgh University Press, Edingburgh, 1948.

C.R. Sarma and S. Rettrup. A Programmable Spin-Free Method for Configuration Interaction. *Theoret. Chim. Acta (Berl.)*, 46:63–72, 1977.

J.C. Slater. Theory of Complex Spectra. *Phys. Rev.*, 34:1293–1322, 1929.

J.C. Slater. Molecular Energy Levels and Valence Bonds. *Phys. Rev.*, 38:1109–1144, 1931.

H. Weyl. *The Theory of Groups and Quantum Mechanics.* Dover, London, 1950. (Originally published in german in 1931).

E.P. Wigner. *Group Theory and Its Application to the Quantum Mechanics of Atomic Spectra.* Academic Press, London, 1959. (Originally published in German 1931).

NEW DEVELOPMENTS IN MANY BODY PERTURBATION THEORY AND COUPLED CLUSTER THEORY

DIETER CREMER and ZHI HE
Department of Theoretical Chemistry
University of Göteborg
Kemigarden 3, S-41296 Göteborg, Sweden

1. Introduction

Many body perturbation theory (MBPT) methods using the Møller-Plesset (MP) perturbation operator [1] are the most popular correlation corrected ab initio methods in Quantum Chemistry for calculating dynamic electron correlation effects. [2-9] The popularity of MP methods results from several reasons: (a) MP theory leads to a hierarchy of well-defined methods, which provide increasing accuracy with increasing order n. (b) Correlation effects are included stepwise in a systematic manner that facilitates their analysis and the understanding of the correlation problem. (c) Most important is the fact that all MP methods are size-extensive. [3,10] (d) Up to fourth order, MP energies can be calculated at relatively small computational cost since calculations involve just single, noniterative evaluation steps.

There are also some disadvantages of MP theory, which have to be mentioned. (a) MP methods are not variational. (b) At a given order n of MP perturbation theory, there exists not a well-defined wave function. (c) One observes often an oscillatory or erratic rather than monotonic convergence behaviour of calculated MPn energies. [11-14] The first two problems are of just minor consequence. For example, it is more important to use a size-extensive rather than a variational method for calculating electron correlation effects. Also, one can calculate molecular properties in form of response properties using analytical energy derivatives without ever referring to a wave function. [11,15] However, the third problem is more serious: One has early observed that the MPn energy can strongly oscillate for small values of n before it converges to the full CI (FCI) energy value, which is identical with the infinite order MP energy. [11-14] Oscillations are also found for other properties such as the internal coordinates of molecular geometries, dipole moments, vibrational frequencies or infrared intensities. [11]

Clearly, these oscillations make the use of MP methods less attractive, which is one of the major reasons why Coupled Cluster (CC) methods have replaced MP method more and more in the nineties. [10,16-21] CC methods are related to MP methods in so far as they are also size-extensive and non-variational. Of course, CC methods are more expensive than MP methods since the CC wave function and, by this, the CC energy has to be calculated iteratively. The gain from the extra cost is increased accuracy that results from the fact that due to the exponential ansatz CC contrary to MP includes infinite order effects. By this, it is guaranteed that oscillations in the CC energy series are excluded and CC energies can effectively compete with those of other high-accuracy ab initio methods. Since the infinite order effects, it is much more difficult to keep track which correlation effects are covered by a given CC method and which not. A solution to this problem can be found by using MP theory to analyze the correlation contributions covered by a given CC method. [22,23]

The last ten years have seen many attempts to improve the repertoire of MP and CC methods for their effective use in Quantum Chemistry. The present account of MP and CC theory does not intend to present a summary of this work. Instead it exclusively concentrates on research carried out at Theoretical Chemistry of the University of Göteborg to develop techniques for including higher order correlation effects into MP or CC theory. [22-32] There is reason to believe that higher order correlation effects will make it possible to successfully apply single determinant theory even in the case of a typical multi-reference problem. Apart from this, the analysis of higher order correlation effects provides a basis for the understanding of the convergence behaviour of the MPn series and, by this, of the electron correlation problem in general. Once the convergence behaviour of the MPn series is well understood, the prediction of reliable FCI energies from MPn energies for low n becomes possible. [27-29]

Knowledge about the MPn methods and their coverage of electron correlation effects can directly be used to predict performance and accuracy of CC methods since it is possible to express correlation contributions covered by CC in terms of MP correlation effects. [22,23] For example, it is one of the key questions of the last years whether approximate CC methods, which do not include all cluster operators or handle part of the correlation problem by perturbation theory, can replace full CC methods. [30-32] We will deal with these questions in this work, which is structured in the following way.

In chapter 2, we will present a procedure, by which MP perturbation methods can be developed to higher orders. This procedure is a combination of the two traditional approaches in perturbation theory, namely the algebraic development procedure applied at lower orders and the diagrammatic development procedure applied at fourth and fifth order MP perturbation theory. We will show that by the combination of the two traditional approaches one avoids their disadvantages and is able to derive fourth order MP (MP4) [6,7] and fifth order MP (MP5) theory [8,9] in a compact form expressed in terms of cluster operators of first and second order. The cluster operator equations can easily be converted into a two-electron integral equations and programmed for use on a computer.

In addition to the discussion of how to develop MPn methods, we will shortly review cost requirements of MP methods where we will use the two-electron integral equations. We will stress the importance of the use of intermediate arrays in the computation of MPn correlation energies since this is the best way of cutting down computational cost for MPn methods. We will also discuss the various correlation effects covered by MP theory at low orders to get a better understanding of the accuracy of MPn methods presently used. [28,29]

In chapter 3, sixth order MP (MP6) perturbation theory [24-29,33] will be developed along the lines discussed in chapter 2, i.e. the development will start from the general MP energy formula, then partition the principal term in connected and disconnected cluster operator contributions, and, finally, extract all those terms that represent linked diagram contributions [34] to the MP6 correlation energy. The final cluster operator equations of the MP6 energy will be transformed term by term into two-electron integral formulas. It will be shown that the most costly terms are those that result from disconnected cluster operators. However, the computational cost of the disconnected cluster operator terms can systematically be reduced by using intermediate arrays. In this way, we will be able to give final two-electron integral formulas that lead to a minimum of computational cost for calculating the MP6 correlation energy. [24,25]

Also in chapter 3, the implementation of the first MP6 computer program for routine calculations will be discussed where special emphasis will be laid on the various ways of testing such a complicated program for programming errors. Some applications of MP6 will be discussed. [25]

In chapter 4, a short summary of CC theory is given. The projection equations of CC theory with single and double excitations (CCSD) [19] will be derived in their connected form and compared with those of the corresponding quadratic CI (QCI) approach, QCISD [35], which represents an approximation to the more complete CCSD theory. It will be pointed out that the QCI approach as developed by Pople and co-workers [35] can not be extended to the triple (T) excitation level because at this level it looses the property of size-extensivity.

In chapter 5, we will analyse CC and QCI methods on the basis of perturbation theory. [22,23] A graphical method will be presented to assess the infinite order effects of CC theory. It will be shown that high accuracy can be expected from CC methods that include in some way T excitations that describe three-electron correlation effects. Clearly, the best method in terms of accuracy is CCSDT [20] while satisfactory results can also be expected from CCSD(T) [36] that includes T effects in a perturbative way. Compared to the corresponding CC methods, QCISD and QCISD(T) lack many energy contributions and, therefore, they are unable to describe T effects in a balanced way. This will be clearly shown on the basis of the perturbation analysis. [23]

In chapter 6, we will use the work on MP and CC methods to develop a hierarchy of size-extensive QCI methods. [30] For this purpose, a systematic procedure of converting the non-size-extensive CI methods into extended CI methods, which are size-extensive. [30] We will show that, if correctly applied, the original QCI concept of Pople and co-workers [35] leads to just two size-extensive extended CI (ECI) methods, namely QCISD = ECISD and ECISDT. [30] At the quadruple (Q) excitation level as well as any higher excitation level, ECI methods merge with the corresponding CC methods, which means that the original QCI concept does not lead to a hierarchy of approximate CC methods. [30] However, using the linked diagram theorem the ECI equations can be converted into a connected form and, then, systematically simplified to projection equations with linear and quadratic cluster operator terms. In this way, a hierarchy of size-extensive QCI methods is developed that are parallel to the CC methods, but have the advantage of a rather simple form that can easily be converted into a computer program. [30]

We will discuss in chapter 6, the development of the first size-extensive QCISDT method and its application to simple electron systems for which FCI results are known. [31,32] We will show that QCISDT leads to the same accuracy as CCSDT, but has the advantage of being much easier to implement on a computer. In addition, QCISDT converges in many cases faster than CCSDT, which leads to time savings. [32]

Finally, in chapter 7 we will summarize the most important aspects of this review and shortly discuss the future of MBPT and CC theory.

2. Møller-Plesset Perturbation Theory

There are two different ways of developing MP methods for use in quantum chemical calculations. The first way can be called the algebraic approach since it is based on an algebraic derivation of matrix elements from general perturbation theory formulas. It works very well for low order perturbation theory [4,5,7], however becomes problematic for higher orders. In the latter case, one can distinguish between a principal term and one or several renormalization terms in the general perturbation theory formula. The linked diagram theorem [34] shows that it is superfluous to evaluate the renormalization terms since these are all cancelled by appropriate parts of the principal term. One realizes this by writing principal and renormalization terms in form of diagrams. The renormalization terms correspond to unlinked diagram contributions to the energy, which are cancelled by the unlinked diagram contributions of the principal term. Only the linked diagram contributions of the principal term determine the nth order MP correlation energy.

Because of the linked diagram theorem it is of advantage to derive the MP energy formulas by diagrammatic techniques which immediately identify those terms that really contribute to the correlation energy. Accordingly, diagrammatic derivations of the third, fourth and even fifth order MP energy have been made, which clearly demonstrated superiority over the algebraic approach. [6,8,37] However, the diagrammatic approach has also its disadvantages. This becomes obvious when considering the increase in linked diagrams contributing to the correlation energy. If one uses Brandow diagrams, there are 1, 3, 39, 840, and 28300 antisymmetrized diagrams at second, third, fourth, fifth, and sixth order, respectively. This means that it is hardly possible to derive the sixth order correlation energy in terms of linked diagrams.

Therefore, we have proposed a third approach for developing higher order perturbation theory formulas. [24] This third approach is based on a combination of algebraic and diagrammatic techniques and comprises the following steps.

1) Principal term and renormalization terms are derived from the general perturbation theory formula.

2) Since it is clear that all renormalization terms will be cancelled by parts of the principal term, derivation of the MPn equations concentrates just on the principal term. This will be dissected into various parts according to the excitations involved at the corresponding order of perturbation theory. The various parts will be written in a cluster operator form.

3) Each part of the principal term characterized by S, D, T, Q, P, H, etc. excitations can be described as representing connected or disconnected energy

diagrams according to the nature of the cluster operators appearing in the energy formula.

4) All connected (closed) energy terms correspond to linked diagram contributions and enter the formula for the correlation energy while the disconnected energy terms represent unlinked diagram contributions which according to the linked diagram theorem can be discarded.

5) The final cluster operator form of the linked diagram contributions is transformed into two-electron integral formulas. This is facilitated by the fact that all those terms that originally involved disconnected cluster parts can be simplified by using intermediate arrays.

The advantages of this approach are that

a) superfluous energy contributions are never determined within the algebraic derivation and

b) a tedious analysis of all linked diagram terms is not necessary.

The latter point will become clear if step 3 as the key step of the procedure 1)-5) is described in more detail. Each cluster operator $\hat{T}$ can be described in terms of simplified Brandow diagrams. [8] Combination of the $\hat{T}$ diagrams with the diagrams of the perturbation operator $\hat{V}$ may lead to closed connected or closed disconnected diagrams, which means that the corresponding matrix elements represent linked or unlinked energy contributions. It is also possible that the combination of $\hat{T}$ and $\hat{V}$ diagrams leads to disconnected open diagrams. In this case, the diagrams correspond to the wave operator and cover both linked and unlinked contributions. One has to combine the wave operator part with further parts of the energy formula to get a separation into connected closed (= linked) and disconnected closed (= unlinked) energy diagrams. In any case, it is possible to identify for each part of the principal term whether it contains just linked or in addition unlinked diagram contributions. The diagrams one has to use for this purpose are rather simple because they correspond to some basic operators and need not to be specified with regard to hole and particle lines. [8]

2.1 DERIVATION OF THE MØLLER-PLESSET CORRELATION ENERGY AT LOWER ORDERS

Using the procedure outlined above, we will derive in the following MP2, MP3, MP4, and MP5 energy formula. For the perturbation expansion, the Hartree-Fock (HF) wave function is used as zeroth order function.

$$\hat{H}_0|\Phi_0\rangle = E_0|\Phi_0\rangle \tag{2.1}$$

with
$$\hat{H}_0 = \sum_p \hat{F}_p = \sum_p (\hat{h}_p + \hat{g}_p), \tag{2.2}$$

In Eq. (2.2), $\hat{h}$ denotes the one-electron part of the Hamiltonian and $\hat{g}$ covers the sum over Coulomb operators $\hat{J}_q$ and exchange operators $\hat{K}_q$, which describe two-electron interactions between electrons p and q.

With these definitions, the eigen value E_0 of Eq. (2.1) and HF energy E_{HF} are given by

$$E_0 = \langle \Phi_0 | \hat{H}_0 | \Phi_0 \rangle \tag{2.3a}$$

$$E_{HF} = \langle \Phi_0 | \hat{H} | \Phi_0 \rangle \tag{2.3b}$$

In the following the HF spin orbitals are denoted by ψ_p. It is assumed that they are eigen functions of the Fock operator $\hat{F}_p$ with eigen value ϵ_p. Following a widespread convention we will use indices $i, j, k, \ldots$ to label occupied spin orbitals and indices $a, b, c, \ldots$ to label unoccupied (virtual) spin orbitals. In cases where the formulas hold for both type of spin orbitals indices $p, q, r, \ldots$ are used.

To solve the non-relativistic electronic Schrödinger equation

$$\hat{H} \Psi = E \Psi \tag{2.4}$$

one considers the true Hamiltonian $\hat{H}$ and the true wave function Ψ as related to the HF Hamiltonian and HF wave function by a perturbation, i.e. $\hat{H}$ splits into unperturbed Hamiltonian $\hat{H}_0$ and perturbation operator $\hat{V}$: [1]

$$\hat{H} = \hat{H}_0 + \hat{V} \tag{2.5}$$

Hence, the perturbation operator $\hat{V}$ is given as the difference between the exact Hamiltonian $\hat{H}$ and the zeroth order Hamiltonian $\hat{H}_0$:

$$\hat{V} = \sum_{p<q} \hat{r}_{pq}^{-1} - \sum_p \hat{g}_p. \tag{2.6}$$

The energy E of Eq. (2.4) can be expanded in a perturbation series

$$E = E_{HF} + E_{MP}^{(2)} + E_{MP}^{(3)} + E_{MP}^{(4)} + E_{MP}^{(5)} + \ldots \tag{2.7a}$$

or

$$E = E_0 + E_{MP}^{(1)} + E_{MP}^{(2)} + E_{MP}^{(3)} + E_{MP}^{(4)} + E_{MP}^{(5)} + \ldots \tag{2.7b}$$

The energy difference $\Delta E = E - E_{HF}$ represents the correlation energy

$$\Delta E = E - E_{HF} = \sum_{n=2} E_{MP}^{(n)} \tag{2.8}$$

that is calculated as the sum of the Møller-Plessent (MP) perturbation contributions at order n.

The MP energy $E_{MP}^{(n)}$ at nth order can be written as

$$E_{MP}^{(n)} = \langle \Phi_0 | \hat{V} \hat{\Omega}^{(n-1)} | \Phi_0 \rangle, \tag{2.9}$$

where the wave operator $\hat{\Omega}$ at nth order is given by Eq. (2.10):

$$\hat{\Omega}^{(n)} = \hat{G}_0 \left[\hat{V} \hat{\Omega}^{(n-1)} - \sum_{m=1}^{n-1} E_{MP}^{(m)} \hat{\Omega}^{(n-m)} \right] \tag{2.10}$$

with $\hat{G}_0$ being the reduced resolvent:

$$\hat{G}_0 = \sum_{k=1}^{\infty} \frac{|\Phi_k\rangle \langle \Phi_k|}{E_0 - E_k}. \tag{2.11}$$

For a given order n, the correlation energy contribution $E_{MP}^{(n)}$ takes the form of Eq. (2.12)

$$E_{MP}^{(n)} = \langle \Phi_0 | \hat{V} (\hat{G}_0 \bar{V})^{n-1} | \Phi_0 \rangle + \text{ renormalization terms} \qquad (n \geq 2) \tag{2.12}$$

with $\bar{V}$ being

$$\bar{V} = \hat{V} - \langle \Phi_0 | \hat{V} | \Phi_0 \rangle \tag{2.13}$$

The first term of Eq. (2.12) is the principal term while all additional terms are renormalization terms. The number of renormalization terms increases rapidly with order n and, therefore, it is rather difficult algebraically to derive the energy formula for increasing order n.

However, in this situation the linked diagram theorem [34] helps, which states that only the linked diagram terms of Eq. (2.12) contribute to the correlation energy. All linked diagram contributions to the energy are contained in the principal term while the renormalization terms represent just unlinked diagram contributions, which are cancelled by the corresponding unlinked diagram contributions of the principal term. Therefore, Eq.(2.12) can be simplified to give Eq. (2.14):

$$E_{MP}^{(n)} = \langle \Phi_0 | \hat{V} (\hat{G}_0 \bar{V})^{n-1} | \Phi_0 \rangle_L \tag{2.14}$$

where the L indicates limitation to "linked" diagrams. This means that a derivation of the energy formula can focus just on the linked diagram contributions of the principal term. Linked diagram contributions to the energy can easily be identified by considering that they have to be closed and connected. If diagrams are not closed, they represent wave operator diagrams. In this case, a linked diagram can be either connected or disconnected, which makes it advisable to close the diagram first to an energy diagram and then to decide whether it is of linked or unlinked nature.

At second order MP (MP2) perturbation theory, there are no unlinked diagram contributions in Eq. (2.12). The energy $E_{MP}^{(2)}$ can be written as

$$E_{MP}^{(2)} = \langle \Phi_0 | \hat{V} \hat{G}_0 \hat{V} | \Phi_0 \rangle \tag{2.15a}$$

$$= \sum_d^D \langle \Phi_0 | \hat{V} | \Phi_d \rangle (E_0 - E_d)^{-1} \langle \Phi_d | \hat{V} | \Phi_0 \rangle \tag{2.15b}$$

in which d corresponds to double (D) excitations. In the following, we will denote single (S), triple (T), quadruple (Q), pentuple (P), and hextuple (H) excitations by subscripts s, t, q, p and h. For general excitations X, Y, etc., we will use subscripts x, y, etc.

In Eq. (2.15b), energies E_d are (in the same way as E_0) eigen values of the zeroth-order Hamiltonian $\hat{H}_0$ corresponding to the eigen functions $|\Phi_d\rangle$ ($|\Phi_0\rangle$).

We can rewrite Eq. (2.15b) by defining a D excitation cluster operator $\hat{T}_2^{(1)}$ at first order according to Eq. (2.16)

$$\hat{T}_2^{(1)} |\Phi_0\rangle = \sum_d^D a_d |\Phi_d\rangle \tag{2.16}$$

where the first order D excitation amplitudes a_d are given by

$$a_d = (E_0 - E_d)^{-1} \langle \Phi_d | \hat{V} | \Phi_0 \rangle \tag{2.17}$$

With the cluster operator $\hat{T}_2^{(1)}$ of Eq. (2.16), the second order energy adopts the simple form of Eq. (2.18):

$$E_{MP}^{(2)} = \langle \Phi_0 | \hat{V} \hat{T}_2^{(1)} | \Phi_0 \rangle \tag{2.18}$$

The analogous expression for the third order energy $E_{MP}^{(3)}$ can be easily obtained:

$$E_{MP}^{(3)} = \langle \Phi_0 | \hat{V} \hat{G}_0 \hat{V} \hat{G}_0 \hat{V} | \Phi_0 \rangle \tag{2.19a}$$

$$= \sum_{d_1,d_2}^D \langle \Phi_0 | \hat{V} | \Phi_{d_1} \rangle (E_0 - E_{d_1})^{-1} \langle \Phi_{d_1} | \bar{V} | \Phi_{d_2} \rangle$$

$$\times (E_0 - E_{d_2})^{-1} \langle \Phi_{d_2} | \hat{V} | \Phi_0 \rangle \tag{2.19b}$$

$$= \langle \Phi_0 | \hat{V} \hat{T}_2^{(2)} | \Phi_0 \rangle \tag{2.19c}$$

The second order D excitation cluster operator $\hat{T}_2^{(2)}$ is given by

$$\hat{T}_2^{(2)}|\Phi_0\rangle = \sum_d^D b_d|\Phi_d\rangle \qquad (2.20)$$

with the second order amplitude b_d being

$$b_d = (E_0 - E_d)^{-1}\langle\Phi_d|\bar{V}\hat{T}_2^{(1)}|\Phi_0\rangle \qquad (2.21)$$

At fourth order MP theory, one encounters for the first time beside the principal term also a renormalization term. Using Eq. (2.14), the contribution $E_{MP}^{(4)}$ is given by

$$E_{MP}^{(4)} = \langle\Phi_0|\hat{V}\hat{G}_0\bar{V}\hat{G}_0\bar{V}\hat{G}_0\hat{V}|\Phi_0\rangle_L \qquad (2.22a)$$

$$= \sum_x^{S,D,T,Q} \left(\langle\Phi_0|(\hat{T}_2^{(1)})^\dagger \bar{V}|\Phi_x\rangle(E_0 - E_x)^{-1}\langle\Phi_x|\bar{V}\hat{T}_2^{(1)}|\Phi_0\rangle\right)_L$$

$$= \sum_{i=1,2,3} \langle\Phi_0|(\hat{T}_2^{(1)})^\dagger \bar{V}\hat{T}_i^{(2)}|\Phi_0\rangle + \langle\Phi_0|(\hat{T}_2^{(1)})^\dagger \left[\bar{V}\frac{1}{2}(\hat{T}_2^{(1)})^2\right]_C|\Phi_0\rangle$$

$$(2.22b)$$

$$= E_S^{(4)} + E_D^{(4)} + E_T^{(4)} + E_Q^{(4)} \qquad (2.23)$$

The renormalization term is associated with the disconnected cluster operator $(\hat{T}_2^{(1)})^2$:

$$\langle\Phi_0|(\hat{T}_2^{(1)})^\dagger\left[\bar{V}\frac{1}{2}(\hat{T}_2^{(1)})^2\right]_D|\Phi_0\rangle = \langle\Phi_0|(\hat{T}_2^{(1)})^\dagger \hat{T}_2^{(1)}(\bar{V}\hat{T}_2^{(1)})_C|\Phi_0\rangle \qquad (2.24a)$$

$$= \langle\Phi_0|(\hat{T}_2^{(1)})^\dagger \hat{T}_2^{(1)}|\Phi_0\rangle E_{MP}^{(2)} \qquad (2.24b)$$

$$= \qquad (2.24c)$$

$\bar{V}$: $\qquad (2.25a)$

$$\hat{T}_1^{(n)}: \quad \vee \qquad (2.25b)$$

$$\hat{T}_2^{(n)}: \quad \vee\vee \qquad (2.25c)$$

$$\hat{T}_3^{(n)}: \quad \vee\vee\vee \qquad (2.25d)$$

This is an unlinked diagram term, which can be disregarded in the MP4 calculation. In Eq.s (2.22) and (2.24) the subscripts C and D denote restriction to connected and disconnected diagrams. Simplified graphical representations of perturbation operator $\bar{V}$ and cluster operators $\hat{T}_i^{(n)}$ (i = 1,2,3) at nth order perturbation theory, are given in (2.25a)-(2.25d). [24] Obviously, unlinked diagram terms result from a disconnected cluster operator and, therefore, terms involving disconnected cluster operators such as $\hat{T}_2^2$, $\hat{T}_1\hat{T}_2$, etc., have to be analyzed.

In Eq. (2.22b), the S and T excitation cluster operators $\hat{T}_1^{(2)}$ and $\hat{T}_3^{(2)}$ are defined by

$$\hat{T}_1^{(2)}|\Phi_0\rangle = \sum_s^S b_s|\Phi_s\rangle \qquad (2.26)$$

and

$$\hat{T}_3^{(2)}|\Phi_0\rangle = \sum_t^T b_t|\Phi_t\rangle \qquad (2.27)$$

where the corresponding amplitudes b_s and b_t are given by

$$b_s = (E_0 - E_s)^{-1}\langle\Phi_s|\bar{V}\hat{T}_2^{(1)}|\Phi_0\rangle \qquad (2.28)$$

and

$$b_t = (E_0 - E_t)^{-1}\langle\Phi_t|\bar{V}\hat{T}_2^{(1)}|\Phi_0\rangle \qquad (2.29)$$

At fifth order, the MP correlation energy contribution $E_{MP}^{(5)}$ can be expressed according to Eq. (2.30):

$$E_{MP}^{(5)} = \langle\Phi_0|\hat{V}\hat{G}_0\bar{V}\hat{G}_0\bar{V}\hat{G}_0\bar{V}\hat{G}_0\hat{V}|\Phi_0\rangle_L \qquad (2.30)$$

Using the expression for $\hat{G}_0$ (Eq.(2.11)) and the cluster operators $\hat{T}_2^{(1)}$ and $\hat{T}_i^{(2)}$ (i = 1, 2, 3), one obtains

$$E_{MP}^{(5)} = \sum_{x_1,x_2}^{S,D,T,Q} \left(\langle\Phi_0|(\hat{T}_2^{(1)})^\dagger \bar{V}|\Phi_{x_1}\rangle (E_0 - E_{x_1})^{-1} \langle\Phi_{x_1}|\bar{V}|\Phi_{x_2}\rangle \right.$$
$$\left. \times (E_0 - E_{x_2})^{-1} \langle\Phi_{x_2}|\bar{V}\hat{T}_2^{(1)}|\Phi_0\rangle \right)_L$$
$$= \sum_{i,j=1,2,3} \langle\Phi_0|(\hat{T}_i^{(2)})^\dagger \bar{V}\hat{T}_j^{(2)}|\Phi_0\rangle + 2\sum_{i=2,3} \langle\Phi_0|(\hat{T}_i^{(2)})^\dagger \bar{V}\frac{1}{2}(\hat{T}_2^{(1)})^2|\Phi_0\rangle_C$$
$$+ \langle\Phi_0|\frac{1}{2}((\hat{T}_2^{(1)})^\dagger)^2 \bar{V}\frac{1}{2}(\hat{T}_2^{(1)})^2|\Phi_0\rangle_C \quad (2.31)$$
$$= E_{SS}^{(5)} + 2E_{SD}^{(5)} + E_{DD}^{(5)} + 2E_{ST}^{(5)}$$
$$+ 2E_{DT}^{(5)} + E_{TT}^{(5)} + 2E_{DQ}^{(5)} + 2E_{TQ}^{(5)} + E_{QQ}^{(5)} \quad (2.32)$$

The 14 terms of MP5 can be reduced to 9 unique terms by considering that $E_{SD}^{(5)} = E_{DS}^{(5)}$, etc., and weighting each term by appropriate factors of one and two in Eq. (2.32). The last three terms of Eq. (2.32), namely $E_{DQ}^{(5)}$, $E_{TQ}^{(5)}$, and $E_{QQ}^{(5)}$, contain the disconnected cluster operator $(\hat{T}_2^{(1)})^2$. In addition, the term $E_{TS}^{(5)}(= E_{ST}^{(5)})$ contains the open disconnected diagram part (2.33):

$$\langle\Phi_t|\bar{V}\hat{T}_1^{(2)}|\Phi_0\rangle : \quad \bigvee\cdots\bigvee \quad \bigvee \quad (2.33)$$

which contributes a linked energy diagram to $E_{MP}^{(5)}$ when closed by the triple cluster operator $(\hat{T}_3^{(2)})^\dagger$. However, it is relevant for the derivation of the MP6 method discussed in chapter 3 that term (2.33) can lead to unlinked diagram contributions at higher orders of perturbation theory and, therefore, it is reasonable to place the term E_{TS} in a separate class, which at higher orders will be associated with disconnected T contributions. In this way, $E_{MP}^{(5)}$ is partitioned into $E(MP5)_1$, $E(MP5)_2$, and $E(MP5)_3$:

$$E_{MP}^{(5)} = E(MP5)_1 + E(MP5)_2 + E(MP5)_3 \quad (2.34)$$

where

$$E(MP5)_1 = E_{SS}^{(5)} + 2E_{SD}^{(5)} + E_{DD}^{(5)} + 2E_{DT}^{(5)} + E_{TT}^{(5)}$$
$$= \sum_{i=1,2} (2-\delta_{1,i})\langle\Phi_0|(\hat{T}_1^{(2)})^\dagger \bar{V}\hat{T}_i^{(2)}|\Phi_0\rangle$$
$$+ \sum_{i=2,3} (2-\delta_{2,i})\langle\Phi_0|(\hat{T}_2^{(2)})^\dagger \bar{V}\hat{T}_i^{(2)}|\Phi_0\rangle$$
$$+ \langle\Phi_0|(\hat{T}_3^{(2)})^\dagger \bar{V}\hat{T}_3^{(2)}|\Phi_0\rangle \quad (2.35)$$

represents the contributions of the connected cluster operators,

$$E(MP5)_2 = 2E^{(5)}_{DQ} + 2E^{(5)}_{TQ} + E^{(5)}_{QQ} \tag{2.36}$$

the contributions from the disconnected Q cluster operators, and

$$E(MP5)_3 = 2E^{(5)}_{TS} \tag{2.37}$$

a contribution, which at higher orders results from disconnected T cluster operators. It is straightforward to evaluate the contributions covered by $E(MP5)_1$, however care has to be taken with regard to all terms associated with disconnected cluster operators since they lead to both linked and unlinked diagrams. For example, in the Q terms of $E(MP5)_2$, one has to identify those disconnected parts of the wave operator that upon closure lead to unlinked diagrams, which can be eliminated.

$$\begin{aligned} E^{(5)}_{DQ} &= \langle \Phi_0 | (\hat{T}^{(2)}_2)^\dagger \bar{V} \frac{1}{2}(\hat{T}^{(1)}_2)^2 | \Phi_0 \rangle_C \\ &= \langle \Phi_0 | (\hat{T}^{(2)}_2)^\dagger \left[\bar{V} \frac{1}{2}(\hat{T}^{(1)}_2)^2 \right]_C | \Phi_0 \rangle \end{aligned} \tag{2.38}$$

As also found for the Q term of MP4, the disconnected diagram part $\left[\bar{V} \frac{1}{2}(\hat{T}^{(1)}_2)^2 \right]_D$ solely leads to unlinked diagram contributions so that $E^{(5)}_{DQ}$ is determined just by the connected diagram part $\left[\bar{V} \frac{1}{2}(\hat{T}^{(1)}_2)^2 \right]_C$ as shown in Eq.(2.38).

The second Q term of Eq. (2.36) can also be partitioned into two parts according to the splitting into $\left[\bar{V} \frac{1}{2}(\hat{T}^{(1)}_2)^2 \right]_C$ and $\left[\bar{V} \frac{1}{2}(\hat{T}^{(1)}_2)^2 \right]_D$. However, closure of the disconnected cluster part by the $\hat{T}^{(2)}_3$ operator leads to a connected contribution.

$$\begin{aligned} E^{(5)}_{TQ} &= \langle \Phi_0 | (\hat{T}^{(2)}_3)^\dagger \bar{V} \frac{1}{2}(\hat{T}^{(1)}_2)^2 | \Phi_0 \rangle_C \\ &= \langle \Phi_0 | (\hat{T}^{(2)}_3)^\dagger \left[\bar{V} \frac{1}{2}(\hat{T}^{(1)}_2)^2 \right]_D | \Phi_0 \rangle_C + \langle \Phi_0 | (\hat{T}^{(2)}_3)^\dagger \left[\bar{V} \frac{1}{2}(\hat{T}^{(1)}_2)^2 \right]_C | \Phi_0 \rangle \\ &= E^{(5)}_{TQ}(I) + E^{(5)}_{TQ}(II) \end{aligned} \tag{2.39}$$

By combining $E^{(5)}_{TQ}(I)$ with $E^{(5)}_{TS}$ of $E(MP5)_3$, one can get rid of the triple cluster operator $\hat{T}^{(2)}_3$ according to

$$E^{(5)}_{TS} + E^{(5)}_{TQ}(I) = \langle \Phi_0 | (\hat{T}^{(2)}_3)^\dagger \bar{V} \hat{T}^{(2)}_1 | \Phi_0 \rangle + \langle \Phi_0 | (\hat{T}^{(2)}_3)^\dagger \hat{T}^{(1)}_2 (\bar{V} \hat{T}^{(1)}_1)_C | \Phi_0 \rangle_C \tag{2.40}$$

$$= \langle \Phi_0 | (\hat{T}^{(1)}_2)^\dagger (\bar{V} \hat{T}^{(2)}_1 \hat{T}^{(1)}_2)_C | \Phi_0 \rangle \tag{2.41}$$

where we have used the fact that [38]

$$\langle \Phi_y | (\bar{V}\hat{T}_m\hat{T}_n)_D | \Phi_0 \rangle = \langle \Phi_y | \hat{T}_m(\bar{V}\hat{T}_n)_C | \Phi_0 \rangle + \langle \Phi_y | \hat{T}_n(\bar{V}\hat{T}_m)_C | \Phi_0 \rangle \qquad (2.42)$$

and have applied the factorization theorem [34],

$$(xy)^{-1} = (x+y)^{-1}(x^{-1} + y^{-1}). \qquad (2.43)$$

The combination of $E_{TS}^{(5)}$ and $E_{TQ}^{(5)}(I)$ leads to a reduction of computational cost from $O(M^7)$ to $O(M^6)$ since the computational requirements for triple amplitudes b_t involve an $O(M^7)$ dependence while the calculation of $E_{TS}^{(5)} + E_{TQ}^{(5)}(I)$ in Eq. (2.41) requires only $O(M^6)$ steps.

The last Q term of Eq. (2.36) can also be split into two parts, which have to be evaluated separately.

$$\begin{aligned} E_{QQ}^{(5)} &= \langle \Phi_0 | \frac{1}{2}((\hat{T}_2^{(1)})^\dagger)^2 \bar{V} \frac{1}{2}(\hat{T}_2^{(1)})^2 | \Phi_0 \rangle_C \\ &= \langle \Phi_0 | \frac{1}{2}((\hat{T}_2^{(1)})^\dagger)^2 \left[\bar{V} \frac{1}{2}(\hat{T}_2^{(1)})^2 \right]_D | \Phi_0 \rangle_C \\ &\quad + \langle \Phi_0 | \frac{1}{2}((\hat{T}_2^{(1)})^\dagger)^2 \left[\bar{V} \frac{1}{2}(\hat{T}_2^{(1)})^2 \right]_C | \Phi_0 \rangle \\ &= E_{QQ}^{(5)}(I) + E_{QQ}^{(5)}(II) \end{aligned} \qquad (2.44)$$

In this way, each of the nine terms of $E_{MP}^{(5)}$ (see Eq.2.32) is expressed in a cluster operator form, which can easily be transformed into appropriate two-electron integral formulas. Any computer program for the calculation of MP2, MP3, MP4 or MP5 correlation energies is based on the two-electron integral equations and, therefore, the transformation into the latter has to be done in the most economic way. This aspect will be discussed in the next section.

2.2 DERIVATION OF MØLLER-PLESSET CORRELATION ENERGIES IN TERMS OF TWO-ELECTRON INTEGRAL FORMULAS

MP methods are practical up to fourth order and become more difficult to apply at higher orders. This becomes obvious when inspecting the two-electron integral formulas of MP2, MP3, MP4, etc. For example, the appropriate expression for the MP2 energy is given by [3,4]

$$E_{MP}^{(2)} = \frac{1}{4} \sum_{ij} \sum_{ab} \langle ij || ab \rangle a_{ij}^{ab} \qquad (2.45)$$

where a_{ij}^{ab} are the D amplitudes that are defined by

$$a_{ij}^{ab} = (\epsilon_i + \epsilon_j - \epsilon_a - \epsilon_b)^{-1}\langle ab||ij\rangle \tag{2.46}$$

The double-bar integrals $\langle ij||ab\rangle$ are antisymmetrized two-electron integrals of the general type $\langle pq||rs\rangle$:

$$\langle pq||rs\rangle = \int\int \psi_p^*(1)\psi_q^*(2)\frac{1}{r_{12}}[\psi_r(1)\psi_s(2) - \psi_s(1)\psi_r(2)]d\tau_1 d\tau_2$$

Eq. (2.45) is obtained by transformation of Eq.s (2.16) and (2.18), respectively, using Slater rules for matrix elements over orthonormal spin orbitals ψ_p. The computational cost for the evaluation of the MP2 energy results just from the transformation of two-electron integrals over basis functions χ_μ into two-electron integrals over spin orbitals ψ_p, which is proportional to $O(M^5)$ where M denotes the number of basis functions. This cost factor is actually much lower than the cost suggested by the transformation Eq. (2.47).

$$\langle ij|kl\rangle = \sum_\mu\sum_\nu\sum_\lambda\sum_\sigma \langle \mu\nu|\lambda\sigma\rangle c_{\mu i}c_{\nu j}c_{\lambda k}c_{\sigma l} \tag{2.47}$$

If one would carry out the one-step transformation of (2.47), then the computational work would be proportional to $O(M^8)$. This can be seen by realizing that about M^4 two-electron integrals over basis functions χ_μ have to be calculated at the SCF level (this the reason why the cost of a HF calculation is proportional to $O(M^4)$), which are transformed into M^4 two-electron integrals over spin orbitals. Yoshimine and co-workers [39] realized that the M^8 transformation could be dissected into a sequence of four M^5 transformations by calculating intermediate arrays $\langle\mu\nu|\lambda l\rangle$, $\langle\mu\nu|kl\rangle$, and $\langle\mu j|kl\rangle$ which represent partially transformed two-electron integrals:

$$\langle\mu\nu|\lambda l\rangle = \sum_\sigma \langle\mu\nu|\lambda\sigma\rangle c_{\sigma l} \tag{2.48}$$

$$\langle\mu\nu|kl\rangle = \sum_\lambda \langle\mu\nu|\lambda l\rangle c_{\lambda k} \tag{2.49}$$

$$\langle\mu j|kl\rangle = \sum_\nu \langle\mu\nu|kl\rangle c_{\nu j} \tag{2.50}$$

$$\langle ij|kl\rangle = \sum_\mu \langle\mu j|kl\rangle c_{\mu i} \tag{2.51}$$

In this way, the integral transformation can be carried out at a cost level which is not so much higher than that of a HF calculation. On the other hand, it is clear that any correlation corrected ab initio calculation involves at least $O(M^5)$ computational steps because of (2.48) - (2.51).

The computational cost for the calculation of the MP3 correlation energy can be determined from the appropriate two-electron integral formula given in Eq. (2.52): [5]

$$E_{MP}^{(3)} = \frac{1}{4} \sum_{ij} \sum_{ab} \langle ij||ab\rangle b_{ij}^{ab} \tag{2.52}$$

where the second order D excitation amplitudes are defined in Eq. (2.53):

$$b_{ij}^{ab} = (\epsilon_i + \epsilon_j - \epsilon_a - \epsilon_b)^{-1} \left[\frac{1}{2} (\sum_{ef} \langle ab||ef\rangle a_{ij}^{ef} + \sum_{mn} \langle mn||ij\rangle a_{mn}^{ab}) \right.$$

$$\left. - \sum_{me} \sum_{P} (-1)^P P(i/j|a/b)\langle mb||je\rangle a_{im}^{ae} \right] \tag{2.53}$$

The calculation of the b_{ij}^{ab} amplitudes requires $O(M^6)$ steps and, accordingly, the calculation of the MP3 correlation energy is an $O(M^6)$ operation.

One could expect that the calculation of the MP4 energy is an $O(M^8)$ procedure because one has to loop over occupied spin orbitals i,j,k,l and virtual spin orbitals a,b,c,d to get the Q term. However, the idea of using intermediate arrays becomes important when calculating $E_Q^{(4)}$. To calculate $E_{MP}^{(4)}$, one has to define first the second order amplitudes for S and T excitations, respectively:

$$b_i^a = -(\epsilon_i - \epsilon_a)^{-1} \left[\frac{1}{2} \sum_{m,ef} \langle ma||ef\rangle a_{im}^{ef} + \frac{1}{2} \sum_{mn,e} \langle mn||ie\rangle a_{mn}^{ae} \right] \tag{2.54}$$

$$b_{ijk}^{abc} = (\epsilon_i + \epsilon_j + \epsilon_k - \epsilon_a - \epsilon_b - \epsilon_c)^{-1} \sum_{P} (-1)^P P(i/jk|a/bc) \left[\sum_e \langle bc||ei\rangle a_{jk}^{ae} \right.$$

$$\left. - \sum_m \langle ma||jk\rangle a_{im}^{bc} \right] . \tag{2.55}$$

Then, one can simply transform matrix elements such as $\langle \Phi_d | \bar{V} \hat{T}_i^{(2)} | \Phi_0 \rangle$ (i = 1, 2, 3) and $\langle \Phi_d | [\bar{V} \frac{1}{2}(\hat{T}_2^{(1)})^2]_C | \Phi_0 \rangle$ of the cluster operator Eq. (2.22) into two-electron integral formulas using the auxiliary arrays $vn(ij.., ab..)$ shown in Table 2.1.

TABLE 2.1. Definition of auxiliary arrays v1 - v9 for use in MP4 and MP5 calculations.

Array	Matrix element	Two-electron integral formula	Cost
v1(ij,ab)	$\langle\Phi_{ij}^{ab}\|\bar{V}\bar{T}_1^{(2)}\|\Phi_0\rangle$	$\sum_e \sum_P (-1)^P P(i/j)\langle ab\|ej\rangle b_i^e + \sum_m \sum_P (-1)^P P(a/b)\langle ma\|ij\rangle b_m^b$	$O(M^5)$
v2(ij,ab)	$\langle\Phi_{ij}^{ab}\|\bar{V}\bar{T}_2^{(2)}\|\Phi_0\rangle$	$\frac{1}{2}\left[\sum_{ef}\langle ab\|ef\rangle b_{ij}^{ef} + \sum_{mn}\langle mn\|ij\rangle b_{mn}^{ab}\right]$	$O(M^6)$
v3(ij,ab)	$\langle\Phi_{ij}^{ab}\|\bar{V}\bar{T}_3^{(2)}\|\Phi_0\rangle$	$-\sum_{me}\sum_P(-1)^P P(i/j)\langle a/b\rangle\langle mb\|je\rangle b_{im}^{ae}$ $-\frac{1}{2}\left[\sum_{me,f}\sum_P(-1)^P P(a/b)\langle bm\|ef\rangle b_{ij m}^{aef}\right.$ $\left.+\sum_{mne}\sum_P(-1)^P P(i/j)\langle mn\|ej\rangle b_{imn}^{abe}\right]$	$O(M^7)$
v4(ij,ab)	$\langle\Phi_{ij}^{ab}\|[\bar{V}\frac{1}{2}(\bar{T}_2^{(1)})^2]_C\|\Phi_0\rangle$	$\frac{1}{2}\sum_{mn}x1(ij,mn)a_{mn}^{ab} - \sum_e\sum_P(-1)^P P(a/b)x2(b,e)a_{ij}^{ae}$ $-\sum_m\sum_P(-1)^P P(i/j)x3(j,m)a_{im}^{ab} + \sum_{nf}\sum_P(-1)^P P(i/j)x4(ia,nf)a_{jn}^{bf}$	$O(M^6)$
v5(ij,ab)	$\langle\Phi_{ij}^{ab}\|[\bar{V}\bar{T}_1^{(2)}\bar{T}_2^{(1)}]_C\|\Phi_0\rangle$	$\sum_P(-1)^P P(i/j)\left[\sum_n y1(n,j)a_{in}^{ab} + \sum_{n,e}\sum_P(-1)^{P'}P'(a/b)y2(a,n,j,e)a_{in}^{be}\right.$ $\left.+\sum_P(-1)^P P(a/b)\left[\sum_f y3(a,f)a_{ij}^{bf} + \sum_{m,f}\sum_P(-1)^{P'}P'(i/j)y4(a,m,i,f)a_{jm}^{bf}\right]\right.$ $-\frac{1}{2}\sum_{ef}y6(ab,ef)a_{ij}^{ef}$	$O(M^6)$
v6(i,a)	$\langle\Phi_i^a\|\bar{V}\bar{T}_1^{(2)}\|\Phi_0\rangle$	$-\sum_{me}\langle ma\|ie\rangle b_m^e$	$O(M^5)$
v7(ijk,abc)	$\langle\Phi_{ijk}^{abc}\|\bar{V}\bar{T}_3^{(2)}\|\Phi_0\rangle$	$-\frac{1}{2}\left[\sum_{ef}\sum_P(-1)^P P(a/bc)\langle bc\|ef\rangle b_{ijk}^{aef}\right.$ $\left.+\sum_{mn}\sum_P(-1)^P P(i/jk)\langle mn\|jk\rangle b_{imn}^{abc}\right]$ $-\sum_{me}\sum_P(-1)^P P(i/jk/a/bc)\langle ma\|ie\rangle b_{jkm}^{bce}$	$O(M^7)$
v8(ijk,abc)	$\langle\Phi_{ijk}^{abc}\|(\bar{V}\frac{1}{2}(\bar{T}_2^{(1)})^2)_C\|\Phi_0\rangle$	$\sum_P(-1)^P P(i/jk/a/bc)\left[\sum_f y7(i,bc,f)a_{jk}^{af} + \sum_m y8(jk,m,a)a_{im}^{bc}\right]$	$O(M^8)$
v9(ijkl,abcd)	$a_{ij}^{ab}a_{kl}^{cd}\langle\Phi_{ijkl}^{abcd}\|[\bar{V}\frac{1}{2}(\bar{T}_2^{(1)})^2]_C\|\Phi_0\rangle$	$-a_{ij}^{ab}a_{kl}^{cd}\sum_m\sum_P(-1)^P P(i/jkl\|ab/cd)\left\{\sum_{P'}(-1)^{P'}P'(j/kl)\left[\frac{1}{2}\sum_n\langle kl\|mn\rangle a_{jn}^{cd}\right]\right.$ $\left.-\sum_e\sum_{P''}(-1)^{P''}P''(c/d)\langle mc\|je\rangle a_{kl}^{de}\right\}a_{im}^{ab}$ $-\frac{1}{2}a_{ij}^{ab}a_{kl}^{cd}\sum_{ef}\sum_P(-1)^P P(ij/kl\|a/bcd)\left[\sum_{P'}(-1)^{P'}P'(b/cd)\langle cd\|ef\rangle a_{kl}^{bf}\right]a_{ij}^{ae}$	$O(M^{10})$

TABLE 2.2. Definition of intermediate arrays xi and yj used in MP4 and MP5 calculations to reduce cost.

Arrays xi and yj	Array vk	Two-electron integral formula	Cost
x1(ij,mn)	v4(ij,ab)	$\frac{1}{2}\sum_{ef}(mn\|ef)a_{ij}^{ef}$	$O(M^6)$
x2(b,e), x3(j,m)	v4(ij,ab)	$\frac{1}{2}\sum_{mn,f}(mn\|ef)a_{mn}^{bf}$, $\frac{1}{2}\sum_{n,ef}(mn\|ef)a_{jn}^{ef}$	$O(M^5)$
x4(ia,nf)	v4(ij,ab)	$\sum_{m,e}(mn\|ef)a_{im}^{ae}$	$O(M^6)$
y1(n,j), y3(a,f)	v5(ij,ab)	$\sum_{m,e}(mn\|je)b_m^e$, $\sum_{m,e}(am\|ef)b_m^e$	$O(M^5)$
y2(a,n,j,e), y4(a,m,i,f)	v5(ij,ab)	$\sum_m(mn\|je)b_a^m$, $\sum_e(am\|ef)b_i^e$	$O(M^5)$
y5(mn,ij)	v5(ij,ab)	$\sum_e \sum_P (-1)^P P(i/j)(mn\|je)b_i^e$	$O(M^5)$
y6(ab,ef)	v8(ijk,abc)	$\sum_m \sum_P (-1)^P P(a/b)(am\|ef)b_m^b$	$O(M^6)$
y7(i,bc,f)	v8(ijk,abc)	$\sum_{me}(am\|ef)a_{jk}^{ef} + \sum_{ne}\sum_P(-1)^P P(b/c)(mb\|ef)a_{im}^{ce} - \frac{1}{2}\sum_{mn}(mn\|if)a_{mn}^{bc}$	$O(M^6)$
y8(jk,m,a)	v8(ijk,abc)	$\frac{1}{2}\sum_{ef}(am\|ef)a_{jk}^{ef}$, $\sum_{ne}\sum_P(-1)^P P(j/k)(mn\|ej)a_{kn}^{ae}$	$O(M^5)$
y9(b,e), y10(j,m)	v9(ijkl,abcd)	$\sum_{ij,a}a_{ij}^{ab}a_{ij}^{ae}$, $\sum_{iab}a_{ij}^{ab}a_{im}^{ab}$	$O(M^5)$
y11(j,b,m,c), y12(i,j,k,n)	v9(ijkl,abcd)	$\sum_{ia}a_{ij}^{ab}a_{km}^{ac}$	$O(M^6)$
y13(a,b,c,f), y14(k,l,e,f)	v9(ijkl,abcd)	$\sum_{ij}a_{ij}^{ab}a_{ij}^{cf}$, $\sum_{ab}a_{ij}^{ab}a_{kl}^{ab}$	$O(M^6)$
y15(c,d,m,n), y16(k,c,m,e)	v9(ijkl,abcd)	$\sum_{kl}a_{kl}^{cd}(mn\|kl)$, $\sum_{cd}a_{kl}^{cd}(cd\|ef)$	$O(M^6)$
y17(k,l,e,f)	v9(ijkl,abcd)	$-\frac{1}{2}\sum_b y9(b,e)a_{kl}^{bf}$	$O(M^6)$
y18(b,d,e,f)	v9(ijkl,abcd)	$\frac{1}{8}y9(b,e)y9(d,f) + \frac{1}{4}\sum_{jb}y11(j,b,k,e)a_{jl}^{bf}$	$O(M^6)$
	v9(ijkl,abcd)	$+\frac{1}{8}\sum_{ac}y13(a,b,c,f)y13(c,d,a,e)$	
y19(c,d,m,n)	v9(ijkl,abcd)	$-\frac{1}{2}\sum_{j,k}y11(j,b,k,e)y11(k,d,j,f)$	$O(M^6)$
y20(m,n,j,l)	v9(ijkl,abcd)	$-\frac{1}{8}\sum_j y10(j,m)a_{jn}^{cd} + \frac{1}{4}\sum_{jb}y11(j,b,m,c)a_{jn}^{bd}$	$O(M^6)$
		$\frac{1}{8}y10(j,m)y10(l,n) + \frac{1}{8}\sum_{ik}y12(i,j,k,n)y12(k,l,i,m)$	
y21(k,c,m,e)	v9(ijkl,abcd)	$\sum_{j,a}y11(j,a,m,c)a_{jk}^{ae} - \frac{1}{4}\sum_j y10(j,m)a_{jk}^{ce}$	$O(M^6)$
		$-\frac{1}{2}\sum_{bc}y11(j,b,m,c)y11(l,c,n,b)$	
y22(m,b,l,e)	v9(ijkl,abcd)	$\frac{1}{2}\sum_{jk}y11(i,b,k,e)y12(k,l,i,m) - \sum_{j,c}y11(j,b,m,c)y11(l,c,j,e)$	$O(M^6)$
		$-\frac{1}{4}y9(b,c)y10(l,m) + \frac{1}{2}\sum_{ad}y13(a,b,d,e)y11(l,d,m,a)$	$O(M^6)$

This leads to the MP4 energy of Eq. (2.56):

$$E_{MP}^{(4)} = \frac{1}{4} \sum_{ij} \sum_{ab} a_{ij}^{ab} \left(v1(ij,ab) + v2(ij,ab) + v3(ij,ab) + v4(ij,ab) \right) \quad (2.56)$$

in which the array $v4(ij,ab)$ is calculated with the help of the intermediate arrays $x1(ij,mn), x2(b,e), x3(j,m)$ and $x4(ia,nf)$ listed in Table 2.2. For the calculation of the intermediate arrays, one needs $O(M^6)$ or less costly computational steps, which means that the calculation of the Q term of $E_{MP}^{(4)}$ is actually not more costly than calculating the S or D term of MP4. According to Table 2.1, the cost for calculating the MP4 correlation energy is determined by the array $v3(ij,ab)$ associated with the T term. Since this is an $O(M^7)$ operation, full MP4 is just one power of M more costly than MP3 while MP4(SDQ) and MP3 are comparable in cost. [5-7]

At MP5 and higher levels of MP perturbation theory, the development of an efficient computer program is directly connected with the derivation of suitable intermediate arrays. By defining the right intermediate arrays, the mathematical algorithms for MP5, MP6, etc. can be executed on a computer in a minimum of time. This is indicated in Table 2.3 for MP5 and MP6, which are reduced from $O(M^{10})$ to $O(M^8)$ and $O(M^{12})$ to $O(M^9)$ procedures by using series of intermediate arrays. One can say that the development of such electron correlation methods focuses a) on how to get rid of unwanted unlinked diagram contributions and b) on how to set up the right intermediate arrays in the two-electron integral equations.

In a similar way as for $E_{MP}^{(4)}$, one can derive two-electron integral formulas for $E(MP5)_1$, $E(MP5)_2$ and $E(MP5)_3$ of $E_{MP}^{(5)}$. The following equations are obtained.

$$E(MP5)_1 = E_{SS}^{(5)} + 2E_{DS}^{(5)} + E_{DD}^{(5)} + 2E_{DT}^{(5)} + E_{TT}^{(5)}$$

$$= \sum_{i,a} b_i^a v6(i,a) + \frac{1}{4} \sum_{ij} \sum_{ab} b_{ij}^{ab} \left[2v1(ij,ab) + v2(ij,ab) + 2v3(ij,ab) \right]$$

$$+ \frac{1}{36} \sum_{ijk} \sum_{abc} b_{ijk}^{abc} v7(ijk,abc) \quad (2.57)$$

$$E_{DQ}^{(5)} = \frac{1}{4} \sum_{ij} \sum_{ab} b_{ij}^{ab} v4(ij,ab) \quad (2.58)$$

$$E_{TS}^{(5)} + E_{TQ}^{(5)}(I) = \frac{1}{4} \sum_{ij} \sum_{ab} [a_{ij}^{ab} v5(ij,ab) + \langle ij||ab\rangle(b_i^a b_j^b - b_i^b b_j^a)] \quad (2.59)$$

TABLE 2.3. Comparison of MPn methods.

MPn	Number of total terms	Number of unique terms	Cost (without intermediate arrays)	Most expensive term	Cost (intermediate arrays included)	Most expensive terms
MP2	1	1	$O(M^5)$		$O(M^5)$	$(ij\|\|ab)$
MP3	1	1	$O(M^6)$		$O(M^6)$	b_{ij}^{ab}
MP4	4	4	$O(M^8)$	$E_Q^{(4)}$	$O(M^7)$	$E_T^{(4)}$
MP5	14	9	$O(M^{10})$	$E_{QQ}^{(5)}$	$O(M^8)$	$E_{TT}^{(5)}$
MP6	55	36	$O(M^{12})$	$E_{QHQ}^{(6)}$	$O(M^9)$	$E_{TQT}^{(6)}, E_{QQQ}^{(6)}(II)$ and $E_{TQQ}^{(6)}(II)$
MP7	221	141	$O(M^{14})$	$E_{QHHQ}^{(7)}$	$O(M^{10})$	$E_{TQQT}^{(7)}$, etc.
MP8	915	583	$O(M^{16})$	$E_{QHOHQ}^{(8)}$	$O(M^{11})$	$E_{TQS_TQT}^{(8)}$, etc.

$$E_{TQ}^{(5)}(II) = \frac{1}{36} \sum_{ijk} \sum_{abc} b_{ijk}^{abc} v8(ijk, abc) \tag{2.60}$$

$$E_{QQ}^{(5)}(I) = \frac{1}{4} \sum_{mn} \sum_{ef} a_{mn}^{ef} Q_{mn}^{ef}(\tilde{b}_d, a \times a) \tag{2.61}$$

$$E_{QQ}^{(5)}(II) = \frac{1}{32} \sum_{ij} \sum_{ab} \sum_{kl} \sum_{cd} v9(ijkl, abcd) \tag{2.62}$$

in which $v5(ij, ab)$, $v6(i, a)$, $v7(ijk, abc)$, $v8(ijk, abc)$, and $v9(ijkl, abcd)$ are taken from Table 2.1 and the array $Q_{ij}^{ab}(\tilde{b}_d, a \times a)$ is defined in Eq. (2.63).

$$\begin{aligned} Q_{ij}^{ab}(\tilde{b}_d, a \times a) = \frac{1}{4} \sum_{mn} \sum_{ef} \tilde{b}_{mn}^{ef} & \left[a_{ij}^{ef} a_{mn}^{ab} - 2 \sum_P (-1)^P P(a/b) a_{ij}^{ae} a_{mn}^{bf} \right. \\ & \left. -2 \sum_P (-1)^P P(i/j) a_{im}^{ab} a_{jn}^{ef} \right. \\ & \left. + 4 \sum_P (-1)^P P(i/j) a_{im}^{ae} a_{jn}^{bf} \right] \end{aligned} \tag{2.63}$$

with

$$\tilde{b}_{mn}^{ef} = (\epsilon_m + \epsilon_n - \epsilon_e - \epsilon_f) b_{mn}^{ef} \tag{2.64}$$

As can be seen from Table 2.1, arrays $v5$, and $v8$ are build up using intermediate arrays $y1, y2, y3, y4, y5, y6, y7$, and $y8$ of Table 2.2. The calculation of each of these arrays does not require more than $O(M^6)$ operations. Hence, the determination of DQ requires just $O(M^6)$ steps, that of TS+TQ(I) also just $O(M^6)$ steps, that of TQ(II) is determined by the cost of evaluating the second order triple amplitudes, which requires $O(M^7)$ steps. If one would not use intermediate arrays, the QQ term would lead to a cost factor of $O(M^{10})$ (Table 2.3), however this is reduced to $O(M^6)$ by using intermediate arrays $y9$ - $y22$, as shown in Eq.(2.65)

$$\begin{aligned} E_{QQ}^{(5)}(II) = & \sum_{kl,ef} y17(k,l,e,f) y14(k,l,e,f) + \sum_{bdef} y18(b,d,e,f) \langle bd||ef \rangle \\ & + \sum_{mncd} y19(c,d,m,n) y15(c,d,m,n) + \sum_{mnjl} y20(m,n,j,l) \langle mn||jl \rangle \\ & + \sum_{kmce} y21(k,c,m,e) y16(k,c,m,e) \\ & + \sum_{lmbe} y22(m,b,l,e) \langle mb||le \rangle \end{aligned} \tag{2.65}$$

Table 2.3 gives an impression on how computational cost increase with the order of MP perturbation theory. Clearly, the reduction of cost by the use of intermediate arrays increases with order n. Accordingly, MP6 is an $O(M^9)$ rather than an $O(M^{12})$ method, MP7 an $O(M^{10})$ rather than an $O(M^{14})$ method, etc. This indicates that higher order perturbation theory methods are no longer feasible for routine calculations, however MP5, MP6, and even MP7 are suitable for important test calculations or investigations of small molecules. For example, there are $O(M^{10})$ methods such as CISDTQ [40] or CCSDTQ [21], which are nowadays available for expert calculations. Accordingly, the development of MP6 or even MP7 is desirable since these methods provide direct information on higher correlation effects and the convergence behaviour of the MPn series.

Inspection of Table 2.3 reveals that it will almost be impossible with the strategies presently available to derive equations for the 141 energy terms of MP7. The MP6 correlation energy is built up from 36 energy contributions, which also requires an enormous amount of work if one follows either the algebraic or the diagrammatic development procedure, however which becomes feasible when using the strategy described at the beginning of this chapter (see also chapter 3 and references 24, 25).

2.3 CORRELATION EFFECTS COVERED AT VARIOUS ORDERS OF MØLLER-PLESSET PERTURBATION THEORY

Second order MP (MP2) theory covers D excitations and, accordingly, describes pair correlation effects. [3,4] There is no coupling between the D excitations at second order and, therefore, each pair correlation correction is determined as if no other electron pairs are present in the molecule. This leads to an overestimation of pair correlation effects, which has been documented in the literature. [11,28,29] At third order MP (MP3) theory, coupling between D excitations is introduced and in this way, an exaggeration of pair correlation effects at MP2 is partially corrected. [5,11,28,29]

At fourth order MP (MP4) theory, D excitations are complemented by single (S), triple (T) and quadruple (Q) excitations. [6,7] Single excitations describe orbital relaxation effects, which are needed to adjust orbitals to the correlated movement of the electrons. This leads to some limited improvements of the spin orbitals, however, these changes cannot be compared with the systematic self-consistent-field type of adjustment of orbitals within a MCSCF calculation.

The (connected) triple excitations cover three-electron correlation effects that are smaller than the pair correlation effects. For a given electronic system, one can normally distinguish a number of core electron pairs, bond electron pairs, and lone electron pairs. [41] Each of these pairs can correlate with any of the other electrons in a three-electron situation, which means that the number of three-electron combinations is much larger than the number of core, bond, and lone pairs. Although the T correlation effects are usually quite small their large number leads to sizeable contributions to the correlation energy. [11,28,29] These T contributions may become even rather large if electron pairs are packed close together in a particular area of a molecule. This occurs for multiple bonds in the bonding region, for atoms with two or more lone pairs in the lone pair region or in the valence region of strongly electronegative atoms. In these cases, T excitations help to keep the electron pairs more separated and, therefore, T correlation corrections increase in magnitude.

Quadruple effects in a MP4 calculation correspond to disconnected quadruples [6,7], i.e. they do not describe the correlation of four electrons, but the simultaneous correlation of two independent electron pairs. These pair-pair correlations essentially represent positive correction terms to the pair correlation energy calculated at MP2. They are quite important to get a balanced description of pair correlation. Of course, in large molecules the simultaneous correlation of 3, 4, etc. electron pairs is also important, but these effects are not introduced to MP theory before sixth order (3 electron pairs), eighth order (4 electron pairs), etc.

Three-electron correlation effects can be exaggerated at MP4 for the same reason pair correlation effects are exaggerated at MP2. [11,28,29] MP5 introduces the coupling between S, D, T, and Q excitations in form of SS, SD, ST, DD, DT, DQ, TT, TQ, and QQ correlation effects and, therefore, MP5 gives a better account of T and Q effects. New correlation effects are introduced at sixth order in form of connected Q, disconnected pentuple (P) and disconnected hextuple (H) excitations, i.e. MP6 introduces for the first time four-electron correlation effects. The latter are important in all those situations, in which electrons cluster in a confined region of atomic space as has been demonstrated by MP6 calculations. [28,29] Again, connected Q and disconnected P or H effects can be exaggerated at MP6 because QQ, PP or HH couplings are introduced not before seventh order. In this way, the MPn series continues by introducing new effects at even orders and correcting them via the appropriate couplings at odd orders.

One can compare MP perturbation theory with a car that is fuelled at even orders but slowed down at odd orders. Cremer and co-workers have considered this basic nature of MP perturbation theory as the reason for an erratic or oscillatory behaviour of the MPn series. [11,15,28,29] Energy and other molecular properties do not converge smoothly to an infinite order limit but very often oscillate in the range n = 1 (HF) to n = 5 (MP5). This oscillatory behaviour of results becomes apparent when comparing MP2 and MP4 with MP1 (= HF),

MP3, and MP5 results. It represents a major drawback of perturbation theory, which can fully be understood when MP6 results are included into the comparison. [28,29] Therefore, we will discuss in the following chapter MP6 theory.

3. Møller-Plesset Perturbation Theory at Sixth Order

The MP6 correlation energy comprises 55 energy contributions of the type $E^{(6)}_{ABC}$, which reduce to 36 terms because of symmetry. These energy terms are given in Figure 3.1 in a graphical way as paths connecting excitations S, D, T, and Q at order n with excitations S, D, T, and Q at order 4 under the constraint that Slater rules for the corresponding matrix terms are obeyed. [22,23] For example, one obtains 14 fifth order paths in this way, namely the SS, SD, ST, DS, DD, DT, DQ, TS, TD, TT, TQ, QD, QT, and the QQ path. At sixth order, one has to consider that T and Q excitations can couple with P and H excitations. Therefore, the diagram extends to the right when the paths go down to levels n-1, etc. However, any allowed path can only start and end at A = S, D, T, Q, which is indicated by (wiggled) separation lines for the starting level n in Figure 3.1.

In the lower half of Figure 3.1 all 55 energy paths of MP6 are listed, 19 of which are equivalent because of symmetry. Hence, there remain 36 unique paths corresponding to 36 unique energy terms $E^{(6)}_{ABC}$, which have to be calculated to determine the MP6 correlation energy.

Our work on a MP6 method for routine calculations was triggered by several reasons.

(1) MP6 is after MP2 and MP4 the next even order method that should be of interest because of the introduction of new correlation effects described by connected Q or disconnected P and H excitations.

(2) With MP6 one has three energies (MP2, MP4, MP6) in the class of even order methods and three in the class of odd order methods (MP1 = HF, MP3, MP5). In this way, one gets a somewhat more realistic basis to test the initial convergence behaviour of the MPn series.

(3) Inspection of Table 2.3 and Figure 3.1 reveals that MP6 is actually the last method that can be developed using traditional techniques. MP7 has already a total of 221 terms, 141 of which are unique. Therefore, setting up MP7 or even higher MPn methods will require some form of automated method development strategy based on computer algebra languages.

(4) The cost of a MP6 calculation is proportional to $O(M^9)$ (see Table 2.3). This is too expensive for calculations on larger molecules, but still gives a chance for systematic studies on small molecules.

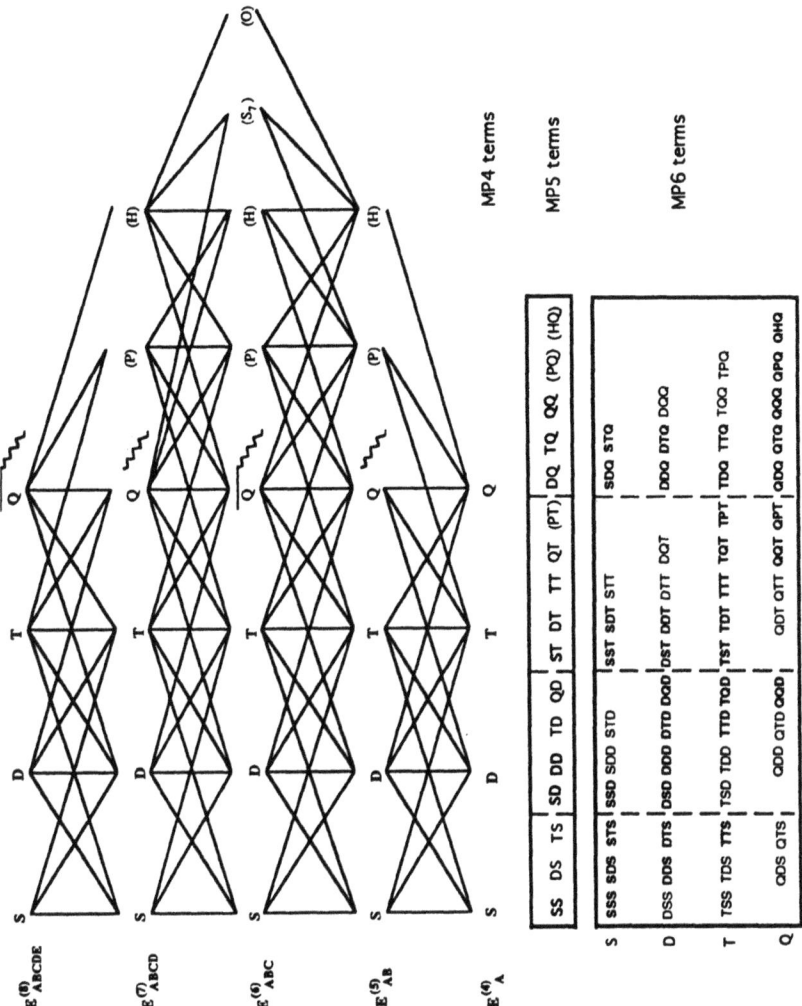

Figure 3.1. Graphical representation of energy contributions $E^{(n)}_{ABC..}$ at nth order many-body perturbation theory (n = 4, 5, 6, 7, 8) (upper part of the figure). A particular energy contribution $E^{(n)}_{ABC..}$ is given by the solid line that starts at A = S, D, T or Q in row $E^{(n)}$ and connects B, C, etc. at row n-1, n-2, etc. until n = 4 is reached. Note that at the n-1, n-2, .., n = 5 level also those excitations are included that can couple with A = S, D, T, Q at level n and level 4 according to Slater rules. They are given in parentheses after a separator (downward directed wiggles) to the right of the S, D, T, Q excitations. - At the bottom of the diagram, 5th order and 6th order energy terms $E^{(5)}_{AB}$ and $E^{(6)}_{ABC}$ respectively, are listed in correspondence to the energy paths shown in the upper half of the diagram. Unique terms are given in bold print. Reprinted with permission from Z. He and D. Cremer, Int. J. Quant. Chem (1996) 59, 15. Copyright (1996) J. Wiley, Inc.

(5) Apart from this, there is the possibility of developing useful approximate MP6 methods, which are less costly than the full MP6 approach because they include just the more important energy contributions $E^{(6)}_{ABC}$ rather than the full set of 36 energy terms.

(6) Various correlation methods, presently in use, have been compared in terms of MP4 and MP5. [8,9,36,42] It is known that many of the methods are accurate up to fourth order while only a few are accurate up to fifth order. Investigation of MP6 provides a basis to extend existing methods in such a way that they become correct up to sixth order. [43]

(7) Utilizing MP6 correlation energies it will be possible to test and improve existing extrapolation formulas, by which reliable estimates of the exact correlation energy can be obtained.

Focusing on these goals, we have applied the general procedure outlined in chapter 2 to derive an appropriate MP6 correlation energy formula that can be programmed for a computer.

3.1 DERIVATION OF A SIXTH ORDER ENERGY FORMULA IN TERMS OF CLUSTER OPERATORS

Since the general MP6 energy formula leads to 42 terms, 41 one of which are renormalization terms that are cancelled by appropriate parts of the principal term, it is out of question to attempt an algebraic derivation of the MP6 energy formula. On the other hand, one has to realize that the number of diagrams dramatically increases with the order n (1 (n=2), 3 (3), 39 (4), 840 (5), and 28300 (6) antisymmetrized Brandow diagrams), which means that an implementation of MP6 on the basis of a diagrammatic approach is also not feasible. Therefore, we avoid to derive the MP6 energy formula in a diagram-by-diagram manner or in an algebraic fashion. Instead, we use the procedure outlined in chapter 2, which represents a balanced mixture of diagrammatic and algebraic approach. This procedure comprises for MP6 the following steps.

(a) Starting from the general formula

$$E^{(n)}_{MP} = \langle \Phi_0 | \hat{V} (\hat{G}_0 \bar{V})^{n-1} | \Phi_0 \rangle_L \qquad (3.1)$$

and the expression for the reduced resolvent $\hat{G}_0$ (Eq.(2.11)), we will develop the MP6 energy equation in terms of S, D, T, Q, P, and H contributions. Contributions of higher excitations will be truncated according to Slater rules.

(b) Cluster operators $\hat{T}_2^{(1)}$ and $\hat{T}_i^{(2)}$ (i = 1, 2, 3) will be introduced into the $E_{MP}^{(6)}$ expression since this helps to identify those MP6 terms that contain disconnected parts of operator products such as $(\hat{T}_2^{(1)})^2$, $(\bar{V}\hat{T}_i^{(2)})_D$, etc.

(c) The terms of $E_{MP}^{(6)}$ will be classified in two categories: one covers terms only resulting from connected diagrams of operator products such as $(\bar{V}\hat{T}_i^{(2)})_C$ called connected operator diagram terms; and another contains terms resulting from disconnected diagram parts of operator products. Contrary to the former, we define the latter as disconnected operator diagram terms.

(d) The disconnected diagram parts of operator products will be combined with other operators to ultimately obtain connected energy terms, i.e. linked diagram contributions to the correlation energy.

According to this procedure, the MP6 energy formula can be derived in the following way. [24].

The general expression of $E_{MP}^{(6)}$ is given in Eq. (3.2):

$$E_{MP}^{(6)} = \langle \Phi_0 | \hat{V}(\hat{G}_0 \bar{V})^5 | \Phi_0 \rangle_L \qquad (3.2)$$

According to Eq.(2.11) and Slater rules, Eq. (3.2) can explicitly be written as

$$E_{MP}^{(6)} = \sum_{x_1,x_2}^{SDTQ} \sum_{y}^{SDTQPH} \Big(\langle \Phi_0 | (\hat{T}_2^{(1)})^\dagger \bar{V} | \Phi_{x_1} \rangle (E_0 - E_{x_1})^{-1} \bar{V}_{x_1 y} (E_0 - E_y)^{-1} \bar{V}_{y x_2}$$
$$\times (E_0 - E_{x_2})^{-1} \langle \Phi_{x_2} | \bar{V} \hat{T}_2^{(1)} | \Phi_0 \rangle \Big)_L \qquad (3.3a)$$

$$= \sum_{X_1,X_2}^{SDTQ} \sum_{Y}^{SDTQPH} \mathcal{A}(X_1, Y, X_2)_L \qquad (3.3b)$$

with $\mathcal{A}(X_1, Y, X_2)$ being

$$\mathcal{A}(X_1, Y, X_2) = \sum_{x_1}^{X_1} \sum_{x_2}^{X_2} \sum_{y}^{Y} \langle \Phi_0 | (\hat{T}_2^{(1)})^\dagger \bar{V} | \Phi_{x_1} \rangle (E_0 - E_{x_1})^{-1} \bar{V}_{x_1 y} (E_0 - E_y)^{-1}$$
$$\times \bar{V}_{y x_2} (E_0 - E_{x_2})^{-1} \langle \Phi_{x_2} | \bar{V} \hat{T}_2^{(1)} | \Phi_0 \rangle \qquad (3.4)$$

By using the cluster operators $\hat{T}_2^{(1)}$ and $\hat{T}_i^{(2)}$ (i = 1, 2, 3), one can partition the MP6 energy into three different $\mathcal{A}(X_1, Y, X_2)$ terms:

$$E_{MP}^{(6)} = \mathcal{A}_1[(\bar{V}\hat{T}_i^{(2)})_C] + \mathcal{A}_2[(\hat{T}_2^{(1)})^2]_L + \mathcal{A}_3[(\bar{V}\hat{T}_i^{(2)})_D]_L \qquad (i=1,2,3) \qquad (3.5)$$

The first part, $\mathcal{A}_1$, covers all connected cluster operator diagrams resulting from $(\bar{V}\hat{T}_i^{(2)})_C$ and fully contributes to to $E_{MP}^{(6)}$ in form of $E(MP6)_1$:

$$\mathcal{A}_1[(\bar{V}\hat{T}_i^{(2)})_C] = E(MP6)_1$$

$$= \sum_{i,j=1,2,3} \sum_y^{S,D} \langle\Phi_0|[(\hat{T}_i^{(2)})^\dagger \bar{V}]_C|\Phi_y\rangle(E_0 - E_y)^{-1}\langle\Phi_y|(\bar{V}\hat{T}_j^{(2)})_C|\Phi_0\rangle$$

$$+ \sum_{i,j=2,3} \sum_t^{T} \langle\Phi_0|[(\hat{T}_i^{(2)})^\dagger \bar{V}]_C|\Phi_t\rangle(E_0 - E_t)^{-1}\langle\Phi_t|(\bar{V}\hat{T}_j^{(2)})_C|\Phi_0\rangle$$

$$+ \sum_q^{Q} \langle\Phi_0|[(\hat{T}_3^{(2)})^\dagger \bar{V}]_C|\Phi_q\rangle(E_0 - E_q)^{-1}\langle\Phi_q|(\bar{V}\hat{T}_3^{(2)})_C|\Phi_0\rangle \quad (3.6)$$

$$= E_{SSS}^{(6)} + 2E_{SSD}^{(6)} + 2E_{SST}^{(6)} + E_{SDS}^{(6)} + 2E_{SDD}^{(6)}$$
$$+ 2E_{SDT}^{(6)} + E_{DSD}^{(6)} + 2E_{DST}^{(6)} + E_{DDD}^{(6)} + 2E_{DDT}^{(6)} + E_{TST}^{(6)}$$
$$+ E_{TDT}^{(6)} + E_{DTD}^{(6)} + 2E_{DTT}^{(6)} + E_{TTT}^{(6)} + E_{TQT}^{(6)} \quad (3.7)$$

The second and the third part, $\mathcal{A}_2$ and $\mathcal{A}_3$, cover all disconnected cluster operator diagram terms resulting from $(\hat{T}_2^{(1)})^2$ or various combinations of $\hat{T}_i^{(2)}$ with $\bar{V}$ as illustrated in Figure 3.2.

$$\mathcal{A}_2[(\hat{T}_2^{(1)})^2]_L = \sum_{X_1}^{S,D,T,Q} \sum_Y^{D,T,Q,P,H} (2 - \delta_{X_1,Q} - \delta_{X_1,S}\delta_{Y,T} - \delta_{X_1,D}\delta_{Y,Q}$$
$$- \delta_{X_1,T}\delta_{Y,P})\mathcal{A}(X_1,Y,Q)_L \quad (3.8)$$

where

$$\mathcal{A}(X_1,Y,Q) = \sum_y^{Y} \langle\Phi_0|(\hat{T}_i^{(2)})^\dagger \bar{V}|\Phi_y\rangle(E_0 - E_y)^{-1}\langle\Phi_y|\bar{V}\frac{1}{2}(\hat{T}_2^{(1)})^2|\Phi_0\rangle$$
$$(i = 1,2,3 \text{ for } X_1 = S, D, T) \quad (3.9)$$

or

$$\mathcal{A}(Q,Y,Q) = \sum_y^{Y} \langle\Phi_0|\frac{1}{2}((\hat{T}_2^{(1)})^\dagger)^2 \bar{V}|\Phi_y\rangle(E_0 - E_y)^{-1}\langle\Phi_y|\bar{V}\frac{1}{2}(\hat{T}_2^{(1)})^2|\Phi_0\rangle \quad (3.10)$$

As indicated in Figure 3.2, the disconnected Q cluster operator $(\hat{T}_2^{(1)})^2$ in $\mathcal{A}_2[(\hat{T}_2^{(1)})^2]_L$ couples with the perturbation operator $\bar{V}$. This leads to disconnected and connected cluster operator diagram parts, which in turn lead to the energy contributions $\mathcal{A}(X_1,Y,Q_D)_L$ and $\mathcal{A}(X_1,Y,Q_C)_L$ of Eq.s (3.11) - (3.14).

$$\mathcal{A}(X_1,Y,Q_D)_L = \sum_y^{Y} \left(\langle\Phi_0|(\hat{T}_i^{(2)})^\dagger \bar{V}|\Phi_y\rangle(E_0 - E_y)^{-1}\langle\Phi_y|\hat{T}_2^{(1)}(\bar{V}\hat{T}_2^{(1)})_C|\Phi_0\rangle\right)_L$$
$$(i = 1,2,3 \text{ when } X_1 = S, D, T; Y = T, Q, P) \quad (3.11)$$

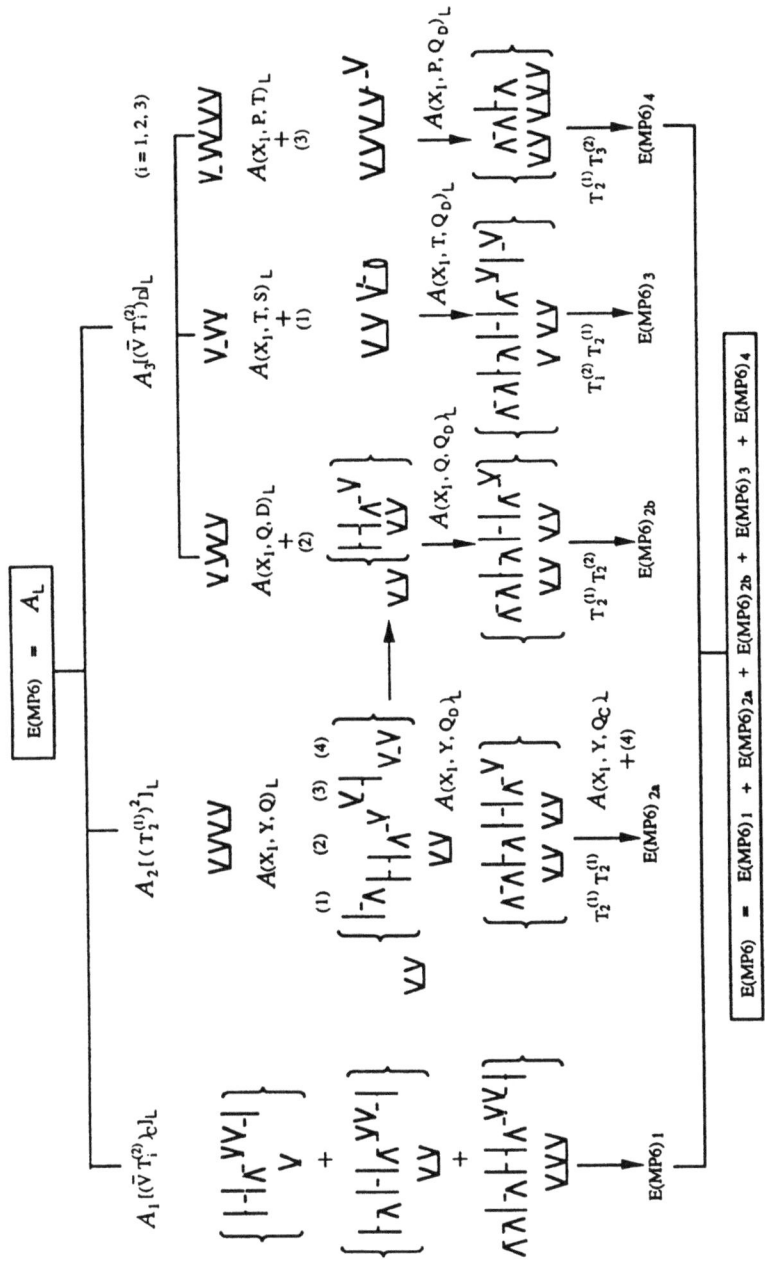

Figure 3.2. Derivation of the MP6 energy formula. For the perturbation operator and the cluster operators simplified Brandow diagrams are used. For more details, see text. *Reprinted with permission from D. Cremer and Z. He, J. Phys. Chem., (1996) 100, 6173. Copyright (1996) American Chemical Society.*

$$A(X_1,Y,Q_C)_L = \sum_{y}^{Y} \langle\Phi_0|(\hat{T}_i^{(2)})^\dagger \bar{V}|\Phi_y\rangle(E_0 - E_y)^{-1}\langle\Phi_y|(\bar{V}\frac{1}{2}(\hat{T}_2^{(1)})^2)_C|\Phi_0\rangle$$
$$(i = 1,2,3 \text{ when } X_1 = S, D, T; Y = D, T, Q) \quad (3.12)$$

$$A(Q,Y,Q_D)_L = \sum_{y}^{Y} \left(\langle\Phi_0|\frac{1}{2}((\hat{T}_2^{(1)})^\dagger)^2 \bar{V}|\Phi_y\rangle(E_0 - E_y)^{-1}\langle\Phi_y|(\bar{V}\frac{1}{2}(\hat{T}_2^{(1)})^2)_D|\Phi_0\rangle\right)_L$$
$$(Y = T, Q, P, H) \quad (3.13)$$

$$A(Q,Y,Q_C)_L = \sum_{y}^{Y} \left(\langle\Phi_0|\frac{1}{2}((\hat{T}_2^{(1)})^\dagger)^2 \bar{V}|\Phi_y\rangle(E_0 - E_y)^{-1}\langle\Phi_y|(\bar{V}\frac{1}{2}(\hat{T}_2^{(1)})^2)_C|\Phi_0\rangle\right)_L$$
$$(Y = D, T, Q) \quad (3.14)$$

In Ref. 24, the sum of $A(X_1,Y,Q_C)_L$ and $A(Q,H,Q_D)_L$ is denoted $E(MP6)_{2a}$, which covers 12 different energy contributions:

$$E(MP6)_{2a} = \sum_{X_1}^{S,D,T} 2A(X_1,D,Q_C) + A(Q_C,D,Q_C) + A(S,T,Q_C)$$
$$+ \sum_{X_1}^{D,T,Q} (2 - \delta_{X_1,Q})A(X_1,T,Q_C) + A(D,Q,Q_C)$$
$$+ \sum_{X_1}^{T,Q}(2 - \delta_{X_1,Q})A(X_1,Q,Q_C) + A(Q,H,Q_D)_L \quad (3.15)$$

$$= 2 \sum_{X}^{S,D,T} E^{(6)}_{XDQ} + E^{(6)}_{QDQ} + E^{(6)}_{STQ}(II) + E^{(6)}_{QTQ}(II)$$
$$+ 2 \sum_{X}^{D,T} E^{(6)}_{XTQ}(II) + E^{(6)}_{DQQ}(II) + E^{(6)}_{QQQ}(II)$$
$$+ 2E^{(6)}_{TQQ}(II) + E^{(6)}_{QHQ} \quad (3.16)$$

In some cases, it is useful to split the terms $E^{(6)}_{X_1YQ}$ into two parts:

$$E^{(6)}_{X_1YQ} = E^{(6)}_{X_1YQ}(I) + E^{(6)}_{X_1YQ}(II) \quad (3.17)$$

which correspond to $A(X_1,Y,Q_D)_L$ and $A(X_1,Y,Q_C)_L$, respectively. For $E^{(6)}_{STQ}(II)$ and $E^{(6)}_{DQQ}(II)$, their Hermitian conjugates $E^{(6)}_{QTS}$ and $E^{(6)}_{QQD}$ are included in $A_3[(\bar{V}\hat{T}_i^{(2)})_D]_L$, which is defined by (compare with Figure 3.2)

$$A_3[(\bar{V}\hat{T}_i^{(2)})_D]_L = \sum_{X_1}^{D,T,Q} A(X_1,Q,D)_L + \sum_{X_1}^{S,D,T,Q} A(X_1,T,S)_L + \sum_{X_1}^{T,Q} A(X_1,P,T)_L$$
$$(3.18)$$

The terms of Eq.(3.18) can be combined with the corresponding terms $A(X_1, Y, Q_D)_L$ of Eq.s (3.11) and (3.13) to lead to simpler formulations in cluster operator form as shown in Eq.s (3.19) to (3.23).

$$E(MP6)_{2b} = \sum_{X_1}^{D,T,Q} (A(X_1, Q, D)_L + A(X_1, Q, Q_D)_L)$$

$$= \langle\Phi_0|(\hat{T}_2^{(2)})^\dagger (\bar{V}\hat{T}_2^{(1)}\hat{T}_2^{(2)})_C|\Phi_0\rangle + 2\langle\Phi_0|(\hat{T}_3^{(2)})^\dagger \bar{V}\hat{T}_2^{(1)}\hat{T}_2^{(2)}|\Phi_0\rangle$$
$$+ \langle\Phi_0|\frac{1}{2}((\hat{T}_2^{(1)})^\dagger)^2 \bar{V}\hat{T}_2^{(1)}\hat{T}_2^{(2)}|\Phi_0\rangle_C \qquad (3.19)$$

$$= \left[E_{DQD}^{(6)} + E_{DQQ}^{(6)}(I)\right] + 2\left[E_{TQD}^{(6)} + E_{TQQ}^{(6)}(I)\right]$$
$$+ \left[E_{QQD}^{(6)} + E_{QQQ}^{(6)}(I)\right] ; \qquad (3.20)$$

$$E(MP6)_3 = \sum_{X_1}^{S,D,T} (A(X_1, T, S)_L + A(X_1, T, Q_D)_L)$$
$$+ A(Q, T, S)_L + A(Q, T, Q_D)_L$$
$$= \sum_{i=1,2,3} (2 - \delta_{i,1})\langle\Phi_0|(\hat{T}_i^{(2)})^\dagger \bar{V}\hat{T}_1^{(2)}\hat{T}_2^{(1)}|\Phi_0\rangle_C$$
$$+ \langle\Phi_0|\frac{1}{2}((\hat{T}_2^{(1)})^\dagger)^2 \bar{V}\hat{T}_1^{(2)}\hat{T}_2^{(1)}|\Phi_0\rangle_C \qquad (3.21)$$

$$= E_{STS}^{(6)} + E_{STQ}^{(6)}(I) + 2\left[E_{DTS}^{(6)} + E_{DTQ}^{(6)}(I)\right] + 2\left[E_{TTS}^{(6)} + E_{TTQ}^{(6)}(I)\right]$$
$$+ E_{QTS}^{(6)} + E_{QTQ}^{(6)}(I) \qquad (3.22)$$

$$E(MP6)_4 = \sum_{X_1}^{T,Q} (A(X_1, P, T)_L + A(X_1, P, Q_D)_L)$$

$$= \langle\Phi_0|(\hat{T}_3^{(2)})^\dagger \bar{V}\hat{T}_2^{(1)}\hat{T}_3^{(2)}|\Phi_0\rangle_C + \langle\Phi_0|\frac{1}{2}((\hat{T}_2^{(1)})^\dagger)^2 \bar{V}\hat{T}_1^{(1)}\hat{T}_3^{(2)}|\Phi_0\rangle_C$$
$$\qquad (3.23a)$$

$$= \left[E_{TPT}^{(6)} + E_{TPQ}^{(6)}\right] + \left[E_{QPT}^{(6)} + E_{QPQ}^{(6)}\right] . \qquad (3.23b)$$

The final MP6 energy expression covers all four energy parts $E(MP6)_1$, $E(MP6)_2$, $E(MP6)_3$, and $E(MP6)_4$:

$$E_{MP}^{(6)} = E(MP6)_1 + E(MP6)_{2a} + E(MP6)_{2b} + E(MP6)_3 + E(MP6)_4 \qquad (3.24)$$

which correspond to the connected cluster operator part ($E(MP6)_1$), the disconnected Q cluster operator part ($E(MP6)_2$), the disconnected T cluster operator part ($E(MP6)_3$), and the disconnected P cluster operator part ($E(MP6)_4$). [24]

3.2 SETTING UP TWO-ELECTRON INTEGRAL FORMULAS

The development of a MP6 computer program implies the transformation of the MP6 cluster operator equations into two-electron integral formulas. Analogue to the auxiliary arrays vn used in chapter 2 for the derivation of the MP4 and MP5 energy, a series of auxiliary arrays un is derived at the MP6 level of theory. For lower n, these arrays actually correspond to the vn arrays used at the MP4 and MP5 level as is indicated in Table 3.1. Table 3.2 shows how each correlation energy contribution is determined with one of the auxiliary arrays un according to equations which are explicitly listed in Ref. 25. Utilizing the auxiliary arrays un, two-electron integral formulas for all MP6 terms listed in Table 3.2 can be readily obtained. However, the computational cost resulting out of this one-to-one transformation would be, as indicated in Table 3.2, $O(M^{12})$ thus leading to a MP6 method that would not be practical even with today's supercomputers.

TABLE 3.1. Relationship between auxiliary arrays vi in MP5 and auxiliary arrays uj in MP6.

MP5	MP6
$v1$(ij,ab)	$u4$(ij,ab)
$v2$(ij,ab)	$u5$(ij,ab)
$v3$(ij,ab)	$u6$(ij,ab)
$v4$(ij,ab)	$u10$(ij,ab)
$v5$(ij,ab)	$u14$(ij,ab)
$v6$(i,a)	$u1$(i,a)
$v7$(ijk,abc)	$u8$(ijk,abc)
$v8$(ijk,abc)	$u11$(ijk,abc)
$v9$(ijkl,abcd)	$u12$(ijkl,abcd)

As mentioned in section 3.1, all energy contributions of $E(MP6)_2$, $E(MP6)_3$, and $E(MP6)_4$ (#17 to #38 in Table 3.2) result from disconnected cluster operators. In Ref. 25, we have shown that correlation energy contributions associated with disconnected cluster operator diagrams can be expressed with the help of intermediate arrays that significantly reduce computational cost. This is indicated in

TABLE 3.2. Two-electron integral formulas for the energy contributions of MP6.[a]

#	Contribution	Auxiliary arrays	Expected cost
	$E(MP6)_1$		
1	$E^{(6)}_{SSS}$	u1(i,a)	$O(M^5)$
2	$E^{(6)}_{SSD}$	u1(i,a),u2(i,a)	$O(M^6)$
3	$E^{(6)}_{SST}$	u1(i,a),u3(i,a)	$O(M^7)$
4	$E^{(6)}_{DSD}$	u2(ij,ab)	$O(M^6)$
5	$E^{(6)}_{DST}$	u2(ij,ab),u3(ij,ab)	$O(M^7)$
6	$E^{(6)}_{TST}$	u3(i,a)	$O(M^7)$
7	$E^{(6)}_{SDS}$	u4(ij,ab)	$O(M^5)$
8	$E^{(6)}_{SDD}$	u4(ij,ab),u5(ij,ab)	$O(M^6)$
9	$E^{(6)}_{SDT}$	u4(ij,ab),u6(ij,ab)	$O(M^7)$
10	$E^{(6)}_{DDD}$	u5(ij,ab)	$O(M^6)$
11	$E^{(6)}_{DDT}$	u5(ij,ab),u6(ij,ab)	$O(M^7)$
12	$E^{(6)}_{TDT}$	u6(ij,ab)	$O(M^7)$
13	$E^{(6)}_{DTD}$	u7(jk,abc)	$O(M^7)$
14	$E^{(6)}_{DTT}$	u7(ijk,abc),u8(ijk,abc)	$O(M^8)$
15	$E^{(6)}_{TTT}$	u8(ijk,abc)	$O(M^8)$
16	$E^{(6)}_{TQT}$	u9(ijkl,abcd)	$O(M^9)$
	$E(MP6)_2$		
17	$E^{(6)}_{SDQ}$	u4(ij,ab),u10(ij,ab)	$O(M^8)$*
18	$E^{(6)}_{DDQ}$	u5(ij,ab),u10(ij,ab)	$O(M^8)$*
19	$E^{(6)}_{TDQ}$	u6(ij,ab),u10(ij,ab)	$O(M^8)$*
20	$E^{(6)}_{QDQ}$	u10(ij,ab)	$O(M^8)$*
21	$E^{(6)}_{STQ}(II) + E^{(6)}_{QTQ}(II)_a$	b_s, a_d, u11(ijk,abc)	$O(M^7)$*
22	$E^{(6)}_{DQQ}(II) + E^{(6)}_{QQQ}(II)_a$	a_d, b_d, u12(ijkl,abcd)	$O(M^9)$*
23	$E^{(6)}_{QTQ}(II)_b$	u11(ijk,abc)	$O(M^9)$*
24	$E^{(6)}_{QQQ}(II)_b$	u12(ijkl,abcd)	$O(M^{10})$*
25	$E^{(6)}_{DTQ}(II)$	u7(ijk,abcd),u11(ijk,abc)	$O(M^9)$*
26	$E^{(6)}_{TTQ}(II)$	u8(ijk,abc),u11(ijk,abc)	$O(M^9)$*
27	$E^{(6)}_{TQQ}(II)$	u9(ijkl,abcd),u12(ijkl,abcd)	$O(M^{10})$*
28	$E^{(6)}_{QHQ}(I)$	Q^{cd}_{kl},u10(ij,ab)	$O(M^8)$*
29	$E^{(6)}_{QHQ}(II)$		$O(M^{12})$*
30	$E^{(6)}_{DQD} + E^{(6)}_{DQQ}(I)$	a_d, b_d, Q^{cd}_{kl}	$O(M^8)$*
31	$E^{(6)}_{TQD} + E^{(6)}_{TQQ}(I)$	a_d,u9(ijkl,abcd)	$O(M^9)$*
32	$E^{(6)}_{QQD} + E^{(6)}_{QQQ}(I)$	a_d, b_d,u12(ijkl,abcd)	$O(M^{10})$*
	$E(MP6)_3$		
33	$E^{(6)}_{STS} + E^{(6)}_{STQ}(I)$	b_s,u13(i,a)	$O(M^6)$*
34	$E^{(6)}_{DTS} + E^{(6)}_{DTQ}(I)$	b_s, b_d,u14(ij,ab)	$O(M^7)$*
35	$E^{(6)}_{TTS} + E^{(6)}_{TTQ}(I)$	u15(ijk,abc),u16(i,a)	$O(M^8)$*
36	$E^{(6)}_{QTS} + E^{(6)}_{QTQ}(I)$	Q^{cd}_{kl}, u4(ij,ab), u17(l,d)	$O(M^8)$*
	$E(MP6)_4$		
37	$E^{(6)}_{TPT} + E^{(6)}_{TPQ}$	u3(i,a),u16(i,a),b_t,u18(ijk,abc)	$O(M^{10})$*
38	$E^{(6)}_{QPT} + E_{QPQ}$	u6(ij,ab),u19(i,a),u20(ijkl,abcd)	$O(M^{10})$*

[a] Ref. 25. The stars in the column "cost" indicate that the expected cost factor can be reduced by using intermediate arrays. See table 3.3.

Table 3.3 where for each of the disconnected cluster operator terms #17 to #38 of Table 3.2 the intermediate arrays $x1 - x20$, $y1 - y17$, and $z1 - z19$ are given that help to reduce computational cost to a minimum (see Ref. 25, for the definition of all intermediate arrays). This is demonstrated in the following using the term $E^{(6)}_{QHQ}(II)$ (#29 in Tables 3.2 and 3.3) as an appropriate example.

If $E^{(6)}_{QHQ}(II)$ would be calculated according to Eq. (36) of Ref. 25, one would have to determine an auxiliary array $u21(ijkl, abcd)$ according to Eq.s (3.25) and (3.26):

$$E^{(6)}_{QHQ}(II) = \frac{1}{32} \sum_{ij,ab} \sum_{kl,cd} a^{ab}_{ij} a^{cd}_{kl} u21(ijkl, abcd) \tag{3.25}$$

with $u21(ijkl, abcd)$ being defined by

$u21(ijkl, abcd)$

$$= -\frac{1}{4} \sum_m \sum_{PP'} (-1)^{P+P'} P(i/jkl|ab/cd) \left[P'(j/kl) \sum_{n,ef} \langle mn||ef\rangle a^{ef}_{kl} a^{cd}_{jn} \right] a^{ab}_{im}$$

$$+ \sum_f \sum_P (-1)^P P(ij/kl|a/bcd) \sum_{mn,e} \langle mn||ef\rangle \left[\sum_{P'} (-1)^{P'} P'(k/l|b/cd) a^{eb}_{mk} a^{cd}_{nl} \right]$$

$$- \frac{1}{4} \sum_{P'} (-1)^{P'} P'(b/cd) a^{cd}_{mn} a^{eb}_{kl} \right] a^{af}_{ij} \tag{3.26}$$

Since $u21$ depends on the eight indices i,j,k,l,a,b,c,d and requires in addition summations over indices m,n,e,f, the cost factor for $u21$ is proportional to $O(M^{12})$. However, this cost factor can be reduced since $u21(ijkl, abcd)$ does not depend on the H energy denominator. This provides the possibility of replacing $u21(ijkl, abcd)$ by a series of much cheaper intermediate arrays such as contractions between double amplitudes and combinations of these contractions (see Eq.s (A5) - (A9) and (A33) -(A42) of Ref. 25). This is outlined in Eq.s (3.27) - (3.36), which start with a dissection of $E^{(6)}_{QHQ}(II)$ into three parts:

$$E^{(6)}_{QHQ}(II) = \sum_{mn} \sum_{ef} \langle mn||ef\rangle \left[\frac{1}{4} w3(mn, ef)_a + w3(mn, ef)_b + \frac{1}{4} w3(mn, ef)_c \right] \tag{3.27}$$

Arrays $w3(mn, ef)_a$, $w3(mn, ef)_b$ and $w3(mn, ef)_c$ are determined with the help of intermediate arrays y12 - y17 according to Eq.s (3.28) - (3.36) (see Ref. 25).

$$w3(mn, ef)_a = -\sum_{kl} a^{ef}_{kl} y16(kl, mn) + \sum_{ab} x5(ab, ef) y17(ab, mn) \tag{3.28}$$

$$w3(mn, ef)_b = \sum_{lb} a^{eb}_{ml} y14(nf, lb) + \sum_{jb} x3(jb, me) y15(nf, jb) \tag{3.29}$$

$$w3(mn, ef)_c = -\sum_{kl} x4(kl, mn) y12(kl, ef) + \sum_{ca} a^{ca}_{mn} y13(ca, ef) \tag{3.30}$$

MANY BODY PERTURBATION THEORY AND COUPLES CLUSTER THEORY 273

TABLE 3.3. Use of intermediate arrays to reduce the cost of a MP6 calculation.[a]

#	Energy contribution	Auxiliary arrays		Intermediate arrays	Cost
	$E(MP6)_2$				
17	$E_{SDQ}^{(6)}$	$u4(ij,ab)$,	$u10(ij,ab)$	$z1 - z4$	$O(M^6)$
18	$E_{DDQ}^{(6)}$	$u5(ij,ab)$,	$u10(ij,ab)$	$z1 - z4$	$O(M^6)$
19	$E_{TDQ}^{(6)}$	$u6(ij,ab)$,	$u10(ij,ab)$	$z1 - z4$	$O(M^7)$
20	$E_{QDQ}^{(6)}$	$u10(ij,ab)$		$z1 - z4$	$O(M^6)$
21	$E_{STQ}^{(6)}(II) + E_{QTQ}^{(6)}(II)_a$	b_i^a,	$w1(i,a)$	$x1$-$x5$, $y1,y2$, etc.	$O(M^6)$
22	$E_{DQQ}^{(6)}(II) + E_{QQQ}^{(6)}(II)_a$	b_{ij}^{ab},	$w2(ij,ab)$	$x11$-$x13$, $y3 - y11$	$O(M^6)$
23	$E_{QTQ}^{(6)}(II)_b$	$u11(ijk,abc)$		$z5,\ z6$	$O(M^7)$
24	$E_{QQQ}^{(6)}(II)_b$	$u12(ijk,abc)$		$z7,\ z8$	$O(M^9)$
25	$E_{DTQ}^{(6)}(II)$	$u7(ijk,abc)$,	$u11(ijk,abc)$	$z5,\ z6$	$O(M^7)$
26	$E_{TTQ}^{(6)}(II)$	$u8(ijk,abc)$,	$u11(ijk,abc)$	$z5,\ z6$	$O(M^8)$
27	$E_{TQQ}^{(6)}(II)_b$	$u9(ijkl,abcd)$,	$u12(ijkl,abcd)$	$z7,\ z8$	$O(M^9)$
28	$E_{QHQ}^{(6)}(I)$	Q_{kl}^{cd},	$u10(ij,ab)$	$z1 - z4$	$O(M^6)$
29	$E_{QHQ}^{(6)}(II)$	$w3(mn,ef)_a,\ w3(mn,ef)_b,\ w3(mn,ef)_c$		$x1 - x5,\ y12 - y17$	$O(M^6)$
30	$E_{DQD}^{(6)} + E_{DQQ}^{(6)}(I)$	a_d,		$z1 - z4$	$O(M^6)$
31	$E_{TQD}^{(6)} + E_{TQQ}^{(6)}(I)$	b_t,	$w4(ijk,abc)$	$z5,\ z6$	$O(M^7)$
32	$E_{QQD}^{(6)} + E_{QQQ}^{(6)}(I)$	b_d,	$w2(ij,ab)$	$y3 - y11$	$O(M^6)$
	$E(MP6)_3$				
33	$E_{STS}^{(6)} + E_{STQ}^{(6)}(I)$	b_s,	$u13(i,a)$	$\sum b_s(mn\|ef)$	$O(M^5)$
34	$E_{DTS}^{(6)} + E_{DTQ}^{(6)}(I)$	b_d,	$u14(ij,ab)$	$\sum b_s(mn\|je)$, etc.	$O(M^6)$
35	$E_{TTS}^{(6)} + E_{TTQ}^{(6)}(I)$	$u15(ijk,abc), u16(i,a)$		$z9,\ z10$	$O(M^7)$
36	$E_{QTS}^{(6)} + E_{QTQ}^{(6)}(I)$	Q_{kl}^{cd}, $u4(ij,ab)$,	$u17(l,d)$	$z1 - z4,\ x14 - x19$	$O(M^6)$
	$E(MP6)_4$				
37	$E_{TPT}^{(6)} + E_{TPQ}^{(6)}$	$u3(i,a)$, $u16(i,a)$, b_t, $u18(ijk,abc)$		$z11 - z17$	$O(M^8)$
38	$E_{QPT}^{(6)} + E_{QPQ}^{(6)}$	$u6(ij,ab)$, $u19(i,n)$, $u20(ijkl,abcd)$		$z18$-$z19$, $x1$-$x10$, $x16,\ x17,\ x20$-$x21$	$O(M^8)$

[a] Ref 25. Compare with table 3.2. Note that the xn and yn are different from the xn and yn in Table 2.2.

$$y12(kl,ef) = \sum_b a_{kl}^{eb} x1(b,f) - \sum_{jd} a_{lj}^{fd} x3(ke,jd) \qquad (3.31)$$

$$y13(ca,e,f) = \sum_{ik} x3(ia,ke)x3(if,kc) - \sum_{bd} x5(ab,fd)x5(eb,cd)$$
$$+ x1(a,f)x1(c,e) \qquad (3.32)$$

$$y14(nf,lb) = -\sum_{kj} x4(kl,jn)x3(jb,kf) + x2(l,n)x1(b,f)$$
$$- \sum_k x3(lb,kf)x2(k,n) + \sum_{kc} x3(lc,kf)x3(kb,nc)$$
$$+ \sum_{ac} x5(ba,cf)x3(lc,na) \qquad (3.33)$$

$$y15(nf,jb) = -\sum_{kc} a_{kj}^{cf} x3(kc,nb) + \frac{1}{2}\sum_{kl} a_{kl}^{fb} x4(kl,nj) - \sum_d a_{nj}^{db} x1(d,f)$$
$$\qquad (3.34)$$

$$y16(kl,mn) = \sum_j x4(kl,jn)x2(j,m) - \sum_{ij} x4(li,mj)x4(kj,ni)$$
$$- x2(l,n)x2(k,m) - \sum_{bc} x3(lb,mc)x3(kc,nb) \qquad (3.35)$$

$$y17(ab,mn) = \sum_{kc} a_{mk}^{ac} x3(kc,nb). \qquad (3.36)$$

In these equations, the intermediate arrays x1 - x5 appear (see Table 3.3 and Ref. 25), which are contractions of double amplitudes, and the intermediate arrays $y12-y17$. In total, 11 intermediate arrays are used to reduce the original $O(M^{12})$ dependence of Eq. (3.25). Inspection of Eq.s (3.28) - (3.36) reveals that the calculation of the 11 intermediate arrays involves just $O(M^6)$ computational steps: The total cost of the calculation of $E_{QHQ}^{(6)}(II)$ has been reduced from $O(M^{12})$ to $O(M^6)$, which means a dramatic decrease in needed computer time.

Manipulations as the one described in the case of the $E_{QHQ}^{(6)}(II)$ term have been applied for all terms associated with disconnected cluster operators so that the reduced cost factors listed in Table 3.3 result. Compared to MP4 where just four intermediate arrays are needed (Table 2.1), the number of intermediate arrays for MP6 (57 in total) increases by a factor of 14. This demonstrates that any development of higher order MP methods has to concentrate on those correlation energy contributions which are associated with disconnected cluster operators.

In Ref. 25, each energy contribution of $E(MP6)$ is expressed in terms of two-electron integral formulas, which can directly be programmed for calculation on a computer. Of course, these formulas look rather complicated because of the manifold of double-bar integral terms, the complexity of summations and the large number of arrays to be formed. Therefore, the implementation of the MP6 method in form of a FORTRAN program on a computer requires a carefully worked out strategy.

3.3 IMPLEMENTATION AND TESTING OF A MP6 COMPUTER PROGRAM

As has been discussed in section 3.2, all disconnected cluster operator terms associated with T, Q or P excitations in $E(MP6)_2$, $E(MP6)_3$, and $E(MP6)_4$ can be calculated with the help of intermediate arrays. However, in some cases it is of advantage to combine the calculation of disconnected cluster operator terms with that of related connected cluster operator terms involving a higher cost factor rather than calculating each MP6 term individually. In this way, superfluous I/O operations are suppressed. We have found that in this way the calculation of E(MP6) becomes much more efficient.

A MP6 computer program can be structured in the following way. [25] First, all needed first order and second order amplitudes are collected, which, of course, are available from lower order MP calculations. Then, the loop over T excitations is carried out, which leads to some of the TTA terms as well as the TTT coupling contribution. In the next step, the Q loop is executed, which is the most expensive part of the program. There is a relatively large number of terms, the calculation of which can be based on existing MP5 programs. Finally, terms are collected to give the MP6 correlation energy.

In Ref. 25, a MP6 program has been set up in a way that many (but not all) correlation energy contributions can be determined individually. This gives the chance to analyze these terms, to investigate the importance of the most expensive terms and to develop partial MP6 methods that cover well-defined excitation and correlation effects.

Although the writing of a MP6 program is a time consuming task, even more time consuming is the testing of a new MP6 program. The question whether more than 3000 lines of FORTRAN code are without errors cannot be answered in a simple way. In this work, three testing strategies were developed to search for programming errors. First, a number of benchmark calculations were carried out for which MPn energies ($n \leq 48$) derived from full CI results are available. [12,13]

Then, parts of the program were reprogrammed by using alternative calculation strategies and, finally, MP5 results were used to test MP6 energies.

Testing with the help of FCI calculations could be done for about 30 atoms and molecules with no more than 10 electrons since FCI calculations with reasonable basis sets can be carried out in these cases. The majority of FCI based MP6 correlation energies agrees with our values within 10^{-6} hartree. This difference is also found for many of the lower order correlation contributions if one compares results from FCI-MPn and MPn calculations. Although the agreement between FCI-MP6 and our MP6 data seems to suggest reliability of the new MP6 program, it does not prove that the latter is without any errors. Since all test molecules are rather small possessing just a limited number of electrons, higher excitations such as P or H do not contribute significantly to the final correlation energy. As a consequence, any errors in these terms do not show up in the comparison between FCI-MP6 and MP6 energies. This also holds for any other low value term and has to be considered in the testing.

Therefore, we tested each term of the MP6 program (see list of terms in Table 3.2) individually by extensive reprogramming. For example, in the case of energy terms associated with disconnected cluster operators, we have programmed alternative evaluation procedures that do not take advantage of intermediate arrays. This leads to a rather simple structuring of the FORTRAN code, but also to program versions that can be used only for testing purposes because they are too expensive for normal use. An energy contribution tested in this way was considered to be correct when the difference in energy values obtained by different program versions is smaller or equal to 10^{-10} hartree.

A third way of efficiently searching for errors in the MP6 program was to replace second order amplitudes by the appropriate first order amplitudes to get the corresponding fifth order energy contributions, which can be directly compared with existing MP5 results. [8,9] This procedure is straightforward and can be extended to (partial) third order amplitudes to be replaced by second order amplitudes or products of first order amplitudes. In each case, it was verified that the energy contributions obtained at MP5 did not differ from the corresponding directly calculated MP5 terms by more than 10^{-10} hartree. After checking all MP6 energy contributions listed in Table 3.2 either by reprogramming or by exchanging amplitudes, we concluded that our MP6 program was without errors and could be used for calculating MP6 correlation energies. The MP6 codes were installed on a CRY Y-MP to be run within the ab initio package COLOGNE94. [44]

3.4 COMPARISON OF MP6 AND FULL CI CORRELATION ENERGIES

FCI energies are known for a number of atoms and simple molecules, which accordingly provide an appropriate basis for a comparison with E(MP6) energies obtained with the same basis at the same geometry. The set of reference systems includes charged and uncharged atoms (F and F^-), different states of molecules ($^3B_2$ and $^1A_1$ state of CH_2, $^2B_1$ and $^2A_1$ state of NH_2) as well as AH_n molecules both at their equilibrium geometry (R_e) and in geometries with (symmetrically) stretched AH bonds ($1.5R_e$, $2R_e$: "stretched geometries"). Calculation of the latter represents a critical test on the performance of a correlation method because wave functions of molecules with stretched geometries possess considerable multireference character. In total, 26 energy calculations have been carried out for the comparison. [26]

MP6 correlation energies cover on the average 98 - 99% of the exact (FCI) correlation energy for atoms and molecules at equilibrium geometries. For molecules with stretched geometries ($1.5R_e$ and $2.0R_e$), this coverage can drop to 80 - 85% because of difficulties in describing a problem with relatively high multireference character by a single determinant approach. There are systems, for which the MP6 correlation energy becomes more negative than the FCI correlation energy thus reflecting the non-variational character of MP theory.

We have investigated the mean absolute deviation between FCI and various MPn energies. [26] If just equilibrium geometries are considered, then there is a slight improvement when going from MP4 $\sim$ MP5 energies (mean absolute deviation 2.12 mhartree) to MP6 energies (mean absolute deviation 1.75 mhartree, [26]). If stretched geometries are included in the comparison, then mean absolute deviations become larger by a factor of 3 and decrease more clearly with increasing order n of MPn perturbation theory (MP4: 7.26, MP5: 6.47, MP6: 4.66 mhartree [26]). This suggests that fifth and sixth order corrections become more important with increasing multireference character of a system and that the relative improvement of energies is larger at the MP6 than the MP5 level of theory.

Because of the $O(M^9)$ dependence of MP6 methods, its application is limited to relatively small atoms and molecules. Therefore, it was interesting to test whether deletion of costly MP6 energy terms leads to useful approximate MP6 methods that are more economic and can be applied to larger molecules. We have checked two alternatives. [26] First, we have deleted the three terms $E^{(6)}_{TQT}$, $E^{(6)}_{QQQ}(II)_b$, and $E^{(6)}_{TQQ}(II)$ that require $O(M^9)$ computational steps. In this way, we have obtained an approximate MP6 method (MP6(M8)) with computational requirements $\leq O(M^8)$. In a second step, we have eliminated all terms that require $O(M^8)$ computational steps. Thus, an approximate $O(M^7)$ method has been obtained (MP6(M7)).

The average errors of MP6(M8) and MP6(M7) are 8 and 13%, respectively, of the total MP6 correlation energy. The difference $\Delta E^{(6)}(M8)$ is with the exception of F^- and the stretched geometry of H_2O considerably smaller than 1 mhartree. This is also true in the case of $\Delta E^{(6)}(M7)$. We have also investigated deviations in relative energies and compared them with those of other MPn methods and FCI. The mean absolute deviation of MP2 relative energies from the corresponding FCI values is rather large (12.5 kcal/mol), which has to do with the fact that the majority of the problems investigated involves systems with multireference character. At MP4, the mean absolute deviation decreases to 8.3 kcal/mol, then to 7.4 at MP5 and, finally, to 5.7 at MP6, i.e. the largest reduction in the mean absolute deviation is obtained at MP4 and MP6, which underlines that MP6 leads to the largest improvements after MP4.

The approximate MP6 methods give about the same mean absolute deviations (5.8 kcal/mol [26]) than MP6, i.e. the three methods MP6, MP6(M8), and MP6(M7) lead to similar relative energies. For example, the singlet-triplet splitting in the case of CH_2 is calculated to be 12.98, 12.99, and 13.06 kcal/mol at MP6, MP6(M8), and MP6(M7), respectively (FCI value 11.97 kcal/mol [26]). A similarly good agreement is obtained for the differences between the $^2A_1$ and the $^2B_1$ state of NH_2 taken at R_e, $1.5R_e$ and $2R_e$ of the NH bond distance. On the other hand, there is a clear improvement of relative energies when going from MP5 to MP6(M8) or MP6(M7). Since the latter method has similar time requirements as MP4, MP6(M7) is an attractive new method for getting higher order correlation corrections for small and medium-sized molecules.

4. Coupled Cluster Theory

Coupled Cluster theory is tightly connected with the Linked Diagram (LD) theorem which states that the exact electronic energy and wave function of the Schrödinger equation can be written as a sum of linked diagrams in field theory language without any contributions from unlinked diagrams. [34] Equivalently, one can say that the wave function is expressed with the help of an exponential of cluster operators, which was first suggested by Coester and Kümmel in physics in the late 1950s [45] and later introduced into Quantum Chemistry by Cizek and Paldus. [16,17] The exponential form of the wave function guarantees correct scaling with the number of electrons, which leads to the important property of size-extensivity of calculated energies. [10] Sometimes one uses also the term size-consistency in connection with the investigation of dissociation or addition reactions. [5] However, the term size-extensivity is more general than the term size-consistency and, therefore, the former is used throughout this work. Contrary to CI methods, MP and CC methods are all size-extensive. [10]

The major difference between MP and CC methods is that MP theory is used to include all contributions resulting from S, D, T, etc. excitations through some finite order while CC theory covers selected contributions to the MP correlation energy to infinite order. Therefore, it does not lead to the oscillations in correlation energy typical of the MPn series. On the other hand, the cost factor of a given CC method is considerably larger than its MP equivalent (defined by the excitations covered) as we will show in the following.

The use of CC methods in ab initio theory was started on a routine basis in the early 1980s due to the development work of the Pople and the Bartlett group. [18] However, little systematic research with the lower CC methods was done in the first years after CCD [18] and CCSD [19] programs became generally available. This has changed in the last five years when CC methods covering T effects were introduced that provided high-accuracy in calculated energies. As an example for many other investigations the CC study of simple H_2 addition reactions carried out by Kraka and co-workers may be mentioned here. [46] In this work, it was demonstrated that with a CC method that includes T effects, more precise activation barriers and reaction energies could be calculated than previously with MRCI, MCSCF or CI methods. Because of its many advantages, CC theory attracts a lot of research efforts and, therefore, one has investigated how currently used CC methods can be improved to get even higher accuracy. Before this work is discussed, we shortly review the CC projection approach because it is hardly discussed in any of the standard Quantum Chemistry text books.

4.1 THE PROJECTION COUPLED CLUSTER APPROACH

The Coupled Cluster (CC) wave function Ψ_{CC} is expressed in terms of the cluster operator $\hat{T}$ as

$$\Psi_{CC} = e^{\hat{T}} |\Phi_0\rangle \tag{4.1}$$

where $\hat{T}$ is the cluster operator for the n electrons of a given electronic system

$$\hat{T} = \hat{T}_1 + \hat{T}_2 + \hat{T}_3 + \ldots + \hat{T}_n \tag{4.2}$$

and

$$\hat{T}_n = \frac{1}{(n!)^2} \sum_{i,j\ldots,a,b,\ldots} a_{ij\ldots}^{ab\ldots} \hat{t}_{ij\ldots}^{ab\ldots} \tag{4.3}$$

In Eq. (4.3), the operators $\hat{t}_{ij\ldots}^{ab\ldots}$ represent elementary substitution operators and the amplitudes $a_{ij\ldots}^{ab\ldots}$ are the corresponding cluster amplitudes. With a wave function of the form (4.1) the expectation value ΔE of the Hamiltonian $\bar{H}(= \hat{H} - \langle \Phi_0|\hat{H}|\Phi_0\rangle)$ can be written as

$$\Delta E = \frac{\langle \Phi_0 | e^{\hat{T}^\dagger} \bar{H} e^{\hat{T}} | \Phi_0 \rangle}{\langle \Phi_0 | e^{\hat{T}^\dagger} e^{\hat{T}} | \Phi_0 \rangle}$$

$$= \langle \Phi_0 | \left(e^{\hat{T}^\dagger} \bar{H} e^{\hat{T}}\right)_C | \Phi_0 \rangle \qquad (4.4)$$

The expectation value ΔE is obtained by standard variational theory where the variational parameters are the cluster amplitudes $(a_{ij\ldots}^{ab\ldots})^*$. In the first step, one obtains

$$\langle \Phi_{ij\ldots}^{ab\ldots} | (e^{\hat{T}^\dagger} \bar{H} e^{\hat{T}})_C | \Phi_0 \rangle = 0 \qquad (4.5)$$

By inserting the identity operator $\hat{I} = e^{\hat{T}} e^{-\hat{T}}$ into Eq.(4.5) and using the fact that

$$e^{-\hat{T}} \bar{H} e^{\hat{T}} = (\bar{H} e^{\hat{T}})_C \qquad (4.6)$$

Eq.(4.5) becomes

$$\langle \Phi_{ij\ldots}^{ab\ldots} | (e^{\hat{T}^\dagger} e^{\hat{T}})(\bar{H} e^{\hat{T}})_C | \Phi_0 \rangle = 0 \qquad (4.7)$$

which is equivalent to a set of projection equations:

$$\langle \Phi_i^a | (\bar{H} e^{\hat{T}})_C | \Phi_0 \rangle = 0 \qquad (4.8a)$$

$$\langle \Phi_{ij}^{ab} | (\bar{H} e^{\hat{T}})_C | \Phi_0 \rangle = 0 \qquad (4.8b)$$

$$\vdots$$

The correlation energy ΔE of Eq.(4.4) is given by

$$\Delta E = \langle \Phi_0 | (\bar{H} e^{\hat{T}})_C | \Phi_0 \rangle \qquad (4.9)$$

The full CC energy ΔE, which is identical to the full CI correlation energy, is size-extensive and variational according to the derivation given above.

In practice, one has to truncate the cluster operator $\hat{T}$ of Eq. (4.2) at a finite level n to obtain a practical method. This, however, leads to the loss of the variational character of the CC method. For example, when

$$\hat{T} \approx \hat{T}_1 + \hat{T}_2 \qquad (4.10)$$

the CCSD wave function is obtained

$$\Psi_{CCSD} = e^{\hat{T}_1 + \hat{T}_2} | \Phi_0 \rangle \qquad (4.11)$$

The CCSD equations are obtained by projecting the Schrödinger equation onto single and double excited determinants:

$$\langle \Phi_i^a | (\bar{H} e^{\hat{T}_1+\hat{T}_2}) | \Phi_0 \rangle = a_i^a \Delta E_{CCSD} \tag{4.12}$$

$$\langle \Phi_{ij}^{ab} | (\bar{H} e^{\hat{T}_1+\hat{T}_2}) | \Phi_0 \rangle = \langle \Phi_{ij}^{ab} | \hat{T}_2 + \frac{1}{2}\hat{T}_1^2 | \Phi_0 \rangle \Delta E_{CCSD} \tag{4.13}$$

$$\Delta E_{CCSD} = \langle \Phi_0 | \bar{H}(\hat{T}_2 + \frac{1}{2}\hat{T}_1^2) | \Phi_0 \rangle \tag{4.14}$$

The left-hand side of Eq. (4.12) can be split into connected part and disconnected part:

$$\langle \Phi_i^a | (\bar{H} e^{\hat{T}_1+\hat{T}_2}) | \Phi_0 \rangle = \langle \Phi_i^a | (\bar{H} e^{\hat{T}_1+\hat{T}_2})_C | \Phi_0 \rangle + \langle \Phi_i^a | (\bar{H} e^{\hat{T}_1+\hat{T}_2})_D | \Phi_0 \rangle \tag{4.15}$$

The disconnected part of Eq. (4.15) can be rewritten according to (4.16), (4.17) and (4.18):

$$\langle \Phi_i^a | (\bar{H} e^{\hat{T}_1+\hat{T}_2})_D | \Phi_0 \rangle = \langle \Phi_i^a | (\bar{H}(\frac{1}{2}\hat{T}_1^2 + \hat{T}_1\hat{T}_2 + \frac{1}{3!}\hat{T}_1^3))_D | \Phi_0 \rangle$$

$$= \langle \Phi_i^a | \hat{T}_1 (\bar{H}(\hat{T}_1 + \hat{T}_2 + \frac{1}{2}\hat{T}_1^2))_C | \Phi_0 \rangle \tag{4.16}$$

$$= \langle \Phi_i^a | \hat{T}_1 | \Phi_0 \rangle \langle \Phi_0 | (\bar{H}(\hat{T}_1 + \hat{T}_2 + \frac{1}{2}\hat{T}_1^2))_C | \Phi_0 \rangle \tag{4.17}$$

$$= \langle \Phi_i^a | \hat{T}_1 | \Phi_0 \rangle \langle \Phi_0 | (\bar{H}(\hat{T}_2 + \frac{1}{2}\hat{T}_1^2))_C | \Phi_0 \rangle$$

$$= a_i^a \Delta E_{CCSD} \tag{4.18}$$

where in Eq. (4.17) the identity operator $\sum_p^\infty |\Phi_p\rangle\langle\Phi_p|$ has been inserted. It becomes clear from Eq. (4.18) that the disconnected part just cancels the term on the right-hand side of Eq. (4.12) so that Eq. (4.12) takes the form of (4.19)

$$\langle \Phi_i^a | (\bar{H} e^{\hat{T}_1+\hat{T}_2})_C | \Phi_0 \rangle = 0 \tag{4.19}$$

or alternatively

$$\langle \Phi_i^a | \bar{H}(\hat{T}_1 + \hat{T}_2 + \frac{1}{2}\hat{T}_1^2 + \hat{T}_1\hat{T}_2 + \frac{1}{6}\hat{T}_1^3) | \Phi_0 \rangle_C = 0 \tag{4.20}$$

In a similar way, the disconnected part of the left-hand side of Eq. (4.13) can be rewritten:

$$\langle \Phi_{ij}^{ab} | (\bar{H} e^{\hat{T}_1+\hat{T}_2})_D | \Phi_0 \rangle$$

$$= \langle \Phi_{ij}^{ab} | \left(\bar{H}(1 + \hat{T}_1 + \hat{T}_2 + \frac{1}{2}\hat{T}_1^2 + \hat{T}_1\hat{T}_2 + \frac{1}{2}\hat{T}_2^2 + \frac{1}{6}\hat{T}_1^3 + \frac{1}{2}\hat{T}_1^2\hat{T}_2 + \frac{1}{24}\hat{T}_1^4) \right)_D | \Phi_0 \rangle$$

$$= \langle \Phi_{ij}^{ab} | \hat{T}_1 \left(\bar{H}(\hat{T}_1 + \hat{T}_2 + \frac{1}{2}\hat{T}_1^2 + \frac{1}{2}\hat{T}_1\hat{T}_2 + \frac{1}{3!}\hat{T}_1^3) \right)_C | \Phi_0 \rangle$$
$$+ \langle \Phi_{ij}^{ab} | \hat{T}_2 \left(\bar{H}(\hat{T}_1 + \hat{T}_2 + \frac{1}{2}\hat{T}_1^2) \right)_C | \Phi_0 \rangle$$
$$+ \frac{1}{2} \langle \Phi_{ij}^{ab} | \hat{T}_1^2 \left(\bar{H}(\hat{T}_2 + \frac{1}{2}\hat{T}_1^2) \right)_C | \Phi_0 \rangle \qquad (4.21)$$

$$= \langle \Phi_{ij}^{ab} | \hat{T}_1 \sum_s^S | \Phi_s \rangle \langle \Phi_s | \left(\bar{H}(\hat{T}_1 + \hat{T}_2 + \frac{1}{2}\hat{T}_1^2 + \frac{1}{2}\hat{T}_1\hat{T}_2 + \frac{1}{3!}\hat{T}_1^3) \right)_C | \Phi_0 \rangle$$
$$+ \langle \Phi_{ij}^{ab} | \hat{T}_2 | \Phi_0 \rangle \langle \Phi_0 | \left(\bar{H}(\hat{T}_1 + \hat{T}_2 + \frac{1}{2}\hat{T}_1^2) \right)_C | \Phi_0 \rangle$$
$$+ \frac{1}{2} \langle \Phi_{ij}^{ab} | \hat{T}_1^2 | \Phi_0 \rangle \Delta E_{CCSD} \qquad (4.22)$$

$$= \langle \Phi_{ij}^{ab} | (\hat{T}_2 + \frac{1}{2}\hat{T}_1^2) | \Phi_0 \rangle \langle \Phi_0 | \left(\bar{H}(\hat{T}_2 + \frac{1}{2}\hat{T}_1^2) \right)_C | \Phi_0 \rangle \qquad (4.23)$$

$$= \langle \Phi_{ij}^{ab} | \hat{T}_2 + \frac{1}{2}\hat{T}_1^2 | \Phi_0 \rangle \Delta E_{CCSD} \qquad (4.24)$$

where Eq. (4.20) has been used in Eq. (4.22). Hence, the unlinked diagram terms in Eq. (4.13) also mutually cancel. The final D amplitude equations take the form of Eq. (4.25)

$$\langle \Phi_{ij}^{ab} | (\bar{H} e^{\hat{T}_1 + \hat{T}_2})_C | \Phi_0 \rangle = 0 \qquad (4.25)$$

which can explicitly be written as

$$\langle \Phi_{ij}^{ab} | \bar{H}(1 + \hat{T}_1 + \hat{T}_2 + \frac{1}{2}\hat{T}_1^2 + \hat{T}_1\hat{T}_2 + \frac{1}{2}\hat{T}_2^2 + \frac{1}{6}\hat{T}_1^3 + \frac{1}{2}\hat{T}_1^2\hat{T}_2 + \frac{1}{24}\hat{T}_1^4) | \Phi_0 \rangle_C = 0 \qquad (4.26)$$

Eq.s (4.20) and (4.26) are the CCSD (Coupled-Cluster Singles and Doubles) projection equations in connected form:

$$\langle \Phi_i^a | \bar{H}(\hat{T}_1 + \hat{T}_2 + \frac{1}{2}\hat{T}_1^2 + \hat{T}_1\hat{T}_2 + \frac{1}{6}\hat{T}_1^3) | \Phi_0 \rangle_C = 0 \qquad (4.20)$$

$$\langle \Phi_{ij}^{ab} | \bar{H}(1 + \hat{T}_1 + \hat{T}_2 + \frac{1}{2}\hat{T}_1^2 + \hat{T}_1\hat{T}_2 + \frac{1}{2}\hat{T}_2^2 + \frac{1}{6}\hat{T}_1^3 + \frac{1}{2}\hat{T}_1^2\hat{T}_2 + \frac{1}{24}\hat{T}_1^4) | \Phi_0 \rangle_C = 0 \qquad (4.26)$$

which have to be solved to obtain ΔE_{CCSD} as an approximation to the true correlation energy.

$$\Delta E_{CCSD} = \langle \Phi_0 | \bar{H}(\hat{T}_2 + \frac{1}{2}\hat{T}_1^2) | \Phi_0 \rangle \qquad (4.27)$$

Since S and D excitation amplitudes a_i^a and a_{ij}^{ab} of Eq.s (4.20) and (4.26) have to be known to set up and solve the CCSD equations, a solution can only be found by a trial-and-error procedure. Once the amplitudes are known, they can be used to evaluate the CCSD energy ΔE_{CCSD} according to Eq. (4.27). All CC correlation energies obtained by truncation of $\hat{T}$ similar to (4.10) are still size-extensive, however they are no longer variational. Obviously, truncation of the cluster operator $\hat{T}$ at various excitation levels leads to a hierarchy of Coupled Cluster equations. For example, by truncating at $\hat{T}_3$, the CCSDT projection equations [20], and by truncating at $\hat{T}_4$, the CCSDTQ projection equations [21] are obtained. Since one has to iterate in each case, CCSD requires $N_{iter}O(M^6)$ computational steps, CCSDT $N_{iter}O(M^8)$ computational steps, and CCSDTQ $N_{iter}O(M^{10})$ computational steps. While CCSD computer programs are generally available, only few groups have developed expert programs that can solve the CCSDT [20,47] or even the CCSDTQ projection equations. [21]

4.2 THE QUADRATIC CI APPROACH
- AN APPROXIMATE COUPLED CLUSTER METHOD

The quadratic CI (QCI) method was suggested by Pople and co-workers. [35] Although it can be considered as a method that results from a simplification of the corresponding CC equations, Pople and co-workers took a different view and considered QCI as a CI method corrected for the size-extensivity error of CI. [35] To achieve size-extensivity for CI, the authors added new terms to the CI projection equations, which is demonstrated in the following for the case of a truncated CI approach with just S and D excitations (CISD) included. [48]

The CISD wave function can be represented by

$$\Psi_{CISD} = (1 + \hat{T}_1 + \hat{T}_2)|\Phi_0\rangle \tag{4.28}$$

Then, the Schrödinger equation in the CISD approximation takes the form

$$\bar{H}(1 + \hat{T}_1 + \hat{T}_2)|\Phi_0\rangle = \Delta E_{CISD}(1 + \hat{T}_1 + \hat{T}_2)|\Phi_0\rangle \tag{4.29}$$

The projection of Eq. (4.29) on S and D excited determinants as well as the reference wave function (i.e. the HF function) leads to Eq.s (4.30), (4.31), and (4.32):

$$\langle\Phi_i^a|\bar{H}(\hat{T}_1 + \hat{T}_2)|\Phi_0\rangle = a_i^a \Delta E_{CISD} \tag{4.30}$$

$$\langle\Phi_{ij}^{ab}|\bar{H}(1 + \hat{T}_1 + \hat{T}_2)|\Phi_0\rangle = a_{ij}^{ab} \Delta E_{CISD} \tag{4.31}$$

$$\Delta E_{CISD} = \langle\Phi_0|\bar{H}\hat{T}_2|\Phi_0\rangle \tag{4.32}$$

which also can be viewed as resulting from a minimization of the expectation value ΔE_{CISD}. In this and the following, intermediate normalization, i.e. $\langle\Phi_0|\Phi_{CISD}\rangle = 1$, is used throughout.

Since in both Eq. (4.30) and Eq. (4.31) unlinked diagram terms appear, the energy ΔE_{CISD} is not size-extensive, which is a general problem of all truncated CI methods. The simplest way of restoring size-extensivity in the CISD equations is to add $\hat{T}_1\hat{T}_2$ on the left-hand side of Eq. (4.30) and $\frac{1}{2}\hat{T}_2^2$ on the left-hand side of Eq. (4.31). These quadratic terms cancel the unlinked diagram terms of the right-hand side of Eq.s (4.30) and (4.31), namely $a_i^a \Delta E_{CISD}$ and $a_{ij}^{ab} \Delta E_{CISD}$. In this way a size-extensive modified CISD method is obtained, which was coined Quadratic Configuration Interaction with Single and Double excitations (QCISD) by Pople and co-workers. [35] The final QCISD projection equations in connected form are given by

$$\langle \Phi_i^a | \bar{H}(\hat{T}_1 + \hat{T}_2 + \hat{T}_1\hat{T}_2) | \Phi_0 \rangle_C = 0 \tag{4.33}$$

$$\langle \Phi_{ij}^{ab} | \bar{H}(1 + \hat{T}_1 + \hat{T}_2 + \frac{1}{2}\hat{T}_2^2) | \Phi_0 \rangle_C = 0 \tag{4.34}$$

$$\Delta E_{QCISD} = \langle \Phi_0 | \bar{H}\hat{T}_2 | \Phi_0 \rangle \tag{4.35}$$

Pople and co-workers have also attempted to develop a QCI method in S, D and T space, which replaces the non-size-extensive CISDT approach. [35] These authors suggested the QCISDT projection equations given in (4.36) - (4.39):

$$\langle \Phi_i^a | \bar{H}(\hat{T}_1 + \hat{T}_2 + \hat{T}_3) | \Phi_0 \rangle = a_i^a \Delta E_{QCISDT} \tag{4.36}$$

$$\langle \Phi_{ij}^{ab} | \bar{H}(1 + \hat{T}_1 + \hat{T}_2 + \hat{T}_3 + \frac{1}{2}\hat{T}_2^2) | \Phi_0 \rangle = a_{ij}^{ab} \Delta E_{QCISDT} \tag{4.37}$$

$$\langle \Phi_{ijk}^{abc} | \bar{H}(\hat{T}_1 + \hat{T}_2 + \hat{T}_3 + \hat{T}_2\hat{T}_3) | \Phi_0 \rangle = a_{ijk}^{abc} \Delta E_{QCISDT} \tag{4.38}$$

$$\Delta E_{QCISDT} = \langle \Phi_0 | \bar{H}\hat{T}_2 | \Phi_0 \rangle \tag{4.39}$$

However, the two quadratic terms added are not sufficient to cancel the unlinked diagram term $a_i^a \Delta E_{QCISDT}$ in Eq. (4.36) and an unlinked diagram term resulting from $\hat{T}_1$ in Eq. (4.38). In addition, the disconnected diagram part of $\hat{T}_2\hat{T}_3$ in Eq. (4.38) leads to a new unlinked diagram term to the energy ΔE_{QCISDT} so that the QCISDT energy defined by Eq.s (4.36) - (4.39) is not size-extensive. Because of this failure, the QCI concept has been criticized. [49] On the other hand, QCISD was for a long time the most often used (approximate) CC method, which simply had to do with the fact that a) a QCISD program was early available to chemists through Pople's ab initio package GAUSSIAN [50] and b) QCISD results seemed to be superior to CCSD results. The criticism with regard to the QCI concept caused us to analyze QCISD on the basis of perturbation theory (chapter 5) and to look for other ways of restoring size-extensivity at the CISDT level of theory (chapter 6). [22,23,30-32]

5. Analysis of Coupled Cluster Methods in Terms of Perturbation Theory

Since CC methods cover infinite order effects, they are more accurate than MP or CI methods based on the same excitations. In general, it is difficult to say which effects are covered by a given CC method and how it compares with MP or CI methods. [8,9,42] Therefore, one analyzes CC methods in terms of perturbation theory, which is particularly useful since the structure of lower order perturbation contributions to the correlation energy is well-known. As shown in chapter 2, MP4 and MP5 correlation energies can be dissected into contributions from specific excitations according to Eq.s (5.1) and (5.2):

$$E_{MP}^{(4)} = E_S^{(4)} + E_D^{(4)} + E_T^{(4)} + E_Q^{(4)} \tag{5.1}$$

$$E_{MP}^{(5)} = E_{SS}^{(5)} + 2E_{SD}^{(5)} + 2E_{ST}^{(5)} + E_{DD}^{(5)} + 2E_{DT}^{(5)} + 2E_{DQ}^{(5)} + E_{TT}^{(5)} + 2E_{TQ}^{(5)} + E_{QQ}^{(5)} \tag{5.2}$$

In a similar way, the MP6 correlation energy can be dissected into 36 contributions:

$$\begin{aligned}E_{MP}^{(6)} =& E_{SSS}^{(6)} + 2E_{SSD}^{(6)} + 2E_{SST}^{(6)} + E_{SDS}^{(6)} + 2E_{SDD}^{(6)} + 2E_{SDT}^{(6)} + 2E_{SDQ}^{(6)} + E_{STS}^{(6)} \\ &+ 2E_{STD}^{(6)} + 2E_{STT}^{(6)} + 2E_{STQ}^{(6)} + E_{DSD}^{(6)} + 2E_{DST}^{(6)} + E_{DDD}^{(6)} + 2E_{DDT}^{(6)} \\ &+ 2E_{DDQ}^{(6)} + E_{DTD}^{(6)} + 2E_{DTT}^{(6)} + 2E_{DTQ}^{(6)} + E_{DQD}^{(6)} + 2E_{DQT}^{(6)} + 2E_{DQQ}^{(6)} \\ &+ E_{TST}^{(6)} + E_{TDT}^{(6)} + 2E_{TDQ}^{(6)} + E_{TTT}^{(6)} + 2E_{TTQ}^{(6)} + E_{TQT}^{(6)} + 2E_{TQQ}^{(6)} \\ &+ E_{TPT}^{(6)} + 2E_{TPQ}^{(6)} + E_{QDQ}^{(6)} + E_{QTQ}^{(6)} + E_{QQQ}^{(6)} + E_{QPQ}^{(6)} + E_{QHQ}^{(6)} \end{aligned} \tag{5.3}$$

Each of these contributions represents a special correlation effect as has been discussed in section 2.3. However, with increasing order n of MP perturbation theory the number of terms increases exponentially and, therefore, it becomes impossible to keep track of each single term and to check whether it is covered by a certain CC method or not. Nevertheless, it will be helpful if one knows that a given CC method is correct up to nth order perturbation theory, which means that all contributions up to this order are contained in the CC approach. In such a case, one can expect that the CC method is as accurate as the corresponding MP perturbation method still contained in the CC method. For example, one has shown that CCSD is correct up to third order and also contains apart from the T contribution all other fourth order terms. One can expect that CCSD calculations are superior to either MP2, MP3 or MP4(SDQ) calculations because

CCSD contains beside the third and fourth order contributions also infinite order contributions not covered by any of the MP methods.

While it is rather easy to compare CC methods with MP methods, the comparison of different CC methods based on the same excitations is much more difficult since it has to be carried to higher orders of perturbation theory. For this purpose, we have developed a graphical method that reveals which contributions to the correlation energy at higher orders of perturbation theory are covered by the CC method in question. [22,23] According to this method, each energy contribution at nth order perturbation theory is described as a path that connects those excitations A, B, C, etc. at orders n, n-1, n-2, etc. down to order 4 that characterize the contribution $E_{ABC...}^{(n)}$. A path can start at one of the excitation levels S, D, T or Q and has to end at one of these levels at order 4. In between, it can leave SDTQ space under the provision that Slater rules for a two-electron operator are fulfilled. [23,24]

For finite MP perturbation theory the possible energy paths form a regular network, which in horizontal direction takes the form of a wedge. This is the direction of increasing excitation levels that can be included for increasing order n. In Figure 5.1, this is shown for MP8. At this level, Q excitations can couple via H excitations to octuple (O) excitations in the sequence QHOHQ according to Slater rules, i.e. MP8 is the first perturbation method that includes correlation effects from septuple (S_7) and O excitations. There are 915 different paths representing the 915 correlation energy contributions $E_{ABCDE}^{(8)}$ of the eighth order MP energy. While it would be very time consuming and of little use to write down all 915 contributions, the diagram in Fig.5.1, gives these contributions in a compact and easy to understand form. With the diagram 5.1, it is straightforward to determine the unique energy paths and to exclude the symmetry equivalent paths so that the 583 unique energy contributions of MP8 can be described. In addition, the diagram makes obvious how the couplings between different excitations grow with the order n forming ladders that stretch in the direction of infinity.

Graphical representations as the one shown for MP8 in Figure 5.1 also have to stop at some finite level. However, even for finite n they give a good impression how the diagram develops for n going to infinity and, therefore, they are well suited to describe the infinite order effects of a CC method. Therefore, they have been used in connection with an algebraic expansion method to analyse the infinite order effects of CC methods.

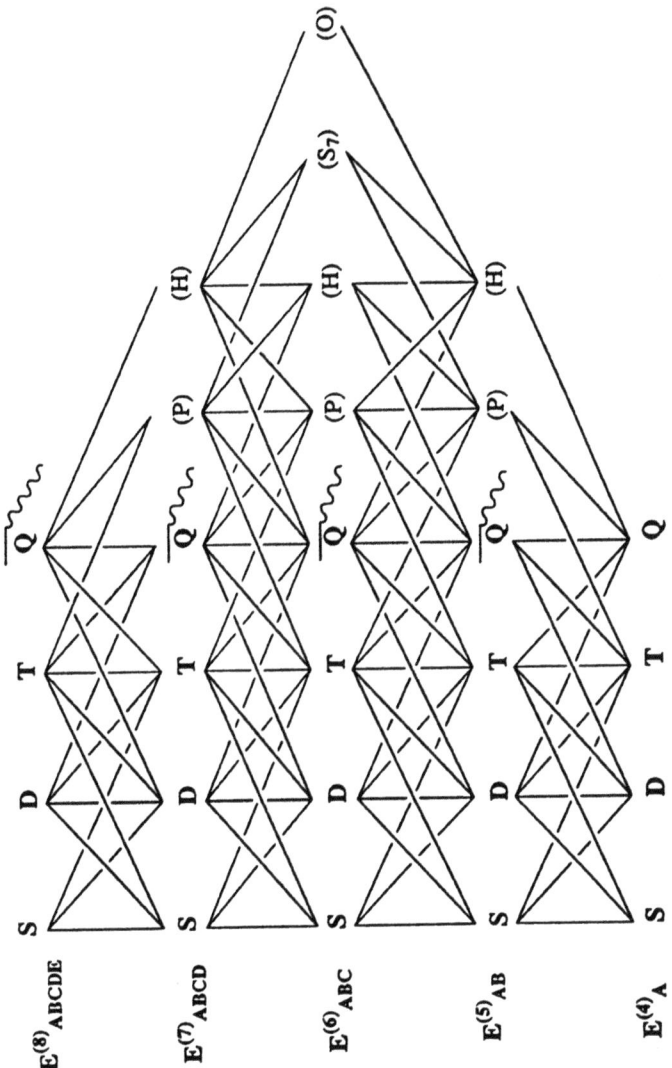

Figure 5.1. Graphical representation of all energy contributions at nth order MP perturbation theory (n= 4, 5, 6, 7, 8). A particular energy contribution $E^{(n)}{}_{ABC...}$ is given by the solid line that starts at A = S, D, T or Q in the $E^{(n)}$ row and connects B, C, etc. at row n-1, n-2, etc. until n=4 is reached. Note that at the nth order level also those excitations are included that arise from energy terms at higher order levels (m > n). They are given in parentheses after a separator to the right of the scheme. Reprinted with permission from Z. He and D. Cremer, Int. J. Quant. Chem. Symp. (1991) 25, 43. Copyright (1991) J. Wiley, Inc.

5.1 EXPANSION OF CC METHODS TO HIGHER ORDERS OF PERTURBATION THEORY

CC projection methods have to be solved iteratively. During the iterations, contributions from higher and higher orders of perturbation theory are successively added to the CC correlation energy. These contributions become smaller and smaller with increasing n for a given molecular system. Depending on the convergence criterion of the convergence procedure, contributions being smaller than a given threshold are cut off, which means that the highest orders of perturbation theory are neglected. Of course, this does not contradict the fact that CC methods cover infinite order effects. It only means that it is hardly useful to have the CC calculation running through an infinite number of iterations to cover even infinitely small energy contributions for n becoming infinitely large.

The contributions from perturbation theory covered by a CC method can be determined by expanding the CC correlation energy to higher orders, which may be demonstrated for the CCD (CC with double excitations) method. The CCD projection equations [18] can be written as

$$\Delta E_{CCD} = \sum_d^D \langle \Phi_0 | \bar{H} | \Phi_d \rangle a_d(CCD) \qquad (5.6)$$

and

$$a_d(CCD) = (E_0 - E_d)^{-1} \langle \Phi_d | \bar{V}(1 + \hat{T}_2 + \frac{1}{2}\hat{T}_2^2) | \Phi_0 \rangle_C \qquad (5.7)$$

Iterative solving of Eq. (5.7) leads in the kth step to the amplitudes $a_d^{(k)}(CCD)$ (also written as $a_d^{(k)}$) given in Eq. (5.8):

$$a_d^{(k)}(CCD) = (E_0 - E_d)^{-1} V_{d0} + (E_0 - E_d)^{-1} \left[\sum_{d_1}^D \bar{V}_{dd_1} a_{d_1}^{(k-1)} \right.$$
$$\left. + \sum_{d_1 d_2}^D \frac{1}{2} \left(\langle \Phi_d | \bar{V} \hat{t}_{d_1} \hat{t}_{d_2} | \Phi_0 \rangle a_{d_1}^{(k-1)} a_{d_2}^{(k-1)} \right)_C \right] \qquad (5.8)$$

For $k = 1$, $a_d^{(1)}$ is given by

$$a_d^{(1)} = (E_0 - E_d)^{-1} V_{d0} = C_{1,d}^{(1)} \qquad (5.9)$$

thus leading to the correlation energy $\Delta E_{CCD}^{(1)}$

$$\Delta E_{CCD}^{(1)} = E_{MP}^{(2)} \qquad (5.10)$$

which is simply the second order MP correlation energy. In the second iteration step, the amplitudes take the form:

$$a_d^{(2)} = C_{1,d}^{(2)} + C_{2,d}^{(2)} + C_{3,d}^{(2)} \tag{5.11}$$

where the terms $C_{i,d}^{(2)}(i = 1,2,3)$ correspond to the cluster operators 1, $\hat{T}_2$, and $\frac{1}{2}\hat{T}_2^2$ of Eq. (5.7). They are given by

$$C_{1,d}^{(2)} = C_{1,d}^{(1)} \tag{5.12}$$

$$C_{2,d}^{(2)} = (E_0 - E_d)^{-1} \sum_{d_1}^{D} \bar{V}_{dd_1} C_{1,d_1}^{(1)} \tag{5.13}$$

and

$$C_{3,d}^{(2)} = (E_0 - E_d)^{-1} \sum_{d_1 d_2}^{D} \frac{1}{2} \left(\langle \Phi_d | \bar{V} \hat{t}_{d_1} \hat{t}_{d_2} | \Phi_0 \rangle C_{1,d_1}^{(1)} C_{1,d_2}^{(1)} \right)_C \tag{5.14}$$

Hence, the correlation energy ΔE_{CCD} is expanded in the second iteration step up to fourth order:

$$\Delta E_{CCD}^{(2)} = E_{MP}^{(2)} + E_{MP}^{(3)} + E_Q^{(4)} \tag{5.15}$$

In the third iteration step, ΔE_{CCD} is expanded up to eighth order and at the kth iteration step, the CCD amplitudes $a_d^{(k)}(CCD)$ and the energy ΔE_{CCD} cover perturbation contributions up to order $2^k - 1$ and 2^k, respectively. In this way, the CCD correlation energy is expanded to higher and higher orders with proceeding iterations.

In the same way, the CCSD correlation energy ΔE_{CCSD} can be expanded in terms of perturbation theory for increasing numbers of iteration steps:

$$\Delta E_{CCSD} = \Delta E_{CCSD}^{(1)} + \Delta E_{CCSD}^{(2)} + \Delta E_{CCSD}^{(3)} + \Delta E_{CCSD}^{(4)} + \Delta E_{CCSD}^{(5)} + \ldots \tag{5.16}$$

In the first three iterative steps, CCSD covers the following energy contributions:

$$\Delta E_{CCSD}^{(1)} = E_{MP}^{(2)} \tag{5.17}$$

$$\Delta E_{CCSD}^{(2)} = E_{MP}^{(2)} + E_{MP}^{(3)} + E_S^{(4)} + E_Q^{(4)} \tag{5.18}$$

$$\Delta E_{CCSD}^{(3)} = E_{MP}^{(2)} + E_{MP}^{(3)} + E_S^{(4)} + E_D^{(4)} + E_Q^{(4)} + E_{CCSD}^{(5)} + E_{CCSD}^{(6)} + \ldots \tag{5.19}$$

with $E_{CCSD}^{(5)}$ and $E_{CCSD}^{(6)}$ being defined by (5.20) and (5.21)

$$E_{CCSD}^{(5)} = E_{SS}^{(5)} + 2E_{SD}^{(5)} + E_{DD}^{(5)} + E_{DQ}^{(5)} + E_{QD}^{(5)} + E_{QQ}^{(5)}(I) + E_{TS}^{(5)} + E_{TQ}^{(5)}(I) \tag{5.20}$$

$$\begin{aligned}E^{(6)}_{CCSD} =& E^{(6)}_{SSS} + 2E^{(6)}_{SSD} + E^{(6)}_{SDS} + 2E^{(6)}_{SDD} + 2E^{(6)}_{SDQ} + E^{(6)}_{STS} + E^{(6)}_{STQ}(I) \\ &+ E^{(6)}_{QTS}(I) + E^{(6)}_{DSD} + E^{(6)}_{DDD} + 2E^{(6)}_{DDQ} + E^{(6)}_{DQD} + E^{(6)}_{DQQ}(I) \\ &+ E^{(6)}_{QQD}(I) + E^{(6)}_{DTS} + E^{(6)}_{DTQ}(I) + E^{(6)}_{QTQ}(I) + E^{(6)}_{QQQ}(I) + E^{(6)}_{QDQ} \\ &+ E^{(6)}_{QHQ}(I) + E^{(6)}_{TSD} + E^{(6)}_{TQD}(I) + E^{(6)}_{TQQ}(I) \\ &+ E^{(6)}_{TSS} + E^{(6)}_{TTS}(I) + E^{(6)}_{TTQ}(I) \end{aligned} \quad (5.21)$$

In these equations, the symbol (I) identifies those terms that are only partially contained in the CCSD correlation energy.

Eq.s (5.20) and (5.21) give those MP5 and MP6 energy contributions that are fully or partially covered by CCSD. The energy $\Delta E^{(3)}_{CCSD}$ of Eq. (5.19) added in the third iteration involves at least 8th order MP contributions.

5.2 COMPARISON OF CCSD AND QCISD

A similar expansion as the one described for CCSD in section 5.1 can also be carried out for the QCISD correlation energy. The resulting equations become rather complicated for n = 6 and even higher orders and, therefore, it is of advantage to present results in a graphical way using the same techniques developed for the graphical representation of MPn correlation energy contributions. In Figure 5.2, the corresponding CCSD diagram is shown. [22]

In Figure 5.2 and the following figures, energy contributions that are fully contained at a particular level of perturbation theory are given by solid lines and those, which are only partially contained, have at least in one part of the total path representing an energy contribution a dotted line. The corresponding diagram for QCISD is shown in Figure 5.3. A combination of the CCSD and the QCISD diagrams is given in Figure 5.4 in order to make the differences between the two methods more obvious.

In the combination diagram, terms that are common to both methods are given in thick solid lines and those, which can only be found in CCSD, in thin solid lines. If a term is just partially contained hashed lines are used for common terms and dotted lines for terms that are just covered by CCSD.

CCSD and QCISD are correct up to third order. Figures 5.2 and 5.3 indicate in addition that both CCSD and QCISD are correct at any order of perturbation theory in the truncated configuration space that is made up from S and D excitations, i.e. within this space all infinite-order effects are covered. This, of course, is trivial since it reflects just the nature of the CC ansatz. More important is that

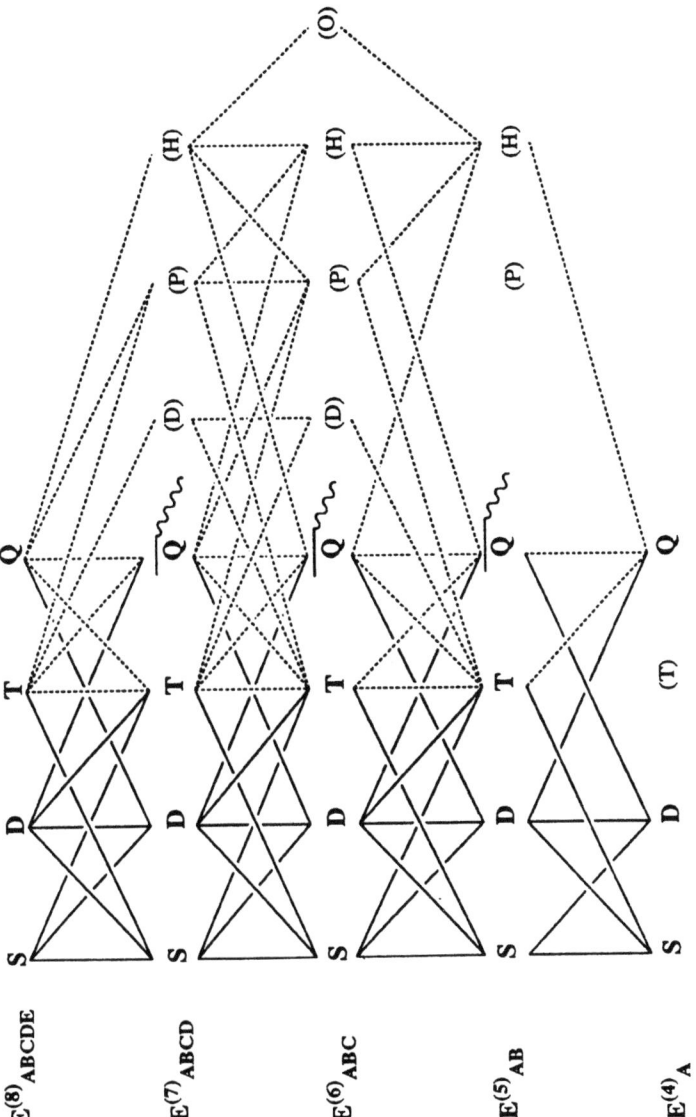

Figure 5.2. Graphical analysis of energy contributions at nth order MP perturbation theory (n= 4, 5, 6, 7, 8) covered by the CCSD correlation energy. See explanations given for Figure 5.1. Note that solid (dashed) lines denote energy terms fully (partially) contained in the CCSD correlation energy. *Reprinted with permission from Z. He and D. Cremer, Int. J. Quant. Chem. Symp. (1991) 25, 43. Copyright (1991) J. Wiley, Inc*

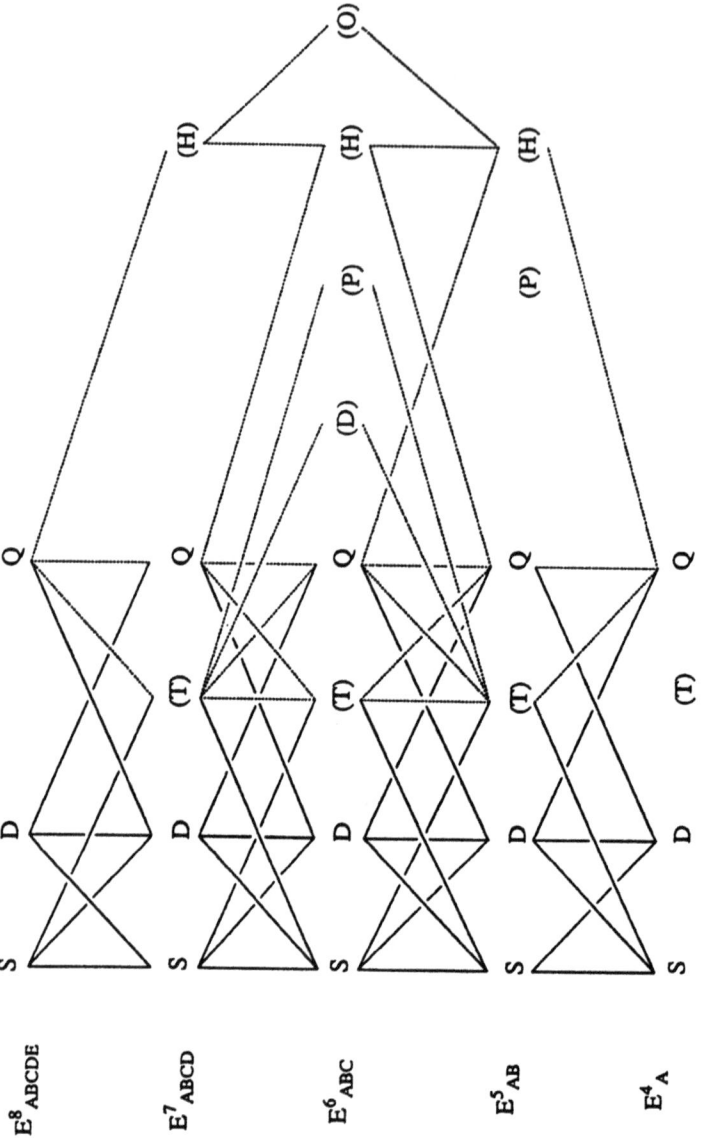

Figure 5.3. Graphical analysis of energy contributions at nth order MP perturbation theory (n= 4, 5, 6, 7, 8) covered by the QCISD correlation energy. See explanations given for Figure 5.1. Note that solid (dashed) lines denote energy terms fully (partially) contained in the QCISD correlation energy. *Reprinted with permission from Z. He and D. Cremer, Int. J. Quant. Chem. Symp. (1991) 25, 43. Copyright (1991) J. Wiley, Inc.*

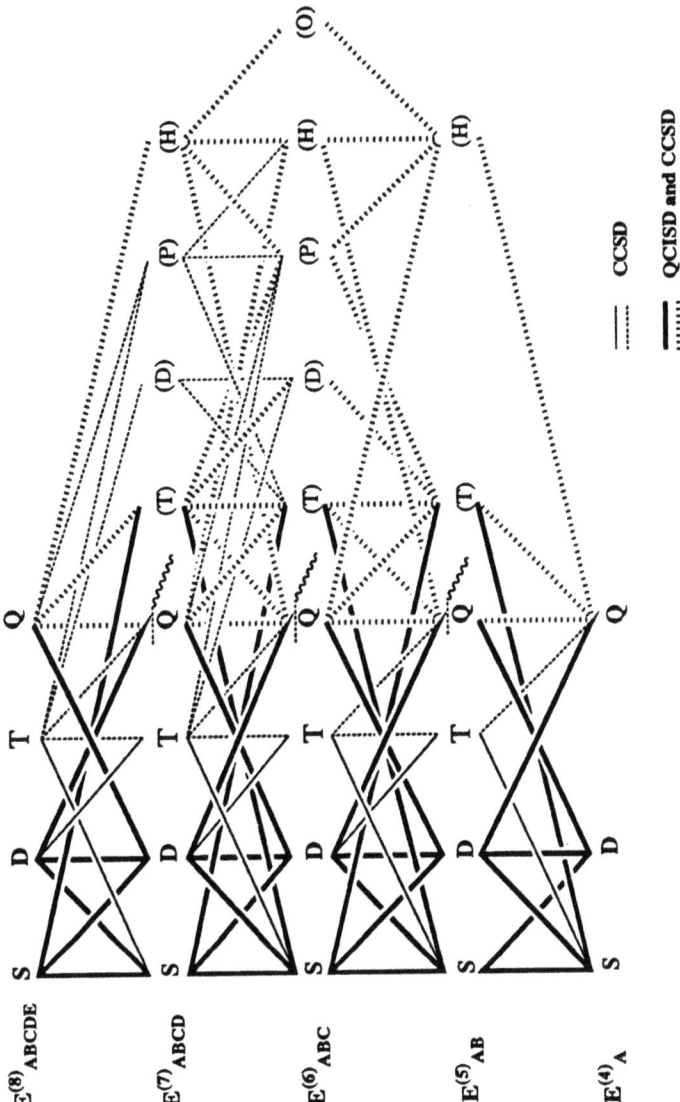

Figure 5.4. Graphical analysis of energy contributions at nth order MP perturbation theory (n= 4, 5, 6, 7, 8) covered by the CCSD and QCISD correlation energy. See explanations given for Figure 5.1. Note that bold solid (hashed) lines denote energy terms that are fully (partially) contained in both the CCSD and the QCISD correlation energy. Those energy terms that are only covered by the CCSD energy terms are denoted by normal solid or dashed lines depending on whether they are fully or partially contained. Reprinted with permission from Z. He and D. Cremer, *Theor. Chim. Acta,* (1994) 85, 305. Copyright (1994) Springer Verlag.

for both methods infinite order effects are also covered in the SDQ space with just the A...QQ terms being partial. In general, the diagrams reveal that CCSD and QCISD are equivalent in any SDQHO...X space where X excitations are generated from disconnected $\hat{T}_2^m$ cluster operators.

The major differences between CCSD and QCISD are in SDTQ space. They result from the omission of the term $\hat{T}_1\hat{T}_2'$ ($\hat{T}_2' = \hat{T}_2 + \frac{1}{2}\hat{T}_1^2$) in the D equations of QCISD and can be summarized by Eq. (5.22):

$$\sum_{n=2}^{7}(\Delta E_{CCSD}^{(n)} - \Delta E_{QCISD}^{(n)}) = E_{TS}^{(5)} + E_{TQ}^{(5)}(I) + E_{TSS}^{(6)} + E_{TSD}^{(6)} + E_{DTS}^{(6)}$$

$$+ E_{DTQ}^{(6)}(I) + E_{TTS}^{(6)}(I) + E_{TTQ}^{(6)}(I) + E_{TQD}^{(6)}(I)$$

$$+ E_{TQQ}^{(6)}(I) + \sum E_{ABCD}^{(7)} \qquad (5.22)$$

where $\sum E_{ABCD}^{(7)}$ denotes a sum of 41 terms (see Ref. 23).

Both CCSD and QCISD cover disconnected T effects, which according Figures (5.2), (5.3), (5.4) and Eq. (5.22) are introduced into CCSD one order of perturbation theory earlier than into QCISD. For example, TS couplings enter CCSD at n = 5, but QCISD at n = 6; similarly, TT couplings CCSD at n = 6, but QCISD at n = 7; and TTT couplings CCSD at n = 7 and QCISD at n = 8. Clearly, *QCISD is limping behind CCSD by one order of perturbation theory.*

We have compared the energy terms covered by CCSD and QCISD up to eighth order. [22] With increasing order n, QCISD covers less and less terms going down from a 50% coverage at MP5 to a 24% coverage at MP8. CCSD covers 64% of all MP5 terms and 43% of all MP8 terms, which means that CCSD contains almost twice as many terms at higher orders.

Of course, each energy contribution at a given order of perturbation theory can have a different importance for the description of a given electronic system. Therefore, one may argue that there is a chance that QCISD covers all important terms and just neglects the unimportant ones. For example, the TS and TQ terms at MP5, that are not covered by QCISD, lead often to positive energy contributions and decrease the absolute magnitude of the correlation energy. As a consequence, QCISD correlation energies are often more negative than CCSD correlation energies, which one could take as indication that the right terms have been neglected in QCISD. On the other hand, the correct description of correlation effects must avoid an exaggeration of certain correlation effects as discussed in chapter 2 in the case of the pair correlation effects. The appearance of positive contributions normally means a correction of correlation effects exaggerated at lower orders and, therefore, these terms should not be neglected.

In general, it is hardly possible to make predictions with regard to the importance of each energy contribution for a given system. Therefore, the simple rule of thumb is that the more complete description which covers more energy contributions should also be the better. Out of this perspective, a statistical comparison of two methods should give some indication on the performance of the two methods. [22,23] For example, in the case of CCSD and QCISD the analysis clearly shows that for molecules, that require higher order effects, CCSD should perform significantly better than QCISD because (a) QCISD covers a much smaller number of energy contributions than CCSD at larger order n; and (b) part of the T, P, S_7, ..., Y contributions (Y is any odd order excitation) generated by the cluster operator $\hat{T}_1 \hat{T}_2'$ are delayed at the QCISD level by one order of perturbation theory.

5.3 COMPARISON OF CCSD(T) AND QCISD(T)

When T correlation effects are included into CCSD and QCISD in a perturbative way, then CCSD(T) [36] and QCISD(T) [35] energies are obtained according to

$$\Delta E_{CCSD(T)} = \Delta E_{CCSD} + \Delta E_T(CCSD) \qquad (5.23)$$

and

$$\Delta E_{QCISD(T)} = \Delta E_{QCISD} + \Delta E_T(QCISD) \qquad (5.24)$$

where the T corrections are given by

$$\Delta E_T(CCSD) = \sum_p^{SD} \sum_d^{D} \sum_t^{T} a_p(CCSD) \bar{V}_{pt}(E_0 - E_t)^{-1} \hat{V}_{td} a_d(CCSD) \qquad (5.25)$$

$$\Delta E_T(QCISD) = \sum_p^{SD} \sum_d^{D} \sum_t^{T} a_p(QCISD) \bar{V}_{pt}(E_0 - E_t)^{-1} \hat{V}_{td} a_d(QCISD)$$

$$+ \sum_s^{S} \sum_d^{D} \sum_t^{T} a_d(QCISD) \bar{V}_{dt}(E_0 - E_t)^{-1} \hat{V}_{ts} a_s(QCISD)$$

$$(5.26)$$

CCSD(T) and QCISD(T) correlation energies were expanded in a similar way as in the case of CCSD and QCISD. [23] Similar methods such as CCSD(TQ) and QCISD(TQ), which were developed to have CC methods that are correct in fifth order perturbation theory [9], were also investigated. We refrain from a lengthy

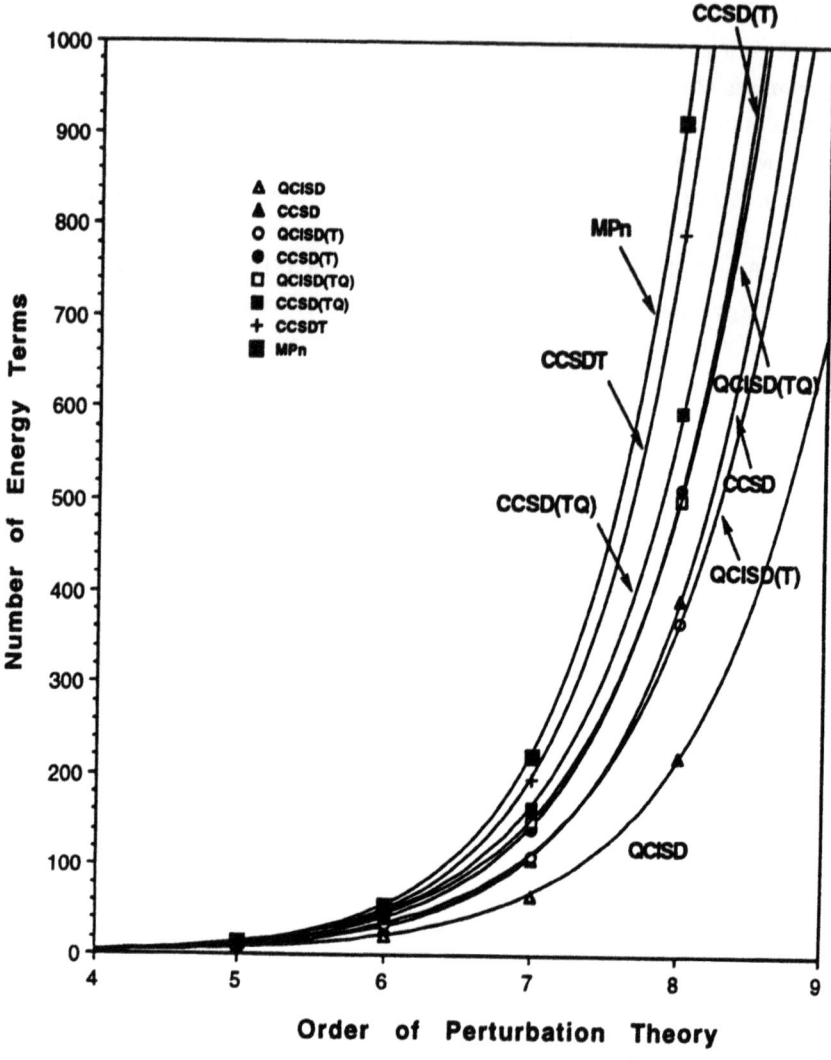

Figure 5.5. Number of energy contributions $E^{(n)}_{ABC...}$ covered by QCISD, CCSD, QCISD(T), CCSD(T), QCISD(TQ), CCSD(TQ), CCSDT, and MPn given as a function of the order n. Numbers are given without considering symmetry. *Reprinted with permission from Z. He and D. Cremer, Theor. Chim. Acta, (1994) 85, 305. Copyright (1994) Springer Verlag.*

discussion of all the results obtained in this work [22,23] and, instead, summarize results in Figure 5.5, where the numbers of energy contributions covered by a particular CC or QCI method at nth (n = 5, 6, 7 and 8) order perturbation theory are given as functions of n. These curves clearly reveal that the discrepancy between the QCI and CC descriptions is gradually reduced from CCSD/QCISD to CCSD(T)/QCISD(T) and CCSD(TQ)/QCISD(TQ).

A noniterative improvement of a SD method by T excitations is more important for QCISD than for CCSD, since $\Delta E_T(QCISD)$ adds more terms to the QCISD correlation energy than $\Delta E_T(CCSD)$ adds to the CCSD correlation energy. As for the total number of energy contributions, QCISD(T) falls back behind CCSD at higher orders of perturbation theory as shown in Figure 5.5. Of course, this does not necessarily imply that CCSD is a better method than QCISD(T).

CCSD(T) should lead to a much better description of T effects than QCISD(T) since it contains more T contributions (including important TT coupling terms) than QCISD(T). Therefore, CCSD(T) is probably the method with the better cost-performance ratio. The difference between QCI and CC is considerably decreased at the CCSD(TQ) and QCISD(TQ) level of theory when considering in particular T correlation effects. QCISD(TQ) should lead to a performance comparable to that of CCSD(TQ).

In molecular investigations that require the inclusion of T effects, the various CC and QCI methods should lead to improved results in the following order:

$$MP4(SDTQ) < QCISD(T) < CCSD(T) < QCISD(TQ), CCSD(TQ) < CCSDT.$$

MP4 that does not cover any TT coupling effects will always exaggerate T effects and some of this exaggeration will be carried over to QCISD(T), which includes the TT coupling effects at a relatively late stage. CCSD(T) should be the method that leads to a relatively balanced description of T effects while CCSDT is certainly a method, which comes close to FCI performance. This discussion clearly shows that, at the presence, applied work should be done with CCSD(T) while future work should concentrate on CCSDT or equivalent T methods within the CC approach. This is further discussed in chapter 6.

6. Coupled Cluster Methods with Triple Excitations

Triple (T) excitations resulting from the cluster operator $\hat{T}_3$ describe three-electron correlation effects. These effects are generally rather small, however due to the large number of these effects in an electronic system, their sum leads to a significant contribution to the total correlation energy (see the discussion in section

2.3). It has been shown that T correlation effects are particularly important in molecules with multiple bonds, nonclassical bonding, hypervalent bonding, and many other electronic situations. [11,15] In addition, one has put some hope into the expectation that CC methods with T excitations help to describe electron systems with significant multireference effects even within a single determinant approach. There are, however, at the moment two obstacles that hinder the general use of CC methods with T excitations such as CCSDT. First, development and programming of CCSDT is a rather tedious procedure which has been solved only by few experts. [20,47, see also 21] Second, the calculation of CCSDT energies involves $O(M^8)$ computational steps and, therefore, is too costly for routine calculations of larger molecules. Because of the complexity of the CCSDT equations and the associated cost factor, the development of the last years has gone in the direction of replacing CCSDT by CC methods that contain the T correlation effects in an approximated way. [35,36,51] To be mentioned in this connection are the CCSDT-n methods [51] and CCSD or QCISD with a perturbative inclusion of T effects such as CCSD(T) [36] and QCISD(T) [35]. For example, CCSD(T) and QCISD(T) are nowadays the most often used CC methods for high-accuracy calculations on nontrivial chemical problems. This has to do with the fact that the perturbative inclusion of T effects into CCSD or QCISD leads to $O(M^7)$ procedures, which can be applied to problems with 200 basis functions and more. [52]

Although CCSD(T) and QCISD(T) are the most often used CC methods for describing T effects, it is also clear from the discussion given in chapter 5 that the perturbative T methods, but in particular QCISD(T) can lead to an unbalanced description of T effects. Figure 5.5 reveals how much CCSD(T) and QCISD(T) differ from CCSDT for higher levels of perturbation theory. CCSDT, on the other hand, comes close to full CI or infinite order MPn results and, therefore, it has to be a major goal in CC theory to extend the existing methods for general use in Quantum Chemistry also to CC methods with a full account of T effects. In the following, the main features of CCSDT and approximate CC methods, that cover $\hat{T}_3$ effects fully, is described.

6.1 IMPLEMENTATION OF A COUPLED CLUSTER SINGLES, DOUBLES, AND TRIPLES METHOD: CCSDT

Truncation of the cluster operator $\hat{T}$ (introduced in chapter 4, Eq. 4.2) at $\hat{T}_3$ leads to the CCSDT method [20,47] that is defined by the projection Eq.s (6.1) - (6.3):

$$\langle \Phi_i^a | \bar{H}(\hat{T}_1 + \hat{T}_2 + \hat{T}_3 + \hat{T}_1\hat{T}_2 + \frac{1}{2}\hat{T}_1^2 + \frac{1}{3!}\hat{T}_1^3)|\Phi_0\rangle = a_i^a \Delta E_{CCSDT} \qquad (6.1)$$

$$\langle \Phi_{ij}^{ab}|\bar{H}(1+\hat{T}_1+\hat{T}_2+\hat{T}_3+\frac{1}{2}\hat{T}_2^2+\hat{T}_1\hat{T}_2+\frac{1}{2}\hat{T}_1^2+\hat{T}_1\hat{T}_3$$
$$+\frac{1}{3!}\hat{T}_1^3+\frac{1}{2}\hat{T}_1^2\hat{T}_2+\frac{1}{4!}\hat{T}_1^4)|\Phi_0\rangle$$
$$=\langle \Phi_{ij}^{ab}|\hat{T}_2+\frac{1}{2}\hat{T}_1^2|\Phi_0\rangle \Delta E_{CCSDT} \qquad (6.2)$$

$$\langle \Phi_{ijk}^{abc}|\bar{H}(\hat{T}_1+\hat{T}_2+\hat{T}_3+\frac{1}{2}\hat{T}_2^2+\hat{T}_1\hat{T}_2+\frac{1}{2}\hat{T}_1^2+\hat{T}_1\hat{T}_3$$
$$+\hat{T}_2\hat{T}_3+\frac{1}{3!}\hat{T}_1^3+\frac{1}{2}\hat{T}_1^2\hat{T}_2+\frac{1}{2}\hat{T}_1^2\hat{T}_3+\frac{1}{4!}\hat{T}_1^4)|\Phi_0\rangle$$
$$=\langle \Phi_{ijk}^{abc}|\hat{T}_3+\hat{T}_1\hat{T}_2+\frac{1}{3!}\hat{T}_1^3|\Phi_0\rangle \Delta E_{CCSDT} \qquad (6.3)$$

In the same way as described for CCSD in chapter 4, one can show that unlinked diagram terms cancel each other so that the S, D, and T equations (6.1) - (6.3) are obtained in connected form:

$$\langle \Phi_i^a|\bar{H}(\hat{T}_1+\hat{T}_2+\hat{T}_3+\hat{T}_1\hat{T}_2+\frac{1}{2}\hat{T}_1^2+\frac{1}{3!}\hat{T}_1^3)|\Phi_0\rangle_C = 0 \qquad (6.4)$$

$$\langle \Phi_{ij}^{ab}|\bar{H}(1+\hat{T}_1+\hat{T}_2+\hat{T}_3+\frac{1}{2}\hat{T}_2^2+\hat{T}_1\hat{T}_2+\frac{1}{2}\hat{T}_1^2+\hat{T}_1\hat{T}_3$$
$$+\frac{1}{3!}\hat{T}_1^3+\frac{1}{2}\hat{T}_1^2\hat{T}_2+\frac{1}{4!}\hat{T}_1^4)|\Phi_0\rangle_C = 0 \qquad (6.5)$$

and

$$\langle \Phi_{ijk}^{abc}|\bar{H}(\hat{T}_2+\hat{T}_3+\frac{1}{2}\hat{T}_2^2+\hat{T}_1\hat{T}_2+\frac{1}{2}\hat{T}_1^2+\hat{T}_1\hat{T}_3+\hat{T}_2\hat{T}_3+\frac{1}{3!}\hat{T}_1^3+\frac{1}{2}\hat{T}_1^2\hat{T}_2$$
$$+\frac{1}{2}\hat{T}_1^2\hat{T}_3+\frac{1}{4!}\hat{T}_1^4)|\Phi_0\rangle_C = 0 \qquad (6.6)$$

The expression for the correlation energy ΔE_{CCSDT}

$$\Delta E_{CCSDT} = \langle \Phi_0|\bar{H}(\hat{T}_2+\frac{1}{2}\hat{T}_1^2)|\Phi_0\rangle \qquad (6.7)$$

keeps the same form as in the CCSD case since higher excitation cluster operators $\hat{T}_n$ ($n \geq 3$) can not directly affect the total energy.

The equation for the CCSDT correlation energy in form of two-electron integrals is given by (6.8)

$$\Delta E_{CCSDT} = \frac{1}{4}\sum_{ij,ab}\langle ij||ab\rangle \tau_{ij}^{ab} \qquad (6.8)$$

where τ_{ij}^{ab} is determined via S and D amplitudes

$$\tau_{ij}^{ab} = a_{ij}^{ab} + a_i^a a_j^b - a_i^b a_j^a \tag{6.9}$$

The S and D amplitudes are calculated by iterative solution of Eq.s (6.10) and (6.11):

$$(\epsilon_i - \epsilon_a)a_i^a = \tilde{u}_i^a + \tilde{v}_i^a + \sum_{l<m}\sum_{d<e}\langle lm||de\rangle a_{ilm}^{ade} \tag{6.10}$$

$$(\epsilon_i + \epsilon_j - \epsilon_a - \epsilon_b)a_{ij}^{ab} = \langle ab||ij\rangle + \tilde{u}_{ij}^{ab} + \tilde{v}_{ij}^{ab} +$$

$$\sum_{l,d<e}(\langle bl||de\rangle a_{ijl}^{ade} + \langle al||de\rangle a_{ijl}^{dbe}) + \sum_{l<m,d}(\langle lm||dj\rangle a_{ilm}^{abd} + \langle lm||di\rangle a_{ljm}^{abd})$$

$$+\frac{1}{2}\sum_{mn}\sum_{ef}\langle mn||ef\rangle \left(2a_m^e a_{ijn}^{abf} + \sum_P (-1)^P P(i/j) a_i^e a_{jmn}^{abf}\right.$$

$$\left.+\sum_P (-1)^P P(a/b) a_m^a a_{ijn}^{bef}\right) \tag{6.11}$$

which via the T amplitudes a_{ijk}^{abc} depend on the T equation (6.12):

$$(\epsilon_i + \epsilon_j + \epsilon_k - \epsilon_a - \epsilon_b - \epsilon_c)a_{ijk}^{abc} =$$

$$-\sum_P (-1)^P P(i/jk|a/bc)\left[\sum_d X_1^{CCSDT}(i,d,b,c)a_{jk}^{ad} + \sum_l X_2^{CCSDT}(j,k,l,a)a_{il}^{bc}\right]$$

$$+\sum_P (-1)^P P(a/bc)\left[\frac{1}{2}\sum_{ef} X_3^{CCSDT}(b,c,e,f)a_{ijk}^{aef} + \sum_f Y_1^{CCSDT}(f,a)a_{ijk}^{fbc}\right]$$

$$+\sum_P (-1)^P P(i/jk)\left[\frac{1}{2}\sum_{mn} X_4^{CCSDT}(m,n,j,k)a_{imn}^{abc} + \sum_n Y_2^{CCSDT}(n,i)a_{njk}^{abc}\right]$$

$$-\sum_P (-1)^P P(i/jk|a/bc)\sum_{me} X_5^{CCSDT}(m,a,i,e)a_{jkm}^{bce} \tag{6.12}$$

For setting up S, D, and T equations, arrays $\tilde{u}_i^a$, $\tilde{v}_i^a$, $\tilde{u}_{ij}^{ab}$ and $\tilde{v}_{ij}^{ab}$, which contain the terms of the CCSD equations, have to be calculated according to Eq.s (6.13) - (6.16):

$$\tilde{u}_i^a = -\sum_{ld}\langle la||id\rangle a_l^d - \frac{1}{2}\sum_{lde}\langle la||de\rangle \tau_{il}^{de} - \frac{1}{2}\sum_{lmd}\langle lm||id\rangle \tau_{lm}^{ad} \tag{6.13}$$

$$\tilde{u}_{ij}^{ab} = \sum_d (\langle ab||dj\rangle a_i^d - \langle ab||di\rangle a_j^d) + \sum_l (\langle la||ij\rangle a_l^b - \langle lb||ij\rangle a_l^a)$$

$$+ \frac{1}{2}\sum_{de}\langle ab||de\rangle \tau_{ij}^{de} + \frac{1}{2}\sum_{lm}\langle lm||ij\rangle \tau_{lm}^{ab}$$

$$- \sum_{ld}\sum_p (-1)^P P(i/j|a/b)\langle lb||jd\rangle(a_{il}^{ad} - a_i^d a_l^a) \quad (6.14)$$

$$\tilde{v}_i^a = \frac{1}{2}\sum_{lm}\sum_{de}\langle lm||de\rangle \left(a_i^d a_{lm}^{ea} + a_l^a a_{im}^{ed} + 2a_l^d(a_{im}^{ae} - a_i^e a_m^a)\right) \quad (6.15)$$

$$\tilde{v}_{ij}^{ab} = \frac{1}{4}\sum_{lm}\sum_{de}\langle lm||de\rangle \left[\tau_{ij}^{de}\tau_{lm}^{ab} - 2(a_{ij}^{ad}\tau_{lm}^{be} + a_{ij}^{be}\tau_{lm}^{ad} + a_{il}^{ab}\tau_{jm}^{de} + \tau_{il}^{de}a_{jm}^{ab})\right.$$

$$\left. + 4\left(a_{il}^{ad}(a_{jm}^{be} - a_j^e a_m^b) + a_{il}^{be}(a_{jm}^{ad} - a_j^d a_m^a) - a_i^d a_j^a a_{jm}^{be} - a_i^e a_l^b a_{jm}^{ad}\right)\right]$$

$$+ \sum_P (-1)^P P(i/j) \sum_{mn,d} \langle mn||jd\rangle (a_m^d a_{in}^{ab} - \frac{1}{2}a_i^d \tau_{mn}^{ab})$$

$$+ \sum_P (-1)^P P(a/b) \sum_{m,ef} \langle am||ef\rangle (a_m^e a_{ij}^{bf} - \frac{1}{2}a_m^b \tau_{ij}^{ef})$$

$$+ \sum_P (-1)^P P(i/j|a/b) \left(\sum_{mn,d}\langle mn||jd\rangle a_m^a (a_{in}^{bd} - a_i^d a_n^b)\right.$$

$$\left. + \sum_{m,ef}\langle am||ef\rangle a_i^e (a_{jm}^{bf} - a_j^f a_m^b)\right) \quad (6.16)$$

In addition, the intermediate arrays X_n^{CCSDT} and Y_m^{CCSDT} listed in Table 6.1 have to be evaluated. They reduce the computational cost of the CCSDT method to $O(M^8)$. The most costly terms in the T equations result from the cluster operators $\hat{T}_3$ and $\hat{T}_2\hat{T}_3$.

6.2 DEVELOPMENT OF A QCI METHOD WITH SINGLE, DOUBLE, AND TRIPLE EXCITATIONS: QCISDT

Pople, Head-Gordon, and Raghavachari (PHR) have suggested a QCISDT method in their original QCI publication [35], which turned out to be not size-extensive although the method was developed to overcome the size-extensivity error of the normal CI approach. [49] This deficiency of the QCISDT method of PHR does not mean that a size-extensive QCISDT method with just quadratic corrections added

TABLE 6.1. Intermediate arrays used in the triple amplitude equations of QCISDT and CCSDT.[a] Reprinted with permission from D. Cremer and Z. He, Chem. Phys. Lett. (1994) 222, 40. Copyright (1994) Elsevier Science B. V.

Array	QCISDT	CCSDT
$X_1(i,d,b,c)$	$(id\|\|bc) + \frac{1}{2}\sum_{mne}(mn\|\|ed)a_{imn}^{ebc}$	$X_1^{QCISDTc} - \sum_P (-1)^P P(b/c)\{\sum_{me}(mb\|\|ed)a_{im}^{ce}$ $-\sum_m Z1(mb,id)a_m^c\} + \sum_{m<n}((mn\|\|id) + \sum_e (mn\|\|ed)a_i^e)\tau_{mn}^{bc}$ $+\sum_e (bc\|\|ed)a_i^e - \sum_{mn,e}(mn\|\|ed)a_m^b a_{in}^{ec}$
$X_2(j,k,l,a)$	$(jk\|\|la) + \frac{1}{2}\sum_{mef}(ml\|\|ef)a_{mjk}^{aef}$	$X_2^{QCISDTc} + \sum_P (-1)^P P(j/k)\{\sum_{me}(ml\|\|ej)a_{km}^{ae} + \sum_e Z2(al,ej)a_k^a\}$ $+ \sum_{e<f}((la\|\|ef) + \sum_m(ml\|\|ef)a_m^a)\tau_{jk}^{ef} + \sum_m(ml\|\|jk)a_m^a$
$X_3(b,c,e,f)$	$(bc\|\|ef) + \frac{1}{2}\sum_{mn}(mn\|\|ef)a_{mn}^{bc}$	$(bc\|\|ef) + \frac{1}{2}\sum_{mn}(mn\|\|ef)\tau_{mn}^{bc} + \sum_P (-1)^P P(b/c)\sum_m(mb\|\|ef)a_m^c$
$X_4(m,n,j,k)$	$(mn\|\|jk) + \frac{1}{2}\sum_{ef}(mn\|\|ef)a_{jk}^{ef}$	$(mn\|\|jk) + \frac{1}{2}\sum_{ef}(mn\|\|ef)\tau_{jk}^{ef} + \sum_P (-1)^P P(j/k)\sum_e(mn\|\|ek)a_j^e$
$X_5(m,a,i,e)$	$(ma\|\|ie) - \sum_{nf}(mn\|\|ef)a_{in}^{af}$	$(ma\|\|ie) - \sum_{nf}(mn\|\|ef)(a_{in}^{af} - a_n^a a_i^f) - \sum_f(ma\|\|ef)a_i^f$ $- \sum_n(mn\|\|ie)a_n^a$
$Y_1(f,a)$	$\frac{1}{2}\sum_{mne}(mn\|\|ef)a_{mn}^{ae}$	$\frac{1}{2}\sum_{mne}(mn\|\|ef)\tau_{mn}^{ae} + \sum_{me}(ma\|\|ef)a_m^e$
$Y_2(n,i)$	$\frac{1}{2}\sum_{mef}(mn\|\|ef)a_{im}^{ef}$	$\frac{1}{2}\sum_{mef}(mn\|\|ef)\tau_{im}^{ef} + \sum_{me}(mn\|\|ie)a_m^e$

[a] The arrays τ_{ij}^{ab}, $Z1(mb,id)$ and $Z2(al,ej)$ are defined by:

$$\tau_{ij}^{ab} = a_{ij}^{ab} + a_i^a a_j^b - a_i^b a_j^a$$

$$Z1(mb,id) = (mb\|\|id) + \sum_e (mb\|\|ed)a_i^e + \sum_{ne}(mn\|\|ed)a_{in}^{be}$$

$$Z2(al,ej) = (al\|\|ej) - \sum_m(ml\|\|ej)a_m^a + \sum_{mf}(ml\|\|ef)a_{jm}^{af}$$

to the linear CISDT terms is impossible. Therefore, the question arises whether and how a size-extensive QCISDT method can be constructed. This problem automatically leads to the more general question whether there exists a hierarchy of size-extensive QCI methods that is in line with the original idea of PHR, namely a simple improvement of CI by just including quadratic correction terms. [30]

One can approach this problem in the following way [30]:

The physically not meaningful terms in the projection equations of truncated CI show up in the diagrammatic description in form of unlinked diagrams. The unlinked diagrams result from disconnected terms in the CI equations of a given truncation level. One has to eliminate all disconnected terms from the CI projection equations in order to obtain a size-extensive CI energy. Based on these considerations, a general procedure for restoring size-extensivity in a CI approach has been developed. [30] This procedure comprises three steps:

1) Analysis of disconnected terms in the CI projection equations. 2) Cancellation of disconnected terms by the addition of appropriate new terms. 3) Final test whether the addition of new terms to the CI equations does not lead to new disconnected terms.

If new disconnected terms appear, one will have to add further terms until all disconnected terms disappear. In the most general case one has to loop several times through the sequence 1) - 3) until a size-extensive method is obtained. Since the methods obtained in this way are size-extensive extended CI approaches that do not necessarily comply with the quadratic CI method of PHR, we have called them extended CI (ECI) methods. [30] Hence, *an ECI method can be considered as a CI method, to which a minimum number of terms have been added to restore size-extensivity*, or alternatively *as an approximated CC method that differs from the corresponding CI method by a minimal number of terms*.

If the CI space is restricted to S and D excitations, the new terms to be added are quadratic as has been shown by PHR.[35] However, if higher excitations are included, e.g. T excitations at the CISDT level, size-extensivity will require the inclusion of both quadratic and cubic terms as will be shown in the following. Accordingly, one would have to speak of cubic CI, quartic CI, etc. However, it is better to speak of ECI methods and to refrain from introducing a new terminology. [30]

The CISDT projection equations are given by

$$\langle \Phi_i^a | \bar{H}(\hat{T}_1 + \hat{T}_2 + \hat{T}_3) | \Phi_0 \rangle = a_i^a \Delta E_{CISDT} \tag{6.17}$$

$$\langle \Phi_{ij}^{ab} | \bar{H}(1 + \hat{T}_1 + \hat{T}_2 + \hat{T}_3) | \Phi_0 \rangle = a_{ij}^{ab} \Delta E_{CISDT} \tag{6.18}$$

$$\langle\Phi^{abc}_{ijk}|\bar{H}(\hat{T}_1+\hat{T}_2+\hat{T}_3)|\Phi_0\rangle = a^{abc}_{ijk}\Delta E_{CISDT} \quad (6.19)$$

$$\Delta E_{CISDT} = \langle\Phi_0|\bar{H}\hat{T}_2|\Phi_0\rangle \quad (6.20)$$

There are three unlinked diagram terms ($a^a_i\Delta E_{CISDT}$, $a^{ab}_{ij}\Delta E_{CISDT}$ and $a^{abc}_{ijk} \times \Delta E_{CISDT}$) in these equations and one disconnected term associated with $\hat{T}_1$ in Eq.(6.19), which also gives an unlinked diagram contribution to the energy ΔE_{CISDT}. In order to cancel those unlinked diagram contributions one has to add quadratic terms $\hat{T}_1\hat{T}_2$, $\frac{1}{2}\hat{T}_2^2$ and $\hat{T}_2\hat{T}_3$ in Eq.s (6.17), (6.18) and (6.19), respectively. In this way, the three unlinked diagram terms are cancelled, however those unlinked diagram contributions resulting from $\hat{T}_1$ are not cancelled. Also, the addition of $\hat{T}_2\hat{T}_3$ in Eq.(6.19) leads to some new unlinked diagram contributions. This can be seen if the term associated with $\hat{T}_2\hat{T}_3$ is written as

$$\langle\Phi^{abc}_{ijk}|\bar{H}\hat{T}_2\hat{T}_3|\Phi_0\rangle = \langle\Phi^{abc}_{ijk}|(\bar{H}\hat{T}_2\hat{T}_3)_C|\Phi_0\rangle + \langle\Phi^{abc}_{ijk}|(\bar{H}\hat{T}_2\hat{T}_3)_D|\Phi_0\rangle \quad (6.21)$$

where the disconnected part can further be partitioned according to

$$\langle\Phi^{abc}_{ijk}|(\bar{H}\hat{T}_2\hat{T}_3)_D|\Phi_0\rangle = \langle\Phi^{abc}_{ijk}|\hat{T}_3(\bar{H}\hat{T}_2)_C|\Phi_0\rangle + \langle\Phi^{abc}_{ijk}|\hat{T}_2(\bar{H}\hat{T}_3)_C|\Phi_0\rangle \quad (6.22)$$

The first term of Eq. (6.22) cancels the unlinked diagram term of the right-hand side of Eq. (6.19) while the second term of Eq. (6.22) adds both linked and unlinked diagram contributions to the correlation energy. This demonstrates clearly that the original QCI concept of PHR [35] that is based on the addition of just quadratic cluster terms does no longer work for CISDT.

Size-extensivity can only be obtained in the case of CISDT by adding further quadratic and even cubic cluster operator terms to the T equation (6.19). [30] Finally, the projection equations (6.23) - (6.26) of size-extensive ECISDT are obtained, which differ considerably from the (non-size-extensive) QCISDT equations of PHR. [30]

$$\langle\Phi_0|\bar{H}\hat{T}_2|\Phi_0\rangle = \Delta E_{ECISDT} \quad (6.23)$$

$$\langle\Phi^a_i|\bar{H}(\hat{T}_1+\hat{T}_2+\hat{T}_3+\hat{T}_1\hat{T}_2)|\Phi_0\rangle = a^a_i\Delta E_{ECISDT} \quad (6.24)$$

$$\langle\Phi^{ab}_{ij}|\bar{H}(1+\hat{T}_1+\hat{T}_2+\hat{T}_3+\frac{1}{2}\hat{T}_2^2)|\Phi_0\rangle = a^{ab}_{ij}\Delta E_{ECISDT} \quad (6.25)$$

$$\langle\Phi^{abc}_{ijk}|\bar{H}(\hat{T}_1+\hat{T}_2+\hat{T}_3+\frac{1}{2}\hat{T}_1^2+\hat{T}_1\hat{T}_2+\hat{T}_1\hat{T}_3+\frac{1}{2}\hat{T}_2^2+\hat{T}_2\hat{T}_3+\frac{1}{2}\hat{T}_1\hat{T}_2^2)|\Phi_0\rangle$$
$$= (a^{abc}_{ijk} + \sum_P(-1)^P P(i/jk|a/bc)a^a_i a^{bc}_{jk})\Delta E_{ECISDT} \quad (6.26)$$

Eq.s (6.24) - (6.26) can be rewritten in connected form:

$$\langle \Phi_i^a | \bar{H}(\hat{T}_1 + \hat{T}_2 + \hat{T}_3 + \hat{T}_1\hat{T}_2)|\Phi_0\rangle_C = 0 \quad (6.27)$$

$$\langle \Phi_{ij}^{ab} | \bar{H}(1 + \hat{T}_1 + \hat{T}_2 + \hat{T}_3 + \frac{1}{2}\hat{T}_2^2)|\Phi_0\rangle_C = 0 \quad (6.28)$$

$$\langle \Phi_{ijk}^{abc} | \bar{H}(\hat{T}_2 + \hat{T}_3 + \frac{1}{2}\hat{T}_1^2 + \hat{T}_1\hat{T}_2 + \hat{T}_1\hat{T}_3 + \frac{1}{2}\hat{T}_2^2 + \hat{T}_2\hat{T}_3 + \frac{1}{2}\hat{T}_1\hat{T}_2^2)|\Phi_0\rangle_C = 0 \quad (6.29)$$

Clearly, ECISDT is not identical with CCSDT since it differs with regard to one cubic term and some quadratic terms. There is not much reason to further investigate a method such ECISDT since it offers little advantage compared to the more complete CCSDT method.

ECI methods in general are not very attractive, which becomes obvious when extending CISDTQ to size-extensive ECISDTQ as described in Ref. 30. It turns out that ECISDTQ is identical with CCSDTQ, i.e. the ECI methods do not represent a hierarchy of independent methods. ECID is identical with CCD, ECISD with QCISD, and ECISDTQ, ECISDTQP, etc. identical with the corresponding CC methods. Hence, the PHR concept of size-extensive QCI methods is not attractive, no matter whether one uses the original recipe [35] or the correct procedure worked out in Ref. 30.

Although the ECI equations may not be useful for any practical purpose, they can be used for deriving a new QCI concept. For this purpose, one starts from the ECI projection equations in their connected form and, in the same spirit as the CCSDT-n methods were developed [51], deletes all terms but the connected linear and certain quadratic terms. In the case of ECISDT (see Eq.s (6.27) - (6.29)), one keeps $\hat{T}_2\hat{T}_n$ with n = 1, 2, and 3 for S, D, and T equations, respectively. In this way, the projection equations of a size-extensive QCISDT method are defined, which we have coined $QCISDT_c$ to emphasize that we start from connected form of the projection equations and to distinguish $QCISDT_c$ from the non-size-extensive QCISDT approach of PHR.

$$\Delta E_{QCISDT_c} = \langle \Phi_0 | \bar{H}\hat{T}_2 | \Phi_0 \rangle \quad (6.30)$$

$$\langle \Phi_i^a | \bar{H}(\hat{T}_1 + \hat{T}_2 + \hat{T}_3 + \hat{T}_1\hat{T}_2)|\Phi_0\rangle_C = 0 \quad (6.31)$$

$$\langle \Phi_{ij}^{ab} | \bar{H}(1 + \hat{T}_1 + \hat{T}_2 + \hat{T}_3 + \frac{1}{2}\hat{T}_2^2)|\Phi_0\rangle_C = 0 \quad (6.32)$$

$$\langle \Phi_{ijk}^{abc} | \bar{H}(\hat{T}_2 + \hat{T}_3 + \hat{T}_2\hat{T}_3)|\Phi_0\rangle_C = 0 \quad (6.33)$$

The $QCISDT_c$ equations differ from the original QCISDT equations (Eq.s 3.36 - 3.39) in three ways. First, only the connected part of the quadratic correction terms is included. Secondly, quadratic corrections $(\bar{H}\hat{T}_2\hat{T}_n)_C$ are added to all

CISDT equations but the energy expression. Finally, the linear term $\langle \Phi_{ijk}^{abc}|\bar{H}\hat{T}_1|\Phi_0\rangle$ disappears since it leads to unlinked diagram contributions that do not add to the correlation energy.

In the same way, one can derive $QCISDTQ_c$ from ECISDTQ = CCSDTQ. [30] Formally, the QCI_c equations can also be obtained by starting at the corresponding CI equations. This can be shown for the projection equations of a truncated CI method that includes up to n-fold excitations:

$$\Delta E_{CI} = \langle \Phi_0|\bar{H}\hat{T}_2|\Phi_0\rangle \tag{6.34}$$

$$\langle \Phi_x|\bar{H}(1+\hat{T}_1+\hat{T}_2+...+\hat{T}_n)|\Phi_0\rangle = c_x \Delta E_{CI} \qquad (x=1,2,...,n) \tag{6.35}$$

or, alternatively, as

$$\langle \Phi_s|\bar{H}(\hat{T}_1+\hat{T}_2+\hat{T}_3)|\Phi_0\rangle = c_s \Delta E_{CI} \tag{6.36}$$

$$\langle \Phi_d|\bar{H}(1+\hat{T}_1+\hat{T}_2+\hat{T}_3+\hat{T}_4)|\Phi_0\rangle = c_d \Delta E_{CI} \tag{6.37}$$

$$\langle \Phi_x|\bar{H}(\sum_{i=x-2}^{min[x+2,n]}\hat{T}_i)|\Phi_0\rangle = c_x \Delta E_{CI} \qquad (n \geq x \geq 3) \tag{6.38}$$

where x is a general excitation indices. For any excitation index x higher than d, there appear just two disconnected terms, namely $\langle \Phi_x|\bar{H}\hat{T}_{x-2}|\Phi_0\rangle (= \langle ab||ij\rangle c_{x-2})$ and $c_x E_{CI}$ in the corresponding projection equation. Introducing $-\bar{H}\hat{T}_{p-2}$ and parts of the term $\bar{H}\hat{T}_2\hat{T}_x$, namely $(\bar{H}\hat{T}_2\hat{T}x)_C$ and $\hat{T}_x(\bar{H}\hat{T}_2)_C$, on the left side of Eq. (6.38), all disconnected terms are cancelled and the QCIc equations are obtained in their general form:

$$\Delta E_{QCI_c} = \langle \Phi_0|\bar{H}\hat{T}_2|\Phi_0\rangle \tag{6.39}$$

$$\langle \Phi_s|\bar{H}(\hat{T}_1+\hat{T}_2+\hat{T}_3+\hat{T}_1\hat{T}_2)|\Phi_0\rangle_C = 0 \tag{6.40}$$

$$\langle \Phi_d|\bar{H}(1+\hat{T}_1+\hat{T}_2+\hat{T}_3+\hat{T}_4+\frac{1}{2}\hat{T}_2^2)|\Phi_0\rangle_C = 0 \tag{6.41}$$

$$\langle \Phi_x|(\sum_{i=x-1}^{min[x+2,n]}\hat{T}_i+\hat{T}_2\hat{T}_x)|\Phi_0\rangle_C = 0 \qquad (n \geq x \geq 3) \tag{6.42}$$

Eq.s (6.39) - (6.42) establish a hierarchy of size-extensive QCI methods that covers the original QCISD method of Pople and co-workers, but can easily be extended to T, Q, and higher excitations.

Since the $QCISDT_c$ projection equations possess a rather simple form compared to the corresponding CCSDT equations, it was attractive to develop $QCISDT_c$ as a new CC method including T effects. Such development is a rather tedious task and, therefore, work has to be based on more than just the assumption that a new method such as $QCISDT_c$ may approximate a more complete method such as CCSDT rather well. Therefore, $QCISDT_c$ and CCSDT are compared in the next section with the help of perturbation theory.

6.3 ANALYSIS OF CCSDT AND QCISDT

Results of the comparison of the CCSDT and $QCISDT_c$ correlation energy in terms of nth ($n \leq 6$) order MBPT as described in chapter 5 are summarized in Figures 6.1 and 6.2. [30] CCSDT covers all contributions directly made up by S, D, and T excitations as is nicely shown by the energy diagram of Figure 6.1. The body of the CCSDT diagram is made up by a ladder of SDT contributions, which reaches to infinite order. Many Q effects are also covered either fully or at least partially where the latter is also true for many higher excitation effects. Figure 6.1 reveals that CCSDT, which is correct up to fourth order, lacks just the QT contribution at fifth order and contains the QQ contribution just partially. At sixth order, CCSDT is also a rather complete method. Out of the 55 sixth order terms, it does not contain five terms, namely QTD, DQT, QQT, QTT, and TQT. Apart from this, CCSDT covers nine other terms just partially.

$QCISDT_c$ is also exact up to fourth order. At higher orders, it lacks TS, TSA, TQ, and TQA energy contributions as is shown by Eq. (6.43), which gives energy differences between CCSDT and $QCISDT_c$ up to 6th order.

$$\Delta E_{CCSDT} - \Delta E_{QCISDT_c}$$
$$= \lambda^{(5)}(E_{TS}^{(5)} + E_{TQ}^{(5)}) + \lambda^{(6)}[E_{DTS}^{(6)} + E_{STQ}^{(6)}(II) + E_{QTS}^{(6)}(II)$$
$$+ E_{DTQ}^{(6)} + E_{TSS}^{(6)} + E_{TTS}^{(6)} + E_{TSD}^{(6)} + E_{TQD}^{(6)} + E_{TST}^{(6)}$$
$$+ E_{TQQ}^{(6)}(I) + E_{TTQ}^{(6)} + E_{TPT}^{(6)}(II)] + O(\lambda^{(7)}) \qquad (6.43)$$

In most cases, contributions such as TS, TQ, TSA or TSQ have positive sign and, therefore, make the correlation energy more positive. This means that $QCISDT_c$ energies will be often lower than CCSDT energies.

According to Figure (6.2) and Eq. (6.44), which gives the difference between $QCISDT_c$ and QCISD(T) up to sixth order, $QCISDT_c$ should be superior to

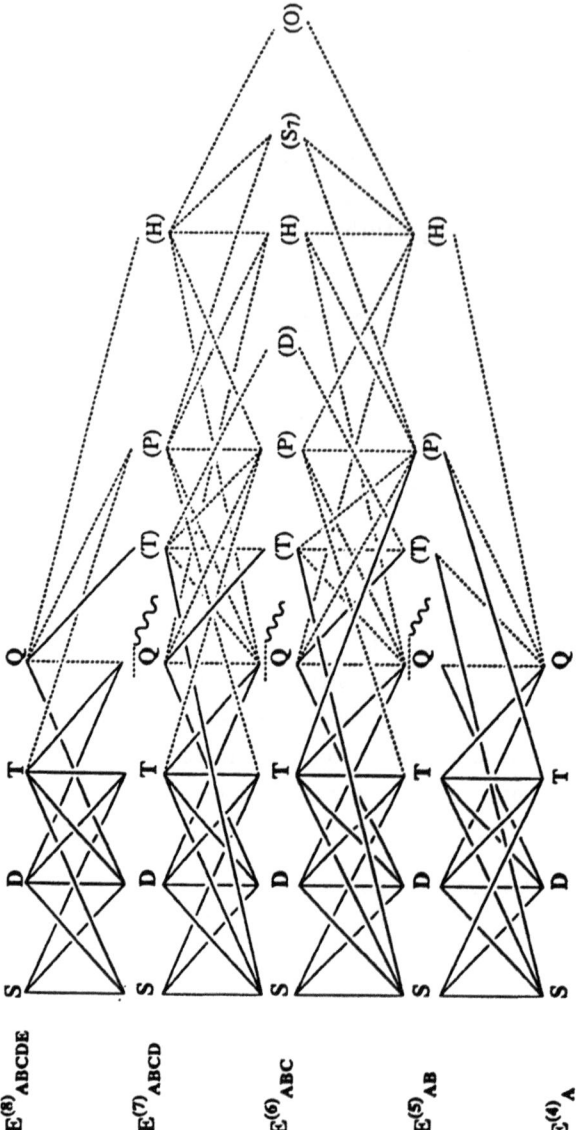

Figure 6.1. Graphical analysis of energy contributions at nth order MP perturbation theory (n= 4, 5, 6, 7, 8) covered by the CCSDT correlation energy. See explanations given for Figure 5.1. Note that solid (dashed) lines denote energy terms fully (partially) contained in the CCSDT correlation energy. Reprinted with permission from Z. He and D. Cremer, Int. J. Quant. Chem. Symp. (1991) 25, 43. Copyright (1991) John Wiley, Inc.

Analysis of Various Methods in Terms of MP6

	QCISD(T)	QCISDTc	ECISDT	CCSDT
SSS	yes	yes	yes	yes
SSD,DSS	y,y	y,y	y,y	y,y
SDS	yes	yes	yes	yes
SDD,DDS	y,y	y,y	y,y	y,y
SDQ,QDS	y,y	y,y	y,y	y,y
DSD	yes	yes	yes	yes
DDD	yes	yes	yes	yes
DQD	yes	yes	yes	yes
DDQ,QDD	y,y	y,y	y,y	y,y
DQQ,QQD	(y),(y)	(y),(y)	(y),(y)	(y),(y)
QDQ	yes	yes	yes	yes
QQQ	(yes)	(yes)	(yes)	(yes)
STS	yes	yes	yes	yes
STD,DTS	y,y	y,r	y,r	y,y
STQ,QTS	(y),(y)	(y),(y)	y,(y)	y,y
DTD	yes	yes	yes	yes
DTQ,QTD			(y),r	y,r
QTQ	(yes)	(yes)	(yes)	(yes)
SST,TSS	y,y	y,r	y,r	y,y
SDT,TDS	y,y	y,y	y,y	y,y
STT,TTS		y,r	y,(y)	y,y
DST,TSD	y,y	y,r	y,r	y,y
DDT,TDD	y,y	y,y	y,y	y,y
DQT,TQD			-,(y)	-,y
DTT,TTD		y,y	y,y	y,y
TDQ,QDT	y,y	y,y	y,y	y,y
TQQ,QQT			(y),r	(y),r
TTQ,QTT			(y),r	y,r
TST				yes
TDT		yes	yes	yes
TQT				
TTT		yes	yes	yes
QPQ		(yes)	(yes)	(yes)
QHQ	(yes)	(yes)	(yes)	(yes)
TPQ,QPT		(y),(y)	(y),(y)	(y),(y)
TPT		(yes)	(yes)	yes

Analysis of Various Methods in Terms of MP5

	QCISD(T)	QCISDTc	ECISDT	CCSDT
SS	yes	yes	yes	yes
DD	yes	yes	yes	yes
TT		yes	yes	yes
QQ	(yes)	(yes)	(yes)	(yes)
SD,DS	y,y	y,y	y,y	y,y
DQ,QD	y,y	y,y	y,y	y,y
ST,TS	y,y	y,r	y,r	y,y
DT,TD	y,y	y,y	y,y	y,y
TQ,QT			(y),r	y,r

Figure 6.2. Analysis of energy contributions at 5th (top) and 6th order MP perturbation theory (bottom) covered by QCISD(T), QCISDTc, ECISDT, and CCSDT correlation energies. Yes or y denote that the particular term is fully contained in the correlation energy while yes or y in parentheses indicates that the term is only partially covered. Reprinted with permission from D. Cremer and Z. He, Theor. Chim. Acta, (1994) 88, 47. Copyright (1994) Springer Verlag.

QCISD(T):

$$\begin{aligned}
\Delta E_{QCISDT_c} &- \Delta E_{QCISD(T)} \\
&= \lambda^{(5)}(E_{TT}^{(5)} - E_{TS}^{(5)}) + \lambda^{(6)}[E_{STT}^{(6)} + 2E_{DTT}^{(6)} + E_{TDT}^{(6)} + E_{TTT}^{(6)} \\
&+ E_{QPQ}^{(6)}(I) + E_{TPQ}^{(6)}(I) + E_{QPT}^{(6)}(I) + E_{TPT}^{(6)}(I) \\
&- E_{DTS}^{(6)} - E_{TSS}^{(6)} - E_{TSD}^{(6)}] + O(\lambda^{(7)})
\end{aligned} \quad (6.44)$$

QCISD(T) suffers from an exaggeration of T effects in molecular calculations due to the fact that TT couplings are totally missing in QCISD(T) at 5th and 6th order. [23] $QCISDT_c$, however, covers 6 of the 11 possible sixth order TAT and TTA coupling terms (partially or totally) and, hence, compares well with CCSDT that covers 9 of these terms. In summary, $QCISDT_c$ should come close to the performance of CCSDT in those cases where TS and TQ contributions are not important. Of course, this improvement is obtained at the cost of going from an $O(M^7)$ method, namely QCISD(T), to an $O(M^8)$ method. However, in view of the simplicity of $QCISDT_c$ it is worthwhile to use this approach as an alternative CC method with full inclusion of T effects.

6.4 IMPLEMENTATION AND APPLICATION OF QCISDT

The $QCISDT_c$ projection equations derived in section 6.2 have to be transformed into two-electron integral expressions in order to set up a $QCISDT_c$ computer program. [31] The $QCISDT_c$ energy expression as well as the S, D, and T equations take a similar form as the corresponding CCSDT equations if expressed in terms of two-electron integrals (in the following $QCISDT_c$ is abbreviated by $QCISDT$): [31]

$$\Delta E_{QCISDT} = \frac{1}{4} \sum_{ij,ab} \langle ij||ab \rangle a_{ij}^{ab} \quad (6.45)$$

$$(\epsilon_i - \epsilon_a)a_i^a = u_i^a + v_i^a + \sum_{l<m}\sum_{d<e} \langle lm||de \rangle a_{ilm}^{ade} \quad (6.46)$$

$$(\epsilon_i + \epsilon_j - \epsilon_a - \epsilon_b)a_{ij}^{ab} = \langle ab||ij \rangle + u_{ij}^{ab} + v_{ij}^{ab} +$$
$$\sum_{l,d<e}(\langle bl||de\rangle a_{ijl}^{ade} + \langle al||de\rangle a_{ijl}^{dbe}) + \sum_{l<m,d}(\langle lm||dj\rangle a_{ilm}^{abd} + \langle lm||di\rangle a_{ljm}^{abd}) \quad (6.47)$$

$$(\epsilon_i + \epsilon_j + \epsilon_k - \epsilon_a - \epsilon_b - \epsilon_c)a_{ijk}^{abc} =$$
$$-\sum_P (-1)^P P(i/jk|a/bc)\left[\sum_d X_1(i,d,b,c)a_{jk}^{ad} + \sum_l X_2(j,k,l,a)a_{il}^{bc}\right]$$

$$+\sum_{P}(-1)^P P(a/bc)\left[\frac{1}{2}\sum_{ef}X_3(b,c,e,f)a_{ijk}^{aef}+\sum_{f}Y_1(f,a)a_{ijk}^{fbc}\right]$$

$$+\sum_{P}(-1)^P P(i/jk)\left[\frac{1}{2}\sum_{mn}X_4(m,n,j,k)a_{imn}^{abc}+\sum_{n}Y_2(n,i)a_{njk}^{abc}\right]$$

$$-\sum_{P}(-1)^P P(i/jk|a/bc)\sum_{me}X_5(m,a,i,e)a_{jkm}^{bce} \qquad (6.48)$$

where u_i^a, v_i^a, u_{ij}^{ab}, and v_{ij}^{ab} contain the terms of the corresponding QCISD equations, namely:

$$u_i^a = -\sum_{ld}\langle la||id\rangle a_l^d - \frac{1}{2}\sum_{lde}\langle la||de\rangle a_{il}^{de} - \frac{1}{2}\sum_{lmd}\langle lm||id\rangle a_{lm}^{ad} \qquad (6.49)$$

$$u_{ij}^{ab} = \sum_{d}(\langle ab||dj\rangle a_i^d - \langle ab||di\rangle a_j^d) + \sum_{l}(\langle la||ij\rangle a_l^b - \langle lb||ij\rangle a_l^a)$$
$$+ \frac{1}{2}\sum_{de}\langle ab||de\rangle a_{ij}^{de} + \frac{1}{2}\sum_{lm}\langle lm||ij\rangle a_{lm}^{ab}$$
$$- \sum_{ld}\sum_{P}(-1)^P P(ij|ab)\langle lb||jd\rangle a_{il}^{ad} \qquad (6.50)$$

$$v_i^a = \frac{1}{2}\sum_{lm}\sum_{de}\langle lm||de\rangle(a_i^d a_{lm}^{ea} + a_l^a a_{im}^{ed} + 2a_l^d a_{im}^{ae}) \qquad (6.51)$$

$$v_{ij}^{ab} = \frac{1}{4}\sum_{lm}\sum_{de}\langle lm||de\rangle[a_{ij}^{de}a_{lm}^{ab} - 2(a_{ij}^{ad}a_{lm}^{be} + a_{ij}^{be}a_{lm}^{ad}) $$
$$+ a_{il}^{ab}a_{jm}^{de} + a_{il}^{de}a_{jm}^{ab}) + 4(a_{il}^{ad}a_{jm}^{be} + a_{il}^{be}a_{jm}^{ad})] \qquad (6.52)$$

The intermediate arrays X_n and Y_n are defined in Table 6.1 and compared with the corresponding CCSDT arrays. It becomes clear from Table 6.1 that the working effort for setting up a CCSDT program is considerably higher than that for setting up a QCISDT program although both CC methods have an $O(M^8)$ dependence. [31]

In the case of an existing CCSDT program, it is rather simple to install the QCISDT projection equations and to check them for possible errors. However in a situation, in which a CCSDT program is not available, it is much easier to set up QCISDT on a computer than to set up CCSDT. In any case, it is advisable to start the programming work from an existing QCISD or CCSD program rather than to program everything from scratch.

To test the accuracy of *QCISDT* energies, we have carried out QCISDT and CCSDT test calculations for 33 different electron systems, for which FCI results are available. [31,32] It turns out that in nearly all cases, QCISDT energies are slightly more negative than the corresponding CCSDT energies. This simply reflects the fact that certain coupling terms, which lead to positive energy contributions, are neglected in QCISDT as has been discussed in section 5.3. As far as the accuracy of QCISDT results is concerned, we determined for the 33 benchmark calculations a mean absolute deviation of 0.568 mhartree from FCI values in the case of QCISDT and 0.436 mhartree in the case of CCSDT, i.e. the two methods hardly differ from each other with slight advantages for the more complete CCSDT method. Apart from this, QCISDT results can be characterized in the following way: [31,32]

(1) *QCISDT* is superior to both QCISD and QCISD(T) with regard to the reproduction of FCI energies.

(2) Compared to QCISD and QCISD(T), QCISDT is more stable in calculations of systems with multireference character.

(3) QCISDT reproduces absolute and relative CCSDT energies for the examples we have studied within 1 mhartree and 0.1 kcal/mol, respectively. Its relative energies are actually slightly better than the corresponding CCSDT energies.

An important observation could be made with regard to the timings of the QCISDT calculations. While one QCISDT iteration step requires about the same time than the corresponding CCSDT iteration step, considerable time savings are obtained in the case of QCISDT calculations because of the faster convergence of this method. In general, time savings because of faster convergence become the larger the more multireference character an electron system possesses. This property of QCISDT becomes understandable if one considers the expansion of the method in terms of perturbation theory. Because of its simpler structure, QCISDT covers considerably less energy terms at higher order of perturbation theory than CCSDT. For example, QCISDT contains 59% of all 915 MP8 terms while CCSDT covers 87% of the MP8 terms. Although most of these terms represent rather small energy contributions, their large number leads to significant additions to the correlation energy at higher orders.

Since during the CC iterations higher and higher perturbation contributions are added to the correlation energy, the method that is more complete (covers more higher order terms) will need more iteration steps to reach convergence. It seems that the number of iteration steps reflects in a way the number of correlation effects covered by a given CC method. Therefore, one can expect that

QCISDT will converge faster than CCSDT in particular in those cases where due to multireference character correlation effects of higher order play an important role.

Of course, one could argue that fast convergence reflects the unphysical nature of an approximate CC method. However, as pointed out in chapter 5 it will hardly be possible to keep track of the multitude of correlation effects and their physical nature if one reaches higher orders. Therefore, one has to make a compromise between economy and performance of a given method, which leads to useful results in the case of QCISDT.

7. Conclusions and Outlook

The most important aspects of this review can be summarized in the following way:

(1) The two traditional methods of deriving MPn perturbation theory methods, namely the algebraic and the diagrammatic approach, can be combined in a new procedure that can handle the development of higher order MP methods. This has been tested for MP4 and MP5, for which final formulas are available in the literature. [6-9] Then, the new approach has been used to derive the first full MP6 method for routine calculations. [24-26] In addition, a dissection of the MP6 correlation energy has been carried out that leads to the development of two approximated MP6 methods with an $O(M^7)$ and an $O(M^8)$ dependence. The former method promises to be used as frequent as MP4 to calculate sixth order correlation effects.

(2) A new way of analyzing and comparing Coupled Cluster methods has been developed that provides an useful basis for predicting the performance of these methods. [22,23] The analysis is based on a graphical representation of energy contributions at various orders of perturbation theory. The compact form of the graphical representations makes a quick comparison of different CC methods possible. It has been used to derive the major differences between CC and QCI methods and to suggest improvements of these methods. [30]

(3) Motivated by the unsuccessful attempts to derive a size-extensive QCISDT method [35], a procedure has been worked out to systematically extend CI methods to size-extensive methods. This investigation revealed that the original QCI concept leads just to one unique QCI method, namely QCISD [35], while all other methods either coincide with the corresponding CC methods or represent ECI

methods with other than just quadratic cluster terms. [30] Based on this result, a new QCI concept was developed that leads to the first size-extensive QCISDT method, which was programmed and tested. [30-32] Size-extensive QCISDT leads to correlation energies that differ on the average from CCSDT correlation energies by just 0.07 mhartree. QCISDT possesses significant advantages with regard to both its implementation on a computer and to its cost requirements (faster convergence in the CC iterations). [31,32]

If one speaks of perturbation theory methods as the ab initio methods of the eighties, one can definitely speak of the Coupled Cluster methods as the methods of the nineties. It is easy to foresee that more and more CC calculations will be carried out to solve pending chemical problems. Already today, CCSD(T) is considered as the method to be used in cases where high accuracy is needed. Certainly, with the next generation of computers, CCSDT will replace CCSD(T). Then, QCISDT will offer an attractive alternative to CCSDT.

MP6 in its MP6(M7) approximation will offer new possibilities of assessing higher order correlation effects at the cost of essentially a MP4 calculation. Systematic application of MP6(M7) will show to which extend this method will complement the frequently used MP2 and MP4 method.

Another important aspect in connection with the use of MPn methods is the analysis of the convergence of the MPn series, which we have not discussed in this review, although important work has been done on this topic recently. [27-29] The MPn series shows for different electronic systems different convergence behaviour. Two cases can be considered, namely one with a monotonic convergence of MPn correlation energies to the FCI value and one with erratic convergence behaviour. Based on a dissection of MP5 and MP6 correlation energies, it is possible to explain the differences in the convergence behaviour and to predict which electronic systems possess monotonous or oscillatory convergence behaviour in the MPn series. [27-29] Utilizing this knowledge, it is possible to predict rather precise values of FCI correlation energies once the MP6 correlation energy is known. [28,29]

8. Acknowledgements

This work was supported by the Swedish Natural Science Research Council (NFR), Stockholm, Sweden. Calculations have been carried out with the CRAY YMP/464 of the Nationellt Superdator Centrum (NSC) in Linköping, Sweden. DC thanks the NSC for a generous allotment of computer time.

9. References

[1] Møller, C. and Plesset, M.S. (1934) *Phys. Rev.* **46**, 618.

[2] For an early application of MBPT to atoms see Kelly, H.P. (1969) *Adv. Chem. Phys.* **14**, 129.

[3] (a) Bartlett, R.J. and Silver, D.M. (1974) *Phys. Rev.* **A10**, 1927; (b) (1974) *Chem. Phys. Lett.* **29**, 199; (c) (1974) *Int. J. Quant. Chem.* **S8**, 271.

[4] Binkley, J.S. and Pople, J.A. (1975) *Int. J. Quant. Chem.* **9**, 229.

[5] Pople, J.A., Binkley, J.S., and Seeger, R. (1976) *Intern. J. Quantum Chem. Symp.* **10**, 1.

[6] (a) Bartlett, R.J. and Shavitt, I. (1977) *Chem. Phys. Lett.* **50**, 190. (b) Bartlett, R.J. and Purvis, G.D. (1978) *J. Chem. Phys.* **68**, 2114. (c) Bartlett, R.J., Sekino, H., and Purvis, G.D. (1983) *Chem. Phys. Lett.* **98**, 66.

[7] (a) Krishnan, R. and Pople, J.A. (1978) *Intern. J. Quantum Chem.* **14**, 91. (b) Krishnan, R., Frisch, M.J., and Pople, J.A. (1980) *J. Chem. Phys.* **72**, 4244.

[8] (a) Kucharski, S. and Bartlett, R.J. (1986) *Adv. Quant. Chem.* **18**, 281. (b) Kucharski, S., Noga, J., and Bartlett, R.J. (1989) *J. Chem. Phys.* **90**, 7282.

[9] Raghavachari, K., Pople, J.A., Replogle, E.S., and Head-Gordon, M. (1990) *J. Phys. Chem.* **94**, 5579.

[10] (a) Bartlett, R.J. (1981) *Ann. Rev. Phys. Chem.* **32**, 359. (b) Bartlett, R.J. (1989) *J. Phys. Chem.* **93**, 1697.

[11] Gauss J. and Cremer, D. (1992) *Adv. Quant. Chem.* **23**, 205.

[12] Handy, N.C., Knowles, P.J., and Somasundram, K. (1985) *Theor. Chim. Acta* **68**, 68.

[13] Handy, N.C. (1994), in G.L. Malli (Edt.), *Relativistic and Electron Correlation Effects in Molecules and Solids*, Nato ASI Series Physics 318, Plenum, New York, p.133.

[14] Olsen, J., Christiansen, O., Koch, H., and Jørgensen, P. (in press) *J. Chem. Phys.*

[15] Kraka E. and Cremer, D. (1992) *J. Mol. Struct.* (THEOCHEM) **255**, 189.

[16] Čížek, J. (1966) *J. Chem. Phys.* **45**, 4256; (1966) *Advan. Chem. Phys.* **14**, 35.

[17] (a) Čížek, J. and Paldus, J. (1971) *Intern. J. Quantum Chem.* **5**, 359. (b) Paldus, J., Čížek, J., and Shavitt, I. (1972) *Phys. Rev. A* **5**, 50.

[18] (a) Pople, J.A., Krishnan, R., Schlegel, H.B., and Binkley, J.S. (1978) *Intern. J. Quantum Chem.* **14**, 545. (b) Bartlett R.J. and Purvis III, G.D. (1978) *Intern. J. Quantum Chem.* **14**, 561.

[19] Purvis III, G.D. and Bartlett, R.J. (1982) *J. Chem. Phys.* **76**, 1910.

[20] Noga J. and Bartlett, R.J. (1987) *J. Chem. Phys.* **86**, 7041.

[21] (a) Oliphant, N. and Adamowicz, L. (1991) *J. Chem. Phys.* **95**, 6645. (b) Kucharski, S.A. and Barlett, R.J. (1992) *J. Chem. Phys.* **97**, 4282.

[22] He, Z. and Cremer, D. (1991) *Int. J. Quantum Chem. Symp.* **25**, 43.

[23] He, Z. and Cremer, D. (1993) *Theor. Chim. Acta* **85**, 305.

[24] He, Z. and Cremer, D. (1996) *Int. J. Quantum Chem.* **59**, 15.

[25] He, Z. and Cremer, D. (1996) *Int. J. Quantum Chem.* **59**, 31.

[26] He, Z. and Cremer, D. (1996) *Int. J. Quantum Chem.* **59**, 57.

[27] He, Z. and Cremer, D. (1996) *Int. J. Quantum Chem.* **59**, 71.

[28] Cremer, D. and He, Z. (1996) *J. Phys. Chem.* **100**, 6173.

[29] Cremer, D. and He, Z. (in press) *Theochem.*

[30] Cremer, D. and He, Z. (1994) *Theor. Chim. Acta* **88**, 47.

[31] Cremer, D. and He, Z. (1994) *Chem. Phys. Lett.* **222**, 40.

[32] He, Z., Kraka, E., and Cremer, D. (1996) *Int. J. Quantum Chem.* **57**, 157.

[33] For the derivation of MP6 energies from CCSDTQ calculations, see Kucharski, S.A. and Bartlett, R.J. (1995) *Chem. Phys. Lett.* **237**, 264.

[34] See, e.g., Lindgren, I. and Morrison, J. (1986), *Atomic Many-Body Theory*, Springer Verlag, Berlin.

[35] Pople, J.A., Head-Gordon, M., and Raghavachari, K.(1987) *J. Chem. Phys.* **87**, 5968.

[36] Raghavachari, K., Trucks, G.W., Pople, J.A., and Replogle, E. (1989) *Chem. Phys. Lett.* **158**, 207.

[37] Wilson, S. (1984) *Electron Correlation in Molecules*, Clarendon Press, Oxford.

[38] Bartlett, R.J., Kucharski, S.A., and Noga, J. (1989) *Chem. Phys. Lett.* **155**, 133.

[39] Bagus, P.S., Liu, B., McLean, A.D., and Yoshimine, M. (1973), in D.W. Smith and W.B. McRae (Edts.), *Energy, Structure and Reactivity: Proceedings of the 1972 Boulder Summer Research Conference on Theoretical Chemistry*, Wiley Interscience, New York, p. 130.

[40] (a) Lee, T.J., Remington, R.B., Yamaguchi, Y., and Schaefer III, H.F. (1988) *J. Chem. Phys.* **89**, 408. (b) Harrison, R.J. and Handy, N.C. (1983) *Chem. Phys. Lett.* **95**, 386.

[41] Cremer, D. (1982) *J. Comp. Chem.* **3**, 165.

[42] He, Z. (1995) Ph. D. Thesis, Göteborg.

[43] Kucharski, S.A. and Bartlett, R.J. (1993) *Chem. Phys. Lett.* **206**, 574.

[44] Gauss, J., Kraka, E., Reichel, F., Olsson, L., He, Z., Konkoli, Z., and Cremer, D. (1994) COLOGNE94, Göteborg.

[45] (a) Coester, F. (1958) *Nucl. Phys.* **7**, 421. (b) Coester, F. and Kümmel, H. (1960) *Nucl. Phys.* **17**, 477.

[46] Kraka, E., Gauss, J., and Cremer, D. (1993) *J. Chem. Phys.* **99**, 5306.

[47] (a) Hoffmann, M. and Schaefer III, H.F. (1987) *Adv. Quantum Chem.* **18**, 207. (b) Scuseria, G.E. and Schaefer III, H.F. (1988) *Chem. Phys. Lett.* **152**, 382.

[48] See, e.g. Shavitt, I. (1977) in H.F. Schaefer III (Ed.) *Methods of Electronic Structure Theory*, Plenum Press, New York, p. 189.

[49] (a) Paldus, J., Čižek, J., and Jeziorski, B. (1989) *J. Chem. Phys.* **90**, 4356. (b) Pople, J.A., Head-Gordon, M., and Raghavachari, K. (1989) *J. Chem. Phys.* **90**, 4635. (c) Scuseria, G.E. and Schaefer III, H.F. (1989) *J. Chem. Phys.* **90**, 3700.

[50] Frisch, M.J., Head-Gordon, M., Trucks, G.W., Foresman, J.B., Schlegel, H.B., Raghavachari, K., Robb, M.A., Binkley, J.S., Gonzales, C., Defrees, D.J., Fox, D.J., Whiteside, R.A., Seeger, R., Melius, C.F., Baker, J., Martin, R.L., Kahn, L.R., Stewart, J.J.P., Topiol, S., Pople, J.A. (1992) GAUSSIAN92, Pittsburgh.

[51] (a) Lee, Y.S. and Bartlett, R.J. (1984) *J. Chem. Phys.* **80**, 4371. (b) Lee, Y.S., Kucharski, S.A., and Bartlett, R.J. (1984) *J. Chem. Phys.* **81**, 5906;

(1985) *J. Chem. Phys* **82**, 5761 (E). (c) Urban, M., Noga, J., Cole, S.J., and Bartlett, R.J. (1985) *J. Chem. Phys.* **83** 4041; (1986) *J. Chem. Phys* **85**, 5383 (E). (d) Scuseria, G.E. and Schaefer III, H.F. (1988) *Chem. Phys. Lett.* **146**, 23.

[52] Stanton, J.F., Gauss, J., Watts, J.D., and Bartlett, R.J. (1991) *J. Chem. Phys.* **94**, 4334.

A PHILOSOPHER'S PERSPECTIVE ON THE "PROBLEM" OF MOLECULAR SHAPE

JEFFRY L. RAMSEY
Department of Philosophy
Oregon State University
Corvallis, OR 97331-3902 USA

I. INTRODUCTION

For more than the last two decades, protagonists have argued there is a problem with classical molecular structure. Woolley (1986, 204) claims "(w)e might expect a deductive account of the behavior and properties of molecules according to quantum theory in which the same notion of [classical] molecular structure is a derived concept. The 'problem' of molecular structure arises because quantum chemistry has not achieved this result." Since shape is irreducible, it is only a "concept for solving chemical problems, not an object of belief" (Woolley 1985, 1083). According to Primas (1981, 250), "the EPR correlations predicted by pioneer quantum mechanics compellingly exclude any classical concept of molecular structure." Woolley and Primas attribute the "problem" to philosophical doctrines held by scientists. "[T]he need to put molecular structure in by hand . . . . entails a conflict with certain *a priori* philosophical stances (classical realism, reductionism) which other [sic] may subscibe [sic] to" (Woolley 1986, 204). Primas (1981, 252) claims "many of the difficulties . . . are created by philosophical preconceptions [such as operationalism, logical monism and realism] which we are free to discard as inconvenient ideologies."

Is there a problem with molecular structure? Are reductionism and realism really responsible? In this paper, I engage in two kinds of projects typical of philosophers of science. First, philosophers typically analyze presuppositions, claiming that what has been presupposed is not necessary. In this vein, I argue that the protagonists in the debate generate much of the "problem" by adopting unnecessarily strong versions of realism, reductionism and theoretical explanation. I argue other more philosophically and physically palatable versions soften the problem of structure considerably. Second, we offer principled motivations for particular views of the world. It has been some time since philosophers have endorsed the old dictum that "philosophy assembles definitions of concepts; science applies those concepts to the empirical world." Quine (1960) holds there is no philosophically defensible demarcation between conceptual and empirical problems, and many philosophers and like-minded scientists have blurred the distinction in practice.[1] Thus, I offer a physical account of shape. This account allows me to agree that there is a "problem" of molecular shape. However, I agree in a way that removes much of the claim's force. Shape is not just a "powerful and illuminating metaphor" (Woolley 1982, 4), a gestalt (Amman 1993), or an aspect of Bohrian complementarity (Primas 1982, 1985). Shape just is physical and real while not a literally true description of the world. This does not entail it is not an object of belief.

I sketch the physical account of shape in Section 2. (Readers wishing more detail should consult Ramsey 1996). Section 3 isolates what sense of shape is being attacked and defended. In Sections 4 and 5 respectively, I argue the versions of realism and reductionism endorsed by Woolley and Primas are philosophically questionable, and, given the physical account provided in the paper, physically suspicious as well.

To preview my arguments, I urge caution about what we should expect from theory. Theory alone is not likely to deliver the kind of reductionism, realism or explanation required for shape to disappear as merely conceptual or, for that matter, be entrenched as a literally referential description of reality. In parallel, philosophical commitments alone are not likely to entail anything definite. A mix of theory, experiment, and suitably formulated philosophical doctrines is preferable.

2. A PHYSICAL ACCOUNT OF SHAPE

Contrary to the pronouncements of Woolley, Sutcliffe and Primas, it is possible to offer a thoroughly physicalist account in which shape is an objective property. However, the account gives up any attempt to derive the concept strictly. It makes shape an approximate rather than an exact concept. This approximate concept refers to a causal property, rigidity, which most molecules display under most experimental conditions. Woolley and Primas are correct, however, in stressing that this property is contingent rather than essential.

Put briefly, the account avoids the paradox of molecular shape by giving up the idea that shape is (or ever was) exactly classical. It retains the notion of the separability of nuclear and electronic motions, which is perfectly "legal" in quantum mechanics (Claverie and Diner 1980, 60; Bohm and Hiley 1993; Stamp 1995, 140). Importantly, however, the account does not impute separability independently of measurement. It makes separability (and thus classical descriptions) conditional on the way in which the system is measured.

If we accept that separability of motions is allowed within quantum mechanics, the question shifts to when separability is warranted. The time-energy form of the uncertainty relation provides the parameters. The momentum-position formulation of the Heisenberg uncertainty relations can easily be transformed into the equivalent energy-time formulation (cf. Löwdin 1988, 13). That is,

$$\Delta p \cdot \Delta x = \Delta E \cdot \Delta t \approx h$$

For most chemical states, the assumption that the nuclear and electronic motions are separable keeps us well within the energy range proscribed by the uncertainty relation. When will we observe states which lie outside the states allowed by the separability-of-motions assumption? Take Cl_2 or C_3H_4 as typical molecules. The lifetime for a classical description of these molecules is on the order of 10^{-10} sec (Patsakos 1976).[2] In infrared spectroscopy (IR), we can resolve two bands corresponding to different sites that are separated by 0.1 cm^{-1}. The lifetimes, Δt, corresponding to this are 5×10^{-11}. Since one needs a process which gives rise to lifetimes of $\sim 10^{-11}$ **or less** to cause broadening -- much less eventual separation--of the bands, we will see only one peak which corresponds to the classical description (Drago 1977, 86-87). We will not observe a delocalized atom in which the nuclear and electronic motions are affecting each other in any significant fashion.

Said alternatively, IR is a good tool for observing some but not all energy states. Let us suppose we are interested in the rotational spectrum, an energy level which explains some important behavior of most molecules. The value of a typical quantum of rotational energy is about 250 wave numbers. If the level splittings caused by nuclear-electronic interaction are around 50 wave numbers, IR will not provide the resolution needed to observe these splittings (Berry 1960). Those processes are occurring on a time scale which is too fast with respect to the instrumentation. Our measuring technique is simply too clumsy.

If we wanted, we could make the hyperfine lines appear. This would mean switching spectroscopic techniques in order to decrease the energy of the incident radiation and increasing the lifetime of localized state. Experimentally, we do so by trapping the molecules in beams. These are the experiments cited by Woolley. In these experiments, extremely fine splittings do appear which cannot be captured by the Hamiltonians constructed with the separability assumption. Thus, Woolley is correct that the separability assumption is not necessary; it can be dispensed with if the conditions warrant. Thus, it is not *necessary* for a molecule to have a (classical) shape *in some experimental situations*.

However, most molecules do have a molecular structure since we are often interested in "higher" level energies (i.e., rotational energy) and, independently, very few of our experimental techniques allow us to make the hyperfine lines appear. With current technology, we can make the lower level lines appear only for very light atomic weight systems. Quite extreme conditions (milli-Kelvins, with the atoms trapped for minutes) are required before it becomes necessary to throw out the separation of nuclear and electronic motions. For heavier atoms and molecules, even more extreme conditions would be needed. For most molecules with even better technology than we have now, there would be no observable difference between treating the system as separated and not separated.

If we had infinite time to measure, we could minimize the energy spread to it theoretical limit. This corresponds to an almost negligible "bump" of the system by the observing instrument. At the limit, we would be measuring eigenstates of a complete molecular Hamiltonian. At any time less than infinity, however, we measure states. We can choose to represent these as states of the complete

Hamiltonian or as eigenstates of approximate Hamiltonians. The fact that we construct the approximate Hamiltonians confers no illicit status on the representation.

On this account I have just offered, separability is not an essential feature of a given molecule. Whether separability is a good assumption depends crucially on the natural energy levels of the molecule *and* the nature of the measuring apparatus. In our world of small energy quanta and machines of non-infinite resolution, most molecules have a shape. It is a factual question whether or not there is a discrepancy between the calculation that employs the fixed nucleus approximation and the actual physical system (where "actual" has to take into account how it is we wish to measure the system). It is in this sense that shape is both physical and real, albeit approximate.

Given that it is possible to argue shape (in the sense of separable motions and the existence of a nuclear frame) is an object of belief and not just a concept, it behooves us to examine why Woolley and Primas argue to the opposite conclusion. As it turns out, they do so because of some prior philosophical commitments regarding reductionism and realism. As noted before, they charge other scientists utilize these doctrines uncritically. It is time to examine carefully the versions employed by Woolley and Primas.

3. INTERLUDE: EXACTLY WHICH CONCEPT OF SHAPE IS BEING CHALLENGED?

Before examining the philosophical presuppositions, a point of clarification is required. Woolley and Primas can respond that their target is the classical version of molecular shape, in which the behavior of molecules is understood "in terms of the relative dispositions of the constituent atoms in the molecules" (Woolley 1978, 1074). I have no wish to defend an absolutely literal version of the ball-and-stick models of chemical education. I have chosen to highlight the issue of separability because they do so. In his (1977, 394), Woolley speaks of the "molecular structure hypothesis (microscopic rigidity)," where rigidity is taken to mean the existence of a nuclear frame of some sort. In other places, Woolley and Sutcliffe charge that all justifications for the separability of electronic and nuclear motion are without theoretical foundation (cf. Sutcliffe 1992, 34). All approximations (clamped nuclei, Born-Oppenheimer, Born-Huang) to the wave equation solve the many-body problem without saying how they do so (Woolley 1978, 1075). Since they have formal validity only (Sutcliffe 1992), they somehow fail to have physical validity, i.e., they are incorrect representations of the physical facts. For Primas (1981), molecules are not part of the ontology of physics; they are created in the abstraction from universal EPR correlations. Woolley (1988a) cites Primas' argument approvingly.

Also, the burden of the argument *should* fall on separability.[3] To the extent that Woolley, Primas and others fail to separate the issue of the representational nature of balls and sticks from the issue of the separability of motions, they seriously confuse the debate.[4] Separate arguments are needed for the two issues. Here is why:

if the ball-and-stick notion is only a concept, separability is not necessarily false. Balls-and-sticks can add representational content which is not warranted but does not invalidate the argument that the motions are separated. At best, if separability is false, the ball-and-stick picture cannot be true.

4. REDUCTION AND EXPLANATION

In this section, I argue Woolley and Primas have assumed a questionable account of reduction and, with it, of theoretical explanation. Specifically, they presume what philosophers term "eliminative reduction." They argue such reductive account will never be found, so shape is not an object of belief. However, this account of reduction is questionable on general grounds. In addition, it entails a view of molecular shape as a sharp, literally referring concept. As the physical account given previously indicates, this is untenable. Shape is an approximate concept. This suggests we find alternate accounts of reducibility and explanation.

Woolley and Primas presuppose an eliminatively reductive account because they suppose shape must be:
1) derivable within mature quantum mechanics, and
2) naturally isolable from the perspective of quantum mechanics.

The quote from Woolley at the beginning of this paper displays his commitment to derivability. Woolley also employs the criterion of natural isolability. In his argument (Woolley 1988b, 1991) that the complete molecular Hamiltonian does not distinguish the various isomers of, say, C_3H_4, he implies the fundamental representation must differentiate them. That he endorses the isolability requirement is also supported by his argument that it is not enough for classical structures to be compatible with the fundamental representation. He rejects the claim that it is enough for classical structures to be identified with a certain subset of states in the quantum mechanical Hilbert space (Woolley 1976; cf. Claverie and Diner 1980, 59). Not only must we know **that** molecules inhabit time-dependent states; we must also know **why** they do so. In sum, Woolley reasons that since shape is neither strictly deducible nor naturally isolable from the perspective of quantum theory, shape is only a concept and not an objective feature of the world.

From a philosophical point of view, all this is very curious for a variety of reasons. First, scientists as well as philosophers of science disagree greatly on what constitutes a reduction. Even if they agree a reduction has occurred or is occurring, they usually disagree on the criteria the reduction should satisfy and the consequences (elimination or preservation or correction) of the reduction. Dupre (1993) echoes a common feeling when he notes there are so many brands of reductionism it is difficult to oppose reductionist claims. In short, rather than trotting out a set of criteria, what one really wants is a justification why that set should be applied in this particular instance. Second, the charge presupposes rather undefended and philosophically questionable notions of separate 'levels' of description or reality. Third, since reductions are supposedly explanatory, it is important to ask whether Woolley and Primas employ a defensible notion of theoretical explanation.

In this instance at least, I think there are good reasons why we should not expect the reduction criteria of deducibility and isolability to be fulfilled. I will discuss isolability first as the argument against it goes rather more quickly. The requirement that events at one level be naturally isolable at another is enormously implausible as an intertheoretic compatibility condition. Consider for a moment analogous problem in the mind-body debate:

> Surely objects like neurons, or events like neuron firings, don't have to be 'naturally isolable' from the perspective of fundamental physics-chemistry in order to be compatible with it; rather, it is enough that they be fully decomposable into naturally isolable parts. The role of a neuron can be typically played by a vast conglomeration of physical-chemical structures and events. This doesn't show that the neurons don't exist. All it shows is that the neuron is a complex structure which has complex neurological parts (Horgan and Woodward 1985, 409-410).

The same applies to separable motions of nuclei and electrons. Separable motions, or even ball-and-stick models, do not have to be naturally isolable from the perspective of quantum mechanics. Even less do we have to be able to identify a term in the equation with some part of the representation. It is enough that they are or can be decomposable. But if we can satisfy ourselves with physical arguments that the motions are separable (and of course they are not always so), nothing follows from the fact that we cannot naturally isolate them from the perspective of the fundamental theory. More to the point of everyday chemical life, we can isolate different isomers. The physical account given above indicates both how and why: the different isomers are vibrating in different deep potential wells. Nothing follows from the fact that we cannot pick out the isomers in the quantum mechanical representation.

The demand for an "emergent derivation," with the higher level concepts appearing naturally within the deduction, is just as implausible. Even Nagel (1979) requires only that a reduction connect antecedently given concepts and laws at two levels. How does the charge of metaphor fare under a more reasonable sense of reduction? Consider Paul Churchland's (1979) model, which turns out to be reasonably plausible when applied to the relation between classical and quantum shape. Even so, it leads to a serious problem with any account, such as Nagel's or "emergent derivation," which shares its general features.

Churchland requires: (1) the truth of the reduced theory need not be preserved, and (2) term by term mapping between the reduced and the reducing theories. (Whether one also is able to identify the principles of the reduced theory as theorems of the reducing theory is an open question which is not important for my purposes here.)

Woolley clearly questions the truth of the classical theory of shape for isolated molecules. The classical theory is a classical realist theory and thus cannot be true on quantum mechanical grounds. Woolley also endorses a version of the mapping requirement: classical "molecular structure has . . . to be associated with those intrinsically time-dependent quantum states for which the identification between classical and quantum configuration spaces can be made, since only then is it valid to relate notions of molecular structure to maxima in the molecular wave function in position representation" (Woolley 1978, 1076). That is, there does exist a set of rules which select out a subset of the available physical states--the intrinsically time-dependent quantum states--and identifies them as chemical structure states.

Given that Churchland's account applies, how can one continue to press the claim of metaphor? Woolley rightfully stresses an important problem: the non-explanatory relation of the identifications or connecting rules. "How this identification [between classical and quantum states] is to be made remains an unsolved problem which is usually glossed over" (Woolley 1978, 1076). The account provides no reason why these two predicates should be so connected. Woolley identifies a scientific problem that philosophical accounts of reduction must respect. Non-explanatory identifications are formal moves that paper over substantive issues involved in why-questions (cf. Hull 1976). We would like to know why chemical structure states occupy only time-dependent quantum states. Churchland's account and, I think, other accounts that focus on structural issues will not answer the question. Woolley's eliminative reductionism likewise fails. He is precluded from explaining shape since his criteria for reduction entail shape is not an objective feature of the world.

Nonetheless, we are forced to say shape is forever unexplainable and thus only metaphorical. We can and should question the structural assumptions embedded in models of reduction as well as the concept of explanation employed.

For instance, models of reduction are most often premised on the idea that one has two theoretical accounts which must be made to coincide. This is true of Nagel's and Churchland's models as well as the model adopted by Woolley and Primas to motivate the claim that molecular shape is irreducible. If one adopts this point of view, one is immediately faced with the issue of whether the reduced theory will be preserved or eliminated in the structural connections achieved in the reduction.[5] Primas (1981) is quite explicit that the new, quantum mechanical theory of shape must recover shape as it has been used in the past; that is, we must save or eliminate *that* concept of shape. Thus, he is forced to argue that shape is an aspect of Bohrian quantum mechanical complementarity (cf. Primas 1982, 1985).

In short, Woolley and Primas employ philosophical models of reduction which force the presumption shape is or was an independent level of discourse. Further, that discourse is supposed to be legitimate rather than in need of correction. But this is a mistake. 'The concepts of classical chemistry were never completely precise. . . . [W]hen we carry these concepts over into quantum chemistry we must be prepared to discover just the same mathematical unsatisfactoryness" (Coulson 1960, 174). On the physical account offered above, the imprecise notion of classical shape became physical, albeit approximately so. There is only one reality. There is not a quantum chemistry separate from quantum physics; there are many levels of approximations which we use to get ourselves from one system to the other. The charge of irreducibility relies on an interpretation of the ball-and-stick model as a separate, intrinsically classical description of the world. This is unwarranted and unnecessary. Furthermore, it is apparently untenable; Jones (1994) argues there are no intrinsically classical domains.

The model of explanation can also be questioned. Woolley and Primas require "ultimate" explanations of molecular shape in which shape, if it is real, is distinguishable in the fundamental representation. In short, we must know on the basis of theory how the world appears and why the world appears as it does. There are good reasons to question such a requirement. Invoking Kant, Sklar (1989) questions the request for ultimate explanations, explanations which posit no contingent particular facts or generalizations. Even if Kant is incorrect that such explanations are unattainable, conditional explanations are familiar and quite useful. No whiff of impropriety should attend them.

Second, ultimate explanations debase explanations in terms of models. As already indicated, Primas and Woolley think classical molecular shape is a powerful and illuminating metaphor but nonetheless somehow not truly descriptive of the world. Inspired by the work of Cartwright (1983) and others, philosophers have begun to articulate a philosophical space for model-based explanations. Some explanations may contain the particulars in the general laws or theoretical equations, but many do not. Cartwright (ibid., pp. 107-118) gives the examples of amplifiers and exponential decay to argue many approximations and models actually improve on the fundamental laws. In other cases, the models "fill in" the structural accounts provided by the laws (cf. Rohrlich 1988, Redhead 1980). Davies (1995) provides an example of this when he notes that the field equations of relativity predict many cosmological models.

This "unhinging" of the laws from the models and from the particulars has led many philosophers away from a conception of explanation as strict derivability. One popular option is to think of explanation as "metaphorical redescription," a phrase originally introduced by Mary Hesse (1966). On this view, an explanation involves a non-literal description of a primary system in terms of a secondary system. For example, we take the transmission of light as our primary system and represent it as wave motion. In this way, we make the behavior understandable. That is, we explain it.

Hughes (1993) uses this notion of explanation to posit the existence of many different types of models which mediate the relationship between the thing to be explained and our available theoretical accounts. Some models are constitutive in that they define how matter is made up. Others are behavioral in that they tell us how the matter acts. Yet others are foundational, drawn from very general theories such as Newtonian or quantum mechanics. Finally, some models are structural in that they employ particular versions of the fundamental terms of the theory.

Hughes uses Pines' (1987, 65) description of a plasma to illustrate these distinctions. A plasma can be depicted constitutively *either* as a fully ionized gas *or* as a collection of ions moving in a background of uniform positive charge.

Behaviorally, the plasma can be treated classically or quantum mechanically. If we treat the plasma classically, it is one of a very general class of Newtonian systems; Hughes calls this system foundational models. Finally, the Newtonian systems presuppose a particular account of space-time. This account is described as a class of models.

This four-fold distinction works well for molecular shape: we can regard a molecule as constituted by its atoms, by a nuclear frame which does not preserve the individuality of the atoms, or by a combined nuclear-electronic frame. Likewise, we can treat the behavior of such systems classically or quantum mechanically, although some constitutive descriptions will not be well served by some behavioral descriptions. Once we have made a choice of "quantum" rather than "classical," we must still decide, for example, whether to include dynamical correlation or not. That is at the level of the foundational models since quantum mechanical descriptions allow one or the other depending on the situation. Finally, we can modify the account of space-time by choosing to include relativistic effects or not.

Overall, the message is that theoretical explanation does not proceed directly and only from the fundamental laws or equations. There are many choices to be made along the way. The physical account I offered in Section 2 is eminently compatible with this notion of theoretical explanation. Since shape is only approximate, we must decide if various constitutive, behavioral, foundational and structural descriptions are compatible with the system being investigated with the experimental apparatus chosen. No one explanation is dictated by the facts. Even more importantly, the existence of a fundamental theoretical level does not mean that all features of the world must be recoverable directly from that level.

5. REALISM, PHYSICALISM AND MATERIALISM

In this section, I argue Woolley and Primas have assumed a particular account of realism and, with it, of physicalism and materialism. As with reduction, many versions of realism exist. Leplin (1984) lists ten theses endorsed by different realists. Also as with reduction, since there are so many versions, one wants a justification why a particular version is being employed. I argue the version utilized is not well justified and, further, is not plausible given the physical account provided earlier.

Stated broadly, Woolley and Primas employ a version of realism in which it is assumed
1) the best current scientific theories are at least approximately true,
2) the central terms of the best current theories are genuinely referential,
3) the theoretical claims of scientific theories are to be read literally, and so read are definitively true or false, and
4) scientific theories make genuine, existential claims.
(These are theses 1,2,7 and 8 listed by Leplin.) While all four are admirable ideals, one can be a realist without endorsing them. The point of view I urge is closest to Leplin's thesis 5: A scientific theory may be approximately true even if referentially

unsuccessful. That is, a scientific theory may be both approximately true and not a literally true description of reality. The terms of the theory may refer only approximately to true causal powers of reality.

Both Woolley and Primas take their criterion of reality from our current, best-developed theory. Both assume there is one level of theoretical description, and this level determines what counts as physically significant. Woolley (1988, 55) claims "only H_{QED} has physical significance" since "matter and radiation can never be completely separated." Elsewhere, he insists that we must be develop our account of molecular behavior from the "stationary state of the isolate [sic] molecule Hamiltonian H_{MOL}" (Woolley 1986, 203; cf. Woolley 1988, 71). "[C]lassical molecular structure cannot be associated with a spectroscopic state of H_{mol} . . ." (Woolley 1988, 60). And again, "I have argued previously that the various different quantum mechanical representations of molecular structure cannot be related to the molecular eigenstates by unitary transformations" (Woolley 1980, 38, emphasis added).[6] Likewise, Primas insists "Observable phenomena are created by abstracting from some Einstein-Podolsky-Rosen correlations. . . . Without such an abstraction there are no phenomena" (1981, 253; cf. also Primas 1990). Objects *must* have a pure state description at every instant *in order to be objects* (ibid.).

Since neither Woolley nor Primas are willing to give up the focus on pure states, both find shape to be a paradox. There are a number of philosophical reasons to question the reliance on pure states. First and foremost, it is only the Schrödinger formulation that has a problem with mixed states. As Bell notes, "(I)n the Heisenberg picture there is no complication in considering mixed rather than pure states" (Bell 1987, 49). Given the presence of an alternative, equivalent scheme, some argument must be given for employing only the scheme that generates the problem. Now, perhaps there is a fundamental problem in quantum mechanics given that one formulation has a problem with mixed states and the other does not, but there seems little rationale for blindly following one rather than the other.

The physical account presented in the previous section provides one rationale for dissolving the paradox by "going Heisenbergian." Chemists usually measure mixed states. In such states, the molecule does not exhibit or exist in the full symmetry appropriate to the stationary states. This implies it is an empirical matter whether the physical processes of spreading and delocalization take place on a timescale which rules out the idea of a rigid molecule.

Second, to focus on pure states is to employ an extremely theory-based criterion of reality. This is dangerous. Even in classical mechanics, the inference from "in the theory" to "real" is quite weak. Jones (1991) notes different formulations of classical mechanics (gravitational forces between particles, gravitational potentials, minimum principles) entail very different ontological commitments. If the inference were stronger, not even spacetime would count as real (Wheeler 1973, 227)! In addition, there are many ways to discriminate the real. Philosophers of science have recently begun to make much of the fact that experimental work has a life of its own quite independent of theoretical work. If experimental work accords shape explanatory status, it should not follow immediately that shape is merely conceptual on the basis of theoretical considerations.

Moreover, even Woolley's own descriptions fail if we employ this reality criterion. He uses models rather than some unalloyed, direct representation of "reality." In frozen molecular beams, we do not observe eigenstates of *complete* molecular Hamiltonians. As he points out in his (1988), Woolley uses Hamiltonians that measure states from the perspective of relativistic quantum mechanics. Thus, the solutions he generates are only eigenstates of approximate Hamiltonians. Using his own criterion of reality, the molecular eigenstates lying outside the traditional classical picture do not have any fundamental physical significance.

Primas fares somewhat better in this regard as he bases his arguments in the abstraction process. However, there are independent reasons to be wary of this argument. In contrast to the idea that objects are created in the process of abstraction, there is every reason to believe the degree of quantum entanglement is an empirical matter which can be determined. Separability, including the separation of nuclear and electronic motion, does occur as a physical process. Further, this process is recoverable in the mathematical formulation. A 'natural truncation' occurs "when the influence of all other states, on the . . . states in question, becomes either adiabatic or just weak" (Stamp 1995, 140). Even Bohm and Hiley (1993, 59) stress the point that quantum wholeness does not preclude factorizable wave functions. Moreover, it is not possible to claim all quantum systems should exhibit quantum behavior such as non-separability or even superposition irrespective of the kind of measurement made on the system. Quantum and classical behavior overlap significantly. Systems which we would expect to exhibit only quantum behavior can exhibit localizable, molecule-like behavior (Ezra and Berry 1982). Likewise, systems which we would expect to exhibit classical behavior can continue to exhibit quantum behavior (Mermin 1990). In short, physical arguments, and not *theory alone*, should be kept in the fore when deciding when a system is correctly described by quantum notions.

On a related point, objects seem to be created from ignoring EPR correlations only if we forget that "Gedanken experiments like EPR are idealizations, i.e. oversimplifications, which focus on certain facets, ignoring (as usual) all the rest" (Groenewold 1995, 554-55). A measurement which *could* recover the EPR correlations is an "idealized observation" (Forrest 1988, 32). But again, as the physical account provided here stresses, measurement processes -- at least the measurement processes chemists utilize -- occur in time. To assert objects are "created" is to forget most chemical measurements are not idealized observations. Chemists measure incoherent mixtures rather than coherent superpositions (Davies 1981, 76-77). Forgetting the idealization is important philosophically since "creation" of objects suggests an underlying reality which is violated or contradicted in the process of abstraction. This in turn gives rise to the notion of "levels" of reality. It is far better to speak in terms of one level of reality examined at different scales of resolution. To be certain, I do not have a complete account of how objects come to display different causal properties in different measuring processes.[7] Nonetheless, this seems a less radical and daunting task than reforming quantum mechanics. Philosophy of chemistry can and should proceed without first having to solve the quantum mechanical measurement problem.

Finally, Woolley and Primas offer their account as a tough-minded materialism, or physicalism. In an early article, Woolley (1982, 4) paraphrases Gertrude Stein by claiming "Matter is matter is matter." Since Stein's original phrasing was "Rose is a rose is a rose is a rose," I suggest we return the additional noun and read it as:

Matter:
Is Matter?
Is.
Matter is:
Matter.[8]

In prose, invoking the mantle of materialism does not resolve the issue. As the characters in Gilbert and Sullivan's <u>Ruddigore</u> might claim, an unarticulated materialism "doesn't really matter, matter, matter, matter, matter . . . "[9] In particular, materialism by itself does not entail an endorsement or rejection of either relationalism or holism. Further, ontological physicalism does not imply methodological physicalism (Brandon 1996). That is, one can easily agree that there is nothing but physical stuff, but this in no way entails that the only fruitful way to study that stuff is from the perspective of theoretical physics. Physical stuff arranges itself in many different ways with operating principles consistent with but not necessarily strictly derivable from the physical laws. As noted previously, an assumption about levels seems to be the culprit here. That undefended assumption generates the idea that shape is just a gestalt, an alternative "picture" of reality.

6. CONCLUSION

To summarize, I have offered a physicalistic account on which it makes sense to speak of microscopic rigidity. The account does not recover directly the notion of a ball-and-stick molecule, but it does preserve the separability of nuclear and electronic motions and thus the existence of a nuclear frame. The account echoes claims made by Woolley and Primas that shape is a contingent rather than an essential property of molecules. It makes sense to speak of shape only conditionally since rigidity (or separability or even a particular shape such as C_{3v} or D_3 symmetry) is dependent on the method of measurement. However, the account contradicts their claim that shape is merely conceptual, metaphorical, or Gestalt-like. It is not a matter of looking at a system from two incompatible vantage points. A better analogue would be "zooming in" on a painting, where the increased resolution causes us to lose the gross features of the representation while discovering new features of the paint surface. We see the same thing from very different perspectives.

I have argued it is not philosophical assumptions *per se* that generate the 'problem' of shape. Rather, specific versions of realism, reductionism, and theoretical explanation generate the puzzle. In particular, Woolley and Primas adopt an extremely stringent theoretical realism when they assume measuring anything less than stationary states is measuring something non-objective. It is this that allows them to claim shape is "not an object of belief." I have argued their philosophical presumptions are not philosophically plausible nor are they demanded by the science. More modest philosophical positions allow shape to be physical, real and contingent even if not perfectly literal.

This is not to suggest the 'problem' of shape evaporates. At the physical level, significant questions remain. Why do molecules inhabit the states they do? Why is it chemists measure mixtures rather than superpositions? Why is it so many systems do exhibit separable motions? In the end, what should we do? Certainly, I cannot prophesy how the science will turn out. Schrödinger may be vindicated over Heisenberg. At best, I can say, "Please do not employ philosophical concepts uncritically. Provide philosophically and physically plausible versions for the version used." Philosophers like to think distinctions matter.

From a philosophical point of view, a host of philosophical issues are raised rather than solved by the perspective urged in this paper. First, as Weininger (1984) highlights, can all the concepts of structure be unified, or are they inequivalent and irreducible? The physical account I have offered rejects the dilemma. It entails an answer that the different concepts account for a complexly realizable property of the world, and those concepts are linked imprecisely rather than being strictly reducible or irreducible. Sober' (1984, 69-70) offers a similar argument about biological fitness. The different theoretical versions all try to account for a single magnitude found in the world. Similar questions arise about force and mass in physics. Whether one philosophical story can be told about all these concepts is an open question; comparative work beckons.

Another issue is when to limit the search for more ultimate explanations. Technically, this is a question about how much information should be carried by the Hamiltonian (cf. Berry 1980). More generally, the question is how much should experience lead the way. I do not know the answer. However, I do not believe the issue can be decided on *a priori* philosophical grounds. In particular, Woolley's and Primas' reductionism is an impoverished philosophy of science. It recognizes only some aspects of good physicalistic or mechanistic explanations.[10]

Finally, a property asserted as real but not interpreted literally poses significant philosophical perplexities. In particular, the problem of molecular shape challenges versions of essentialism held by many philosophers and scientists. Essentialists often claim objects have some properties such as shape independently of the way the object is picked out. However, molecules can lack an orientation in three-dimensional space, and a particular shape is dependent on the way the molecule is picked out in measurement. Thus, physical reasoning indicates the question "Does x possess F or not?" is not well-posed, at least for shape. This is philosophically revolutionary.

I would like to thank R. Stephen Berry, Arthur Fine and Eric Scerri for helpful conversations about and insight into quantum mechanics and quantum chemistry. I would like to thank Mary Jo Nye for the same regarding the historical tradition in chemistry. All errors are, of course, solely my responsibility.

ENDNOTES

1. Examples here are too numerous to list inclusively. Philosophers whose work falls into this category include: Fine (1982, 1986) and Shimony (1993) in the philosophy of physics; Scerri (1989, 1994) in the philosophy of chemistry; and Sober (1984a), Hull (1988), Brandon (1990, 1996) and Wimsatt (1995) in the philosophy of biology. The anthology by Sober (1984b) is indicative of the mutual interest of biologists and philosophers in conceptual problems with empirical consequences.

2. The chlorine molecule has 70 protons and neutrons; the C_3H_4 has 40. Thus, an isolated molecule of the chlorine molecule weighs around 1×10^{-22} grams. Since the bond length in the molecule is about 2 angstroms, the time scale would be reduced to about 5×10^{-10} seconds. The C_3H_4 molecule has 40 protons and neutrons, so it weighs around 7×10^{-23} grams. Roughly, the molecule is about twice as long as the chlorine molecule, so the time scale is of the same order of magnitude.

3. Today's classical picture is predicated on the notion of separability. The electrons move very quickly relative to the protons and neutrons so that the nucleus effectively "sees" an electron "cloud." The balls and sticks are recognized as merely representational devices; they only correspond to without being a literal picture of a "rigid" molecule with largely independent electronic and nuclear motion.

4. The confusion appears most strikingly in their "historical" argument. To the extent that Woolley and Sutcliffe in particular and also Primas treat van't Hoff's structural chemistry on a par with the later justification for the separability of nuclear and electronic motions, they do not recognize that the concept of a "rigid" molecule is historically conditioned. Even though van't Hoff's molecule was referred to as the "static" molecule, that notion changed over time. G. N. Lewis referred to the dynamic conception of moving atoms stemming from the Bohr atomic model. Eventually, the theoretical representation became the separated motions of the modern molecule. In short, the theoretical representation of the molecule has changed throughout history.

Of course, not even van't Hoff was a naive ball-and-stick guy; he worked out a dynamical conception using Newtonian concepts, and he worked on tautomerism and the movement of the hydrogen atom with the molecule.

5. This "epistemological" conception of reduction is separate from an "ontological" conception. On the latter, reality comes in levels. For example, there is the physical level and then the chemical and then the biological; one has the physical basis of the brain and the separate level of the mental. One can endorse epistemological reductionism while eschewing ontological reductionism. That is, one can think there is only one level of stuff (the physical) described by various vocabularies. One can endorse ontological reductionism while eschewing epistemological reductionism. That is, one can think reality is made up of various parts and wholes, all of which are somehow physical, but not think that there should be any connections between the languages

used for the different levels. And of course, one can endorse or reject both reductionisms simultaneously.

6. Here, he is referring to the argument offered in Woolley (1978a).

7. For some initial thought on this, see Ramsey (forthcoming). My remarks there are heavily indebted to the philosopher of biology William Wimsatt (1995).

8. Just in case it is not clear in the text, this is how I intend my reading to be interpreted:

> MATTER for your consideration:
> IS MATTER?
> Yes, it IS.
> The MATTER IS:
> No one is quite sure what MATTER (or materialism) implies.

9. This is the patter-trio sung by Robin, Despard, and Mad Margaret. It occurs in Act II of the operetta. Cf. Sullivan 1976, p. 391.

10. Brandon (1996, 196) makes a similar point about reductionism vs. mechanism in biology.

REFERENCES

Amman, A. (1993), "The Gestalt Problem in Quantum Theory: Generation of Molecular Shape by the Environment," *Synthese* 97: 124-156.

Bell, J. S. (1987), "On Wave Packet Reduction in the Coleman-Hepp model," in his *Speakable and unspeakable in Quantum Mechanics*. New York: Cambridge University Press, pp. 45-51. Originally published in *Helvetica Physica Acta* 48 (1975): 93-98.

Berry, R. S. (1960), "Time-Dependent Measurements and Molecular Structure: Ozone," *Reviews of Modern Physics* 32: 447-454.

_____ (1980), "A Generalized Phenomenology for Small Clusters, However Floppy," in R. G. Woolley (ed), *Quantum Dynamics of Molecules: The New Experimental Challenge to Theorists*. New York: Plenum Press.

Bohm, D. and B. Hiley (1993), *The Undivided Universe: An Ontological Interpretation of Quantum Theory*. London: Routledge.

Brandon, R. (1990), *Adaptation and Environment*. Princeton: Princeton University Press.

_____ (1996), *Concepts and Methods in Evolutionary Biology*. New York: Cambridge University Press.

Claverie and Diner (1980), "The Concept of Molecular Structure in Quantum Theory: Interpretation Problems," *Israel Journal of Chemistry* 19: 54-81.

Churchland, P. (1979), *Scientific Realism and the Plasticity of Mind*. Cambridge, England: Cambridge University Press.

Davies, P. (1981), "Time and Reality," in R. Healey (ed), *Reduction, Time and Reality*. New York: Cambridge University Press, pp. 63-78.

_____ (1995), *About Time*. New York: Simon and Schuster.

Dupre, J. (1993), *The Disorder of Things*. Cambridge, MA: Harvard University Press.

Drago, R. (1977), *Physical Methods in Chemistry*. Philadelphia: W. B. Saunders.

Ezra, G. and R. S. Berry (1982), "Correlation of two particles on a sphere," *Physical Review A* 25: 1513-1527.

Fine, A. (1982), "Hidden variables, joint probability and the Bell inequalities," *Physical Review Letters* 48: 291-295.

_____ (1986), *The Shaky Game: Einstein, Realism and the Quantum Theory*. Chicago: University of Chicago Press.

Forrest, P. (1988), *Quantum Metaphysics*. New York: Basil Blackwell.

Groenewold, H. (1995), "Field or Print," *Synthese* 102: 1-56.

Hesse, M. (1966), "The Explanatory Function of Metaphor," in her *Models and Analogies in Science*. Notre Dame, IN: University of Notre Dame Press, pp. 157-177.

Horgan, T. and J. Woodward (1985), "Folk Psychology is Here to Stay," *Philosophical Review* 94: 197-226.

Hull, D. (1976), "Informal Aspects of Theory Reduction," in R. S. Cohen et. al. (eds.), *PSA 1974*. Dordrecht, pp. 653-670.

_____ (1988), *Science as a Process*. Chicago: University of Chicago Press.

Hughes, R. I. G. (1993), "Theoretical Explanation," in P. French, T. Uehling and H. Wettstein (eds), *Midwest Studies in Philosophy of Science, Vol. XVIII: Philosophy of Science*. Notre Dame, IN: University of Notre Dame Press, pp. 132-153.

Jones, K. R. W. (1994), "Exclusion of intrinsically classical domains and the problem of quasiclassical emergence," *Physical Review A* 50: 1062-1070.

Jones, R. (1991), "Realism about What?" *Philosophy of Science* 58: 185-202.

Leplin, J. (1984), "Introduction," in J. Leplin (ed), *Scientific Realism*. Berkeley: University of California Press.

Löwdin, P-O. (1988), "The Mathematical Definition of a Molecule and Molecular Structure," in J. Maruani (ed.), *Molecules in Physics, Chemistry and Biology, vol. II*. Boston: Kluwer Academic Publishers, pp. 3-60.

Mermin, D. (1990), "Extreme Quantum Entanglement in a superposition of macroscopically distinct states," *Physical Review Letters* 65: 1838-1840.

Nagel, E. (1979), *The Structure of Science*, 2nd ed. Indianapolis: Hackett.

Patsakos, G. (1976), "Classical Particles in Quantum Mechanics," *American Journal of Physics* 44: 158-166.

Pines, D. (1987), "The Collective Description of Particle Interactions," in B. Hiley and F. Peat (eds), *Quantum Implications: Essays in Honour of David Bohm*. London, pp. 65-84.

Primas, H. (1981), *Chemistry, Quantum Mechanics and Reductionism*. New York: Springer-Verlag.

_____ (1982), "Chemistry and Complementarity," *Chimia* 36: 293-300.

_____ (1985), "Kann Chemie auf Physik reduziert werden?" *Chemie in unserer Zeit* 19: 109-119, 160-166.

_____ (1990), "Mathematical and Philosophical Questions in the theory of Open and Macroscopic Quantum Systems," in A. Miller (ed.), *Sixty-Two Years of Uncertainty*. New York: Plenum Press, pp. 233-257.

Quine, W. V. O. (1960), *Word and Object*. Cambridge, MA: MIT Press.

Ramsey, J. (1996), "Molecular Shape, Explanation, Reduction and Approximate Concepts," forthcoming in *Synthese*

Redhead, M. (1980), "Models in Physics," *British Journal for the Philosophy of Science* 31: 145-163.

Rohrlich, F. (1988), "Pluralistic Ontology and Theory Reduction in the PHysical Sciences," *British Journal for the Philosophy of Science* 39: 295-312.

Scerri, E. (1989), "Transition Metal Configurations and Limitations of the Orbital Approximation," *Journal of Chemical Education* 66: 481-483.

_____ (1994), "Has Quantum Chemistry Been At Least Approximately Reduced to Quantum Mechanics?" in D. Hull, M. Forbes, and R. Burian (eds), *PSA 1994, vol. 1*. East Lansing, MI: Philosophy of Science Association, pp. 160-170.

Shimony, A. (1993), *Search for a Naturalistic World View*. Cambridge: Cambridge University Press.

Sklar, L. (1989), "Ultimate Explanations: Comments on Tipler," in A. Fine and J. Leplin (eds), *PSA 1988, vol. 2*. East Lansing: Philosophy of Science Association, pp. 47-55.

Sober, E. (1984a), *The Nature of Selection: Evolutionary Theory in Philosophical Focus*. Cambridge, MA: MIT Press.

_____ (ed) (1984b), *Conceptual Issues in Evolutionary Biology*. Cambridge, MA: MIT Press.

Stamp, P. (1995), "Time, Decoherence, and 'Reversible' Measurements," in S. Savitt (ed.), *Time's Arrows Today: Recent Physical and Philosophical Work on the Direction of Time*. New York: Cambridge University Press, pp. 107-154.

Sullivan, A. S. (1976), *The Complete Plays of Gilbert and Sullivan*. New York: W. W. Norton and Co.

Sutcliffe, B. (1992), "The Chemical Bond and Molecular Structure," *Journal of Molecular Structure (Theochem)* 259: 29-58.

Weininger, S. (1984), "The Molecular Structure Conundrum: Can Classical Chemistry be Reduced to Quantum Chemistry?" *Journal of Chemical Education* 61: 939-944.

Wheeler, J. (1973), "From Relativity to Mutability," in J. Mehra (ed), *The Physicist's Conception of Nature*. Boston: D. Reidel.

Wimsatt, W. (1995), "The Ontology of Complex Systems: Levels of Organization, Perspective, and Causal Thickets" in M. Matthen and R. X. Ware (eds), *Biology and Society* (Canadian Journal of Philosophy, supplementary volume 20), pp. 207-274.

Woolley, R. (1976), "Quantum Theory and Molecular Structure," *Advances in Physics* 25: 27-52.

_____ (1978a), "Further remarks on Molecular Structure in Quantum Theory," *Chemical Physics Letters* 55: 443.

_____ (1978b), "Must a Molecule Have a Shape?" *Journal of the American Chemical Society* 100: 1073-1078.

_____ (1980), "Quantum Mechanical Aspects of the Molecular Structure Hypothesis," *Israel Journal of Chemistry* 19: 30-46.

_____ (1982), "Natural Optical Activity and the Molecular Hypothesis," *Structure and Bonding* 52: 1-35

Woolley, R. (1985), "The Molecular Structure Conundrum," *Journal of Chemical Education 62*: 1082-1084
_____ (1986), "Molecular Shapes and Molecular Structures," *Chemical Physics Letters 125*: 200-205.
_____ (1988a), "Quantum Theory and the Molecular Hypothesis," in J. Maruani (ed.), *Molecules in Physics, Chemistry and Biology, vol.. 1.* Boston: Kluwer Academic Publishers, pp. 45-89.
_____ (1988b), "Must a Molecule Have Shape?" New Scientist 120 (Oct. 22, 1988), 53-58
_____ (1991), "Quantum Chemistry beyond the Born-Oppenheimer Approximation," *Journal of Molecular Structure (Theochem) 230*: 17-46.

VAN DER WAALS INTERACTIONS FROM DENSITY FUNCTIONAL THEORIES:

The He-CO system as a case study

F.A.GIANTURCO and F.PAESANI
Department of Chemistry, University of Rome
Città Universitaria, 00185 Rome, Italy

Abstract

The weak dispersion forces and the short-range repulsion effects which are currently considered to be responsible for the overall shape and anisotropy of the potential energy surfaces that exist between simple molecules and rare gas atoms are known not to be easily amenable to direct quantum chemical calculations. In the present study we examine in detail the possibility of employing some of the most commonly used density functionals which allows us to generate very rapidly the short-range and intermediate range regions of the interaction, while relying on perturbative schemes for the long-range forces. The special heteronuclear CO molecule interacting with an helium atom is used as working example because of its recent revival in the current litterature.

1 Introduction

The last ten years or so have witnessed quite a substantial development in the study of the structural and dynamical properties of those weakly interacting complexes which go under the name of Van der Waals (VdW) gas phase complexes. For the purpose of the present analysis, we define a VdW complex as a bound species formed in the gas phase, under very low collision frequency condition (e.g. in a molecular beam) between two stable neutral monomers. In other words, we are interested in analysing the various aspects of the interaction between the monomers when no conventional 'chemical' bonds are formed between them. In this special situation, on the other hand, the quantum nature of the intermolecular forces indeed manifest itself rather dramatically and therefore a very broad variety of theoretical, computational and experimental tools can be employed to reveal its effects [1-5] and to help us understand their consequences on the structural and dynamical properties of the weak complex. We all know that a modern, molecular description of condensed phase properties requires the full characterization of pairwise interactions, as well as the detailed knowledge of many-body forces. The VdW dimers of closed-shell species, either atoms or molecules, have therefore served traditionally as an important source of information on pair interactions, whereby the primary goal has been to understand the nature of the intermolecular potential over the entire configuration space, the so called potential energy surface (PES). As the partners acquire internal structure, e.g. in the case of polyatomic molecules, the intermolecular degrees of freedom increase from those of the simple atom-atom case and additional, important feature becomes the determination of the PES relative anisotropy.

From the experimental point of view, the insights into the anisotropic forces are best gained from measurements which can sample rather directly extended regions of the PES, and on coupling them to some mathematical scheme which inverts the experimental measurements to yield the underlying, anisotropic multidimensional energy surface. Recent advances in far-IR laser spectroscopy, for instance, allowed for direct measurements of the low-frequency vibrations of Van der Waals bonds [2,3]. Such spectroscopic data are believed to probe rather extensively the well region of the PES, both below and above the barriers to internal motion and overall rotation, and for this reason such a technique is often called vibration-rotation-tunneling (VRT) spectroscopy [6,7]. At the same time, a great deal of progress has been achieved in computational approaches which directly determine the multidimensional PESs and the associated intermolecular dynamics from what is being measured in VRT spectroscopy [4,8,9].

From the fundamental point of view of its theoretical origins, the shape of any given PES, in the weak interaction situation of VdW complexes, results from the interplay of the four basic components of the total interaction energy, i.e. the electrostatic, exchange, induction and dispersion components. Each of these contributions helps to determine the effects the anisotropies of the individual components have on the entire surface one wants to evaluate. If the calculations can be carried out, then the dominant contributions at any given range of geometries can be easily identified and therefore a relationship could be established between the interaction energy and the intrinsic properties of the costituent fragments. This relation can then provide a more complete understanding of the whole interaction phenomenon [10,11].

2 An outline of interaction contributions

As mentioned before, and since the early papers on this subject [12-14] the VdW energies could be construed as made up of four fundamental building blocks which help us to better understand the full interaction: electrostatic, induction, dispersion and exchange components. The first three can be traced back to each of the monomers properties: permanent multipole moments and polarizabilities (static and dymamics). Thus, the electrostatic energy results from permanent electric multipole interactions; the induction energy arises from the interactions of permanent multipole moments of one monomer with the multipole moments induced in the other monomer; the dispersion energy originates in the mutual polarisation of the electronic charge distributions of the interacting partners, whereby the interaction betweeen instantaneous multipoles related to dynamic multipole polarisabilities causes the additional dispersion contribution [15]. On the other hand, the exchange interaction energy is a repulsive effect, a consequence of the Pauli principle which forbids the electrons of one monomer to penetrate the occupied space of the partner. On a qualitative basis, one could conceptually relate it to the tendency of the electron charge densities of the interacting partners to avoid each other. A further type of contribution is sometime invoked as a 'charge transfer' energy [16]. From a rigorous treatment such a contribution is primarily contained within the induction component, altough it is also related to exchange contributions [11]. One should also be aware of the fact that the above four components are not strictly additive and therefore any exact treatment must also reveal the further presence of coupling terms, e.g. induction-dispersion or exchange-dispersion contributions. On the other hand, the pictorial power of the above classification is such that it remains still very useful to analyze VdW potentials as resulting from the relative role played by the above contributions [17].

The ab initio, direct calculation of the PES using the 'supermolecule' approach is obviously both expensive and prone to underestimate some parts of the interaction, especially the long-range part controlled by the dispersion forces. Furthermore, as we shall further see below, even when a potential energy surface is calculated it is usually necessary to fit the calculated points to some parametrized functional form as it is not feasible to perform the ab initio calculations for all the possible geometries which would be needed in further dynamical calculations. In conclusion, the individual contributions to the full PES are still useful indicators of the surface behaviour, may be calculated more cheaply and provide a better insight into the origin of the potential. The above partition, however, is sometimes redefined in terms of potential features in different spatial region and given as:

$$V^{VdW} = V^{sr} + V^{es} + V^{ind} + V^{disp} \qquad (1)$$

The short-range contribution, V^{sr}, includes all 'chemical' forces and contains the repulsive exchange interaction. It decays exponentially with intermolecular distance. The electrostatic term, V^{es}, takes account of the interaction between permanent dipole moments of the partners and it is therefore zero when one of them is a sherical atom. The induction and dispersion terms, V^{ind} and V^{disp}, have been mentioned before and describe the region wich corresponds to net attraction between the neutral partners. It is usually the most difficult region to evaluate since it corresponds to effects where the original monomer charge distributions do not overlap.

The V^{ind} term, for instance, results from the interaction of the induced electric moments of each molecule with the permanent charge distribution of its partners [11]. Since it is caused essentially by a relaxation of charge clouds, it is always negative for molecules in their ground states and thus produces an attractive contribution to the PES.

However, it is rarely the dominant attractive term and can be often neglected in comparison with the dispersion energy contributions. The later contribution is in fact very important for VdW systems, particularly in the region of the PES minimum and of the onset of the long-range attractive forces.

The dispersion energy between a pair of spherical atoms, for instance, may be written in the following form at long-range distances R [1,15]:

$$V^{disp} = - \sum_{n=6}^{\infty} C_n R^{-n} \qquad (2)$$

if relativistic effects are neglected, the atomic interactions contain no odd-order terms before $C_{11}R^{-11}$. As one move to short-range, however, this behaviour is modified and the true dispersion energy does not diverge as $R \sim 0$. A better representation of it is therefore obtained by using a damped dispersion series:

$$V^{disp} = - \sum_{n=6}^{\infty} C_n R^{-n} \mathcal{D}_n(R) \qquad (3)$$

In anisotropic systems, therefore, the dispersion coefficients C_n and the damping functions $\mathcal{D}_n$ are also explicit functions of the relative orientations and of the internal vibrational coordinates of the monomers. The damping functions are unity at long-range but fall to zero as $R = 0$, thereby taking into account the overlap effects and the finite size of the charge distributions. They also correct for the presence of the other short-range terms. In case of using ab initio methods in the inner region of interaction, then the dispersion forces should be given by such calculations in the well region and in the intermediate range of distances while the dispersion coefficients as in eq.(2) are still needed at long range. The calculations that we will show in the following section deal mostly with the contributions that originate from the first three terms of eq.(1), while the dispersion interactions will be introduced by the expansion series of eq.(2). In that case the existing evaluations of the C_n coefficients, as we shall see, will be employed in the calculation.

3 The density functional approach

Since the early days of quantum mechanics, several attempts have been made to reduce the complicated, many-body problems to effective one-body problems. Some of the most popular approaches have shown that a many-body system can be dealt with statistically as a one-body system by relating the local electron density, $\rho(\underline{r})$, to the total average potential, $V(\underline{r})$, felt by the electron in the many-body case. These approaches produced two mean-field equations known as Hartree-Fock-Slater (HFS) and Thomas-Fermi-Dirac (TFD) equations [18,19]. It was further thought that an alternative formulation based on a density theory instead of on a wavefunction theory would avoid the full solution of the eigenvalue problem and aim instead at a global knowledge of the nature of at least the electronic ground state of the system [20].

A more rigorous mathematical foundation of a density-based theory, known as the density functional theory (DFT), was established when Hohenberg and Kohn [21] provided a basic theorem stating that the ground state properties of an inhomogeneous many-body system can be expressed as a unique functional of its particle density. In the last two decades the DFT approach has therefore emerged as a powerful tool for the analysis of a large variety of atomic and molecular systems [22-24] and it is also beginning to be applied in direct scattering problems [25,26].

As a consequence of this dramatic increase on the applications of the LDA and of the DFT methods to deal specifically with molecular problems, a great number of effects has also been aimed at examining the performance of such methods. However, it still remains rather difficult to make general statements concerning the level of accuracy and reliability achieved by them, because the magnitude of the errors in the various DFT approaches depend, as expected, on the specific exchange and correlation functionals which are being used. A good review of this area, and a general status of DFT calculations versus more conventional Hartree-Fock (HF) method, was provided in 1987 [27]. It was pointed out there that the quality of the results acutely depends on wheter, and at what stage of the calculations, some spherical averaging are invoked. Thus, the LDA approximation, for instance, generally presents a reasonable semi-quantitative picture and interprets trends of properties correctly along series of compounds. On the other hand, it can yield quantitative errors in relative energies as large as 1 eV or so [28].

When briefly outlining the standard derivation of the applied theory of a density functional, for the sake of better clarity of the discussion that will follow, we can start by reminding ourselves that the DFT total energy functional of an N-electron system can be written as [23]:

$$\begin{aligned} E[\rho] &= T_s[\rho] + J[\rho] + E_{xc}[\rho] + \int V(\underline{r})\rho(\underline{r})d\underline{r} \\ &= \sum_i^N \sum_s \int \psi_i^*(\underline{r},s)(-\frac{1}{2}\nabla^2)\psi_i(\underline{r},s)d\underline{r} + J[\rho] \\ &+ E_{xc}[\rho] + \int V(\underline{r})\rho(\underline{r})d\underline{r} \end{aligned} \quad (4)$$

where $\rho(\underline{r})$ is the spinless total electron density, T_s the sum of single-particle kinetic-energy operators, J is the coulomb interaction between electrons and E_{xc} the non classical exchange and correlation energy contribution coming from the chosen form of the functional of the total density $\rho(\underline{r})$. The external potential $V(\underline{r})$ depends on the problem at hand (e.g. for an isolated system it could represent the coulomb nuclear potential) and the $\psi_i(\underline{r},s)$ are the single-particle orbitals given by the Kohn-Sham equations, with $\underline{r}$ and s being the spatial and spin coordinates respectively [29].

The above equation leads naturally to the effective potential which appears in each of the exact, single-particle equations for each Kohn-Sham orbital

$$\tilde{h}_{eff}\psi_i = \left[-\frac{1}{2}\nabla^2 + \tilde{V}_{eff}\right]\psi_i = \epsilon_i\psi_i \quad (5)$$

where

$$\tilde{V}_{eff}(\underline{r}) = V(\underline{r}) + \int \frac{\rho(\underline{r}')}{|\underline{r}-\underline{r}'|}d\underline{r}' + \tilde{V}_{xc}(\underline{r}) \quad (6)$$

$$\rho(\underline{r}) = \sum_{i=1}^N \sum_s |\psi_i(\underline{r},s)|^2 \quad (7)$$

By solving the above nonlinear equations in an iterative form one obtains the Kohn-Sham density of eq.(7) and therefore one recovers a new expression for the total energy:

$$E_{tot} = E[\rho] = \sum_i \epsilon_i - \frac{1}{2}\int \frac{\rho(\underline{r})\rho(\underline{r}')}{|\underline{r}-\underline{r}'|}d\underline{r}d\underline{r}' + E_{xc}[\rho] - \int \bar{V}_{xc}(\underline{r})\rho(\underline{r})d\underline{r} \quad (8)$$

where

$$\sum_i^N \epsilon_i = T_s[\rho] + \int V_{eff}(\underline{r})\rho(\underline{r})d\underline{r} \quad (9)$$

The above result is indeed exact provided we know the functional form in eq.(9) and the E_{xc} functional form in eq.(8). It is toward the solution of this specific aspect that many computational and theretical efforts have been directed in recent years [30,31]. In the present study, therefore, we intend to show how succeful such treatments can be for weakly interacting, anisotropic VdW systems and we will use the $He - CO$ system as a paradigmatic, case study of the problem.

The simplest way of treating E_{xc} is to resort to what we have previously defined as the local density approximation (LDA), whereby the Slater exchange functional in local form is combined with a further electron correlation functional also based on the properties of an homogeneous electron gas [31]. The most significant problem with this type of simplification is the tendency of the LDA to overestimate the molecular binding energies (in 'conventional' molecules) sometimes by as much as 100%. As we shall see, this particular aspect of the LDA approximation becomes even more acute for such weakly bound species as those given by VdW complexes between light atoms and molecules.

A substantial breakthrough in this regard has therefore been the development of more complicated gradient corrected density functionals that are sometimes referred to as nonlocal functional forms [32-46] whereby the treatment of dynamical correlation contributions and nonlocal exchange effect is done numerically by refined quadrature schemes which can markedly simplify both the evaluation of the Kohn-Sham orbitals of eq.(5) and of the variationally optimized total energy of eq.(8)[48]. We therefore have available today several ways of evaluating gradient-corrected formulations of E_{xc} contributions in eq.(8) and recent study [32], for instance, have found that geometries, vibrational frequencies, dipole moments and other properties of small first-row molecules are in reasonable agreement with experiments. A qualitative rule often mentioned in this context is that these approaches are comparable in accuracy to Moeller-Plesset second order (MP2) theory. We shall further discuss this point below, when examining the present case study. The gradient-corrected DFT approaches have one clear advantage, however. This is that the computed heats of atomization for simple molecules are remarkable both in their accuracy and for the fairly modest basis sets required to achieve this accuracy. It seems apparent from the above studies [28,32] that the final electron density converges much more rapidly with the one-electron basis than does the correlation energy computed from Hartree-Fock (HF)-based treatments.

In conclusion, given the broad variety of DFT approaches which can be used, and have been used, to treat overall energy structures in neutral molecules and ionic clusters [50,51], it becomes of interest to extend such models to study orientational anisotropy and strength of bonding in nonchemical systems which exibit weak VdW coupling. One of the most thoroughly studied systems of this type has been the interaction between a CO molecule in its ground electronic ground state and the Helium atom, also in the ground electronic state. The following section will therefore briefly review what has been already done to determine the anisotropic VdW interaction of $He - CO$.

4   An overview of earlier studies on the He-CO interaction

It is well known that the CO molecule (or functional group) plays a rather important role in several areas of chemistry, either as a separate molecular species, as a ligand in transition metal complexes or as a carbonyl group with its ubiquitous presence in organic compounds. More specifically, the weak complex that CO in its $^1\Sigma^+$ electronic state forms with an He atom has been of great interest for interstellar physics and chemistry for a very large time [52] and therefore the determination of the full PES, and of its dependence on the vibrational state of CO, has been the subject of many experimental studies [53-55].

Correspondingly, several ab initio calculations of the full interaction, or of part of it, have been tried over the years and have employed increasingly more sophisticated methods of computation. The earliest attempt was an evaluation of the rigid rotor (RR) surface with the CO internuclear distance kept at its experimental equilibrium value and the electron-gas method being employed to take correlation forces into account [52]. This will be called the $V_{GT}(R,\gamma)$ interaction potential. Here (R,γ) are the two residual Jacobi coordinates, i.e. the He distance from the CO center-of-mass (R) and the angle that the latter forms with the r_{CO} bond distance (γ). Later ab initio calculations [56], which considered an extended Configuration Interaction (CI) expansion which however did not include the possible consequences of the Basis-Set-Superposition-Error (BSSE) [57], provided an entirely different PES. They yelded the first extended, fully ab initio evaluation of the interaction. The ensuing potential will be called henceforth the V_{TKD} PES.

The main differences between the above surfaces are located in the region of the potential minimum: the V_{TKD} interaction turns out to be more repulsive at all orientations while the strength of the interaction, i.e. the depth of the well and the attractive region, is always larger from the electron-gas calculations of the V_{GT}. If one uses the conventional multipolar expansion:

$$V(R,\gamma) = \sum_{\lambda=0}^{\lambda_{max}} V_\lambda(R) P_\lambda(\cos\gamma) \qquad (10)$$

then the V_λ coefficients with $\lambda \neq 0$, odd and even, correspond to a measure of the strength of the anisotropic coupling. They are invariably larger in the V_{GT} expansion than in the expansion of the V_{TKD} interaction [56].

Modifications of the V_{TKD} interactions were suggested by a series of calculations, classical and quantum, of the transport properties for this system and of the behaviour of its total differential cross section measured in beam experiments [58,59]. It was found there that a judicius scaling of the first three coeffcents in eq.(10) was capable of yielding rather good accord with the existing experiments without marked modifications of the original V_{TKD} calculations.

An entirely different, empirical potential surface has been recently proposed [60] by using a new set of parameters which had been obtained from the fully resolved infrared spectra of the VdW dimer [61], which had suggested the inadequacy of the well region description as provided by the earlier TKD [56] calculations. However, such a new interaction could say nothing on the quality of its description of the repulsive wall anisotropy and location, which is the main region of the PES sampled by the transport properties and the molecular beam experiments. At any rate, the empirical potential suggested above will be compared with our results and labelled $V_{XC_{fit}}$ in the following.

Spurred by this renewed interest of experiments and of the theorists on this special system, several new ab initio calculations have also been completed in recent years. Thus, the rigid rotor PES has been computed at the fourth order of the Moeller-Plesset perturbation theory and further analysed using perturbation theory for the intermolecular forces [62] (the V_{MP4} potential) while a similar approach was employed in another recent pubblication [63] where the MP fourth-order calculations were carried out and the rotovibrational energy levels of the complex were evaluated with the collocation method. Both the above calculations turned out to provide the attractive region and the well depth in good accord with that determined from the potential surface given by the infrared spectra of ref.[61], the surface which was called there the V_{333} potential. Furthermore, both the above calculations have included the effects (found to be rather small) of the BSSE correction in the region of the well.

A very recent comparison with the same infrared spectra was carried out [64] starting from a theoretical calculation which used the symmetry-adapted perturbation theory. It showed how the interaction is dominated by first-order exchange contributions and by the dispersion energy. The bound rotovibrational spectra were used to generate the infrared spectrum [60] and virial coefficients were also computed and compared with experiments. The ensuing V_{SAPT} also turned out to be in good accord with the empirical V_{333} potential obtained from the spectra of ref.[60]. An even more extensive comparison between the two different potentials that were obtained from fitting the experimental infrared spectrum, the V_{333} and the $V_{XC_{fit}}$ potential, has been recently carried out by computing second virial, binary diffusion and shear viscosity (both interaction and mixture) coefficients for both of them [65]. The results indicate a better performance of the $V_{XC_{fit}}$ potential function that will be therefore selected as the one further discussed below.

In conclusion, the above summary shows that several aspects of the $He-CO$ interaction are beginning to be confirmed from a variety of calculations and from the use of experimental data, while some discrepancies still exist, however, on specific features of its repulsive interaction. It becomes of interest, therefore, to call on the DFT approach to see how well and how reliably the various methods described in the previous Section 2 can provide in the present system a description of dispersion forces in the regions of the anisotropic well and of the onset of the repulsive interaction in the present system.

4.1 A COMPARISON OF THE LATEST PES

In order to give a more quantitative feeling to the previous discussion on computed and empirical potential energy functions, it is useful to compare the most recent of them in the region of their strongest interaction. This means that one should look at their behaviour in the region of the potential well as a function of relative orientation and in the region of the onset of the repulsive wall at the same orientational geometries.

The calculations of Figure 1 show therefore a comparison between three ab initio calculations: (1) the TKD calculations of ref.[56] using an extended CI approach; (2) the MP4 calculations of ref.[62], given by a dashed line and (3) the MP4 calculations of ref.[63] given by a dotted line.

One clearly sees there that the three repulsive walls are indeed very similar in location, shape and orientational dependence. On the other hand, the well region differs quite markedly in the sense that the more recent MP4 calculations show both more than 20% deeper wells and also a slightly stronger angular dependence.

Furthermore, if one compares the earliest ab initio potential, the TKD [56] calculation, with the more recent empirically determined potential functions, the results are again rather significant. We report them in the following Figure 2 where the TKD surface is compared with the empirically corrected TKD from an extensive multiproperty analysys, the POT11 of ref.[59], and the empirical potential from infrared spectra, the XC_{fit} of ref.[60].

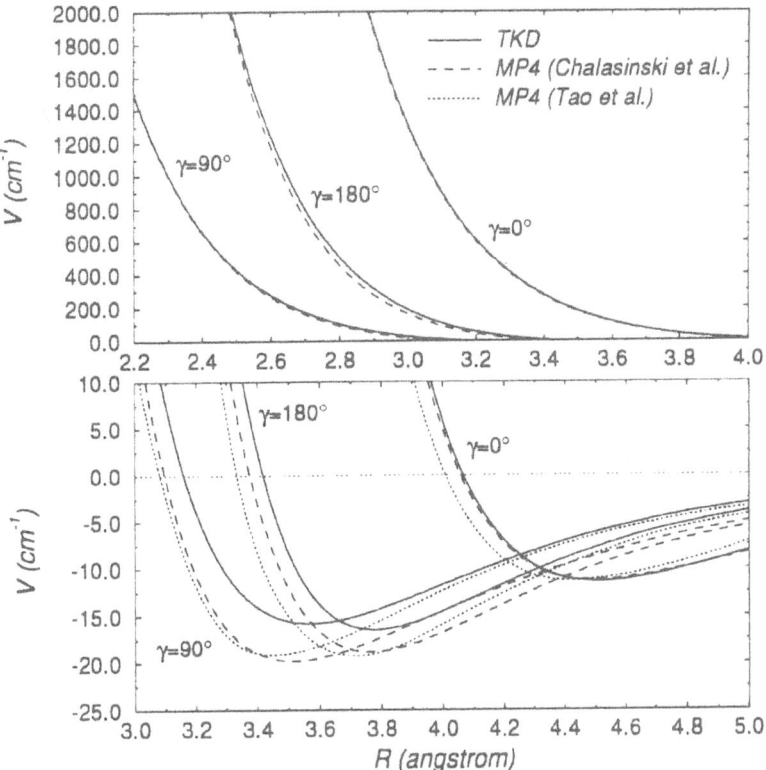

Fig.1: Comparison between recently computed ab initio PES as function of distance and for three different orientation. Top: onset of the repulsive wall; bottom: the region of the potential well. See text for the meaning of symbols.

One clearly identifies now that the two empirical potentials, altough very close to the TKD results in the repulsive part of the surface, show differences between each other and with the latter. In fact, the POT11 is the closest to TKD and only shows a stronger repulsion for $\gamma = 90°$, while the XC_{fit} indicates harder potential walls but slightly weaker wall anisotropy. In the well region, on the other hand, the XC_{fit} has markedly the deepest wells, as expected, while all three potentilas appear to show very similar well anisotropy and only small changes in their sigma values, i.e. their radial distances $\sigma(\gamma)$ where the potentials vanish. We will see below that such small differences will be of significance, however, when comparing such potentials with the best DFT calculations discussed in the next section.

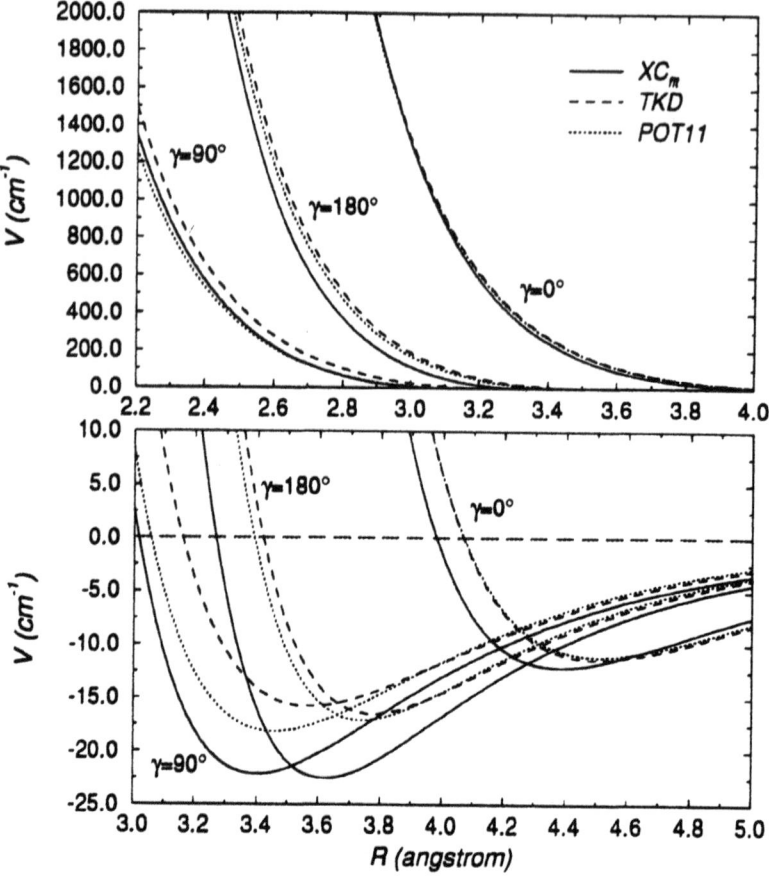

Fig.2: Comparison of computed and empirical PES in the region of the repulsive walls (top) and of the attractive well (bottom). See text for the meaning of symbols.

5 DFT calculations of the interaction

The use of density functionals for the tretments of either hydrogen bonded or Van der Waals systems has been much less widespread than the study of thermochemical data or of molecular equilibrium geometries [66,67]. However, because of the computational difficulties present in the evaluation of such weak interaction forces the avenue of selecting some reliable DFT method appears as a very tempting one and makes it interesting to explore the applicability of such theory to the study of geometry optimization, binding energy determination and of the overall shape of the extended potential energy surfaces. We know, in fact, that the computational time for DFT methods grows as the third power of the number of basis functions [23], while the HF-CI methods grow as N^4 or more. Furthermore, the inclusion of electron correlation occurs rather directly within DF methods while it is brought rather slowly when using CI methods, where necessary truncations of the CI expansion can often jeopardize the final results.

We have already found that for a special class of ionic systems, i.e. those formed with rare gases, the use of DF methods is indeed a very reliable alternative to large CI expansions and can provide us with reasonably good descriptions of, say, the Ar_n^+ trimer and tetramer potential energy surfaces [68] and also turned out to yield accurate estimates of equilibrium geometries, binding energies and relative stabilities of small helium ionic clusters [50]. It therefore becomes of interest to further extend the above analysis to the present interaction: the 'indicators' that have been chosen to evaluate the quality of the calculations are:

(i) the depth of the attractive well and its dependence on orientations;

(ii) the onset of the repulsive walls as a function of the approaching helium atom;

(iii) the relative strength of the well in the three chief configurations of $He - CO$ $\gamma = 0°$, $CO - He$ ($\gamma = 180°$) and the perpendicular approach of He ($\gamma = 90°$).

The quality of the Gaussian basis set expansion employed to obtain either the HF or the Kohn-Sham orbitals of the 'supermolecule' has been of the triple-zeta (TZ) type [69] with (6s,2p,1d) centered on the He and contracted to [3s,2p,1d]. The functions centered on the C atom were (10s,5p,2d,1f), contracted to [4s,3p,2d,1f] while on the O atom the functions were (10s,5p,2d,1f) also contracted to [4s,3p,2d,1f]. The CO equilibrium geometry was kept at its experimental value of 2.1323 a_0. The effect of BSSE corrections was not included in the calculations insofar as one is employing here a very extended basis set to d scribe the supermolecule and little bond polarization effects are caused by the weak VdW interaction f interest here. The total density employed within the various DFT modellings was therefore taken to be the sum of the two separate densities since this turns out to be a reasonable approximation for the present system where no chemical bonds are formed during the interaction.

5.1 THE EFFECTS OF DFT EXCHANGE AND CORRELATION

As mentioned before, the DF calculation of the total energy given as $E(r_{eq}, R, \gamma)$ for the interacting system requires the evaluation of the $E_{xc}[\rho]$ contribution to eq.(8).

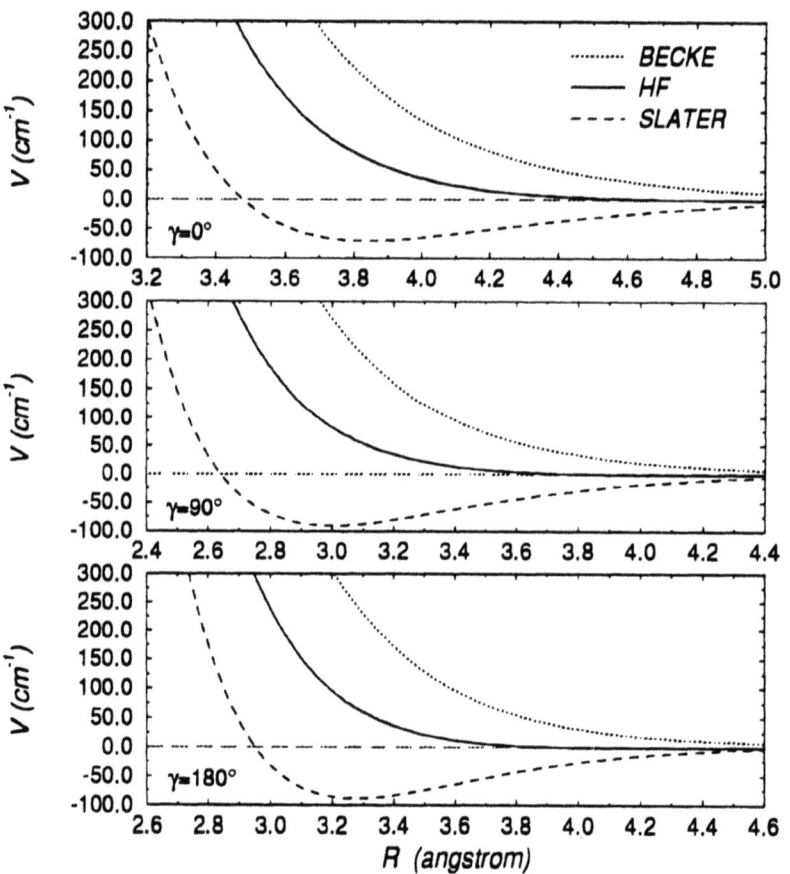

Fig.3: Approximate treatments of exchange at the Hartree-Fock (HF) level of calculation in eq.(8).

On the other hand, one may consider a simpler problem where correlation effects are initially disregarded and therefore the total energy in question is simply, when correctly evaluated, the HF energy of the system. The corresponding DF energy is therefore given by an approximate treatment of the exchange part left after disregarding correlation corrections. A comparison between various evaluations of the interaction potential:

$$V(R, r_{eq}, \gamma) = E(R, r_{eq}, \gamma) - E_{CO}(r_{eq}) - E_{He} \quad (11)$$

when correlation is not included, and where different form of exchange forces are employed, is shown in Figure 3 at the three orientations mentioned before.

The calculations labelled HF correspond to exact Hartree-Fock calculations with correct exchange contributions, while the dashed line indicates calculations using the Slater local exchange approximation [18]. At the same time, the dotted line labelled BECKE corresponds to DFT calculations where the exchange in an LDA scheme was improved with a gradient correction [33]. As expected, the HF calculations always give repulsive potentials since they do not include any dispersion contribution, the latter being the chief cause for the VdW binding in question. The approximate Becke exchange also confirms this result and also gives rather good value for the energy of the isolated partners, as shown in Table 1.

Table 1: Calculated fragment energies (a.u.) and geometry (Å) with the methods of Figure 3.

	E_{He}	E_{CO}	r_{CO}
HF	-2.86115334478	-112.781812561	1.1045
$SLATER$	-2.72271177735	-111.517519633	1.1405
$BECKE$	-2.86247049492	-112.859114798	1.1490

On the other hand, the SLATER approximation incorrectly yields a rather marked minimum without dispersion forces while the BECKE correction yields repulsive regions which are still too strong. One could therefore try to improve on the static-exchange treatments shown before by adding the various corrections which can be obtained through the DFT approach briefly outlined in Section 3. Thus, we report in Figures 4 and 5 the results from the same treatments of the static-exchange interaction discussed before but adding to each of them two different forms of correlation energy ΔE_c.

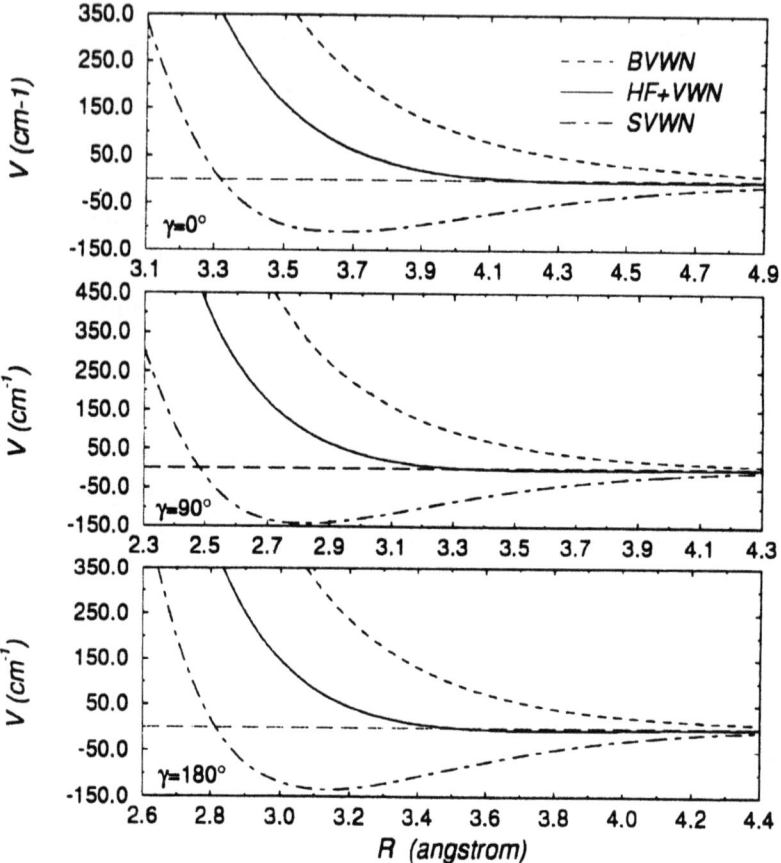

Fig.4: Calculations including correlation energy corrections from the VWN formula. See main text for details.

The correlation correction labelled VWN comes from the treatment described in ref.[47], while the one labelled VWN5 is a further improvement which followed the numerical results of Ceperley and Alder [72]. In the earlier formulation [47] the authors followed the LDA approximation to derive an expression for the correlation energy of an electron gas within the random phase approximation (RPA), while the improved version VWN5 uses a Padè approximant to interpolate the quantum Montecarlo results of ref.[70].

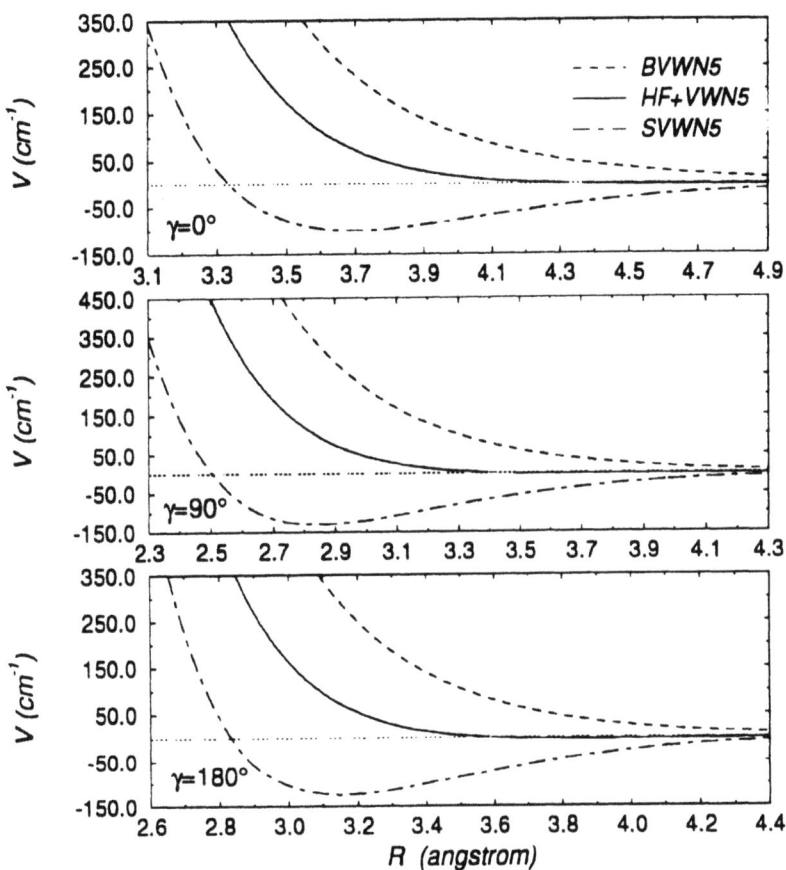

Fig.5: Calculations including correlation energy corrections from the VWN5 formula. See main text for details.

One clearly sees that both correlation corrections improved the results obtained within the HF scheme, which is now providing a small attractive well around 4.0 Å in the two orientations with $\gamma = \pi/2$ and $\gamma = \pi$ (CO-He geometry). On the other hand, the Slater approximation still yields very deep wells at both orientations while the Becke gradient correction remains repulsive at all geometries.

A rather succeful empirical formula which includes both exchange and correlation effects via the use of a DFT approach has been introduced a while ago also by Becke [40], who wrote down the E_{xc} contribution as:

$$E_{xc} = E_{xc}^{LSDA} + a_0(E_x^{HF} - E_x^{LSDA}) + a_x \Delta E_x^{B88} + a_c \Delta E_c \qquad (12)$$

where the LDA scheme is the one mentioned before [18] and the ΔE_x^{B88} is the gradient corrected exchange formula also used before [33].

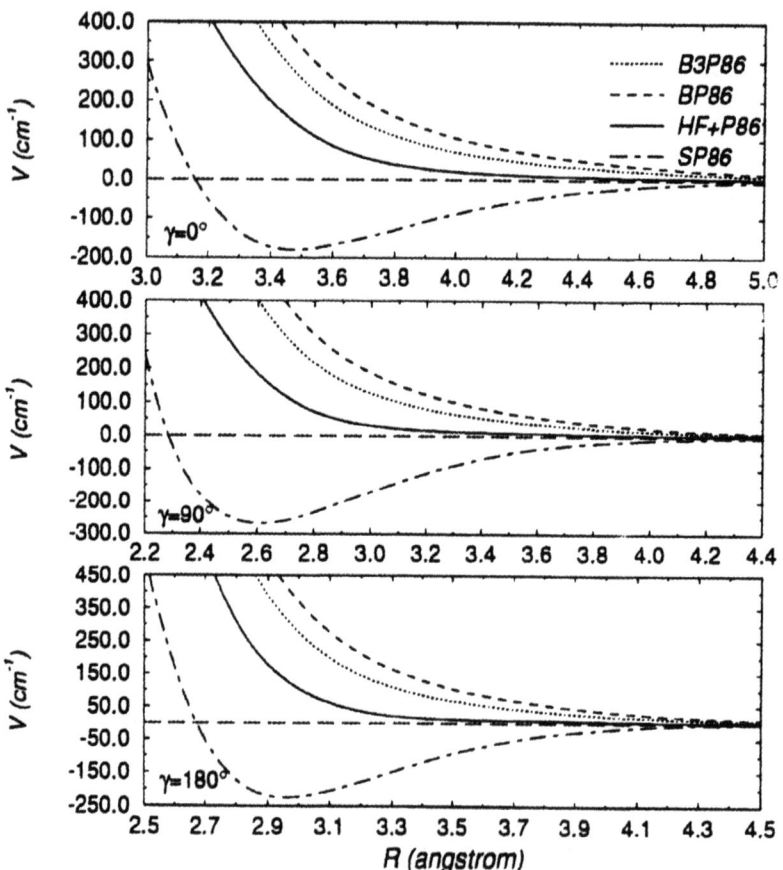

Fig.6: Calculations including correlation energy corrections from the P86 functional. See the main text for details.

The last term on the r.h.s. of the eq.(12) corresponds to different choices for the correlation correction and can be given by: (1) the generalized gradient approximation introduced by Perdew [42,43], who expanded in a Taylor series the functional dependence on the radius enclosing the unit charge density. We will call this exchange correction the P86 correction; (2) the correction to the above formula as introduced later by Perdew and coworkers [44] and which will be called here the PW91 approximation; (3) the gradient corrected correlation as discussed before and in the work of Parr and coworkers [45]. It will be called here the LYP correlation correction.

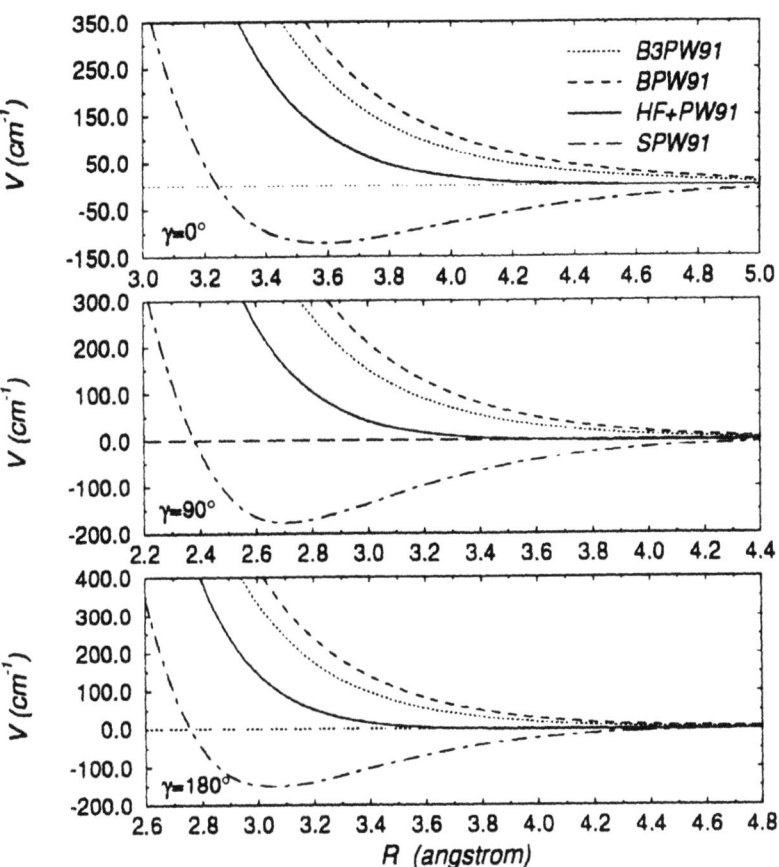

Fig.7: Calculations including correlation energy corrections from the PW91 functional. See the main text for details.

We show in Figures 6 and 7 the changes in the behaviour of the potential curves as obtained from the above approximations and compared with the earlier static-exchange methods described in Figures 3,4 and 5. Each set of curves therefore shows the exact HF treatment, the LDA exchange from the Slater approach, the gradient corrected exchange treatment of Becke [33] and the three-parameter formula of eq.(12) with the values:

$$a_0 = 0.20, \quad a_x = 0.72, \quad a_c = 0.81$$

suggested by their original derivation. The exchange corrections are included using the P86 and the PW91 formula discussed before.

Table 2: Computed total energies (a.u.) and CO equilibrium geometries (Å) using the methods labelled by the various acronyms discussed in the main text.

	E_{He}	E_{CO}	r_{CO}
SVWN	-2.87143716971	-112.739928466	1.1282
SVWN5	-2.83407879697	-112.465220126	1.1291
SPW91	-2.76762354797	-112.002163825	1.1293
SP86	-2.76591794214	-112.011973286	1.1302
SLYP	-2.76619101534	-112.001997905	1.1298
BVWN	-3.01236548982	-114.083125932	1.1362
BVWN5	-2.97490467603	-113.808296923	1.1372
BPW91	-2.90755290527	-113.343983564	1.1373
BP86	-2.90559119697	-113.353783319	1.1381
BLYP	-2.90621768863	-113.343750136	1.1379
B3PW91	-2.90856266736	-113.305086929	1.1260
B3P86	-2.94496220188	-113.593662316	1.1253
B3LYP	-2.91450655369	-113.357254350	1.1262

One sees clearly the following:

(i) the Becke exchange approximation and the B3 interpolation formula behave rather similarly. However the B3 results are, correctly, providing a 'softer' repulsive potential;

(ii) the HF calculations remain always repulsive when corrected with the P86 correlation and yield only a very small well, at too large a distance, when the PW91 correlation functional is employed;

(iii) the Slater, LDA exchange scheme still produces potential wells which are much too deep and located too close to the molecule. This exchange scheme therefore always appears to produce unphysical results.

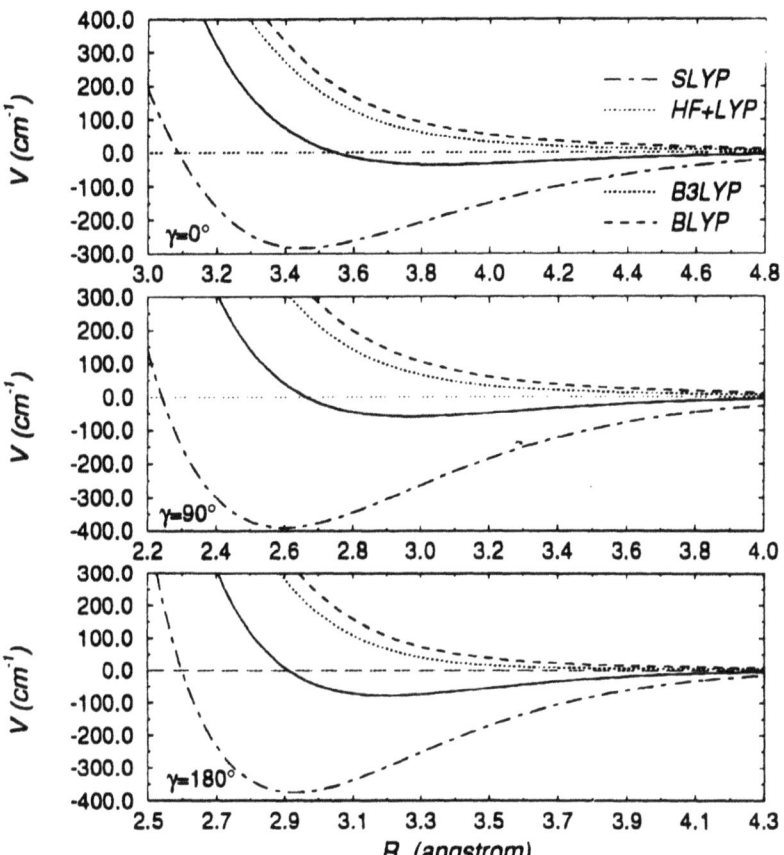

Fig.8: Computed interaction using a nonlocal gradient correction correlation functional [45]. See text for the meaning of symbols.

In Figure 8 we finally show the same comparison but using the gradient corrected density functional labelled as LYP before [45].

The major difference that one notices here is that the nonlocal correlation term now produces marked potential wells when the HF scheme is employed: as we shall see, however, they are much deeper than those suggested by the empirical potentials from experiments since they are of the order of 100 cm^{-1}, to be compared with the 20 to 25 cm^{-1} indicated by experiments [60,61].

In conclusion, the above comparisons strongly indicate that the use of correct exchange becomes an essential ingredient for evaluating the repulsive part of the weak VdW interactions and the DFT approach, given its easier computational implementation, can help with the additional, very important, inclusion of correlation effects. As a further set of numerical comparisons we report in Table 2 the computed total energies and geometries schemes discussed before.

It is clear from the foregoing discussion that the type of computational scheme most likely to succeed is the inclusion of DFT correlation corrections after an exact treatment of the exchange interaction. This approach will therefore be presented in the following Section.

5.2 THE POST-HARTREE-FOCK TREATMENTS

As we shall further analyse before, it is of interest to directly compare the effects of the five DFT schemes for local and non-local correlation corrections once they are used to evaluate the repulsive region and the region of the potential well in the interaction discussed in this work.

Their best use, from the previous results appears to be the one where the static-exchange mean-field interaction is already modelled at the Hartree-Fock level. The results shown in Figure 9 report this analysis. One can clearly see the following behaviour:

> (i) the use of gradient-corrected, nonlocal correlation functionals like the LYP [45] invariably produces well depths that are too large (about a factor 2 or more) and which are located too close in with respect to those suggested experimentally [60];

(ii) the alternative nonlocal form of correlation employed by the P86 scheme [42] generally does not provide the attractive behaviour required by this interaction and yields no well region;

(iii) the further corrected PW91 density functional [44] behaves very similarly to the P86 and still yields the wrong well behaviour;

(iv) the best results appear to originate instead by using the local forms of correlation corrections, where the treatment of the 'molecular' electrons as an homogeneous electron gas seems to approximate most closely the behaviour of the dispersion interaction in a VdW bonding situation. Thus, both the VWN and VWN5 schemes [47] provide the well position and depth closer to what is suggested by the experiments.

A further possibility also come from a parameter-controlled mixture of the use of Kohn-Sham, non interacting orbitals and of the fully interacting system as given by a DFT modelling of the interaction [41]. This scheme goes under the name of Becke (Half-Half) approach since one writes:

$$E_{xc} \approx \frac{1}{2}U_{xc}^0 + \frac{1}{2}U_{xc}^1 \qquad (13)$$

where the U_{xc}^0 term corresponds to the Kohn-Sham contribution and the U_{xc}^1 indicates the correct energy of real system as given by various possible DFT functionals. The expression actually used in our calculation, given the success of LDA methods for correlation corrections, was therefore the following:

$$E_{xc} = 0.5 E_x^{HF} + 0.5 E_{xc}^{SVWN5} \qquad (14)$$

with the meaning of the acronyms already discussed before.

The full calculations of the $He - CO$ rigid-rotor interaction were therefore carried out by fixing the r_{CO} distance as 1.1282 Å and by evaluating ten different orientations between $\gamma = 0°$ and $\gamma = 180°$ with $\Delta\gamma = 20°$. Here again the $\gamma = 0°$ geometry was chosen to the collinear $He - CO$ system.

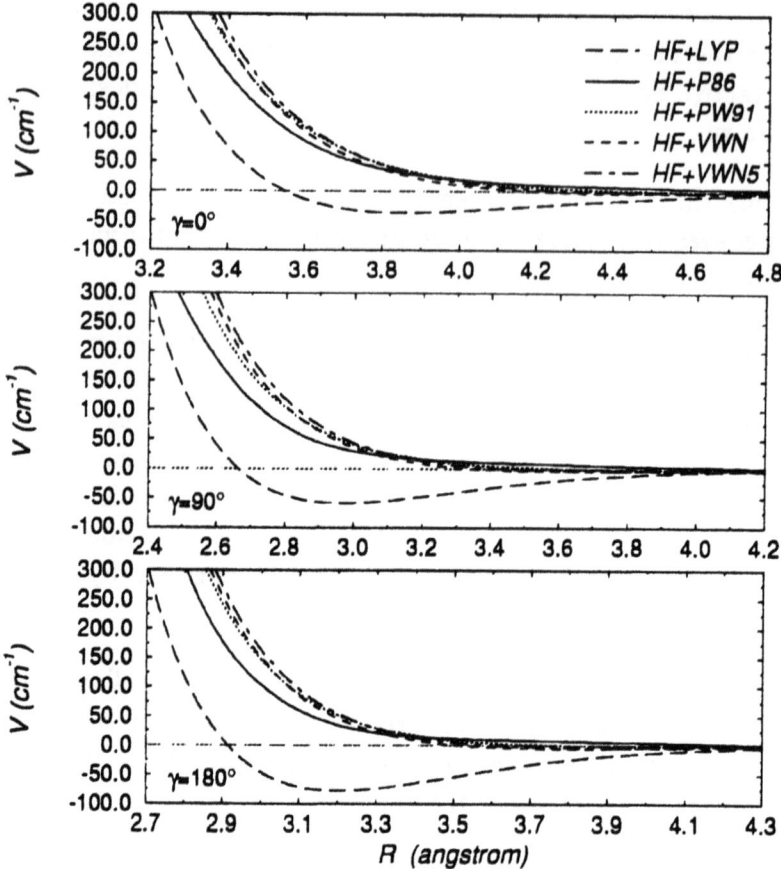

Fig.9: Computed interaction at the Hartree-Fock (HF) level and many different forms of DFT correlation. See text for the meaning of symbols.

We then carried out the post-HF calculations by using three different local correlation schemes: the VWN, the VWN5 and the HALF-HALF suggested by Becke [41] and implemented via eq.(14).

The results presented in Figure 10 show the behaviour of the multipolar coefficients:

$$V(R,\gamma) = \sum_{\lambda=0}^{\lambda_{max}} V_\lambda(R) P_\lambda(\cos\gamma) \qquad (15)$$

for the first four terms of the above expansion. Given each of the 10 angles, 17 values of R have been considered and then the long-range dispersion was included via:

$$V_{LR}(R,\gamma) = -\frac{C_6(\gamma)}{R^6} - \frac{C_7(\gamma)}{R^7} - \frac{C_8(\gamma)}{R^8} \qquad (16)$$

where the coefficients were taken from the experimental data as used by Parker and Pack [71]. At each angle they were then smoothly connected with the present calculations. The differences between the three computed PES are seen via their lower four multipolar coefficients shown in Figure 10.

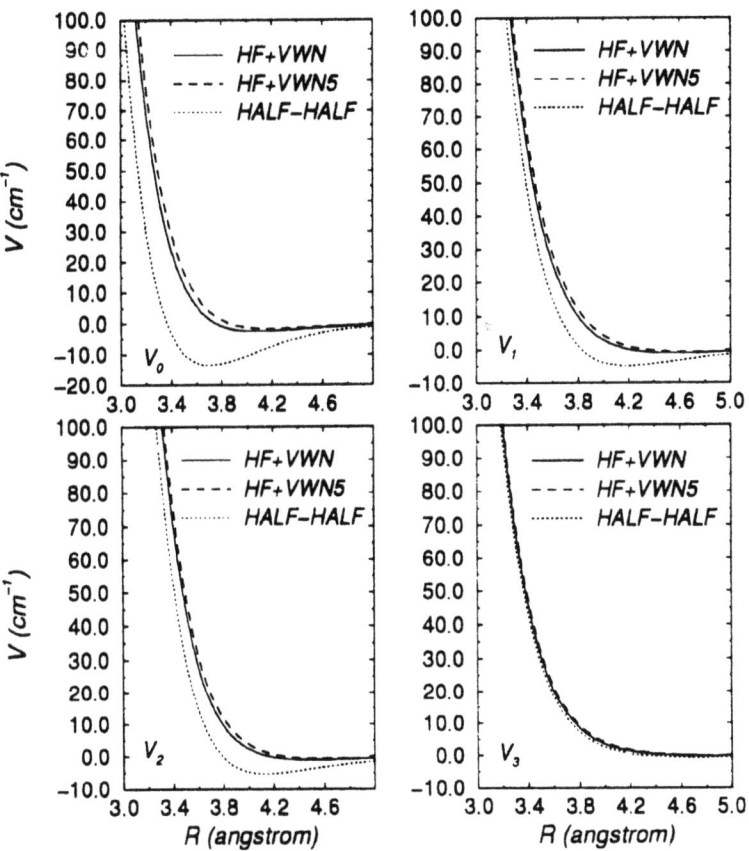

Fig.10: Comparison of computed multipolar coefficients using the post-HF treatment of the DFT approaches discussed in the main text. See there for the meaning of symbols.

One clearly sees once more that the Half-Half approach is now producing the deeper wells, very close to those suggested by the experiments [60]. The other two local treatments, on the other hand, are very close to each other and yield repulsive walls which are 'harder' than the Half-Half for V_0, V_1 and V_2 but more close to it as λ increases: the V_4 coefficients, in fact, are nearly all coincident with each other.

5.3 COMPARING DFT AND AB INITIO TKD RESULTS

As mentioned before, the ab initio mean-field, SCF-CI calculations of ref.[56] constitute the earliest attempt at evaluating the full Rigid-Rotor PES for the present system. Furthermore, the additional testing

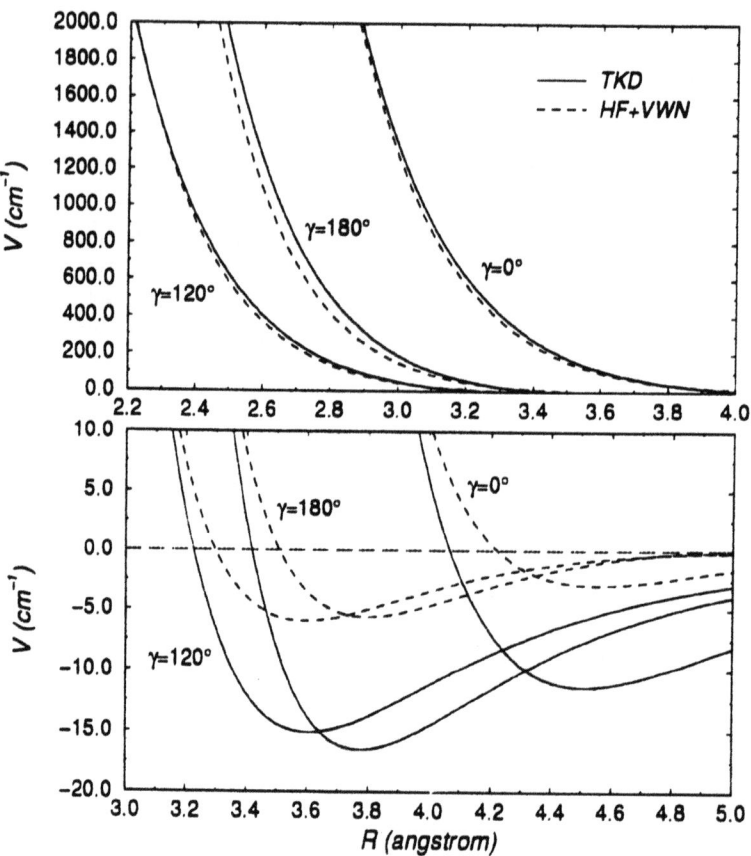

Fig.11: Comparison between SCF-CI calculations of the interaction (TKD) and the post-HF DFT corrections (HF+VWN). See text for the meaning of symbols.

using scattering data [55] and a multiproperty analysis that includes transport properties [59,60] showed that only fairly minor alterations were needed of the ab initio results in order to obtain a good description of those properties. As a consequence, we therefore felt that it still made sense to directly compare the TKD results with the present, post-HF calculations with DFT correlation corrections and with long-range dispersion coefficients.

The results of Figure 11 make the comparison with the correlation calculations which used the LDA treatment via the VWN formula [47]. One clearly sees that both PES are very close to each other in the repulsive region. Furthermore, we report the orientation with $\gamma = 120°$, which is very close to the minimum geometry of the bound system, to show how closely the two PES follow each other along that special 'cut' of the repulsive wall.

On the other hand, the region of the attractive wells (lower part of Figure 11) shows that the ab initio results produce well depths that are more than a factor of three larger than those given by the simpler DFT treatment from the VWN modelling. The situation remains essentially unchanged when the corrected DFT, the VWN5 [47] model is employed and therefore we are not showing here these results. Obviously the treatment of dispersion forces simply at the RPA level for an LDA model with an electron gas is not sufficient to describe the delicate balance of forces which give rise to the behaviour of the VdW binding in neutral systems. On the other hand, when the previously described Half-Half empirical approach is used, one finds the results reported in Figure 12.

One clearly sees that now the repulsive walls have changed with respect the TKD results and correspond to a repulsive potential which is softer for the collision energies up to 800-1,000 cm^{-1}, altough maintaining a repulsive anisotropy very similar to the TKD potential. In the region of the attractive well, as expected from the previous results, the hybrid approach provides now well depths that are larger than those given by the TKD: the combined use of an LDA scheme for exchange and correlation, but mitigated by the correct HF mean-field contribution, is producing well depths which are 20% larger but are still located fairly close to those predicted by the TKD results. Furthermore, the DFT values of $\sigma(\gamma)$ are invariably about 10% smaller than those given by the SCF-CI calculations.

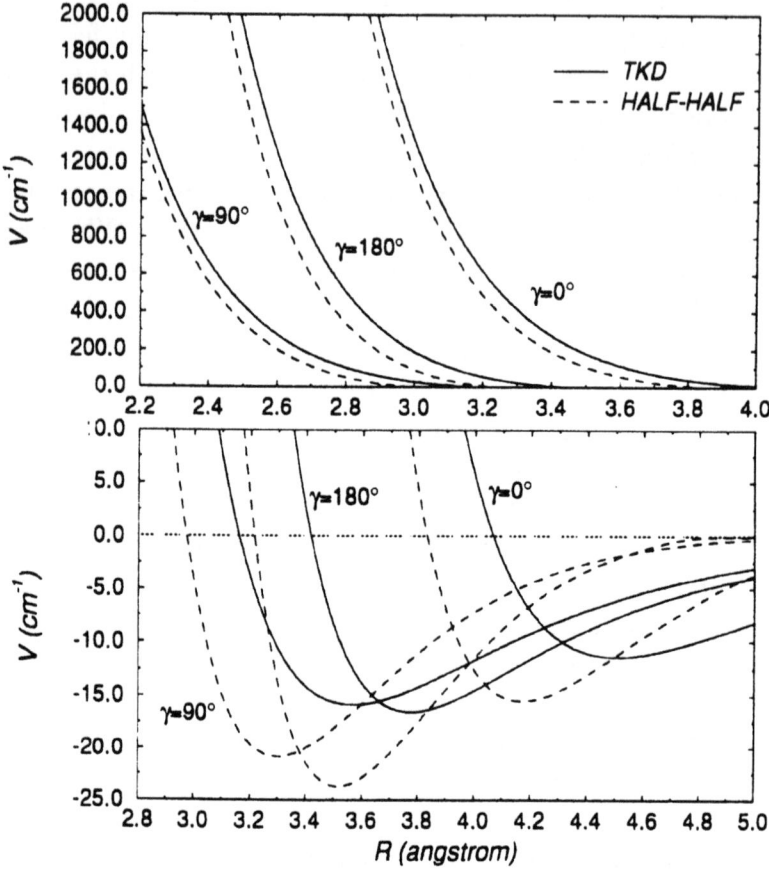

Fig.12: Comparison between ab initio computed potential curves at three different orientations (TKD) and the Half-Half calculations using DFT correlation [41].

In conclusions, we can say that using DFT methods reduces the computational times by about two orders of magnitude with respect to extended CI expansions but still allows us to obtain PES that are reasonably close to the ones produced from ab initio methods. The quite accurate description of the repulsive region, and of its orientational anisotropy, by DFT methods is very good indicator for its possible application for computing collision dependent properties, as the latter largely depend on the repulsive part of the potential. On the other hand, the well region turns out to be well described as far as its anisotropy is concerned, while the DFT models produce well depths which differ markedly from the earlier ab initio results. However, since

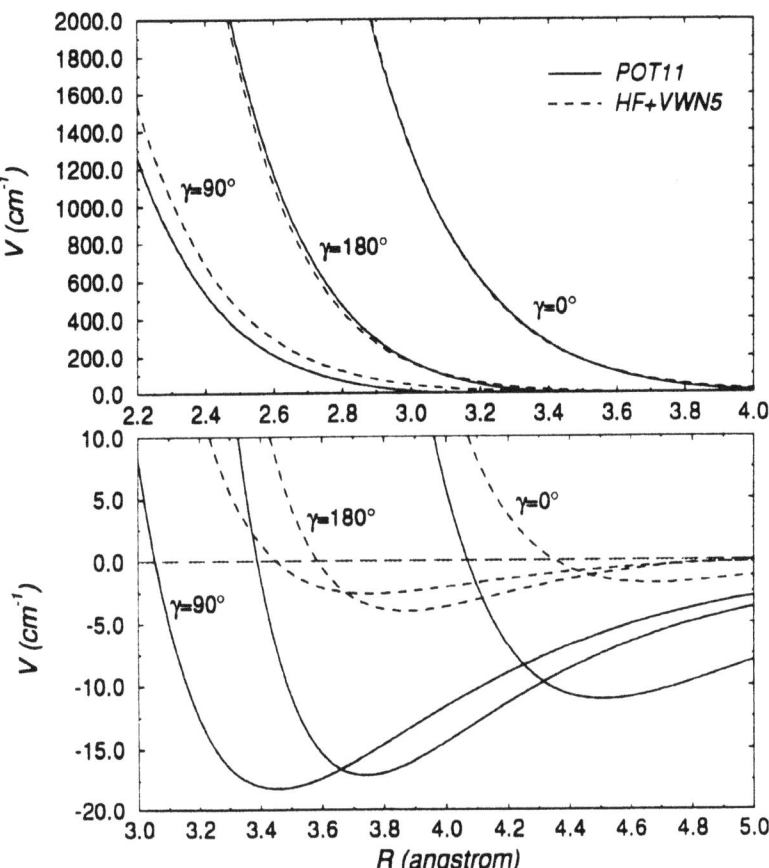

Fig.13: Comparison between computed post-HF potential surfaces (HF+VWN5) and the multiproperty best fit from transport coefficients (POT11) of ref.[59].

the TKD data turned out, as we saw in Section 4, to be unsufficiently converged in terms of both basis set and CI expansions [62,62], and to provide a well depth that is possibly too small, then we still need to further clarify the performance quality of the present DFT modelling of the $He - CO$ interaction by comparing its results with other, available empirical or ab initio PES.

5.4 DFT RESULTS VERSUS A MULTIPROPERTY POTENTIAL

The results shown in Figure 13 consider now the otimized potential that we obtained from several transport property calculations (POT11) [59] and the post-HF calculation with the corrected LDA treatment of correlation [47]. The angular anisotropy of the fitted po-

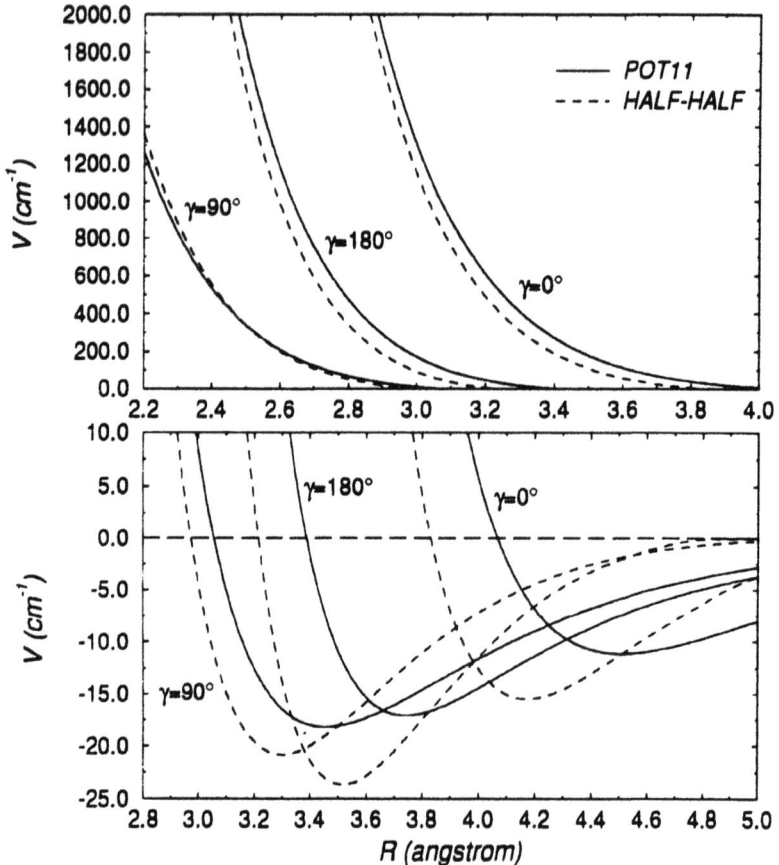

Fig:14. Comparison between computed post-HF potential surfaces (HF+VWN5) and the multiproperty best fit from transport coefficients (POT11) of ref.[59]

tential is larger than the DFT results and therefore a small discrepancy appears at $\gamma = 90°$.

Furthermore, the well depths of the empirical fit are still stronger than the DFT calculations and exhibit even larger differences in the $\sigma(\gamma)$ behaviour, the latter being in general smaller for POT11 than it was in TKD.

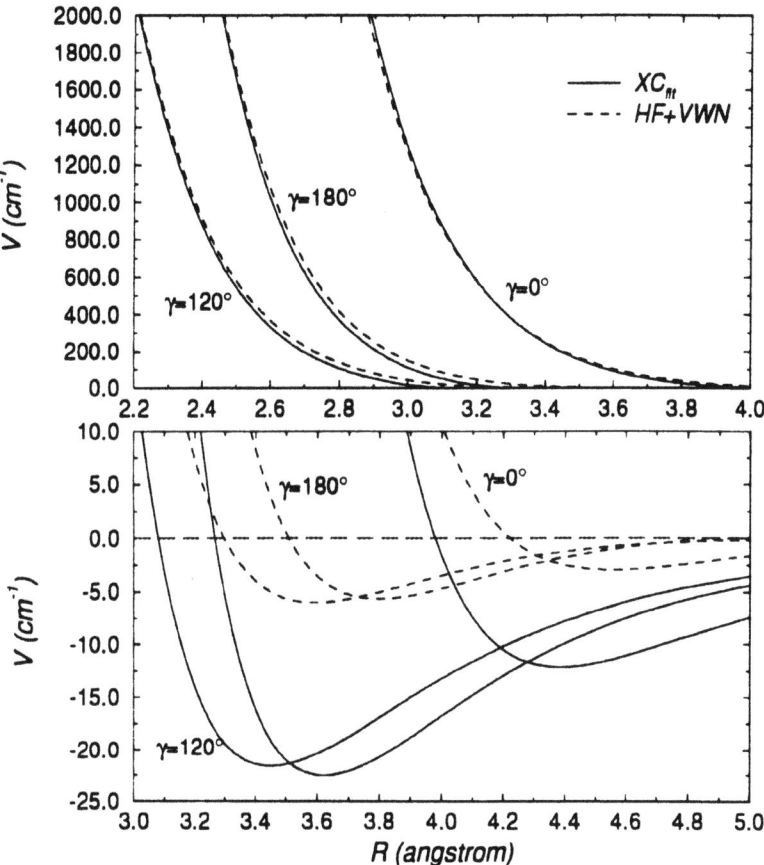

Fig.15: A comparison between the computed PES using the post-HF DFT treatment of correlation (HF+VWN) and the potential given by fitting IR spectra (XC_{fit}) [65]

The further comparison of the multiproperty potential with the Half-Half DFT calculations from eq.(14) is shown in Figure 14. We see clearly there that the repulsive region does not change much, both in location and in anisotropic character, with respect to that from the VWN5 calculations and also follows closely the one proposed by the

POT11 empirical fitting. On the other hand, the well region is very much different now since the well depth from the DFT calculations are larger than those suggested by the fitting potential POT11.

Furthermore, the $\sigma(\gamma)$ values from the dashed curves in the figure clearly show that the DFT potential is now softer than the same potential which had no hybrid corrections and that was shown in the previous figure. This means that a judicious use of both HF meanfield and more global DFT corrections can provide a realistic picture for these weak VdW interactions.

5.5 DFT RESULTS VERSUS A SPECTROSCOPIC POTENTIAL

As we mentioned before, the V_{333} potential energy surface was obtained chiefly by fitting the computed bound states given by the experimental far infrared spectrum of the bound Van der Waals complex [61]. In a further analysis, an empirical modelling of long-range and short-range, exchange-coulomb forces was employed by fitting its parameters to reproduce the same infrared data of the complex [60]. The ensuing XC_{fit} potential energy surface was further employed in a multiproperty study [65] which suggested that the latter PES may well be the best available description for the $He - CO$ interaction.

It is therefore of interest to analyse this new PES and to see how well it agees with the DFT calculations, at least with those done at the post-HF level or which use the hybrid approach of eq.(14).

A comparison with DFT calculations, obtained using the VWN modelling of correlation forces [47] in addition to the HF interaction, is shown in Figure 15.

We report in that figure also the angular orientation ($\gamma = 120°$) that corresponds to the region of the absolute minimum energy structure in order to show how well this delicate part of the interaction is being described by DFT models. The comparison clearly tells us the following:

> (i) the repulsive wall is given by the XC_{fit} surface chiefly by extrapolating the shape and position of the well adjusted to the spectra. We see that the DFT results reproduce it remarkably well and down to very low energies;

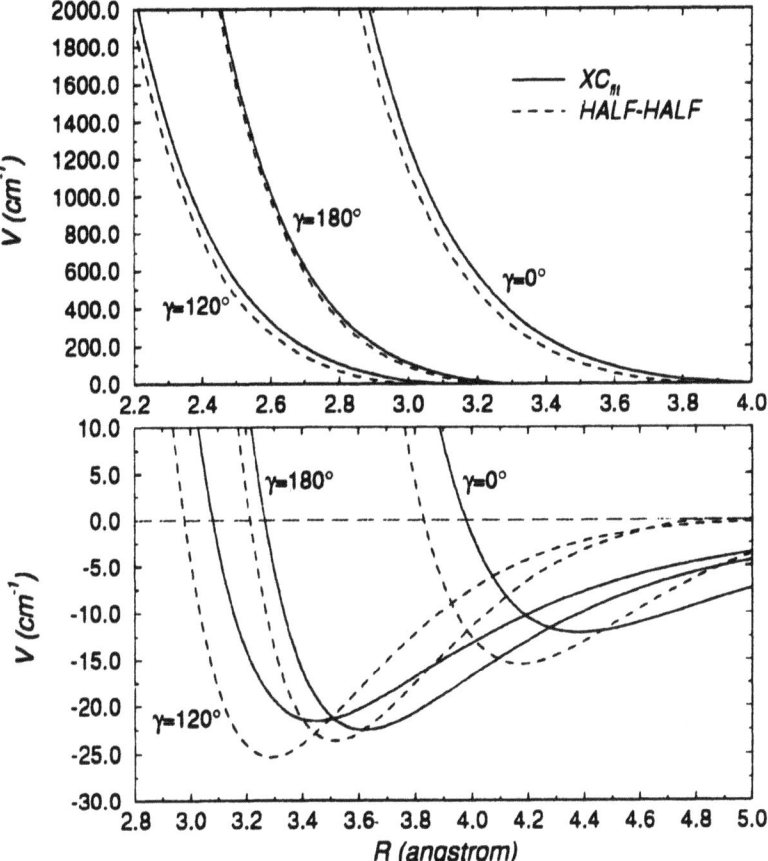

Fig.16: A comparison between the computed post-HF potential curves with the Half-Half hybrid approach [41] and the empirical, fitted potential from infrared spectra [65].

(ii) the $\sigma(\gamma)$ values are smaller than those given by DFT calculations, in agreement with the POT11 findings, but their γ-dependence seems to be well described by the DFT results. This is an important aspect for using the PES in very low energy inelastic dynamics;

(iii) the well positions are also very close between the two treatments, showing that the $R_m(\gamma)$ are nearly the same at all orientations;

(iv) the main difference occurs in the values of the well depths: the ones obtained from the XC_{fit} potential are about a factor of five larger than those given by the DFT calculations.

As one moves to the hybrid treatment that we have mentioned before, it is interesting to see how markedly the modified exchange interaction changes the overall picture. Figure 16 reports, in fact, both PES as obtained from the Half-Half hybrid treatment [41] and the XC_{fit} potential described above.

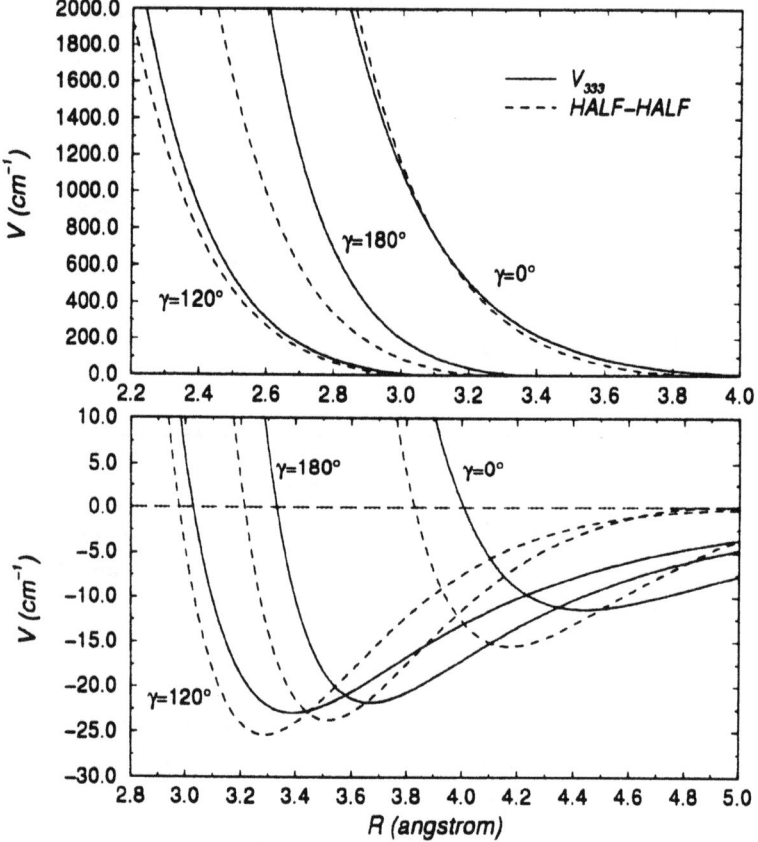

Fig.17: A comparison between the purely empirical potential of ref.[61], V_{333}, and the present DFT calculations. See text for the meaning of symbols.

Here again the repulsive region of the potential is remarkably well described and the hybrid approach, in spite of its partly approximate exchange contributions, also yields the anisotropy of the repulsive wall reasonably well. Furthermore, the well region is now markedly improved since the DFT calculations give each well depth only within a few cm^{-1} from that given by the XC$_{fit}$ empirical potential. Considering that the latter chiefly constructs its fits by using data sensitive to the shape and depth of the anisotropic well, one sees that the hybrid DFT method is doing very well in reproducing as best as possible such small quantities. It is also of interest to compare the present hybrid correction to HF calculation with the pure empirical potential that was produced by IR spectra fitting only, i.e. the V_{333} surface of ref.[61]. We report therefore in Figure 17 such a comparison, done as in the previous cases. One clearly sees that the lack of information on the repulsive region of the interaction that exists in the V_{333} fitting prevents it from reliably describing that part of the surface and therefore makes it to be rather different from the DFT results. However, it has been shown by the present analysis that the repulsive region is invariably well described, both in shape and location, by the present DFT calculations.

6 Summary and conclusions

In the present study we have endeavoured to analyse, as carefully as possible, the relative merits and performance capabilities of several different modellings of exchange and correlation forces using a density functional approach. In particular, we have looked into the possibility that such methods may have to treat several, important, regions of the full interactions between chemical systems which only form weak, Van der Waals bonds with each other. The case study which we have chosen therefore involved the Helium atom interacting with the ground electronic state of the CO molecule, a rather typical example of a simple VdW system. The interaction between these partners, in fact, is a very weak interaction that, at its strongest, is never more than about 25 cm^{-1} but shows a rather marked orientational dependence due to the polar nature of the CO target and to the different polarisabilities (static and dynamic) of the Carbon and Oxygen-atom sides of the molecule. This means that any ab initio calculation needs to balance very carefully, and as a function of both relative distance and relative orientation, the repulsive exchange effects in the regions of charge overlap and the attractive effect due to dispersion forces and induction forces (see Section 2 for further details). For these reasons, then, the ab initio calculations are indeed computationally intensive and require the use of extensive, and slowly converging, Configuration Interaction schemes. The possibility of using instead a global functional form for treating Exchange-Correlation forces becomes therefore rather attractive if it implies a substantial reduction of computational effort. The present case study has thus analysed in detail which of the most currently used density functionals, both those at the local spin density approximation (LDA) level and those employing gradient-corrected functionals, could be used profitably and reliably to yield a realistic description of the strength and the orientational anisotropy of the weak VdW forces. The preceding discussion has highlighted the following aspects of the various DFT approaches:

(i) the use of exact exchange forces plays a crucial role in the sense that the LDA approximations of E_x provided shapes of the potential angular cuts which had the wrong location of the onsets of the repulsive wall, often presented well depth much too large and which were located too close to the target molecule, and finally showed the wrong anisotropy of the attractive regions;

(ii) the best approach is therefore to treat VdW interactions at the post-Hartree-Fock (post-HF) level, thereby adding the correlation corrections using an extensive array of DFT formulations;

(iii) within this context, the RPA, linear response treatment of VWN [47] turns out to provide a very good description of the repulsive regions of interactions. In other words, the shape, location and angular anisotropy of the PES from 50 cm^{-1} up to more than 2000 cm^{-1} is given in very good agreement with the best calculations and with the best empirical potentials produced from fits to several experimental properties of the system. Even the use of an hybrid approach, the Half-Half model of Becke [41], turns out to yield a reasonably good description of that region of the interaction;

(iv) the region of the well, on the other hand, is given by DFT calculations with greater difficulties: the simpler functionals (VWN, VWN5, P86, PW91) usually produce wells positions reasonably close to those suggested by the empirical potentials but the corresponding depths are much too small: usually no deeper than about 5 cm^{-1}, to be compared with the best estimates of about 20 cm^{-1};

(v) the hybrid approach, on the other hand, exhibits a better accord for the latter quantity in the sense that it produces well depths of the order of 25 cm^{-1} for the largest interactions, it gives the correct orientational dependence of the well depth and also provides $\sigma(\gamma)$ values which are the closest to experimental estimates.

Considering the very small values of the potential wells one can therefore say that the DFT calculations provide a very reasonable estimate of such quantities at great savings of computational time: each point, in fact, can be obtained within a time scale which is at least two orders of magnitude faster than a typical CI calculation [62]. Furthermore, any use of the interaction potentials to obtain properties which mainly sample the repulsive region of that interaction is expected to yield quantitative agreement since the present results show that region to be of the quality of the best empirical PES (see Section 5). On the other hand, properties that strongly depend on the shape and location of the well region (e.g. the bound states of the infrared spectrum of the complex [61]) may only be given as qualitatively correct when are computed with the majority of the DFT used to treat correlation forces, while the situation should be markedly better when the hybrid approach [41] is employed.

In conclusion, we feel that to explore the capabilities of DFT approaches when treating interactions chiefly controlled by dynamical correlation effects is certainly a useful exercise, and something which has not yet been done systematically across the range of the best known VdW systems. Furthermore, considering the fact that the attractiveness of any DFT approach increases with the number of the electrons involved in these interactions, we think that it may be very useful to test the most successful of the functionals discussed in the preceding sections on those VdW complexes for which calculations with ab initio methods are still out of range. Thus systems like $Kr - CO$, $Xe - CO$ and the like should provide among the best applications of the previous methods.

7 Acknowledgements

The financial support of the Italian National Research Council (CNR) and the Italian Ministery for University and Research (MURST) is gratefully acknowledged. We are also grateful to Nico Sanna, CASPUR, for his generous help when running GAUSSIAN-DFT.
Finally we warmly thank Professor R.J. Le Roy for sending us the routines for the V_{xc} and V_{333} potential functions and Professor Chalasinski for sending us further information on their calculations.

References

[1] Maitland,G.C., Rigby,M., Smith,E.B., and Wakeham,W.A. (1981) *Intermolecular forces*, Clarendon Press, Oxford.

[2] Weber,A.(ed.) (1987) *Structure and Dynamics of Weakly Bound Molecular Complexes*, Reidel, Dordrecht.

[3] Halberstadt,N. and Janda,K.C. (eds.) (1990) *Dynamics of Polyatomic Van der Waals complexes*, Plenum, New York

[4] Hutson, J.M. (1990) Van der Waals Molecules and intermolecular forces, *Annu. Rev. Phys. Chem* 41, 123-145.

[5] Pullman,B. (ed.) (1981) *Intermolecular Forces*, Reidel, Dordrecht.

[6] Cohen,R.C. and Saykally,R.J. (1991) High resolution spectoscopy of Van der Waals complexes, *Annu. Rev. Phys. Chem.* **42**, 369-387.

[7] Saykally,R.J. and Blake,G.A. (1993) Spectroscopy of weak molecular complexes, *Science* **259**, 1570-1576.

[8] Miller,R.E. (1991) Spectroscopy of molecular dimers, *Adv. Mol. Vib. Coll. Dyn.* **1**, 83-106.

[9] Hutson,J.M. (1991) Intermolecular forces from the far-IR spectra, *Adv. Mol. Vib. Coll. Dyn.* **1**, 1-25.

[10] Buckingham,A.D., Fowler,P.W. and Hutson,J.M. (1988) Theoretical studies of Van der Waals molecules and intermolecular forces, *Chem. Rev.* **88**, 963-1013.

[11] Chalasinski,G. and Szczesniak,M. (1994) Origins of structure and energetics of Van der Vaals clusters from ab initio calculations, *Chem. Rev.* **94**, 1723-1765.

[12] London,F. (1930) *Z. Phys. Chem.* **11**, 491.

[13] Eisenchitz,R. and London,F. (1930) *Z. Phys* **60**, 491.

[14] London,F. (1937) *Trans. Faraday Soc.* **33**, 8.

[15] Margenau,H. and Kestener,N.R. (1971) *Theory of Intermolecular Forces*, Pergamon, Oxford.

[16] Morokuma,K. and Kitaura,K. (1982) *Molecular Interactions*, Ratajczk,H. and Orville-Thomas,W.J. (eds.), Wiley, New York.

[17] Maksic,Z.B. (ed.) (1991) *Theoretical Models of Chemical Bond*, Springer-Verlag, Berlin.

[18] e.g. see: Slater,J.C. (1974) *Self-consistent Field for Molecules and Solids*, McGraw-Hill, New York.

[19] March,N.H. (1975) *Self-Consistent fields in atoms*, Pergamon, Oxford.

[20] Gombas,P. (1949) *Die Statistische Theorie aes Atoms und ihre Anwendungen*, Springer-Verlag, Berlin.

[21] Hohenberg,P.C. and Kohn,W. (1964) *Phys. Rev. B* **136**, 864.

[22] March,N.H. and Deb,B.M. (eds.) (1987) *The Single Particle Density in Chemistry and Physics*, Academic, New York.

[23] Parr,R.G. and Yang,W. (1989) *Density Functional Theory of Atoms and Molecules*, Oxford Univeristy Press, Oxford.

[24] Lundquist,S. and March,N.H. (eds.) *Theory of Inhomogeneous Electron Gas*, Plenum, New York.

[25] Gianturco,F.A. and Rodriguez-Ruiz,J.A. (1992) Correlation forces in electron scattering from atoms and molecules, *J. Mol. Structure* **260**, 99.

[26] Gianturco,F.A. and Rodriguez-Ruiz,J.A. (1993) Correlation forces in electron scattering processes, *Phys. Rev. A* **47**, 1075.

[27] Salahub,D.R. (1987) *Advances in Chem Phys.* **LXIX**, 47

[28] Russo,T.V., Martin,R.L. and Hay,P.J. (1994) Lensity Functional calculations of firs-row transition metals, *J. Chem. Phys.* **101**, 7729.

[29] Kohn,W. and Sham,L.J. (1965) *Phys. Rev. A* **140**, 1133.

[30] e.g. see: Ziegler,T. (1991) *Chem. Rev.* **91**, 651.

[31] Labanowski,J. and Andzelm,J. (eds.) (1991) *Density Functional Methods in Chemistry*, Springer-Verlag, Heidelberg.

[32] Johnson,B.G., Gill,P.M.W. and Pople,J.A. (1993) The performance of a family of density functional methods, *J. Chem. Phys.* **98**, 5612.

[33] Becke,A.D. (1988) Density-functional exchange-energy approximation with correct asymptotic behaviour, *Phys. Rev* **38**, 3098.

[34] Becke,A.D. (1988) A multicenter numerical integration scheme for polyatomic molecules, *J. Chem. Phys.* **88**, 2547.

[35] Becke,A.D. (1988) Correlation energy of an inhomogeneous electron gas: a coordinate-space model, *J. Chem. Phys.* **88**, 1053.

[36] Becke,A.D. and Edgecombe,K.E. (1990) A simple measure of electron localization in atomic and molecule systems, *J. Chem. Phys.* **92**, 5397.

[37] Becke,A.D. (1992) Density functional Thermochemistry.I. The effect of the eschange-only gradient correction, *J. Chem. Phys.* **96**, 2155.

[38] Becke,A.D. (1993) A new mixing of Hartree-Fock and local density functional theories, *J. Chem. Phys.* **98**, 1372.

[39] Becke,A.D. (1992) Density functional Thermochemistry.II. A grdient correction for dynamical correlation, *J. Chem. Phys.* **97**, 9173.

[40] Becke,A.D. (1993) Density functional Thermochemistry.III. The role of exact exchange, *J. Chem. Phys.* **98**, 5648.

[41] Becke,A.D. (1993) A half-half theory of density functionals, *J. Chem. Phys.* **98**, 1372.

[42] Perdew,J.P. and Wang,Y. (1986) *Phys. Rev. B* **33**, 8800.

[43] Perdew,J.P. (1986) *Phys. Rev. B* **33**, 8822.

[44] Perdew,J.P., Chevary,J.J., Vosko,S.H., Jackson,K.A., Pederson,M.R., Singh,D.J. and Fiolhais,C. (1992) *Phys. Rev B* **46**, 6671.

[45] Lee,C., Yang,W. and Parr,R.G. (1988) Development of the Colle-Salvetti correlation-energy formula into a functional of the electron density, *Phys. Rev. B* **37**, 785.

[46] Truong,T.N. and Duncan,W. (1994) A new direct ab initio method for calculating thermal rate constants from DFT, *J. Chem. Phys.* **101**, 7408.

[47] Vosko,S.H., Wilk,L. and Nusair,M. (1980) Accurate spin-dependent electric liquid correlation energies for local spin density calculations, *Can. J. Phys.* **58**, 1200.

[48] Barone,V. (1994) Inclusion of Hartree-Fock exchange in density functional methods, *J. Chem. Phys.* **101**, 6834.

[49] Dunlap,B.I., Connolly,J.W.D. and Sabin,J.R. (1979) *J. Chem. Phys.* **71** 3396,4993.

[50] Gianturco,F.A. and de Lara-Castells,M.P. (1996) Stability and structure of rare-gas ionic clusters using density functional methods, *Int. J. Quantum Chem.* **60**, 593.

[51] Gianturco,F.A. and de Lara-Castells,M.P. (1996) Structure and anisotropy of ionic argon clusters using density functional models, *Chem. Phys.* **208**, 25

[52] Green,S. and Thaddeus,P. (1976) *Astrophys. J.* **205**, 766.

[53] Wickham-Jones,C.T., Williams,H.T. and Simpson,C.J.S.M. (1981) Experimental and theoretical studies of CO vibrational relaxation by the He atoms, *J. Chem. Phys.* **87**, 5294.

[54] Ellis,J., Toennies,J.P. and Witte,G. (1995) Helium atom scattering study of the frustrated translation made of CO adsorbed on the Cu(001) surface, *J. Chem. Phys.* **102**, 5059.

[55] Dilling,W. (1985) PhD Thesis, University of Göttingen, unpublished results.

[56] Thomas,L.D., Kraemer,W.P. and Dierksen,G.H.F. (1980) *Chem. Phys.* **51**, 131.

[57] Boys,S.F. and Bernardi,F. (1970) *Mol. Phys.* **19**, 553.

[58] Gianturco,F.A., Sanna,N. and Serna-Molinera,S. (1993) Quantum and classical calculation of transport and relaxation cross sections in He-CO mixtures, *J. Chem. Phys.* **98**, 3833.

[59] Gianturco,F.A., Sanna,N. amd Serna-Molinera,S. (1994) An improved He-CO interaction from a multiproperty analysis, *Mol. Phys.* **81**, 421.

[60] Le Roy,R.J., Bissonnette,C., Wu,T.H, Dham,A.K. and Meath,W.J. (1994) Improved modelling of atom-molecule potential-energy surface: illustrative application to He-CO, *Faraday Discuss.* **97**, 81.

[61] Chuaqui,C.E., Le Roy,R.J. and McKellar,A.R.W. (1994) Infrared spectrum and potential energy surface of He-CO, *J. Chem. Phys.* **101**, 39.

[62] Kukawska-Tarnawska,B., G.Olzewiski,K. and Chalasinski (1994) Structure and energetics of Van der Waals complexes of Carbon Monoxide with rare gases, *J. Chem. Phys.* **101**, 4964.

[63] Tao,F.M., Drucker,S., Cohen,R.C. and Klemperer,W. (1994) Ab initio potential energy surface and dynamics of He-CO, *J. Chem. Phys.* **101**, 8680.

[64] Moszynski,R., Korona T., Wormer,P.E.S. and Van der Avoid,A. (1995) Ab initio potential energy surface, infrared spectrum and second virial coefficient of the He-CO complex, *J. Chem. Phys.* **103**, 321.

[65] Dham,A.K. and Meath,W.J. (1996) Multi-property predictions from recent He-CO potential energy surfaces, *Mol. Phys.* **88**, 339.

[66] Mele F., Minerva,T., Russo,N. and Toscano,M. (1995) Hydrogebonded and the Van der Waals complexes studied by a density functional method, *Theor. Chim. Acta* **91**, 169.

[67] Corongiu,G., Estrin,D., Murgia,G., Paglieri,L., Pisani,L., Suzzi-Valli,G., Watts,J.D. and Clementi,E. (1996) Revisiting energy surface for NH_3HCl, *Int. J. Quantum Chem.* **59**, 119.

[68] Gianturco,F.A. and de Lara-Castells,M.P. (in preparation) On the concerted mechanism for Ar_4^+ fragmentation.

[69] Woon,D.E. and Dunning Jr,T.H. (1994) *Chem. Phys.* **100**, 2975.

[70] Ceperley,D.M. and Alder,B.J. (1980) *Phys. Rev. Lett.* **45**, 566.

[71] Parker,G.A. and Pack,R.T. (1978) *J. Chem. Phys.* **69**, 3268.

DIFFERENT LEGACIES AND COMMON AIMS : ROBERT MULLIKEN, LINUS PAULING AND THE ORIGINS OF QUANTUM CHEMISTRY

Ana Simões, Department of Physics, University of Lisbon,

Campo Grande, C1, 1700 Lisboa, Portugal

and

Kostas Gavroglu, Department of History and Philosophy of Science,

University of Athens,

John Kennedy 37, Ilisia, Athens 16121, Greece

When referring to the different approaches to the question of atomic bonding nearly all textbooks and research papers project two such methods: the Heitler-London-Slater-Pauling valence bond method and the Hund-Mulliken method of molecular orbitals. Elsewhere we have argued that the views of these protagonists about theory building and the role of theory in chemistry form a set of criteria which justifies a different classification: the Heitler-London approach versus the Pauling-Mulliken approach.[1] Walter Heitler (1904-1981) and Fritz London (1900-1954) shared a common approach to the problem of chemical bonding. The theory they attempted to develop was, in effect, an instantiation of Dirac's reductionist view so clearly expressed as an agenda for chemists in 1929:[2] the underlying laws governing the behavior of electrons were known and, hence, to do chemistry meant to deal with equations which were in principle soluble even though in practice they may only produce approximate solutions. Linus Pauling (1901-1993) and Robert Sanderson Mulliken (1896-1986) thought differently on how the newly developed quantum mechanics could, in practice, be applied to problems of chemistry and, more specifically, to the problem of the chemical bond. They felt that a reductionist agenda was, in practice, useless to the chemist, and by making ample use of semi-empirical methods they developed their respective approaches and whose only criterion for acceptability was their practical success. And, most significantly, they both shared a common outlook on how to construct their theoretical schemata, on the character of the constitutive features of their theories, on what the relation of physics to chemistry should be and on the discourse they developed to legitimate their respective theories.

We have discussed the approach of Fritz London and Walter Heitler quite analytically in ref. 1. In this paper we discuss a number of issues associated with the theoretical outlook shared by Pauling and Mulliken. Firstly, Pauling's valence bond and Mulliken's molecular orbital methods were not simply two practical methods to solve valence problems. They were part of two different conceptual schemata, which can be explicated in terms of two different legacies--that of physics in the case of Mulliken, and that of chemistry in the case of Pauling. Secondly, their theories, and especially Pauling's, raised questions as to the ontological status of theoretical entities very similar to the *problématique* associated with the philosophical discussions about scientific realism.

I

Pauling's valence theory was conceived as an extension of classical valence theory. It was a chemical structural theory which envisioned molecules as aggregates of atoms bonded together along privileged directions. Following Pauling's belief that the task of the chemist should be "to attempt to make every new discovery into a general chemical theory"[3] a comprehensive theory of the chemical bond based on the concept of resonance emerged out of the "The Nature of the Chemical Bond" series[4]. The concept of resonance played a fundamental role in the discovery of the hybridization of bond orbitals, the one-electron and the three-electron bond, and the discussion of the partial ionic character of covalent bonds in heteropolar molecules. Furthermore, the idea of resonance among several hypothetical bond structures explained in "an almost magical way"[5] the many puzzles that had plagued organic chemistry. Resonance established the connecting link between Pauling's new valence theory and the classical structural theory of the organic chemist which Pauling classified as "the greatest of all theoretical constructs." Resonance---originally a physical concept---became decisive in the formulation of a chemical theory.

Linus Pauling received his doctorate from the California Institute of Technology in 1925 on the determination of the structure of crystals by means of X-ray diffraction. He spent 1926 in Europe first at Sommerfeld's Institute for Theoretical Physics at Munich and, then, he visited Schrödinger at Zürich, Born at Göttingen and Bohr at Copenhagen. During his stay in Europe he immersed himself into wave mechanics, made calculations of the values of parameters with the new quantum mechanics, and compared them to the values obtained by the old quantum theory. In 1927 he was offered a position at the California Institute of Technology. Pauling's remarkable ease with quantum mechanics did not intimidate him in subordinating chemistry to physics. Physics to him was a means to a --chemical-- end. He never considered himself a physicist, nor did he consider the study of complex atoms, ions or molecules as belonging to physics. During his stay in Munich, he wrote to his mentor, A.A. Noyes.[6]

> I feel that the theoretical (and experimental) study of the properties of complicated atoms and ions (i.e., with many electrons) and of molecules could really be classed with chemistry; for usually a study of the different phenomena exhibited by different chemical substances is called chemical, and the study of phenomena without especial reference to the chemical substances exhibiting them is called physical. However, we can hardly class this with Physical Chemistry, as ordinarily understood; possibly Molecular Chemistry would be a suitable designation... Most people seem to think that work such as mine, dealing with the properties of atoms and molecules, should be classed as physics; but I (as I have said before) feel that the study of chemical substances remains chemistry even though it reached the state in which it requires the use of considerable mathematics. The question is more than an academic one, for the answer really determines my classification as a physicist or chemist.

Robert Mulliken followed a different path to formulate his theory of chemical bonding. He had completed his doctorate at the University of Chicago in 1921 working on the separation of isotopes. He, then, moved to Harvard where he worked on molecular spectroscopy with F.A. Saunders and E.C. Kemble and assisted in their preparation of the report on the spectra of diatomic molecules for the National Research Council. He went to Europe in the summer of 1927 and visited Göttingen, Zürich and Geneva. He met Schrödinger, Heitler, London and had extensive discussions with Friedrich Hund --who had first proposed the molecular orbital approach in 1926.[7] In 1928 he moved to the Physics Department of the University of Chicago where A.H. Compton and the spectroscopist H.A. Gale promised him a new spectroscope with high-resolution grating. The work on band spectra, and namely, the clarification of the relations between the electronic states and the band spectra structure led Mulliken to propose an entirely different approach to the question of molecule formation and chemical bonding. It amounted to dispense altogether with classical valence theory. He rejected the accepted notion of chemical structure, which pictured a molecule as an aggregate of atoms bonded together. For him the molecule was a unit by itself. To counteract the received view of a molecule as an assembly of atoms he liked to say in the manner of Gertrude Stein that "a molecule is a molecule is a molecule."[8] At the same time he proposed to abandon the classical notions of bonds and valence, and to analyze the phenomena of molecule formation in terms of the electronic structure of molecules. This implied, after all, to analyze a molecule in terms of its nuclei and electrons, or in terms of kernels and outside electrons. New auxiliary concepts were introduced such as promoted and unpromoted electrons, bonding, non-bonding and anti-bonding electrons, and varying bonding power of electrons. His approach to valence theory had been "induced" through a painstaking analysis of an impressively large amount of band spectra data. His was a largely phenomenological theory whose conceptual schema---the extension of Bohr's *Aufbauprinzip* to molecules[9], with the concomitant assignment of quantum numbers to electrons---came from physics and not from chemistry.

Pauling from the very start tried to develop a schema for dealing with problems related to chemical valence by developing a *set of rules*. These rules, whose rationale was only very loosely based on quantum mechanics, in his view "codified" the theoretical approach to questions concerning chemical bonding. In 1926, according to extant manuscripts, he tried unsuccessfully to find such a set of rules in the framework of molecular orbitals.[10] As Pauling confessed[11], the three years elapsed between 1928 when he first sent a short note to the Proceedings of the National Academy of Sciences[12] and his first paper on "The Nature of the Chemical Bond" were used to find a simple and reliable enough method for the computation of bond lengths, bond directions, and resonance energies. Mulliken, on the contrary, was not associated with any equivalent computational schema until the adoption of the Lennard-Jones linear combinations of atomic orbitals (LCAO) after 1929.[13] Although it was clearly recognized by Mulliken that the LCAO method was not necessarily implied by the conceptual framework of his own theory, such an association could be, and was indeed, made. In the manuscript of the book on *Quantum Mechanics of Organic Molecules* which Pauling and his former student and collaborator George Wheland planned to write, in the chapter on "The Quantum Mechanical Treatment of the Electron-Pair Bond" it was stated that the use of atomic orbitals was fundamental both to the valence-bond method as well as to the molecular-orbital method. It was possible to outline a molecular-orbital method which consists in first finding one-electron molecular orbitals by the solution of the wave equation for one electron in the field of the nuclei and other electrons, and then using these molecular orbitals in the construction of suitable wave-functions. "This procedure, however, has never actually been followed; in all numerical applications of the molecular orbital method { EMBED Equation.2 } the molecular orbitals are themselves first built out of atomic orbitals".[14]

Mulliken's assessment of the developments led him to the conclusion that two different descriptive chemical theories provided alternative methods for the assignment of molecular electron configurations, one of them relying entirely on the use of atomic orbitals, whereas, in the other, molecular orbitals of some sort were used to describe shared electrons. In the paper "Electronic structure of polyatomic molecules and valence. VI. On the method of molecular orbits," he considered that[15]

> The first method follows the ideology of chemistry and treats every molecule, so far as possible, as composed of definite atoms or ions. The electron configuration is then the sum of the configurations of these atoms or ions... [This method] has had notable success as a qualitative conceptual scheme for interpreting and explaining empirical rules of valence and in semiquantitative, mostly semiempirical

> calculations of energies of formation and other properties of molecular electronic states, particularly the normal states. ... Departing from chemical ideology, the second method treats each molecule, so far as possible, as a unit... It is the writer's belief that the present... [method of nonlocalized molecular orbitals] may be the best adapted to the construction of an exploratory *conceptual scheme* within whose framework may be fitted both chemical knowledge and data on electron levels from molecular spectra. A procedure adapted to a broad survey and interpretation of observed relations is here aimed at, rather than (at first) one for quantitative calculation, which logically would follow later. Given an observed molecule or ion of known shape and size, what is its electronic structure in terms of an electron configuration using, in general, non-localized orbitals for shared electrons?

Mulliken's quite radical approach to valence theory enables one to understand the importance he attributed to the clarification of its conceptual foundation as well as his constant preoccupation in contrasting the conceptual framework of his theory to the conceptual frameworks of other theories. Mulliken thoroughly discussed Heitler and London's theory[16] outlined in their famous 1927 paper.[17] He, then, proceeded to a detailed analysis of Pauling's work, to such an extent that Mulliken's papers could, in fact, be considered as a sort of response to Pauling's.[18] In these papers Mulliken expressed his conviction that Pauling's work provided a *possible* but *worse* conceptual scheme than his own.[19]

Mulliken's chemical wonderings bore the mark of a physicist. By 1929 he had come to the conclusion that it was crucial to win the attention of chemists. He attempted to blur what was perceived as the discontinuity between his theory and former chemical theories. In Mulliken's papers which appeared in *Chemical Reviews*,[20] and were purposefully addressed to chemists, the relation between the Lewis-Langmuir theory and molecular orbitals was emphasized as part of an argument to win the chemists' support. In a following stage, he went so far as to argue that by comparison with the valence bond method, the molecular orbital method was even closer to G.N.

Lewis's ideas, in the sense that localized molecular orbitals accounted for all degrees of polarity.[21]

In the United States, quantum chemistry was developing simultaneously in two forms, which were the result of two different legacies---that of physics in the case of Mulliken, and that of chemistry in the case of Pauling.[22] Mulliken developed his valence theory in the context of the structural work on molecular spectroscopy which was also studied by a large number of physicists in America.[23] Pauling followed the tradition of chemical theories of valence, which had already been translated by Lewis --and propagated by Langmuir-- in the language of the electronic make-up of molecules.

Chemical considerations were not, of course, altogether absent from Mulliken's work. In his first papers,[24] I. Langmuir's model of the nitrogen molecule, translated in the framework of the old quantum theory, played a role in suggesting that electrons might be orbiting two nuclei, and Langmuir's suggestion that compounds with the same number of atoms and the same number of electrons possessed similar physical properties led Mulliken to search for analogies in the spectroscopic behavior of such molecules. Later on, the relation to Lewis's theory was emphasized in order to facilitate the adoption of his ideas by the chemical community.[25] However, the conceptual core of Mulliken's theory was suggested by physical considerations: the conception of a molecule as an assembly of electrons moving around the nuclei; the idea of studying the electronic structure by extending Bohr's building up principle to molecules; the idea of assigning quantum numbers to electrons in molecules; and the painstaking charting of the energies of the transitions of molecular electrons. The new quantum mechanics did not play a role in the genesis of Mulliken's theory and much of what characterized Mulliken's conceptual scheme of molecular orbitals as the foundation for a radically new approach to valence theory was dependent almost entirely on the old quantum theory. Only after 1929 did quantum mechanics provide the tools for the development of a computational scheme for calculating molecular orbitals as linear combination of atomic orbitals. It was the Pauli exclusion principle -- which Mulliken quickly extended to the electrons of molecular orbitals-- that played the crucial role in his phenomenological approach to molecular structure and chemical binding.

Much in the same way that chemistry played this auxiliary role in Mulliken's thinking, physics crept in more than one way in the otherwise chemical legacy followed by Pauling. It was physics which provided the empirical source of knowledge on structural features. Together with the evidence from X-ray crystallography and electron diffraction, evidence from band spectra was used as a supplementary source of information. Physics, and specifically the new quantum mechanics, suggested new ideas such as hybridization and resonance, and guided Pauling in the formulation of new rules. Yet, there was a strong *rapport* between the conceptual core of the new theory and that of chemical structural theory, and ideas such as resonance and hybridization were used to explain valence properties long known to chemists. The concept of hybridization was used to explain the directional properties of bonds, and resonance was used to explain why sometimes it was impossible to attribute one single Lewis formula to a chemical compound.

The two theories were not received in the same manner by the chemical community. Pauling's theory was relatively easy to accept, since, in effect, it was a recasting of the old structural theory. Mulliken's theory was harder to accept due to its drastic break with the conceptual framework of former chemical theories of valence. Other sorts of factors played a role in the reception of both theories. Differences in scientific styles played their part. Pauling was able to present his theory in lucid texts published in journals read by chemists and developed his ideas in a book form as soon as possible.[26] Mulliken did not have the same ease in communicating his ideas. At the beginning his ideas were presented in a highly technical and rigid manner in physical journals, delaying their acceptance by the chemical community. He never published them in a book form and did not appear to have been a good lecturer. One of Mulliken's former students even claimed that the contrast between Mulliken and Pauling's scientific styles "may be the best example we have of the major role that personality may play in the development of science."[27] And their different personalities only served to accentuate a process that depended above all on the characteristics of the conceptual schemata of their valence theories.

II

By 1935, it was clear that both the valence bond and the molecular orbital methods provided equally effective approaches to untangle valence-related problems. In that year, J.H. Van Vleck and Albert Sherman wrote a long review article where they assessed the situation by comparing and commenting upon the two methods. Concluding their article they posed the question whether it was justified to talk about a quantum theory of valence if Schrödinger's equation had not yet been integrated with any sufficient accuracy for any system more complex than the hydrogen molecule. They proposed to adopt the mental attitude and procedure of an optimist rather than a pessimist. They charcterised the pessimist as the theoretical chemist who demands a rigorous postulational theory, and calculations devoid of any questionable approximations or of empirical appeals to known facts. The pessimist is continuously worried because the omitted terms in the approximations are usually rather large, so that any pretense of rigor should be lacking. The optimist, they suggested, is satisfied with approximate solutions of the wave equation and believes that the approximate calculations give one an excellent "steer" and a very good idea of "how things go," permitting the systematization and understanding of what would otherwise be a maze of experimental data codified by purely empirical valence rules.[28]

Van Vleck and Sherman did not name who they considered to be the optimists or the pessimists, and they, themselves, claimed to have attempted to adopt a middle ground between the two opposite points of view. However, one cannot but wonder at the striking similarities between the praxis of some Americans like Mulliken, Pauling, J.C. Slater or Van Vleck himself, and the optimistic persona, and the praxis of some Germans like Heitler, London or Hund and the pessimistic persona depicted in the review.

In contrast to the received view, it was not the case that everyone thought that the approaches by Mulliken and Friedrich Hund were equally acceptable. Van Vleck did not miss the opportunity to state their differences in his correspondence with Mulliken. For Van Vleck, Hund made a "lot of abstract suggestions without working over the experimental data carefully" whereas Mulliken "as a man thoroughly conversant with the experimental data" was able "to bridge the gap between theory and experiment."[29] In fact, it was to the credit of Hund to have introduced quantum mechanics in spectroscopy.[30] His work gave theoretical support to Mulliken's hint at what he later called "electron promotion,"[31] was used in Mulliken's "correlation diagrams"[32] which relate the state of a molecule to the separated atoms and the united atom descriptions, and paved the way for Mulliken's assignment of individual quantum numbers to electrons in diatomic and polyatomic molecules,[33] as well as for the consideration of the conditions which favor or inhibit the formation of molecules. Hund, in a way, provided a legitimizing framework for Mulliken's phenomenological approach whereas Mulliken provided a justification for Hund's very abstract suggestions, translated them in a language more acceptable by chemists, and went further than Hund's by means of a "systematic study of relevant empirical data in the light of the theory."[34]

By the end of the 1930s, Mulliken became fully convinced that the efficient way to get practical results in attacking the complex problems of molecular structure and bonding depended on the use of semi-empirical methods. He articulated what he meant as follows.[35]

> At present we depend wholly on the quantum mechanics only in a very few cases. In more complicated cases, we make partial use of quantum mechanics in the form of qualitative principles or rules. Into this qualitative theoretical framework we then try to fit all relevant experimental data. By this *semi-empirical method*, we are able to reach many conclusions concerning electronic structures of molecules. In doing this, we may use many kinds of physical and chemical data, including spectroscopic data.

It was in this respect that Pauling appeared to agree wholeheartedly with Mulliken. In the opening paragraph of the first paper of "The Nature of the Chemical Bond" series, Pauling assessed the situation concerning work on the chemical bond as well as the method he planned to follow.[36]

> During the last four years the problem of the nature of the chemical has been attacked by theoretical physicists, especially Heitler and London, by the application of quantum mechanics. This work has led to an approximate theoretical calculation of the energy of formation and of other properties of simple molecules... and has also provided a formal justification of the rules set up in 1916 by G.N. Lewis for his electron bond. In [this] paper it will be shown that many more results of chemical significance can be obtained from the quantum mechanical equations, permitting the formulation of an extensive and powerful set of rules for the electron-pair bond

supplementing those of Lewis. These rules provide information regarding the relative strengths of bonds formed by different atoms, the angles between bonds, free rotation or lack of free rotation about bond axes, the relation between the quantum numbers of bonding electrons and the number and spatial arrangement of the bonds.

Texts of this sort are, in a way, pace setting texts; they are texts exerting a profound influence on chemists and contributing to the formation of the "chemists' culture." According to Pauling, the theoretical physicists were the first to apply quantum mechanics to a chemical problem. At the same time Pauling considered his own work as an extension of their program, his applications provided many more results which could be obtained in the *form of rules* supplementing other rules. In the paper, a sketch of the reasoning leading to the rules was presented. Many approximations and arbitrary assumptions, which might have inhibited a theoretical physicist, were made in the process. They had been obtained as generalizations of the quantum mechanical rigorous treatment of simple systems such as the hydrogen molecule, the helium atom and the lithium atom, together with generalizations supported by Pauling's systematic study of the behavior and properties of chemical compounds, for which rigorous quantum mechanical proofs were inaccessible. Even if the rules were not derivable from quantum mechanical principles, their usefulness in providing information regarding bond types in chemical substances was enough as an empirical criterion for their acceptance.[37]

In 1936, when the series of "The Nature of the Chemical Bond" was already completed, in a talk given to the American Chemical Society, Pauling expressed his belief that there is more to chemistry than an understanding of general principles and that the chemist is, perhaps even more, interested in the characteristics of individual substances---that is, of individual molecules.[38] The work on the nature of the chemical bond had been instrumental for the development of a program of attack on molecular structure along such general guidelines. It was, as Pauling often said, a pragmatic approach to chemistry, a semi-empirical treatment of the problems, and an overall attitude which was so dear to the chemists' traditions.

Earlier in 1930, Pauling wrote a letter to A.A. Noyes discussing the possible changes to the course work in Freshman Chemistry after the successful application of quantum mechanics to chemical problems. Pauling advocated that students should become familiar with descriptive and general chemistry, as a prerequisite for further study. They should know the behavior of chemical substances, their properties and their reactions, before proceeding to a stage where general explanations and laws would be sought for. Only in this way would students develop a "feeling for chemistry" which Pauling deemed necessary to the understanding of chemical theory.[39]

> A knowledge of descriptive chemistry is necessary for the study of theoretical chemistry... I know of no chemist who was attracted to this field of knowledge because of theoretical chemistry. Instead, it is an interest in chemicals and their reactions which has first attracted the chemist, who may (and usually does) become keenly interested in attempting to account for unusual observations, and later become excited over the explanations given by theoretical chemistry...I feel that the freshman year should give the student, through laboratory work as well as lectures, a good familiarity with the chemicals and their reactions. It should not be a bare recital of facts. Simple theories... should be woven in, but all richly illustrated with examples, and not too difficult experiments. The good student at the end of the year should have chemical feeling or intuition---which means that he will usually know what will happen. Then, but not before, he is ready to treat these topics quantitatively and more rigorously. He will be able to handle problems of equilibria leading to quartic equations, because he will know what quantities are small: and these problems will be *chemical* problems to him. A freshman student said, "Our chemistry course is just an extra course in mathematics." That is just what it shouldn't be.

Pauling's description of how students should learn chemistry is all the more interesting because it mirrored what he considered to be the path to discovery in chemistry. According to Pauling physicists, like Heitler and London, lacked in general chemical feeling, and therefore, their efforts to extend the theory to more complicated molecules than the hydrogen molecule were unsuccessful. Their method lacked a proper empirical backup whereas his method had an adequate empirical content and still the same rigorous physical basis.[40]

Heitler and London's incursion in valence theory had fallen short of their initial goals. As Heitler later confessed in an interview his perhaps too ambitious aim had been[41]

> to understand the whole of chemistry. This is perhaps a bit too much to ask, but it was to understand what the chemists mean if they say an atom has a valence of two or three or four; what the chemists mean if they put down a formula with so many bonds here and so many bonds there in a multi-atomic molecule. Both London and I believed that all this must be now within reach of quantum mechanics.

Wigner was amused by Heitler's naiveté.[42]

> I often teased Heitler, I must admit, because Heitler sort of felt that he had explained the whole of chemistry, and I was a little skeptical of that. I asked him, "Well now, what chemical compounds would you predict between nitrogen and hydrogen?" And of course since he did not know any chemistry he could not tell me.

In the same vein one might ask why is it that notions such as covalent bonds or tetrahedral carbon atoms were not arrived at as pure predictions derived solely from quantum mechanics? Even the concept of resonance, usually presented as the example of a quantum concept *par excellence*, was partly anticipated in the theories of Robinson, Ingold and Arndt.

Differences in the assessment of the methodological and ontological status of resonance were the object of a dispute between Pauling and Wheland, who more than anyone else worked towards the extension of resonance theory to organic molecules. Wheland, in his book *The Theory of Resonance and Its Applications to Organic Molecules* dedicated to Pauling, argued that the resonance concept was a "man-made-concept"[43] in a more fundamental way than in most other physical theories. This was his way to counter the widespread view that resonance was "a real phenomenon with real physical significance," which he classified as one example of the nonsense that especially organic chemists were prone to.[44]

> What I had in mind was, rather, that resonance is not an intrinsic property of a molecule that is described as a resonance hybrid, but is instead something deliberately added by the chemist or the physicist who is talking about the molecule. In anthropomorphic terms, I might say that the molecule does not know about resonance in the same sense in which it knows about its weight, energy, size, shape, and other properties that have what I call real physical significance. Similarly ... a hybrid molecule does not know how its total energy is divided between bond energy and resonance energy. Even the double bond in ethylene seems to me less "man-made" than the resonance in benzene. The statement that the ethylene contains a double bond can be regarded as an indirect and approximate description of such real properties as interatomic distance, force constant, charge distribution, chemical reactivity, and the like; on the other hand, the statement that benzene is a hybrid of the two Kekulé structures does not describe the properties of the molecule so much as the mental processes of the person who makes the statement. Consequently, an ethylene molecule could be said to know about its double bond, whereas a benzene molecule cannot be said, with the same justification, to know about its resonance...Resonance is not something that the hybrid *does*, or that could be "seen" with sufficiently sensitive apparatus, but is instead a description of the way that the physicist or chemist has arbitrarily chosen for the approximate specification of the true state of affairs.

Pauling could not disagree more. For him, the double bond in ethylene was as "man-made" as resonance in benzene. Pauling summarized their divergent viewpoints by saying that Wheland seemed to believe that there was a "quantitative difference" in the man-made character of resonance theory when compared to ordinary structure theory---but he could not find such a difference. He asserted that Wheland made a disservice to resonance theory by overemphasizing its "man-made character."[45] Wheland conceded that resonance theory and classical structural theory were qualitatively alike, but he still defended, contrary to Pauling, that there was a "quantitative difference" between the two. He viewed his disagreement with Pauling as a result of different value-judgements on what he classified as philosophical, rather than scientific matters.

Nevertheless, acknowledging or denying the existence of differences between resonance theory and classical structural theory was dependent on their different assessments of the role of alternative methods to study molecular structure. Wheland equated resonance theory to the valence-bond method and viewed them as alternatives to the molecular-orbital method. Pauling conceded that the valence-bond method could be compared with the molecular orbital method, but not with the resonance theory which was largely independent of the valence-bond method. For Pauling the theory of resonance was not merely a computational scheme. It was an extension of the classical structure theory, and as such it shared with its predecessor the same conceptual framework. If one accepted the concepts and ideas of classical structure theory one had to accept the theory of resonance. And, how could one reject their common conceptual base if they had been largely induced from experiment?[46]

> I think that the theory of resonance is independent of the valence-bond method of approximate solution of the Schrödinger wave equation for molecules. I think that it was an accident in the development of the sciences of physics and chemistry that resonance theory was not completely formulated before quantum mechanics. It was, of course, partially formulated before quantum mechanics was discovered; and the aspects of resonance theory that were introduced after quantum mechanics, and as a result of quantum mechanical argument, might well have been induced from chemical facts a number of years earlier.

This discussion with Wheland prompted Pauling to make his position about these issues public. More than the question of the artificiality of the resonance concept, to which he alluded briefly in his Nobel lecture,[47] he wanted to state as clearly as possible his views on theory building. A revised version of the arguments brought about in the discussion with Wheland appeared in *Perspectives in Organic Chemistry*[48] and later on, in the third edition of *The Nature of the Chemical Bond*.[49] In the preface, Pauling pointed out that the theory of resonance involves the "same amounts of idealization and arbitrariness as the classical valence-bond theory." Pauling added a whole section in the new edition to discuss this question. His manifesto was called "The Nature of the Theory of Resonance." There, he argued that the objection concerning the artificiality of concepts applied equally to resonance theory as to classical structure theory. To abandon the resonance theory was tantamount to abandoning the classical structure theory of organic chemistry. Were chemists willing to do that? According to Pauling chemists should keep both theories because they were chemical theories and as such possessed "an essentially empirical (inductive) basis."[50]

> I feel that the greatest advantage of the theory of resonance, as compared with other ways (such as the molecular-orbital method) of discussing the structure of molecules for which a single valence-bond structure is not enough, is that it makes use of structural elements with which the chemist is familiar. The theory should not be assessed as inadequate because of its occasional unskillful application. It becomes more and more powerful, just as does classical structure theory, as the chemist develops a better and better chemical intuition about it... The theory of resonance in chemistry is an essentially qualitative theory, which, like the classical structure theory, depends for its successful application largely upon a chemical feeling that is developed through practice.

In 1947 Charles Coulson, a mathematician turned chemist who taught for many years at Oxford and whose book *Valence* became a standard textbook, wrote an article in a semi-popular magazine on what he thought was resonance.[51]

> Is resonance a *real* phenomenon? The answer is quite definitely no. We cannot say that the molecule has either one or the other structure or even that it oscillates between them...Putting it in mathematical terms, there is just one full, complete and proper solution of the Schrödinger wave equation which describes the motion of the electrons. Resonance is merely a way of dissecting this solution: or, indeed, since the full solution is too complicated to work out in detail, resonance is one way --and then not the only way-- of describing the approximate solution. It is a "calculus", if by calculus we mean a method of calculation; but it has no physical reality. It has grown up because chemists have become used to the idea of localized electron pair bonds that they are loath to abandon it, and prefer to speak of a superposition of definite structures, each of which contains familiar single or double bonds and can be easily visualizable.

The question as to the ontological status of resonance was not an issue which was confined to this exchange between Pauling and Wheland. Pauling's theory of resonance was visciously attacked in 1951 by a group of chemists in the Soviet Union in their Report of the Commission of the Institute of Organic Chemistry of the Academy of Sciences.[52] As they, themselves, stressed their main objection was methodological. They could not accept that by starting from conditions and structures which did not correspond to reality one could be led to meaningful results. Of course, they discussed analytically the work of A.M. Butlerov who in 1861 had proposed a materialist conception of chemical structure: this was the distribution of the action of the chemical force, known as affinity, by which atoms are united into molecules. He insisted that any derived formula should express a real substance, a real situation. According to the Report, Pauling was moving along different directions. For him a chemical bond between atoms existed if the forces acting between them were such as to lead to the formation of an aggregate with sufficient stability to make it convenient for the chemist to consider it as an independent molecular species. To these chemists Pauling's operational definition was totally unacceptable[53]

> In this treatment the objective criterion of reality of the molecule and of the chemical bond vanishes. Since the definition of the molecule and the chemical bond given by Pauling is methodologically incorrect, it naturally leads, when logically developed, to absurd results.

It is interesting to note the initiative of the New York Chapter of the National Council of Arts, Sciences and Professions to organize a meeting on the subject. It was proposed that the meeting have the form of a debate where N.D.Sokolov from Moscow, Charles Coulson and Linus Pauling would each contribute a paper and there would follow a discussion of the points raised in the communications. Coulson felt that the best way would be for Sokolov and Pauling to present their viewpoints and that himself would make a series of comments. Each party would be asked to provide answers to the following questions: What is the resonance theory? What is the evidence in proof or disproof of the resonance theory? Is the convenience of the theory a proof or a corroboration of the theory? Is the resonance theory essentially a theory with physical meaning, or a mathematical technique or both? Has the resonance theory a basis in related sciences, such as physics? Is the resonance theory applicable in all aspects of chemical valence or is it in conflict? [54] The meeting did not take place basically because of the unwillingness of the Soviets, but the points that each party would have had to address were indicative of the uncertainties involved as to the methodological significance and the ontological status of resonance in quantum chemistry.

The differences between a chemical and a physical approach to valence were assessed by Lewis in his last paper on the chemical bond written in 1933, ten years after the publication of his book on valence.[55] Lewis presented a critical evaluation of his earlier ideas and proposals in view of the completion of the first stage in the development of quantum mechanics. He felt that the idea of electronic structure, and, in fact, all the structural ideas which are used by chemists, were obtained by a method which he called analytical, in the sense that from a large body of experimental material the chemist attempts to deduce a body of simple laws which are consistent with the known phenomena. According to Lewis the mathematical physicist postulates laws governing the mutual behavior of particles and then attempts to synthesize an atom or a molecule. When the attempt is successful the physicist has a weapon of extraordinary power which enables answers to questions quantitatively which at best could be answered qualitatively by the other method. Of course, an inaccuracy in a single fundamental postulate may completely invalidate the synthesis, while the results of the analytical method can never be far wrong, since they depend crucially on numerous experimental results. [56]

> It was a recognition of this fact, when I first deduced the idea of the electron pair bond from an analysis of chemical facts, that emboldened me to accept this deduction; although it was obviously incompatible with the then accepted laws of electromagnetics and mechanics. The qualitative principles of molecular structure were presented... as the minimum demands of the chemist which must eventually be met by the more far-reaching and quantitative work of the mathematical physicist. They have not been altered during the recent unprecedented period of discovery in mathematical and experimental physics; but it is possible now to amplify a little, and to express more nearly in the language of modern physics, these minimum demands, or qualitative specifications, to which we have reason to believe the quantitative results of mathematical physics must eventually conform.

For Lewis it was the interplay between the qualitative interventions of chemistry and the quantitative interventions of physics that expressed the constitutive methodology of quantum chemistry. One should note that quality and quantity are not used in their usual sense when they are associated with chemistry and physics, respectively. Chemistry's phenomenological rules acquire their qualitative reliability because of the quantitative exactness of chemistry, whereas the quantitative results of mathematical physics have their particular significance because of the qualitative character of the initial hypotheses of physics. Articulating the methodological novelties of quantum chemistry in such a manner, brings forth a whole series of factors contributing to the becoming of quantum chemistry. This subdiscipline owes its emergence to many complicated and subtle factors and not simply to the solutions proposed for the technical difficulties met with the application of quantum mechanical methods to the problems of chemistry.

Many reasons contributed to the successful way in which quantum chemistry developed in the United States. Mulliken summarized them well when he called himself a middle man between theory and experiment, and between physics and chemistry. A particular kind of institutional atmosphere accounted for the appearance of this new type of scientist, whose definition as a chemist or physicist was in many instances a matter of chance, personal preferences or of institutional affiliation. The institutional ties between chemistry and physics were stronger in the United States than in Europe. In universities like Berkeley and Caltech chemistry students were often learning as much physics as chemistry, and thus were more apt to learn and accept quantum mechanics than their European conterparts. Pauling's knowledge of physics was impressive and Mulliken was an expert on the quantum theory of molecules. Besides Berkeley, Caltech, Harvard and MIT, more universities were promoting the cooperation between physics and chemistry departments. Examples were Princeton, Chicago, Michigan, Minnesota, and Wisconsin. But before Mulliken, Pauling, Slater and Van Vleck, the preceding generation of chemists and physicists---chemists like Lewis and Noyes, R. Tolman, and W. Harkins, and physicists like E.C. Kemble and R.T. Birge---planted the seeds which blossomed into quantum chemistry.

The ability of scientists to be at ease in theory and in experiment might as well account for the successful development of quantum chemistry. Mulliken started as an experimentalist but shifted into theory owing to the delay in getting the high-resolution spectrograph he had been promised when he moved to Chicago in 1928. Pauling's determinations of crystal structures were instrumental as a source of practical information on bond angles and bond lengths to be used in his future more theoretical endeavors.

An altogether different situation occurred in Europe. In England, as Douglas Hartree wrote London, physics meant experimental physics and theoretical physics was in general done by mathematicians.[57] There was also a sharp division between theory and experiment in the German physical community. As to the German chemists they were in general ill-prepared to cope with the challenges of quantum mechanics. One example was Kasimir Fajans, a Professor of Physical Chemistry in Munich when the young Pauling was there in 1926-1927. He was relatively young at the time but old enough to be already one of the leading physical chemists in the world. Many years later, in 1987, Pauling remembered that Fajans' inability to get a good grasp of quantum mechanics was a problem that bothered him the rest of his life.[58]

The differences between Heitler, London and Hund on the one hand, and Pauling and Mulliken on the other hand can best be understood in terms of two diverging cultures for doing quantum chemistry. To the Germans, the American pragmatism looked quite messy; to the Americans, the German's obsession with first principles appeared an unnecessary torture of the chemist's mind. In the 1930s, quantum chemistry developed an autonomous language with respect to physics, and what appeared to be disputes over methods were, in fact, discussions concerning the collective decision of the chemical community about methodological priorities and ontological commitments. These were issues that largely transcended the question of the application of quantum mechanics to chemical problems.

The genesis and development of quantum chemistry as an autonomous subdiscipline owed much to those scientists who successfully managed to escape from the "thought forms of the physicist"[59] by implicitly or explicitly addressing issues such as the role of theory in chemistry, and the methodological status of empirical observations. These were the kinds of reflections which helped inaugurating the creation of a new space for chemists to go about practicing their discipline.

Notes

[1] A. Simões, *Converging Trajectories, Diverging Traditions: Chemical Bond, Valence, Quantum Mechanics and Chemistry, 1927-1937*, Ph.D. Dissertation, University of Maryland at College Park, 1993; K. Gavroglu, A. Simões, "The Americans, the Germans and the beginnings of quantum chemistry," *Historical Studies in the Physical and Biological Sciences*, 25:1 (1994), 47-110; K. Gavroglu, *Fritz London. A Scientific Biography*, (Cambridge: Cambridge University Press, 1995).

[2] P.A.M. Dirac, "Quantum mechanics of many electrons," *Proceedings of the Royal Society of London*, A123 (1929), 714-733.

[3] Pauling Papers, Box 242, Popular Scientific Lectures 1925-1955, "Recent Work on the Structure of Molecules," Talk given to the Southern Section of the American Chemical Society, 1936.

[4] L. Pauling, "The Nature of the chemical bond. Application of results obtained from the quantum mechanics and from a theory of paramagnetic susceptibility to the structure of molecules," *Journal of the American Chemical Society*, 53 (1931), 1367-1400; "The Nature of the chemical bond. II. The one-electron bond and the three-electron bond," 53 (1931), 3225-3237; "The Nature of the chemical bond. III. The transition from one extreme bond type to another," 54 (1932), 988-1003; "The Nature of the chemical bond. IV. The energy of single bonds and the relative electronegativity of atoms", 54 (1932), 3570-3582; L. Pauling, G. Wheland, "The Nature of the chemical bond. V. The quantum mechanical calculation of the resonance energy of benzene and naphthalene and the hydrocarbon free radicals", *Journal of Chemical Physics*, 1 (1933), 362-374; L. Pauling, J. Sherman, "The Nature of the chemical bond.

VI. The calculation from thermochemical data of the energies of resonance of molecules among several electronic structures," *Journal of Chemical Physics*, 1 (1933), 606-617; L. Pauling, J. Sherman, "The Nature of the chemical bond. VII. The calculation of resonance energy in conjugated systems," 1 (1933), 679-686.

[5] Pauling Papers, Box 242, Popular Scientific Lectures 1925-1955, "Resonance and Organic Chemistry," 1941.

[6] Pauling Papers, Box 71, Noyes, A.A. Correspondence 1921-1938, Letter Pauling to Noyes, started 17 December 1926, and finished 28 December 1926.

[7] H.Kragh "Quantum Interdisciplinarity:Friedrich Hund and Early Quantum Chemistry", invited paper delivered at symposium in Gottingen, the Georg-August Universitat, February 6, 1996, in honour of the 100th birthday of Friedrich Hund.

[8] In D.A. Ramsay, J. Hinze, eds, *Selected Papers of Robert S. Mulliken* (Chicago: Chicago University Press, 1973), 13.

[9] Archives for the History of Quantum Physics. Interview with Mulliken.

[10] Pauling Papers, Box 210, Linus Pauling Notes and Calculations vol.III, 1926-1927 "Work on Molecular Orbits," 1926.

[11] Private communication to Ana Simões.

[12] L. Pauling, "The shared-electron chemical bond," *Proceedings of the National Academy of Sciences*, 14 (1928), 359-362.

[13] J.E. Lennard-Jones, "The electronic structure of some diatomic molecules," *Transactions of the Faraday Society*, 25 (1929), 668-686.

[14] Pauling Papers, Box 371, Chapters of Manuscript for Book on Quantum Mechanics of Organic Molecules. Never Published. Emphasis not in original.

[15] R.S. Mulliken, "Electronic structure of polyatomic molecules and valence. VI. On the method of molecular orbits," *Journal of Chemical Physics*, 3 (1935), 375-378, on 376. Emphasis in original.

[16] R.S. Mulliken, "Bonding power of electrons and theory of valence," *Chemical Reviews*, 9 (1931), 347-388; R.S. Mulliken, "Electronic structure of polyatomic molecules and valence. II. General considerations," *Physical Review*, 41 (1932), 49-71.

[17] W. Heitler, F. London, "Wechselwirkung neutraler Atome und homöopolare Bindung nach Quantenmechanik," *Zeitschrift für Physik*, 44 (1927), 455-472.

[18] The paper on the "Bonding Power of Electrons and Theory of Valence", *Chemical Reviews*, 9 (1931), 347-388, as well as the two first papers of the series on the "Electronic Structure of Polyatomic Molecules and Valence," *Physical Review*, 40 (1932), 55-62; 41 (1932), 49-71, appeared in the same period in which Pauling was publishing the first four papers of "The Nature of the Chemical Bond" series.

[19] R.S. Mulliken, "Electronic structure of polyatomic molecules and valence. VI. On the method of molecular orbits," *Journal of Chemical Physics*, 3 (1935), 375-378.

[20] R.S. Mulliken, "Band spectra and chemistry," *Chemical Reviews*, 6 (1929), 503-543; "Bonding power of electrons and theory of valence," *Chemical Reviews*, 9 (1931), 347-388.

[21] R.S. Mulliken, "Electronic structure of polyatomic molecules and valence," *Physical Review*, 40 (1932), 55-62; "Electronic structure of polyatomic molecules and valence. VI. On the method of molecular orbits," *Journal of Chemical Physics*, 3 (1935), 375-378.

[22] A. Russo, "Mulliken e Pauling: Le due vie della chimica-fisica in America," *Testi & Contesti*, 6 (1982), 37-59. Russo explores the idea, first stated by Mulliken on the two different "ideologies," his own and Pauling's, to attack the problem of molecule formation. In this paper, new light is thrown on this question.

[23] A. Assmus, "The molecular tradition in early quantum theory," *Historical Studies in the Physical and Biological Sciences*, 22 (1992), 209-231; " The Americanization of molecular physics," *Historical Studies in the Physical and Biological Sciences*, 23 (1993), 1-33; A. Simões, *Converging Trajectories, Diverging Traditions: Chemical Bond, Valence, Quantum Mechanics and Chemistry, 1927-1937*, Ph.D. Dissertation, University of Maryland at College Park, 1993.

[24] Starting with R.S. Mulliken, "The isotope effect in bad spectra. II. The spectrum of boron monoxide," *Physical Review*, 25 (1925), 259-294.

[25] R.S. Mulliken, "Electronic structure of polyatomic molecules and valence," *Physical Review*, 40 (1932), 55-62; R.S. Mulliken, "Electronic structure of polyatomic molecules and valence. VI. On the method of molecular orbits," *Journal of Chemical Physics*, 3 (1935), 375-378.

[26] L. Pauling, *The Nature of the Chemical Bond and the Structure of Molecules and Crystals: An Introduction to Modern Structural Chemistry*, (New York: Cornell University Press, 1939).

[27] H.C. Longuet-Higgins, "Robert Sanderson Mulliken", *Biographical Memoirs of the Fellows of the Royal Society*, 35 (1990), 329-354, on 337.

[28] J.H. Van Vleck, A. Sherman, "Quantum theory of valence," *Reviews of Modern Physics*, 7 (1935), 167-227, on 169.

[29] Archives for the History of Quantum Physics, Reel 49, Van Vleck Correspondence, Letter from Van Vleck to Mulliken, 22 February 1927.

[30] F. Hund, "Zur Deutung einiger Erscheinungen in den Molekelspektren," *Zeitschrift für Physik*, 36 (1926), 657-674; "Zur Deutung der Molekelspektren. I," 40 (1927), 742-764; "Zur Deutung der Molekelspektren. II," 42 (1927), 93-120; "Zur Deutung der Molekelspektren. IV," 51 (1928), 759-795.

[31] R.S. Mulliken, "Systematic relations between electronic structure and band spectrum structure in diatomic molecules. III. Molecular formation and molecular structure," *Proceedings of the National Academy of Sciences*, 12 (1926), 338-343.

[32] R.S. Mulliken, "The interpretation of band spectra. I, II, III," *Reviews of Modern Physics*, 2 (1930), 60-115; 3 (1931), 89-155; 4 (1932), 1-86.

[33] R.S. Mulliken, "The assignment of quantum numbers for electrons in molecules. I," *Physical Review*, 32 (1928), 186-222; "The assignment of quantum numbers for electrons in molecules. II. Correlation of molecular and atomic electron states," 761-772; "The assignment of quantum numbers for electrons in molecules. III. Diatomic hydrides," 33 (1929), 730-747.

[34] Mulliken Papers, Box 72, Folder 14, Talk 24 February 1928.

[35] Mulliken Papers, Box 74, Folder 6, Talk at the Chicago Meeting of the National Academy of Sciences, 1937 "Electronic Structure of Molecules." Emphasis in original.

[36] L. Pauling, "The Nature of the chemical bond. Application of results obtained from the quantum mechanics and from a theory of paramagnetic susceptibility to the structure of molecules," *Journal of the American Chemical Society*, 53 (1931), 1367-1400, on 1367.

[37] Pauling Papers, Box 212, LP Berkeley Lectures: Quantum Mechanics 1929-1933. Conclusion to the lectures on "The Nature of the Chemical Bond."

[38] Pauling Papers, Box 242, Popular Scientific Lectures 1925-1935, "Recent Work on the Structure of Molecules." Part II. Section "Pauling's Response to the Heitler and London Paper."

[39] Pauling Papers, Box 157, CIT General Files 1922-1964, Letter Pauling to A.A. Noyes, 18 November 1930, emphasis in original.

[40] Private Communication to Ana Simões.

[41] Archives for the History of Quantum Physics. Interview with Heitler.

[42] Archives for the History of Quantum Physics. Interview with Wigner.

[43] G.W. Wheland, *The Theory of Resonance and Its Applications to Organic Molecules* (New York: John Wiley & Sons, 1944).

[44] Pauling Papers, Box 115, Letter Wheland to Pauling, 20 January 1956.

[45] Pauling Papers, Box 115, Letters Pauling to Wheland, 26 January and 8 February 1956.

[46] Pauling Papers, Box 115, Letters Pauling to Wheland, 26 January and 8 February 1956.

[47] L. Pauling, "Modern structural chemistry. Nobel lecture, December 11, 1954," in *Nobel Lectures in Chemistry 1942-1962* (Amsterdam: Elsevier Publishing Company, 1964), 134-148.

[48] L. Pauling, "The nature of the theory of resonance," in Sir A. Todd, ed., *Perspectives in organic chemistry, dedicated to Sir Robert Robinson* (New York: Interscience publishers, 1956).

[49] L. Pauling, *The nature of the chemical bond and the structure of molecules and crystals. An introduction to modern structural chemistry* (New York: Cornell University Press, 1967), third edition.

[50] L. Pauling, "The nature of the theory of resonance," in Sir A. Todd, ed., *Perspectives in organic chemistry, dedicated to Sir Robert Robinson* (New York: Interscience publishers, 1956), 6-7; L. Pauling, *The nature of the chemical bond and the structure of molecules and crystals. An introduction to modern structural chemistry* (New York: Cornell University Press, 1967), third edition, 219-220.

[51] C.A.Coulson, "The meaning of resonance in quantum chemistry" *Endeavour*, 6 (1947), 42-47, on 47.

[52] D.N. Kursanov, M.G. Gonikberg, B. Dubinin, M.I. Kabachnik, E.D. Kaveraneva, E.N. Prilezhaeva, N.D. Sokolov, R.Kh. Freidlina, "The present state of the chemical structural theory". Translation in English by I.S.Bengelsdorf published in *Journal of Chemical Education*, January 1952, 2-13; See also V.M. Tatevskii, M.I. Shakhparanov, "About a Machistic theory in chemistry and its propagantists." Translation by I.S. Bengelsdorf in *Journal of Chemical Education*, January 1952, 13-14. It was not, of course, the case that such sentiments were shared by all the chemists of the community. Characteristic of the differences is the editorial note to the first article where it is stressed that any particular way of dealing with chemical phenomena should not be excluded on *a priori* grounds, but should be first closely studied. In the same article Ya. Syrkin and M. Dyatkina were also attacked. They were the authors of the excellent book *The Chemical Bond and the Structure of Molecules* and had translated Pauling's book into Russian. See also I. Moyer Hunsberger, "Theoretical chemistry in Russia," *Journal of Chemical Education*, October 1954, 504-514.

[53] D.N. Kursanov et al., "The present state of the chemical structural theory". Translation in English by I.S. Bengelsdorf published in *Journal of Chemical Education*, January 1952, 2-13, on 5.

[54] Pauling Papers, Box 261. Letter M.V. King to Pauling, 23 January 1953; letter Coulson to Pauling, 9 October 1953; letter Coulson to King, 18 January 1954; letter King to Pauling, 9 February 1954.

[55] G.N. Lewis, *Valence and the Structure of Atoms and Molecules* (New York: The Chemical Catalog Company, 1923).

[56] G.N. Lewis, "The chemical bond," *Journal of Chemical Physics*, 1 (1933), 17-28, on 17-18.

[57] London Archives, Letter Hartree to London, 16 September 1928.

[58] Letter Pauling to Reymond Holmen, March 1987, as quoted in R. Holmen "Kasimir Fajans" *Bulletin for the History of Chemistry*, 6 (1990), 7-15.

[59] C.A. Coulson, "Recent developments in valence theory," *Pure and Applied Chemistry*, 24 (1970), 257-287, on 259.

POTENTIAL ENERGY HYPERSURFACES FOR HYDROGEN BONDED CLUSTERS (HF)$_N$

MARTIN QUACK AND MARTIN A. SUHM
Laboratorium für Physikalische Chemie
ETH Zürich (Zentrum)
CH-8092 Zürich, Switzerland

Abstract.
Hydrogen fluoride clusters (HF)$_n$ serve as prototype systems for the study of hydrogen bonding on the basis of accurate potential energy hypersurfaces. We discuss conceptual aspects of calculating, sampling, decomposing and representing such multidimensional surfaces and present the current status of molecular one-, two- and three-body potentials for HF as well as results for potential hypersurfaces of (HF)$_n$, $n=2$–6.

1. Introduction

Hydrogen bonds are of great importance for many physical-chemical processes in our everyday life, as they have just about enough strength to generate well defined structure and moderately stable aggregates at ordinary temperatures. They generate the structure of liquid water – an essential environment for life, as well as the structure of DNA and the 3-dimensional structure of proteins. Yet these bonds are weak enough to also allow for the flexibility which is so characteristic for life. Surprisingly, very little is known in quantitative terms about the quantum structure, thermodynamics and reaction dynamics even of the simplest hydrogen bonded systems. Such simple questions as the quantitative kinetics and thermodynamics of homogeneous condensation of a hydrogen bonded liquid have no answer to this day – and even less the question of the quantitative hydrogen bond reaction dynamics in DNA. From a theoretical point of view this lack of knowledge is in part due to a lack of adequate concepts and quantitative knowledge of potential energy hypersurfaces for hydrogen bonding. This situation is not unique to hydrogen bonding but it is particularly obvious in this case.

The concept of the hydrogen bond has reached maturity [1], but hydrogen bonding continues to be intensely investigated due to its prominent role in nature [2] and its intriguing properties [3–5]. The electronic structure calculation of hydrogen bonds is rather straightforward, as the elements (H,N,O,F) are light, the bonding is strong, and electronically excited states are distant. Interpretation of the bond mechanism is more controversial. Electrostatic contributions dominate at long distances [6], but their success for equilibrium structures builds partly on error compensation [7]. For quantitative work, polarisation, exchange, dispersion and other contributions have to be included, and their interplay is not easily cast into analytical form. Thus, there is a widespread need for accurate analytical potential energy surfaces (PES, strictly speaking *hyper*surfaces) [8–12], in which the quantum dynamics of the hydrogen bond [13–17] can be studied.

Experimentally, IR spectroscopy plays a dominant role in the elucidation of the hydrogen bond dynamics, which has a strongly coupled, strongly anharmonic and strongly quantized, multidimensional nature [18,19], which usually prevents a simple inversion strategy [20] to obtain the potential hypersurfaces. It is more advantageous to derive the detailed shape of the potential from state of the art electronic structure calculations and to apply semiempirical corrections where spectroscopic data prove a need for it.

For such an ambitious program, it is desirable to select a simple, yet characteristic prototype system. Hydrogen fluoride is an ideal candidate, as its dimer is the most completely characterized hydrogen bond species incorporating a donor and an acceptor molecule [21], and its abundant cyclic oligomers [22, 23] allow for a systematic study of size effects, also with respect to cooperativity [24–26]. Experimental and theoretical studies of these clusters carried out in Zürich (e.g. [21–23, 25, 27–41]) and elsewhere (e.g. [42–88]) have culminated in the development of various generations of potential energy hypersurfaces. In the present contribution, the strategies behind these developments and the resulting potentials are briefly summarized, with emphasis on the conceptual trends which can be applied to hydrogen bonding in general.

We expect that the hydrogen fluoride PES presented here will find widespread applications in inelastic scattering [89–97], in low [35,39,98–100] and high resolution IR [22, 27–29, 31, 40, 46, 53, 70–73, 76, 84–86, 101, 102], microwave [42, 103–106], and NMR spectroscopy [107–109], in the interpretation of matrix spectra [110–113], electron diffraction data [114], predissociation results [61, 82, 83, 115–123], thermodynamic [23, 50, 124–130], kinetic [131], and nuclear dynamical modelling [39, 59, 66–68, 79–81, 87, 132–141], molecular dynamics [142, 143], classical [47, 69, 144], and quantum

Monte Carlo simulations [32, 34, 38, 63, 145], assessment of quantum chemistry benchmarks [21, 25, 43, 48, 49, 55, 58, 60, 64, 74, 75, 77, 78, 146–172], fitting approaches [30, 36, 37, 41, 45, 51, 52, 54, 56, 57, 62, 65, 173, 174] and the development of empirical refinement strategies [32, 44, 88].

2. Sampling and Representation Methodology

How can we ever hope to obtain a reliable PES of – say – $(HF)_5$, a typical and important constituent of the HF vapor phase? Its $3N-6=24$ internal degrees of freedom contain a vast amount of energetic detail, even if we characterize each coordinate by only, say, 9 displacements from a given reference configuration in order to characterize the important part of configuration space, which is clearly an overly optimistic assumption for most degrees of freedom. The bulk of information is not in the separate coordinates – that only amounts to $1+9\times24=215$ configurations, reduced by up to a factor of 5 due to the 5-fold symmetry in the minimum structure for our numerical example. The overwhelming amount of information lies in the coupling of the coordinates to each other, whose description would require on the order of $(1+9)^{24}$ points for 9 displacements. Again, this number is reduced by a factor of up to $5!=120$ due to symmetry (although the full symmetry saving is only approached for more than 9 displacements), i.e. we require on the order of 10^{23} points.

Obviously, the problem grows exponentially with cluster size [175] and special strategies are needed in order to cope with it [176]. The ultimate goal would be to obtain a complete, quantitatively satisfactory understanding of the analytical structure of such multidimensional potential hypersurfaces. We shall discuss some of the strategies towards this goal and their limitations with the example of HF cluster potential energy surfaces, although the conclusions apply to other hydrogen bonded systems as well.

2.1. FREEZING THE STRUCTURE OF THE FLEXIBLE MONOMERS?

A tempting and frequently used simplification of the configuration space problem is the freezing of all internal monomer degrees of freedom [43, 65, 176, 177]. After all, there is a significant energy gap between chemical and hydrogen bonds (a factor of ≈ 45 in the case of HF dimer [61, 178]). Staying with our $(HF)_5$ example, 10^{18} instead of 10^{23} configurations would be sufficient due to the reduction to 19 dimensions, with more significant savings for larger clusters of more complex monomers. To obtain consistent dissociation results, it is necessary to freeze the monomer coordinate at equilibrium, or even better, at the vibrationally averaged geometry.

What price is to be paid for this simplification? Obviously, the dynamics of the monomer degrees of freedom is lost. In experimental cluster spectroscopy, these high frequency degrees of freedom usually offer the easiest approach to hydrogen bond interactions [19, 22, 46]. With classical nuclear dynamics applications in mind, one may argue that freezing the highly quantized high frequency monomer coordinates is better than approximating them classically [176]. But what are the effects of rigid monomers on the *energetics* of the hydrogen bond? This is quickly estimated in the case of HF clusters. As we will elaborate in section 5.3, the monomer elongation due to complexation is up to 4 pm [109, 114] in the larger clusters. This corresponds to an energy of about 4 kJ/mol per hydrogen bond which is gained upon relaxation of the monomers. Thus, up to 10% of the cluster stabilization energy is missing in a rigid monomer treatment. This error is already larger than the uncertainty of state-of-the art ab initio calculations for these systems [25, 41, 77, 78, 109], but an additional error arises in the zero point energy contribution. Hydrogen bonding induces sizeable negative frequency shifts of the directly involved monomer XH stretching modes [35, 39]. These frequency shifts stabilize the cluster through the reduction of its zero point energy content, thus adding to the relaxation effect. For HF clusters, the zero point motion effect is of the same order of magnitude as the elongation effect, i.e. up to 10% [39]. For DF clusters, it is reduced by one third. Rigid monomer approximations, which can simplify the quantum dynamical treatment very significantly [179], may be acceptable in weak van der Waals complexes [180] and in qualitative investigations of HF dimer, but they are inappropriate for the strong hydrogen bonds in large HF clusters. We have thus adopted a flexible monomer approach from the very beginning [30, 32, 34] and have sought other ways to simplify the problem.

2.2. LOCAL OR GLOBAL SURFACES?

One way to reduce the configurational vastness available to a cluster such as $(HF)_5$ is the restriction to a more or less local environment of the minimum energy structure. Concessions to the globality of an analytical potential energy surface are nearly always made. There is little point in representing an united atom limit for $(HF)_5$, such as Ne_5 or even Sn (not to mention the loss of isotopic invariance...). More realistic, but still very high-lying reaction channels such as those involving F_2 and H_2 [181] are also of minor relevance to the hydrogen bond dynamics in our present applications. Desirable reactive channels would be those which lead to a concerted exchange of hydrogen atoms between neighbouring fluorine atoms [109, 166, 168] and those which lead to ionic products such as FHF^- [182] and HFH^+ [183], perhaps also in applications to condensed phases. However, the constraint of monomer integrity offers significant simplifications of the analytical fitting problem and has normally been applied in our analytical potential energy surfaces (see, however, section 5.5), which may thus be characterized as semi-global. Such a constraint does not affect the hydrogen bond dynamics of the smaller clusters, even at rather high vibrational energies. However, it is a significant approximation in the larger clusters for certain dynamical processes, which have not yet been observed in the IR spectra (see section 5.5). Such a locally constrained dynamics can be made to be the starting point for a more global dynamics by combining regions of space connected by low energy concerted reaction paths in a second step.

The majority of theoretical hydrogen bond studies characterizes the multidimensional potential energy surfaces through their local curvature at selected stationary points [64, 165], nuclear configurations with vanishing potential energy gradient. This is a dramatic approximation compared to global or semiglobal approaches, but it proves to be extremely fruitful for qualitative and semi-quantitative work. The computational effort to locate stationary points on multidimensional PES is only few (say 1-3) orders of magnitude higher than that of single point calculations for many standard quantum chemistry approaches. The same holds for the local curvature, i.e. the force field, with the cluster sizes discussed here. While direct comparison to experimental structures and dynamics is masked by anharmonic quantum effects, the stationary points and their force fields still provide reliable *trends* with cluster size, if electron correlation is accounted for in an adequate way.

Here, a double strategy is clearly indicated. Local characterization of the potential energy surfaces with a multitude of electron correlation approaches provides a zeroth order picture of the structural, energetical and dynamical cluster trends up to relatively large clusters [109]. Semiglobal PES for small and intermediate clusters provide the anharmonic aspects which are required for comparison to experiment, particularly for hydrogen motion [32].

2.3. 'ON-THE-FLY' OR ANALYTICAL REPRESENTATIONS?

The large number of potential energy values which are required to define a multidimensional PES such as that of $(HF)_5$ (on the order of 10^{23} in our numerical example), suggests a radical approach: The number of PES evaluations for a typical dynamical simulation will often be much less, perhaps of the order of 10^6–10^9. Thus, one could evaluate the PES (and its derivatives) 'on-the-fly', precisely at those points which are requested by the nuclear dynamics. In the Car-Parrinello technique [184], this has been turned into practice within the density functional framework through an ingenious algorithm which yields a (classical) dynamics step at much less computational effort than a single potential energy and gradient evaluation. More recently, 'on-the-fly' approaches have also been extended to more conventional quantum chemistry techniques and dynamical algorithms [185, 186]. In both cases, compromises are necessary on the quantum chemistry level due to the large number of nuclear configurations involved. In the density functional case, the deficiencies of the zeroth order local density approximation (LDA), which are formidable for intermolecular interactions, have to be overcome. It is now accepted that gradient corrections [187, 188] can achieve this to a certain extent, but hybrid approaches involving admixture of exact exchange [189] are clearly superior and should be incorporated into the Car-Parrinello framework. In conventional quantum chemistry, where systematic convergence towards accurate PES is now impressively well mastered for hydrogen bonded systems [41, 77, 78, 190], any saving in the number of energy evaluations can be translated into improved energy values. 'On-the-fly' approaches are intrinsically wasteful (but memory-saving), as they do not accumulate knowledge on the potential energy surface. If the dynamical trajectory returns to a previously visited region, this is not reflected in computational saving.

At the other extreme we have analytical representations (fits and interpolations) of the PES to pre-selected configurations. For multidimensional problems, interpolations are less common [191, 192] and we have usually concentrated on fits, using the Levenberg-Marquardt algorithm [175, 193]. If analytical fits are carried out carefully and with the physical nature of the dominant interaction terms in mind, they can be very economic in their requirement of configurations, as we will see below. This is true particularly for problems in relatively low (up to about 5–10) dimensions. However, the human labour involved in finding appropriate analytical representations is formidable, often a more serious problem than the computational labour involved in the electronic structure calculations. As the rapid growth in computer power is not quite matched by that of our brains, the bottle-neck is more and more in the analytical fitting. Furthermore, the selection of appropriate configurations is non-trivial for complicated PES topologies. Also, there is usually a systematic fitting bias due to insufficient flexibility of the analytical model [37]. Nevertheless, an important aspect of a 'good' analytical representation of the PES always remains unmatched by any other approach: It provides a mathematical and sometimes physical, global *understanding of the structure of the PES.*

2.4. COMPROMISE 'ON-THE-FLY' STRATEGIES: VORONOI STEP REPRESENTATIONS

The dilemma between laborious analytical representations and wasteful 'on-the-fly' approaches has triggered some recent midway strategies. The general maxim of these strategies is:

Let the dynamics do the job of sampling the essential regions of configuration space, but let an algorithm automatically collect these points on the PES and construct a systematically improving global representation for the relevant configuration space.

Automatic algorithms are particularly suitable for those contributions to the PES which are not easily cast into analytical form. Wherever the analytical form of a physical contribution to the PES is accurately known, such as for electrostatic model interactions [6,194], this can be subtracted from the total energy prior to the application of self-refining strategies.

Probably the most elementary version of such an automated midway approach is the Voronoi step representation (VSR) [36], which has been successfully applied to classical, path-integral, and diffusion Monte Carlo techniques [36, 37].

At the heart of this method is a distance norm $\mathcal{D}(L, R)$, which defines the proximity between two given points L, R in configuration space. Distance in this context may be viewed as a mixed geometric and energetic quantity. Two configurations which are similar in geometry but strongly different in energy, e.g. due to their location on a steep repulsive wall, should be regarded as more distant than two configurations with similar energy despite significant geometry difference (energy may be replaced by other properties of interest in this context). A simple heuristic choice of the distance norm $\mathcal{D}$ for PES was proposed in [36], namely

$$\mathcal{D}(L,R) \sim \sum_{ij} \left(\frac{1}{\lambda_{ij}^2} - \frac{1}{\rho_{ij}^2} \right)^2 \qquad (1)$$

Here, i and j denote atoms, ρ_{ij} is the cartesian $i-j$ atom-atom distance in configuration R and λ_{ij} is the same quantity in configuration L. Summing over *all* atom-atom distances becomes redundant for molecules with more than 4 atoms, but it remains a simple and universal measure of configurational similarity. The reciprocal dependence of $\mathcal{D}$ on cartesian distance exploits the simple fact that the interaction energy ultimately drops to zero with increasing distance, while the square avoids computation of square roots.

Based on such a distance norm, the Voronoi step representation of a PES is straightforward. Given a set of reference configurations R with known potential energy $V(R)$, the value of the potential energy at an arbitrary configuration L is simply chosen to be that of the closest reference configuration R. The name of the technique is derived from the configurational domains in which the closest reference configuration suddenly changes, the Voronoi polyhedra [195]. The steps contained in this simple surface representation do not pose a serious problem to most kinds of Monte Carlo simulations, suggesting the following dynamical scheme:

The nuclear dynamics starts at a single configuration R_1, which has a simple plane as its Voronoi step representation. As soon as a given threshold distance from R_1 is surpassed in the Monte Carlo random walk, a second reference configuration R_2 is evaluated, i.e. its potential energy is calculated using an appropriate electronic structure approach. $R_{1,2}$ define a new Voronoi step representation, in which the random walk proceeds until the threshold distance is again exceeded. A third reference configuration R_3 is added, and the random walk continues in the VSR of $R_{1,2,3}$, etc. . In this way, successively refined representations of the surface are generated exactly in those regions relevant to the dynamics. In clusters with identical nuclei, all feasible permutations of a reference configuration are added to the set. After some time, the rate of generation of new reference configurations drops, the relevant configuration space becomes saturated, and the distance threshold can be reduced for further refinements, if required. In addition to distance thresholds, energetic similarity criteria may also be introduced. Beyond the potential energy itself, its gradients may be evaluated and used in the Voronoi step representation. For molecular dynamics calculations, where the surface has to be smooth and differentiable, more sophisticated interpolation schemes have been developed independently [196,197]. These can be viewed as generalizations of the basic Voronoi concept. Some earlier approaches are described in [198,199].

Once the dynamic VSR has converged to a set of optimal nuclear configurations, one can proceed in different ways. For Monte Carlo applications, the VSR may be sufficient by itself. If a smooth surface is required, fitting or interpolation can be applied, using either traditional approaches [9,12,190] or neural networks [200–202] (see also the discussion in [23]). VSR is thus seen to bridge the extremes of 'on-the-fly' simulation and analytical fitting to selected configurations, by introducing memory to the one and automated optimal selection to the other. However, in order to be applicable to large systems such as $(HF)_5$, the VSR has to be complemented by the concept of many-body decomposition.

2.5. MANY-BODY DECOMPOSITION

A fundamental assumption implicit in the concept of intermolecular interactions is the retention of monomer identity in the cluster. Although we will see that this assumption is not entirely adequate for larger $(HF)_n$ clusters, it is still a good starting point. In a first approximation, the interaction potential for an aggregate of n HF monomers can then be written as the sum over all $(n-1)n/2$ ordered pair interactions V^2. The total potential energy $V_{i_1 i_2 \cdots i_n}$ ($i_k = k$; $k = 1, 2, \ldots, n$) for $(HF)_n$ is obtained by including the one-dimensional (1-d) potential energy curve V^1 of the monomer for each of the HF molecules:

$$V_{i_1 i_2 \cdots i_n} \approx \sum_{k=1}^{n} V_{i_k}^1 + \sum_{k=1}^{n} \sum_{l=k+1}^{n} V_{i_k i_l}^2 \quad (2)$$

In this approximation, the 24-d potential energy hypersurface of $(HF)_5$ is reduced to a sum over 5 identical 1-d monomer potentials and $5 \cdot 4/2 = 10$ identical 6-d pair potentials (see fig. 1). As illustrated earlier, this amounts to an enormous saving in the required number of configurations. However, such a pairwise additive approximation is not satisfactory in most hydrogen bond systems and particularly inappropriate in HF clusters. There are contributions to the total energy which depend explicitly on the simultaneous coordinates of three, rather than two monomer units. These three-body terms V^3, and more generally m-body terms V^m ($m \le n$) have to be added to give

$$\begin{aligned} V_{i_1 i_2 \cdots i_n} &= \sum_{k=1}^{n} V_{i_k}^1 + \sum_{k=1}^{n} \sum_{l=k+1}^{n} V_{i_k i_l}^2 + \sum_{k=1}^{n} \sum_{l=k+1}^{n} \sum_{m=l+1}^{n} V_{i_k i_l i_m}^3 + \cdots \\ &+ \sum_{k=1}^{n} \sum_{l=k+1}^{n} V_{i_1 i_2 \cdots i_{k-1} i_{k+1} \cdots i_{l-1} i_{l+1} \cdots i_n}^{n-2} \\ &+ \sum_{k=1}^{n} V_{i_1 i_2 \cdots i_{k-1} i_{k+1} \cdots i_n}^{n-1} + V_{i_1 i_2 \cdots i_n}^{n} \end{aligned} \quad (3)$$

From this, the definition of m-body contributions $V_{j_1 j_2 \cdots j_m}^m$ to the total energy $V_{j_1 j_2 \cdots j_n}$ in terms of the total energy of the fragments $V_{i_1 \cdots i_k}$ is (see also [9, 26, 203, 204]):

$$\begin{aligned} V_{j_1 j_2 \cdots j_m}^m &= V_{j_1 j_2 \cdots j_m} + (-1)^1 \sum_{k=1}^{m} V_{j_1 j_2 \cdots j_{k-1} j_{k+1} \cdots j_m} \\ &+ (-1)^2 \sum_{k=1}^{m} \sum_{l=k+1}^{m} V_{j_1 j_2 \cdots j_{k-1} j_{k+1} \cdots j_{l-1} j_{l+1} \cdots j_m} \\ &+ \cdots + (-1)^{m-2} \sum_{k=1}^{n} \sum_{l=k+1}^{n} V_{j_k j_l} + (-1)^{m-1} \sum_{k=1}^{m} V_{j_k} \end{aligned} \quad (4)$$

Obviously, an m-body term $V_{j_1 j_2 \cdots j_m}^m$ has to vanish as soon as at least one of the m monomers j_k involved attains an infinite distance from the others, and it can be calculated from a maximum of $2^m - 1$ separate electronic structure calculations. We note that many-body expansions are often carried out with respect to atoms [9, 181], whereas retention of the monomer

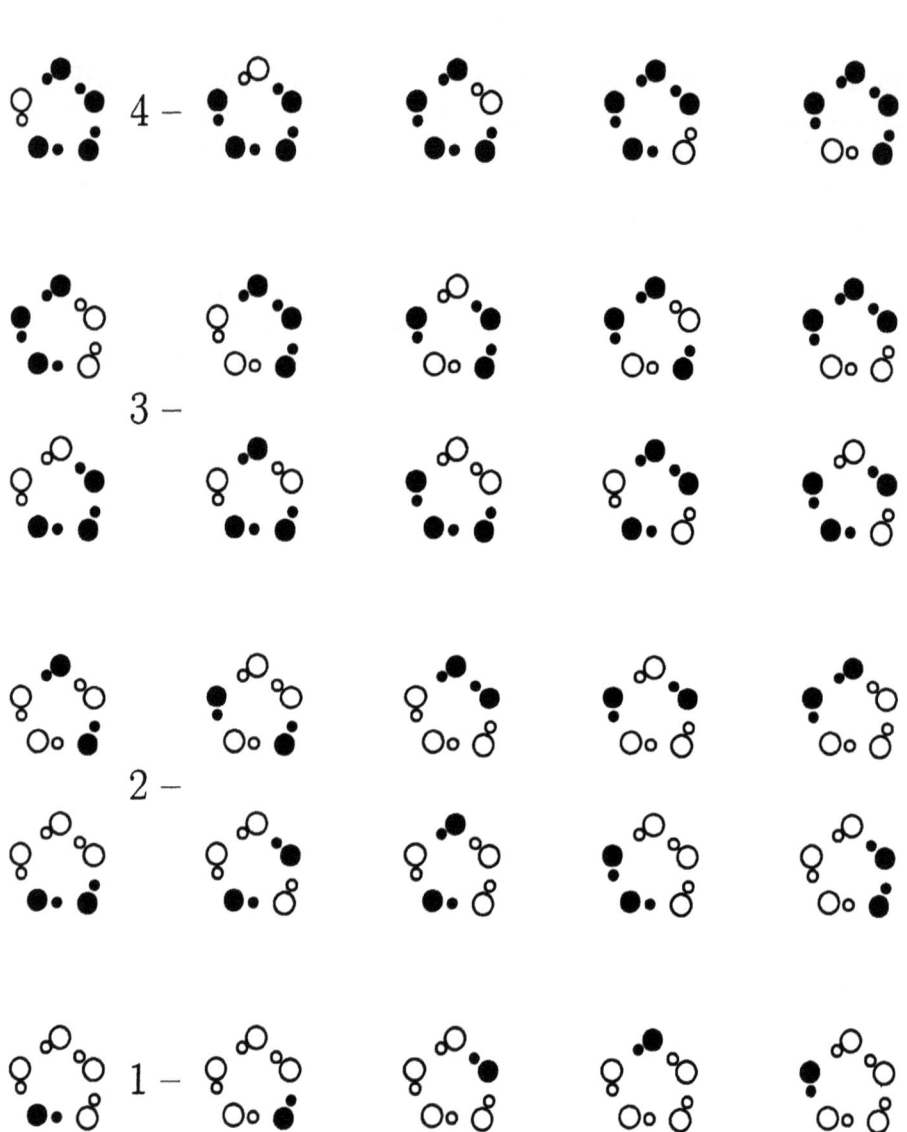

Figure 1. Illustration of 1- to 5-body contributions to the potential energy of a pentamer structure. Filled circles represent the molecules involved in a given n-body contribution.

identity in our PES suggests the HF *molecule* as an appropriate expansion base.

A very important finding for hydrogen fluoride clusters and many other sequentially bound hydrogen bond complexes is the rapid convergence of this many-body expansion of the energy beyond the three-body term [23,25, 76,203,205,206]. Four- and higher-body forces are roughly an order of magnitude smaller than the three-body interactions for typical hydrogen bond geometries. Figure 2 shows a typical result for two pentamer displacement coordinates – concerted hydrogen motion and ring breathing. In this case, all energy contributions beyond the one-body term are attractive. While the three-body term is seen to be very significant even for the qualitative behaviour of the potential, the four-body contribution is much smaller and the five-body effect is hardly visible on the shown scale. Note that properties which depend on energy derivatives, such as vibrational frequencies, may be more sensitive to higher-body contributions, as these clearly depend on the geometry. A many-body decomposition such as that shown in fig. 2 immediately reveals the importance of higher-body contributions for a given property, whereas inspection of cluster size trends [206–208] alone often remains inconclusive. It should be stressed, that in an empirical determination of the relevant potential contributions, say from spectroscopy, it is important to include data, which also provide 'unusual' conformations or configurations at higher than minimal energy. This may not be easy in a purely experimental approach, but causes no problems if ab initio data are used as well.

For HF clusters beyond the pentamer, some higher-body contributions start to cancel among themselves and between each other near equilibrium. This encouraging result brings nearly quantitative PES for arbitrarily large clusters into reach. It suggests that a complete characterization of the dimer, the trimer and (for very accurate work and further away from typical hydrogen bond conformations) the tetramer may prove to be sufficient for an understanding of the energetics of condensed phases of HF [209–212], whereas a pure pair potential would have to rely on severe error compensation for success [177, 213].

2.6. EFFECTIVE OR TRUE PAIR POTENTIALS?

Condensed phases are often modelled by a simplified approach, in which many-body contributions are somehow absorbed into the pair potential [214]. From the preceeding section, it is clear that this is only possible in an averaged, mean-field sense. The resulting effective pair potentials cannot be compared to true pair potentials, as derived from the spectroscopy of molecular dimers. Thus, the direct connection between electronic structure theory and experiment is disrupted, and systematic strategies for improvement are

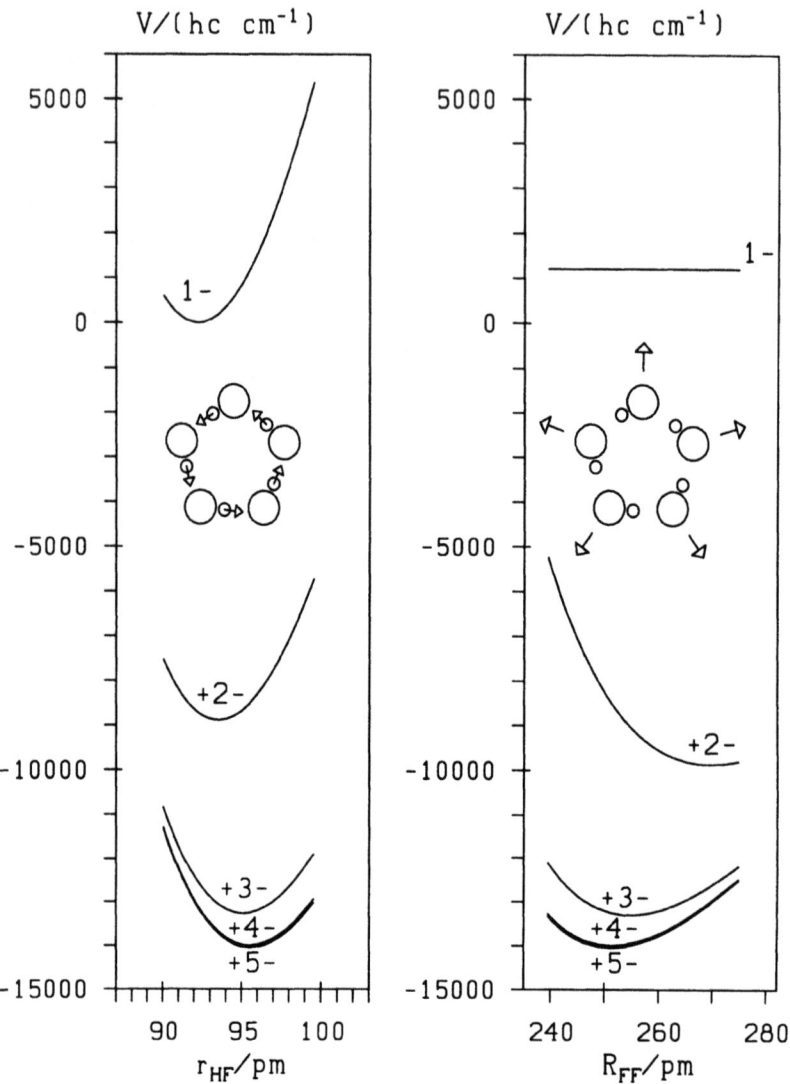

Figure 2. Many-body decomposition of (HF)$_5$ at the 6-311+G** B3LYP level. From top to bottom, the curves marked 1-, +2-, +3-, +4-, +5- represent the potential energy including up to 1-, 2-, 3-, 4-, and 5- body contributions along the totally symmetric HF stretching coordinate r_{HF} (left) and the totally symmetric FF stretching mode, characterized by the nearest FF distance R_{FF} (right), see fig. 6. Convergence is seen to be satisfactory upon 3-body inclusion and essentially complete upon 4-body inclusion for energies, while the shape (curvature) converges somewhat more slowly.

hindered by the introduction of effective pair potentials. Effective pair potentials can be rather successful for weakly interacting systems with rapidly converging many-body interactions. For hydrogen bonded systems, where three-body interactions are very significant, effective pair potentials are little more than parametrizations of bulk phase data in restricted pressure and temperature ranges, with limited extrapolation capabilities.

In a wider sense, all potential energy surfaces discussed here are effective surfaces to some degree. Wherever an approximation is made in the electronic structure approach or in the nuclear dynamics treatment, or wherever the experimental data base remains incomplete, agreement between experiment and theory will always have a fortuitous component. However, these fortuitous aspects can be systematically reduced, at least within the Born-Oppenheimer approximation. Therefore, we prefer an approach in which construction of the pair potential is strictly based on dimer information, while the three-body contributions are developed from trimer data and the rapid convergence of the many-body expansion is tested in the larger aggregates.

2.7. AB INITIO AND EMPIRICALLY ADJUSTED POTENTIALS

As we will see below, state-of-the-art quantum chemistry calculations are capable of semiquantitative, but not quantitative predictions of the HF dimer PES. There will usually be accurate spectroscopic data which even go beyond the fundamental accuracy limits of the potential energy surface concept [215]. Any adjustment of a multidimensional potential energy surface to experiment is faced with a severe difficulty: How should the adjustment be performed without introducing unwanted side effects? The choice of appropriate basis sets and quantum chemical methodology is already a very important empirical adjustment. It should be done carefully, taking into account the anharmonic nature of the PES. As this is only rigorously possible for small systems, the HF dimer is an ideal hydrogen bond prototype for which the ab initio methodology can be tested. Once the method is chosen and the PES has been sampled and fitted, empirical adjustments should be restricted to simple, transparent properties such as the rotational constants or the cluster binding energy. The latter is a notoriously difficult quantity for ab initio theory as it is the difference between two large numbers with large absolute error bars and usually relies on nearly perfect error cancellation. While it is possible to adjust such simple quantities on the analytical surface [32], the adjustment of more complex properties like tunneling splittings [21, 25, 88] or vibrational polyads [21, 41] already runs the risk of deteriorating the performance of the PES for other than the adjusted properties, because the fitting parameters are usually highly correlated. To alleviate the problem, which greatly restricts the predicting power of a modified PES, we have started to use a simple mixing strategy for empirical adjustment [41]. Instead of scanning the PES with one ab initio approach, we use two or more approaches simultaneously [41]. The additional computational effort is often minor, as the most advanced approach typically dominates the computation and lower level results are sometimes obtained at no extra cost (e.g. SCF in MP2 calculations or MP2,3 in MP4 calculations). The different approaches should have a quality ordering. Ideally they might even suggest an extrapolation to the true energy [41,77,216]. Initially, the pointwise potential energies for each approach are separately fitted to the same analytical form, yielding different parameter sets. For each of these parameter sets, experimental quantities such as the spectroscopic binding energy D_0 can be predicted. In order to achieve agreement with experiment, an appropriate combination or extrapolation of the pointwise energies resulting from different ab initio approaches is taken and refitted to the same analytical form. This ensures that the global features of the PES remain largely unaffected while particular properties can be adjusted.

3. Approximations in ab initio electronic structure theory

Electronic structure calculations for weakly bound complexes in general and hydrogen bonds in particular have been reviewed several times (e.g. [217, 218]). Here, we summarize briefly a few key results relevant to HF clusters.

3.1. HARTREE-FOCK BASED METHODS (INCLUDING ELECTRON CORRELATION)

The self-consistent field Hartree-Fock approach itself provides an excellent starting point, as the ground state potential energy surface is well separated from electronically excited states, thus obviating the need for multireference calculations. Within the Hartree-Fock approach, large HF clusters (e.g. knotted rings [25] or extended chains [172]) and periodic systems [152,219] can be treated with moderately-sized basis sets. For more quantitative work, dynamic electron correlation has to be included [150,190], although its contribution in $(HF)_2$ is only a third of that in $(H_2O)_2$ [220] because of the exceptionally small polarizability of HF [221, 222], which enters dispersion interactions quadratically. This suggests 2nd order Møller-Plesset perturbation theory (MP2) as a good compromise between accuracy and efficiency [25,41,75,77,208,223]. The main reason for going beyond the MP2 level lies in the overestimation of the coupling between the monomer and hydrogen bond degrees of freedom [41]. Investigations at MP4 [52,57,77], coupled pair functional [55,64] and coupled cluster levels [41,77] have been carried out, but the currently most successful potential energy surface is still based on MP2 calculations with empirical corrections [41]. The reduction of basis set superposition error (BSSE) [77, 223, 224] is equally important (and computationally demanding at the correlated level) as convergence of the electron correlation contribution. BSSE and other basis set and electron correlation deficiencies often tend to cancel each other for some near-equilibrium properties of hydrogen bonds [77]. While this can be exploited to obtain an overview over cluster size trends at minimum energy structures [25, 64], one should not rely on such a cancellation when scanning multidimensional PES away from the minimum structure, where more subtle aspects of intermolecular interaction come into play [25, 41]. In order to reduce the BSSE and to approach the MP2 basis set limit more efficiently, the R12 approach [216, 225] has been employed successfully [41].

3.2. DENSITY FUNCTIONAL METHODS

The poor scaling of conventional quantum chemistry methods with increasing system size justifies a search for more economic approaches, apart from the many-body decomposition discussed in section 2.5. In this context, the ability of density functional methods [226] to describe hydrogen bonding in HF clusters has been studied recently [75, 109, 169, 208]. The great number of functionals and the lack of systematic improvement strategies renders a judgement quite laborious. While some of the deficiencies are easily recognized through comparison with Hartree-Fock based methods, we have stressed the importance of comparison to experimental data [32, 109]. This is not always simple due to the anharmonic and coupled nature of the PES and the incompleteness of the experimental data base, particularly for larger clusters. We conclude [21, 109] that among the pure gradient-corrected density functionals, the combination of Becke's exchange functional [187] with the correlation functional due to Lee, Yang and Parr [188] (BLYP) performs best, although it has deficiencies which are qualitatively similar to those of the MP2 method (namely exaggerated coupling of the monomer to the hydrogen bond), but much more pronounced. Introduction of exact exchange contributions via Hartree-Fock admixture in the B3LYP functional [188, 189] improves the situation significantly, although this hybrid functional also falls short of the MP2 approach in terms of accuracy [41, 77, 78, 109], while it is computationally less expensive. One should note that an acceptable performance in the hydrogen bonding region does not warrant a comparable quality of density functional PES in other regions and we will discuss such an aspect below. We conclude that density and hybrid functional approaches have powerful exploratory value for system sizes which are not easily amenable to Hartree-Fock based electron correlation methods. This is certainly true for condensed phases [227], for which Car-Parrinello results are now available [143], and to some extent for larger HF clusters. For quantitative, spectroscopic work however, an MP2/CCSD(T) approach [41, 77, 228] seems preferable, since it builds on an excellent zeroth order (Hartree-Fock) approach and provides a systematic improvement strategy.

3.3. QUANTUM MONTE CARLO METHODS

We have briefly discussed the role of Monte Carlo methods in sampling the nuclear configuration space [36, 37] and in determining selected rovibrational quantum states [32, 229]. The very nature of these techniques as well as analytical fitting approaches tolerate a fair amount of statistical scattering in the electronic energies. Hence, electronic structure methods which are possibly less precise but more accurate than conventional basis set methods would be desirable. In this context, quantum Monte Carlo (QMC) algorithms [230–232] may become useful. They scale quite favourably with system size, but not with nuclear charge, so that extended first row element clusters provide a favourable application. QMC methods recover a large amount of electronic correlation energy by finding the lowest eigenvalue consistent with a given fermion nodal structure and by allowing for an explicit introduction of the proper electronic cusps into the wavefunction. While approximation-free applications [233] to extended electronic systems are not yet available and treatments of hydrogen fluoride have not yet gone beyond the monomer [234, 235], electronic quantum Monte Carlo techniques are of great promise for future years due to the foreseeable increase in computing power.

4. Potentials

This section summarizes the current availability of analytical potential energy hypersurfaces for HF clusters, before results for different cluster sizes are given in section 5.

4.1. MONOMER POTENTIAL

A key advantage of HF clusters for the study of hydrogen bonding is the simple structure of the monomer. The potential and dipole functions for HF are very well characterized, both experimentally [178, 236, 237] and from calculations [77, 216, 238–243]. In principle, this knowledge can be fully exploited in the cluster potential energy surfaces [63]. However, the modularity of the many-body decomposition suggests that approximate monomer potentials are suitable as long as the accuracy of pair- and three-body potentials does not match that of the HF monomer potential. The existence of analytical wavefunctions for HF stretching excitations has some advantages in our applications [32]. Therefore we have opted for realistic potential curves which have this feature. A popular one is the Morse oscillator,

$$V(r) = D_e(1 - \exp(-a(r - r_e)))^2 \quad (5)$$

which contains the well depth D_e, the equilibrium bond length r_e and the force constant $f = a^2/(2D_e)$ as adjustable parameters and which we have used in different parametrizations. It can model the first 4 vibrational levels of HF within <0.05% of their experimental value. In order to model rotational excitations at the same level of accuracy, we have recently used [41] a 4-parameter (a, b, A, B) generalized Poeschl-Teller oscillator,

$$V(r) = \exp(-2ar)\left(\frac{B}{(1 - b^2\exp(-2ar))^2} - \frac{A}{(1 + b^2\exp(-2ar))^2}\right) \quad (6)$$

whose vibrational eigenfunctions are also known analytically [244].

4.2. PAIRWISE INTERACTION POTENTIAL

The earliest quantum chemical calculations on HF dimer were reported in 1970 [146, 245] at Hartree-Fock level. Electronic binding energies obtained from crude manual minimization deviate by less than 2% from the most recent best estimate of 19.1 kJ/mol [41]. This is due to fortuitous error cancellation between incomplete minimization, insufficient basis size, BSSE, and lack of correlation treatment. It is not the last error cancellation occurring in this field over the following quarter century...

First systematic scans of the pairwise interaction potential of HF at Hartree-Fock level were restricted to the four intermolecular degrees of freedom [43, 246, 247]. The difficulties in finding suitable analytical representations are well documented in these early attempts [43, 246–248]. A breakthrough was achieved by Alexander and DePristo [173], who expanded the potential in terms of Legendre polynomials. This approach led to the very successful empirical pair potential of Barton and Howard [44] and is still used in the most recent PES [32, 41, 56]. There were numerous attempts to more or less heuristically extend the various 4-d PES to flexible HF monomers [63, 90, 91, 119, 249–252] or to add dispersion terms [45, 248, 249, 252–254], which are missing in the Hartree-Fock approach. Empirical pair potentials, often obtained through calibration to the liquid, have also been obtained [47, 89, 119, 213, 255].

Explicit inclusion of electron correlation effects [48, 150] was applied to restricted scans of the dimer configuration space [49, 54] and to wide meshes which emphasize the repulsive walls [52, 57], leading to analytical fits which typically perform poorly in the hydrogen bond region. All these surfaces have more or less serious deficiencies and limitations, but they provide a rich source of different approaches to the analytical fitting problem. Many of them are reviewed in some more detail in [62].

The only pair potentials which we will discuss in more depth are those building on a coupled pair functional (CPF) approach by the Vienna group, Kofranek, Lischka and Karpfen (KLK) [55], which has been extended over the years [66,162], and those constructed from a recent MP2-R12 approach (KQS) [41] (see fig. 3).

The KLK approach consists of over 10^3 HF dimer energies including electron correlation at the CPF level and a relatively large basis set. The BSSE has not been determined, but it is likely to approach or even exceed 10% of the binding energy. Although this BSSE has not been corrected for in the electronic structure calculation [55], we could show that the binding energy D_e of the KLK PES is still underestimated by about 6% [32,41]. Taken together, this translates into a significantly underestimated hydrogen bond strength, which is also reflected in an overestimated hydrogen bond length (188 pm [55] instead of 183 pm [41] for the H···F distance), and underestimated libration frequencies [41,86]. As these deficiencies are comparable in size to the anharmonic effects in the lowest quantum state of the dimer, they can only be detected via multidimensional quantum dynamics treatments in an analytical representation. Our quantum Monte Carlo simulations [30] in the analytical fit given in [56] (BKLK) revealed artifacts in regions of insufficient ab initio point coverage, which we could avoid in our own analytical representation [30] (QSKLK). As 2/3 of the KLK points lie in the 4-d hyperplane corresponding to rigid equilibrium monomers and only 1/3 correspond to HF elongation, the HF stretching part of the surfaces based on the KLK data set is relatively poorly characterized. Note that according to the simple arguments from section 2, the fraction of points in the 4-d hyperplane should be closer to 1/10 than to 2/3. This deficiency is somewhat alleviated in later fits by [66, 162] (BJKLK), for which the data set outside the 4-d hyperplane has been extended at the averaged coupled pair functional level.

Our efforts [32] have concentrated on removing the most pronounced KLK deficiencies in the hydrogen bond region, namely the overestimated hydrogen bond length and the underestimated hydrogen bond strength. For this purpose, the well depth was increased until the exactly calculated binding energy D_0 agreed with the best experimental value [61,122], leading to a first refined PES called SQSDE [32]. In a second step, the monomer distance (R_{ab}) was adjusted to reproduce the value obtained from a microwave analysis [103] (SQSRAB) or to reproduce the experimental ground state rotational constant B_0 of the dimer (SQSBDE) [32,33]. The latter adjustment was based on a rotationally adiabatic quantum Monte Carlo approach [32] with a precision of ≈1 pm, which neglects small Coriolis corrections of the

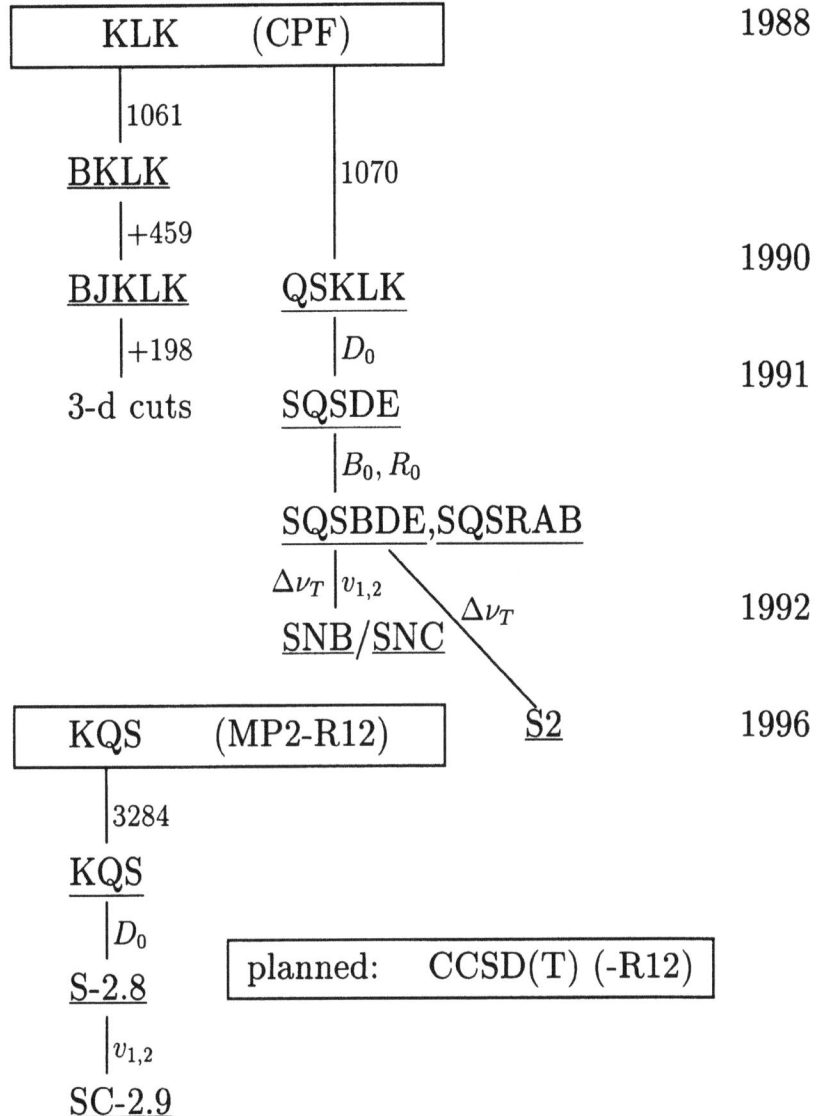

Figure 3. Summary of the most recent analytical 6-d HF pair potentials (underlined) in chronological order (right column). The number of ab initio configurations used in an analytical fit is indicated along the connecting lines. The upper potentials are based on the CPF study by Kofranek, Lischka, and Karpfen [55] (BKLK [56], BJKLK [162], extended 3-d cuts [66], QSKLK [30]) and are semiempirically refined through comparison with the experimental D_0 (SQSDE [32]), and the FF-distance R_0 (SQSRAB [32]), or B_0 rotational constant (SQSBDE) [32,33]). Further refinement attempts include the ground state tunneling splitting $\Delta\nu_T$ and the HF stretching vibrational manifold $v_{1,2}$ (SNB, SNC [21,25,38,41]), or $\Delta\nu_T$ alone (S2 [88]). The lower potentials derive from an MP2-R12 study by Klopper, Quack, and Suhm (KQS [41]) and are adjusted to D_0 (S-2.8 [41]) and additionally to $v_{1,2}$ (SC-2.9 [41]).

order of 1.5 pm [21,44]. Agreement of experimental (HF)$_2$ fundamental vibrations with those predicted on the ab initio and semiempirically refined surfaces was found to be satisfactory, although in one case (ν_6) the comparison relied on the smallness of a-type Coriolis resonances [27, 32] and the underlying experimental data base [27, 29] was (and still is [41, 86, 256]) incomplete. Within such and other limitations, particularly for the coupling of HF stretching modes to the hydrogen bond, the performance of the $\cdots$QS$\cdots$ surfaces was found to be satisfactory in a large number of dynamical calculations [25, 32, 34, 79–81, 139]. The improved BJKLK surface is also successful [59, 67, 257] for quantities which are not critically influenced by hydrogen bond energy and length errors in the 3-6% range. The dynamically most interesting feature of (HF)$_2$, a tunneling splitting due to concerted exchange of the hydrogen bond donor and acceptor units [42, 103] (see section 5.1), was underestimated by 1/3 in all surfaces based on the KLK data set. (Note that there fallaciously appears to be no discrepancy in a 1-d vibrational treatment [59]. This is another example for fortuitous error compensation, as could be shown in an adiabatic path analysis [34] and in explicitly multidimensional treatments [32, 67, 68].) While this discrepancy may not be dramatic in view of the subtle interplay between barrier height and width for such a tunneling process, we have attempted to remove it through empirical adjustment. At the same time, the HF stretching fundamental [46, 116] and overtone transitions [22, 70] required adjustment, as discussed above. The SNB/SNC family of PES superficially solved both discrepancies [21, 25, 38, 41] but it became clear that the adjustment is far from unique. E.g., SNB and SNC differ widely in their direct HF to HF coupling but agree closely in their experimentally available HF stretching spectra. Furthermore, lowering of the tunneling barrier *height* is not the only manipulation consistent with the known experimental spectrum of the dimer. The barrier *width* is at least as important. Interestingly, a later, independent manipulation of the SQSBDE surface with a partially identical goal [88] obtained an even more pronounced lowering of the tunneling barrier. We now have strong evidence [41] that it is the *width* of the SQSBDE barrier which requires most adjustment.

These ambiguities have led to a new effort to scan the pair potential, using the MP2-R12 approach and a more extended basis set [41]. The lower level of electron correlation treatment compared to the earlier CPF study [55] is more than compensated by the reduction of BSSE and other basis set limitations as well as a more complete coverage of configuration space (3284 points). Nevertheless, the resulting analytical potential required adjustment for the binding energy (S-2.8) [41] and the MP2 level led to an overestimated coupling of the monomers to the hydrogen bond, which had to be reduced through scaling (SC-2.9) [41].

The analytical form of this and related pair potentials consists of an angular expansion

$$\sum_i \rho_i(R, r_1, r_2)\alpha_i(\Theta_1, \Theta_2, \tau) \qquad (7)$$

in terms of the angles $\Theta_{1,2}$ which the HF vectors $\vec{r}_1, \vec{r}_2$ form with an intermolecular axis $\vec{R}$ and the dihedral angle τ between the HF vectors [32,41,59] (figure 4). The radial part ρ depends on the monomer separation R and

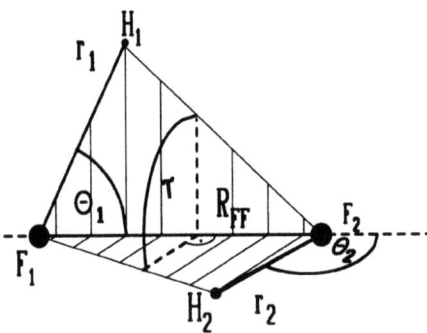

Figure 4. Fluorine centered coordinates for the HF pair potential. $r_{1,2}$ denote the monomer bond lengths, R_{FF} the F–F distance and $\theta_{1,2}, \tau$ the orientation and torsion angles of the HF bond vectors with respect to the FF axis [21, 41]. Some analytical potentials are expanded around the monomer centers of mass instead of the fluorine atoms [30, 32, 44, 56], resulting in center of mass distance R_{ab} and angles $\Theta_{a,b}$ [32].

on the monomer bond lengths $r_{1,2}$. The R dependence can be modelled by sums over exponentials [56], gaussians [32], polynomials, or products thereof [41,57]. Such an angular expansion proves to be sufficiently flexible for long and intermediate monomer distances and it can accommodate the leading electrostatic, induction, and dispersion terms [6] through appropriate inverse power laws in R. For shorter monomer separations, the potential anisotropy becomes too pronounced for low order angular expansions. Direct atom-atom terms and angular expansions around other centers, e.g. the hydrogen atoms have to be added to overcome this limitation [30, 32, 41]. Coupling to the monomer degrees of freedom is subtle. For the leading long range terms, it can be expressed in terms of derivatives of monomer properties such as dipole moment and polarizability with respect to the bond length [258]. For intermediate and short range configurations, ab initio calculations [41] and combination band spectroscopy [86, 259] are important sources for monomer-intermolecular coupling and rather flexible modelling is required to accommodate these effects into the analytical pair potential.

4.3. THREE-BODY POTENTIAL

The leading three-body contribution for polar molecules at long distances is due to dipole polarization. Polarization models are popular in modelling non-pairwise additivity at intermediate distances [51, 69, 174, 260]. At least for first row hydrides, this picture is usually too simple [58]. Hydrogen bonding leads to significant wavefunction distortions due to exchange contributions from the Pauli exclusion principle, which in turn act back on monomer properties such as the polarizability. Thus, the total three-body potential for polar hydrides is due to a subtle interplay between exchange and polarization forces. While difficult to cast into analytical form, this interplay is rather well captured by low level electronic structure approaches [25, 58]. Electron correlation is less important than for pair interactions and basis set requirements are rather modest [25].

A few three-body energies of $(HF)_3$ have been calculated and analyzed in [58]. The first systematic sampling and fitting of the 12-d three-body potential has been reported in [25]. It is based on 1004 configurations at MP2 level, using a DZP basis set and counterpoise correction for the BSSE. Most of these $(HF)_3$ configurations are sampled by the ground state wavefunction. Therefore, the resulting analytical potential HF3BL emphasizes the regions relevant for the $(HF)_3$ dynamics at low excitation. In combination with analytical monomer and pair potentials [25, 32], it led to important structural, energetical, and dynamic predictions [25], which were later supported by experiment [71, 76]. To enhance the applicability to more energetic configurations and to larger clusters, we have extended the sampling with the help of VSR techniques to currently 3000 points. An analytical fit to these points (HF3BG) has an *unweighted* rms deviation of 150 cm^{-1}, in the range of expected ab initio errors and the neglected higher-body contributions. The functional form includes the leading dipole polarization term as well as a sum of exponentials in terms of various atom-atom distances.

The HF3BG potential is among the first semiglobal analytical three-body potentials formulated for flexible molecular monomers. In combination with the most recent accurate pair potential (SC-2.9 [41]), it can now be tested rigorously against experiment [35, 39, 53, 71, 76, 118] through appropriate dynamical treatments. For a review of three-body potentials see also [26, 218].

5. Clusters

5.1. HF DIMER

The hydrogen bond exchange tunneling process between equivalent global minima of the HF dimer

$$H^{(1)}\diagdown_F \cdots H^{(2)}\!\!-\!\!F \;\rightleftharpoons\; F\!-\!H^{(1)} \cdots F\diagdown_{H^{(2)}} \qquad (8)$$

is among the most interesting topological features of the dimer potential energy surface. Although this process is of a multidimensional nature, it can be described qualitatively by its minimum energy path motion in the Θ_1, Θ_2 plane, as shown in figure 5. Segments in which motion is dominated by the

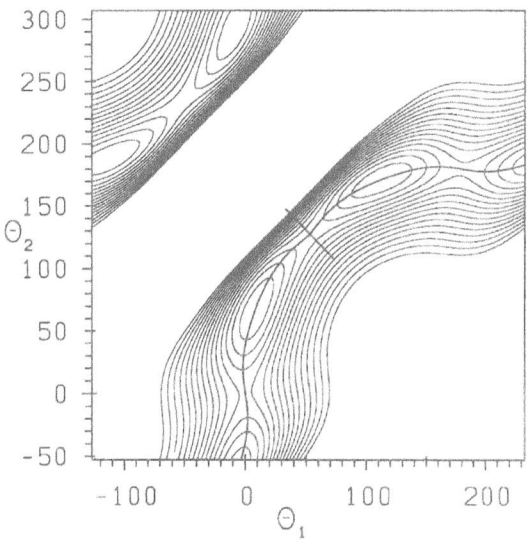

Figure 5. Cut through the SC-2.9 hypersurface [41] at the C_{2h} saddle point ($\tau=180°$) for hydrogen bond interconversion along the two bending angles Θ_1, Θ_2 (in degrees) with contour interval $100\,\mathrm{cm}^{-1}$. The minimum energy path for hydrogen bond exchange and the interconversion ridge are indicated.

free (non-hydrogen bonded) HF molecule are followed by more synchronous motion of the two HF molecules in the internal rotation process. Other coordinates are strongly coupled to this tunneling path. E.g., the saddle point at the center of figure 5 becomes a minimum for short intermolecular distances [32, 55]. The currently best estimate for the electronic barrier height connecting the global minima of the PES is 4.2(2) kJ/mol, but zero

TABLE 1. Best estimates (from experiment and theory combined, including error estimates) for some stationary point properties of (HF)$_2$. At the minimum, each internal coordinate can be qualitatively associated with a normal mode (see [32,41,85] for a more detailed analysis), whose harmonic wavenumber estimate ω_i at the minimum of the potential is given in the last row (accurate within $\pm 20\,\text{cm}^{-1}$ or better). See [77,78] for the best available direct ab initio predictions.

stationary pt.	r_1/pm	r_2/pm	R/pm	$\Theta_1/°$	$\Theta_2/°$	$\tau/°$	$(V_{rel})^a$
C_s minimum	92.3±0.1	92.0±0.1	273.5±1.	7±1	68±2	180	-19.1±0.2
C_{2h} saddle	92.0±0.1	92.0±0.1	271±2	53±2	127±2	180	-14.9±0.2
$C_{\infty v}$ linear	92.0±0.1	91.9±0.1	283±2	0	0	-	-15.2±0.2
ω_i/cm^{-1}	4030	4100	155	550	210	465	

$^a\ V_{rel} = (V_{(\text{HF})_2} - 2V_{\text{HF}})/(\text{kJ mol}^{-1})$

point energy contributions are very significant [34]. Table 1 summarizes some best estimates for local properties of the (HF)$_2$ PES including the minimum and two stationary points along the tunneling path, which are both well below dissociation. The estimates are the result of a comparison to experiment through multidimensional quantum dynamical calculations and can serve as benchmarks for the assessment of ab initio methods in the field of hydrogen bonding [77, 78].

5.2. HF TRIMER

The trimer is the smallest HF cluster in which intermolecular three-body effects can be studied. It is therefore instructive to compare results on a purely pairwise additive PES with those including the three-body contribution. Table 2 gives some structural and energetic predictions on analytical potential energy surfaces. Comparison of the SNB and SNB+HF3BL results reveals sizeable three body effects for structural and energetic properties. The PES curvature at the trimer minimum is rather sensitive to the quality of the pair potential due to the strong deviation of the hydrogen bond from linearity ($\Theta_{\text{HFF}}=0$). The predictions for D_0 are in good agreement with tentative experimental bounds of $D_0 \leq 44.4\,\text{kJ/mol}$ [76,118] and $D_0^{(\text{DF})_3} \geq 46.6\,\text{kJ/mol}$ [71, 76], but only with inclusion of three-body and anharmonicity terms [25, 76].

Figure 7 shows various hydrogen bond rearrangement processes in (HF)$_3$. The most easily feasible process is ring opening, due to the strain inherent in the bent hydrogen bonds of the cyclic structure. Reversal of the hydrogen bond pattern from clockwise to anticlockwise (in the molecular plane) is much less easy than in the dimer, as there is no geared (disrotatory) path-

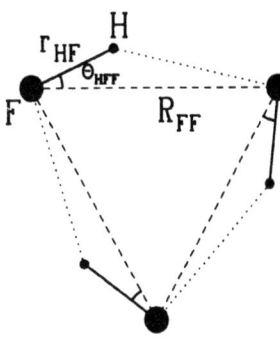

Figure 6. Fluorine centered coordinates for planar cyclic C_{nh} HF cluster structures with the example of the trimer. r_{HF} denotes the monomer bond length, R_{FF} the nearest F–F distance and Θ_{HFF} the angle of the HF bond vectors with respect to the associated FF axis [109]. The structural data for the trimer are given in table 2.

TABLE 2. HF trimer minimum structures (C_{3h} symmetry, FF distance R_{FF}, HF bond length r_{HF} and hydrogen bond angle Θ_{HFF}, see figure 6) and dissociation energies with respect to three isolated monomers (electronic (D_e), harmonically zero point corrected (D_0^h) and fully anharmonic (D_0), also for (DF)$_3$ ($D_0^{(DF)_3}$) all in kJ/mol), on various PES. The D_0 value in parentheses was obtained by replacing the analytical HF3BG surface by its Voronoi step representation (VSR). For the SC-2.9+HF3BG surface, harmonic wavenumbers for HF stretching (ω_{HF}), FF stretching (ω_{FF}), and HF libration (ω_{lib}) are also given (d=degenerate, *italic writing*=IR active). The wavenumbers ω_{FF} and ω_{lib} are 12-27% higher than the corresponding harmonic wavenumbers for SQSBDE+HF3BG and SNB+HF3BL. For the latter, anharmonicity effects (($\nu - \omega)/\omega$) of the order of −15 to −20% have been obtained via QMC calculations [25]. SC-2.9 anharmonicities may be even larger, as in the dimer [41] (see also [53])

PES	r_{HF}/pm	R_{FF}/pm	$\Theta_{HFF}/°$	D_e	D_0^h	D_0	$D_0^{(DF)_3}$
SNB	92.3	273	27	50.9	34.3		
SNB+HF3BL [25]	92.9	258	24	60.6	41.8	43.3	47.0
SQSBDE+HF3BG	93.3	255	24	61.3	42.0	42.6	46.8
SQSBDE+VSR						(41.9)	
SC-2.9+HF3BG	93.3	257	23	64.3	41.1	43.6	48.1
	ω_{HF}	ω_{FF}	ω_{lib}				
SC-2.9+HF3BG	3807	*209d*	500d				
	3884d	228	*625d*				
			719				
			980				

way for an odd number of rotors in a cyclic arrangement. The barrier for

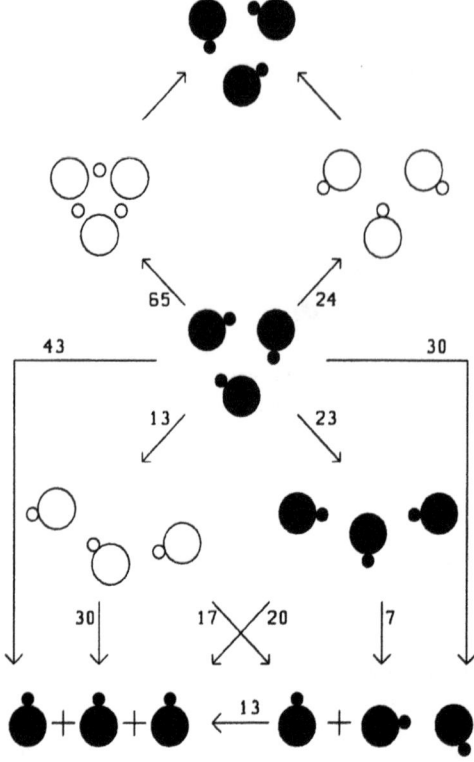

Figure 7. Schematic representation of minimum energy and transition structures for possible rearrangements of the cyclic HF trimer with indication of approximate energy barriers and energy differences in kJ/mol including zero point energy contributions. Filled circles indicate minima or dissociation channels, empty symbols transition structures on the 12-d surface [23, 25, 109]

concerted hydrogen transfer between different F atoms is even higher, but we will see in section 5.5 that this process requires less energy for the larger clusters. For polyatomic monomers such as H_2O, the number of possible hydrogen bond rearrangement patterns is considerably larger [261].

5.3. LARGER HF CLUSTERS - STRUCTURE AND ENERGETICS

A key advantage of the hydrogen fluoride system is the experimental availability of HF oligomers with 4-8 units, even at thermodynamic equilibrium [23, 39]. It is therefore particularly instructive to derive size trends for cluster properties from the multidimensional potential energy surfaces. Figure 8 summarizes the geometrical trends. In this size range, simple, unbranched ring structures are by far the most stable species. From $n=3$ to

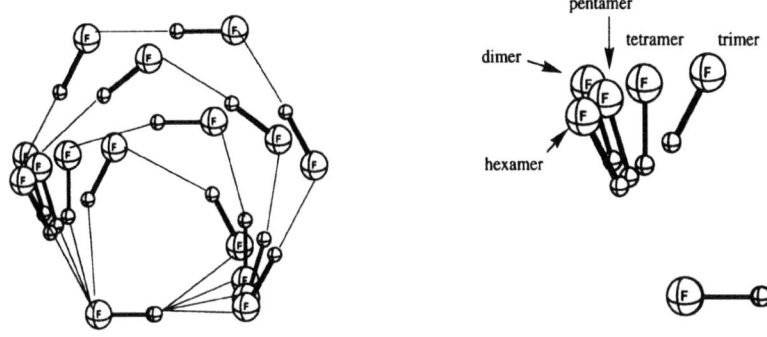

Figure 8. Overlaid minimum energy structures for (HF)$_n$ (n=2–6) at B3LYP 6-311+G** level with a common F atom and HF bond vector in the lower left corner [109]. The minimum energy structure of (HF)$_6$ is slightly puckered at this and other levels of theory, leading to S_6 symmetry. The enlarged fragment on the right illustrates the trend and saturation of the hydrogen bond lengths and angles with cluster size. Note the dimer-like, but significantly contracted local environment in the large clusters.

6, the hydrogen bond becomes nearly linear and the local environment approaches that of the dimer, with the important exception of the hydrogen bond *length*, which is significantly contracted due to non-pairwise additive contributions. As soon as the ring strain is completely relaxed (near n=6), the equilibrium ring structures start to pucker and fold to maintain this local environment, which is also present in the zig-zag chains of the solid [262]. The residual (non-hydrogen bonded) interactions which determine the details of this puckering and folding structure are quite subtle due to the exceedingly small dispersion forces in these clusters. Topological aspects come into play [263]. We have investigated sandwich and knot structures of larger clusters as well as 2-d and 3-d branching [23,25], but will concentrate on the structures up to n=6 in the present work. Table 3 summarizes best estimates of structural and energetic trends [109]. It is seen that most of the systematic structure changes level off beyond about (HF)$_4$. The energetics is best studied by dividing the total binding energy by the number of hydrogen bonds [64] or by taking the difference ΔD between the total binding energies of adjacent cluster sizes n and $n-1$ [23, 25, 39]. The latter corresponds to the evaporation of a monomer from the cluster, and is spectroscopically particularly relevant [39], as it represents the lowest dissociation threshold for these intermediate cluster sizes (see table 3). This threshold is highest for the tetramer due to significant ring strain in the product (HF)$_3$. Upon inclusion of anharmonic zero point energy effects, we

TABLE 3. Structural and energetic trends as a function of cluster size n for $(HF)_n$ minimum structures with $n=2-6$ (see figs. 6 and 8). HF bond lengths r_{HF}, nearest FF distances R_{FF}, hydrogen bond angles Θ_{HFF}, electronic (D_e, $\Delta D_e = D_e(n) - D_e(n-1)$), harmonically zero point corrected (D_0^h, ΔD_0^h) and fully anharmonic (D_0, ΔD_0) dissociation energies into monomers, also with neglected four- and higher-body contributions (D_0^{1+2+3}, ΔD_0^{1+2+3}), and barrier heights for concerted hydrogen exchange (electronic (E_B^e) and harmonically zero point energy corrected (E_B^h)) are listed as best estimates with approximate error bars from a combination of spectroscopy, analytical potential energy surfaces and ab initio calculations [41, 109]. Error bars for differences between neighbouring clusters are significantly smaller.

$n \rightarrow$	2	3	4	5	6
r_{HF}/pm	92.3(1)	93.3(3)	94.4(5)	94.8(6)	94.9(6)
	92.0(1)				
R_{FF}/pm	273.5(10)	259(2)	251(3)	248(4)	247(5)
Θ_{HFF}/ °	7(1)	24(3)	12(2)	6(2)	3(2)
D_e/ (kJ mol^{-1})	19.1(2)	63(3)	117(4)	161(5)	199(6)
ΔD_e/ (kJ mol^{-1})	19.1(2)	44(2)	54(3)	44(4)	38(5)
D_0^h/ (kJ mol^{-1})	11.7(2)	41(3)	83(4)	116(5)	145(6)
ΔD_0^h/ (kJ mol^{-1})	11.7(2)	29(2)	42(3)	33(4)	29(5)
D_0/ (kJ mol^{-1})	12.70(2) [122]	43(3)	84.5(4)	117(5)	146(6)
ΔD_0/ (kJ mol^{-1})	12.70(2) [122]	30(2)	41.5(3)	32.5(4)	29(5)
$D_0^{(1+2+3)}$/ (kJ mol^{-1})	12.70(2) [122]	43(3)	81(4)	108(5)	132(6)
$\Delta D_0^{(1+2+3)}$/ (kJ mol^{-1})	12.70(2) [122]	30(2)	38(3)	27(4)	24(5)
E_B^e/ (kJ mol^{-1})	170(10)	80(10)	55(10)	50(10)	60(10)
E_B^h/ (kJ mol^{-1})	160(10)	65(10)	35(10)	25(10)	30(10)

predict this threshold to be slightly above the highest available IR active excitation near 3445 cm^{-1} [39], although the error bars do not permit a safe statement. Larger clusters have progressively less ring strain, so that the threshold for monomer evaporation drops for the pentamer and the hexamer. Trimer dissociation is more facile than tetramer dissociation despite the fact that two hydrogen bonds are lost in the former process and only one in the latter. This may be rationalized by considering the strength of the hydrogen bonds. In the trimer, there is only one attractive three-body term for *three* hydrogen bonds, whereas in the tetramer there are four such terms for four hydrogen bonds. In addition, trimer ring strain is much larger than in the tetramer, as evidenced by a hydrogen bond angle which is twice as large (table 3).

The importance of isomeric structures deserves a brief discussion, as one might assume such locally stable structures to be quite numerous in these multidimensional cluster PES. We have investigated isomerism from the

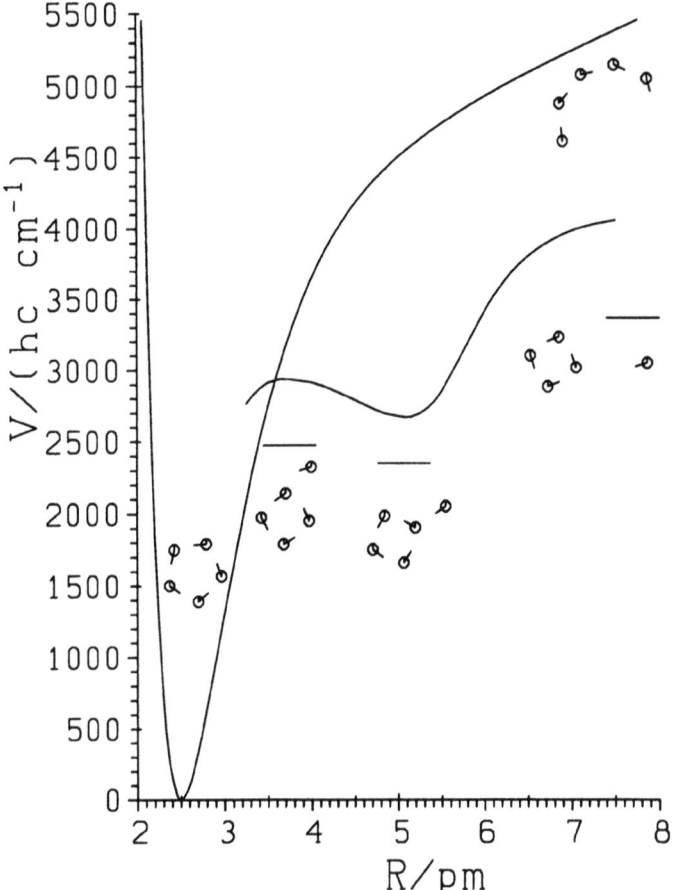

Figure 9. Potential energy V of two planar isomerization paths in $(HF)_5$ at the MP2 DZP level as a function of a selected FF distance R with relaxation of the other coordinates [23]. The steepness of the curves is exaggerated by BSSE. Ring opening requires more energy than evaporation of one monomer unit to form a tetramer. Recombination of the tetramer with the monomer can proceed through an intermediate tailed structure which represents a local minimum. The low planar barrier separating it from the global cyclic minimum is further lowered by out-of-plane motion and by inclusion of zero point energy contributions at the harmonic level (horizontal bars above the snapshots, relative to the zero point energy level of the ring structure). This makes a stabilization of tailed structures during supersonic jet expansions unlikely.

very beginning, but were unable to locate any *thermodynamically competitive* alternatives to the simple rings for $n=4$-7. This is in agreement with early experimental evidence for unpolar clusters beyond the dimer [42] in molecular beams and is easily understood in terms of the potential energy surfaces of these clusters. While the pair interaction energy is dominated by the number of hydrogen bonds and their strain, the three-body potential favours a sequential arrangement of neighbouring hydrogen bonds over any branching at a single F-atom. Figure 9 illustrates this in a semiquantitative way for the recombination of $(HF)_4$ and HF [23]. The intermediate branched structure has a very low barrier towards conversion into the much more favourable ring pentamer. While small amounts of branched clusters under non-equilibrium conditions cannot be rigorously excluded, it is very unlikely that these isomers contribute significantly to the thermodynamics and spectroscopy of HF [39], although this possibility has been suggested on the basis of simplified models [264]. The potential energy surfaces presented here will allow detailed studies of the interconversion dynamics of such isomers to the simple, cyclic ground state.

5.4. SPECTROSCOPIC PROPERTIES

Infrared spectroscopy can provide very detailed insights into the shape of cluster potential energy surfaces, but a reliable connection has to be based on multidimensional quantum dynamical treatments in the case of hydrogen bonds. The availability of such techniques for excited states is usually restricted to small clusters [79–81, 88], small subsets of the configuration space [39], or suitable approximations [32, 34]. The details of these nuclear dynamical approaches are beyond the scope of this review, but we want to summarize some of the results, which are only exploratory for the larger cluster sizes, in table 4. In terms of hydrogen bond libration and stretching, there is a pronounced tightening from the rather floppy dimer towards the larger ring clusters due to the increase in hydrogen bond strength. Thus, the harmonic approximation for hydrogen bond modes may be expected to gain in validity with increasing cluster size and hydrogen bond tunneling effects become less important. The opposite is true for internal hydrogen transfer [109], as we will see in the next section, and for ring puckering modes, which are less likely to exhibit pronounced quantum effects and only become IR active for larger, less symmetric clusters.

High frequency HF stretching modes [39], which are at the focus of current experimental investigations [35, 39, 76, 83], are characteristically shifted towards lower frequency with increasing hydrogen bond strength. This shift is usually estimated in the harmonic approximation [55, 64], but it is subject to two types of large anharmonic effects. On one hand, the strong hydrogen bond coupling and the proximity of hydrogen exchange induces large anharmonic effects within the HF stretching manifold [39]. These contribute on the order of 20–30% to the frequency shift relative to the free monomer. On the other hand, HF stretching motion is embedded in a bath of medium frequency hydrogen bond modes, mainly librations. The zero point energy associated with these bath modes weakens the hydrogen

TABLE 4. IR active (HF)$_n$ hydrogen bond fundamentals and wavenumber shifts relative to free HF as a function of cluster size $n=2$-6. Best estimates of harmonic wavenumbers ω and wavenumber shifts with respect to the monomer $\Delta\omega$ ($= \omega_{\text{HFstretching}} - \omega_{\text{HF}}(\text{monomer})$) as well as integrated molar IR band strengths S are given, but remain rather uncertain for the larger cluster sizes. (S is given within the double-harmonic approximation for convenient comparison to literature data, it can be converted to the more fundamental squared transition dipole moment $\langle\mu_{01}\rangle^2$ and the integrated absorption cross section G via 41.624 $\langle\mu_{01}\rangle^2/D^2 = G/\text{pm}^2 = 16.6054 \, (S/(\text{km mol}^{-1}))/(\omega/\text{cm}^{-1})$ [265]) The band notation is as in table 2, with the additional specification of in plane (ip) and out of plane (op) motion for $n < 6$. For $n=6$, the wavenumber range and total intensity of all librations are given in parentheses. Experimental anharmonic results $\tilde{\nu}$ are given for comparison, where available. For a more complete summary of dimer results see [21], for larger clusters, also [64, 65, 109, 208].

$n \rightarrow$	2	3	4	5	6
$\omega_{\text{FF}}/\text{cm}^{-1}$	155	200	280	270	250
$\tilde{\nu}_{\text{FF}}/\text{cm}^{-1}$	125 [32, 85]		$n=4\rightarrow 6$	$\approx 260\rightarrow 230$ [23]	
$S/(\text{km mol}^{-1})$	25 [78]	40	90	120	130
$\omega_{\text{lib}}^{ip}/\text{cm}^{-1}$	550	600	850	950	(700–1000)
$\tilde{\nu}_{\text{lib}}^{ip}/\text{cm}^{-1}$	475? [86]		$n=5$-6	≈ 600-900 [23]	
$S/(\text{km mol}^{-1})$	160 [78]	800	900	900	(2000)
$\omega_{\text{lib}}^{op}/\text{cm}^{-1}$	465 [41]	700	800	800	
$\tilde{\nu}_{\text{lib}}^{op}/\text{cm}^{-1}$	420? [41]		$n=5$-6	≈ 600-900 [23]	
$S/(\text{km mol}^{-1})$	190 [78]	600	700	700	
$-\Delta\omega_{\text{HF}}/\text{cm}^{-1}$	108 [41, 78]	250	470	600	660
$-\Delta\tilde{\nu}_{\text{HF}}/\text{cm}^{-1}$	93 [46]	249 [118]	516	661	716 [35, 39]
$S/(\text{km mol}^{-1})$	430 [78, 109, 123]	1200	3300	5600	7600

bond, and thus reduces the shift considerably compared to a harmonic calculation. In the dimer, the latter effect dominates [41], whereas for larger clusters, the HF stretching anharmonicity may even overcompensate the zero point energy weakening [39, 41]. As a net effect, harmonic approximations are remarkably successful in the prediction of HF stretching frequency shifts [25, 35, 39, 64, 109, 208] and – for a similar reason – also in the prediction of cluster binding energies D_0 (see table 3) [23, 25]). Rigorous, accurate, full-dimensional predictions of these modes for $n > 3$ are not yet feasible due to methodological limitations in electronic structure approaches, analytical representation and nuclear dynamics. A more viable approach up to $n \approx 5$ is to scan the n-d stretching subspace at an appropriate ab initio level and to carry out n-d variational calculations of the stretching states [39, 190]. These results have to be scaled to experiment for $n=3$ due to limitations in the electronic structure approach and the zero point motional effects mentioned above [39, 41].

5.5. BEYOND MONOMER INTEGRITY - HYDROGEN EXCHANGE

As the concerted hydrogen exchange barrier is rather high for the dimer and the trimer and higher-body contributions become significant in the barrier region for larger clusters, we have not yet fully implemented the appropriate exchange symmetry into our analytical potential energy surfaces, but we have explored some possibilities. A simple strategy would be to treat the existing unsymmetrized pair- (and three-body-) potentials as diabatic surfaces, which are then coupled to each other through an adjustable matrix element. In the case of the dimer, this results in a simple 2×2 problem, whose lower energy eigenvalue can mimic the symmetrized (adiabatic) potential for a suitable coupling matrix element $V_c(\mathbf{x})$. Let $\mathbf{x}$ be a set of 12 cartesian coordinates for $H_1F_1H_2F_2$ and let $\mathcal{P}(\mathbf{x})$ be the same set with H_1 and H_2 interchanged (i.e. $H_2F_1H_1F_2$). Then,

$$V_s = \frac{V(\mathbf{x}) + V(\mathcal{P}(\mathbf{x}))}{2} - \sqrt{\left(\frac{V(\mathbf{x}) + V(\mathcal{P}(\mathbf{x}))}{2}\right)^2 + V_c(\mathbf{x})^2 - V(\mathbf{x})V(\mathcal{P}(\mathbf{x}))} \quad (9)$$

will be a suitably symmetrized potential, provided that $V_c(\mathbf{x})$ has the following properties:

i) $V_c(\mathcal{P}(\mathbf{x})) = V_c(\mathbf{x})$

ii) $V_c \to 0$ if $R_{ij} \to \infty$, where R_{ij} is the separation between any two atoms

iii) V_c should not change the potential significantly in the lower energy range.

For many-body contributions, the procedure becomes more involved, and figure 10 suggests that HF clusters may be a particularly challenging example for its application, as the convergence of the many-body expansion is rather slow in the barrier region.

In order to obtain a semiquantitative overview of the concerted exchange process, which occurs rapidly in HF vapor [107, 108], we have carried out a local investigation of the saddle point properties as a function of cluster size using various quantum chemistry approaches [25, 35, 109] (see also [64, 75, 166, 168]. The results depend quite strongly on electronic structure level and basis size, but table 3 contains tentative estimates of the barrier heights. Note that beyond the trimer, the concerted exchange barrier falls well below the threshold for complete dissociation into monomers and becomes competitive with the lowest available dissociation thresholds. At the same time, at thermal equilibrium the larger clusters possess substantial average internal energy. In this sense, $(HF)_n$ clusters with $n > 3$ may be regarded as molecules with several isomers, rather than merely as

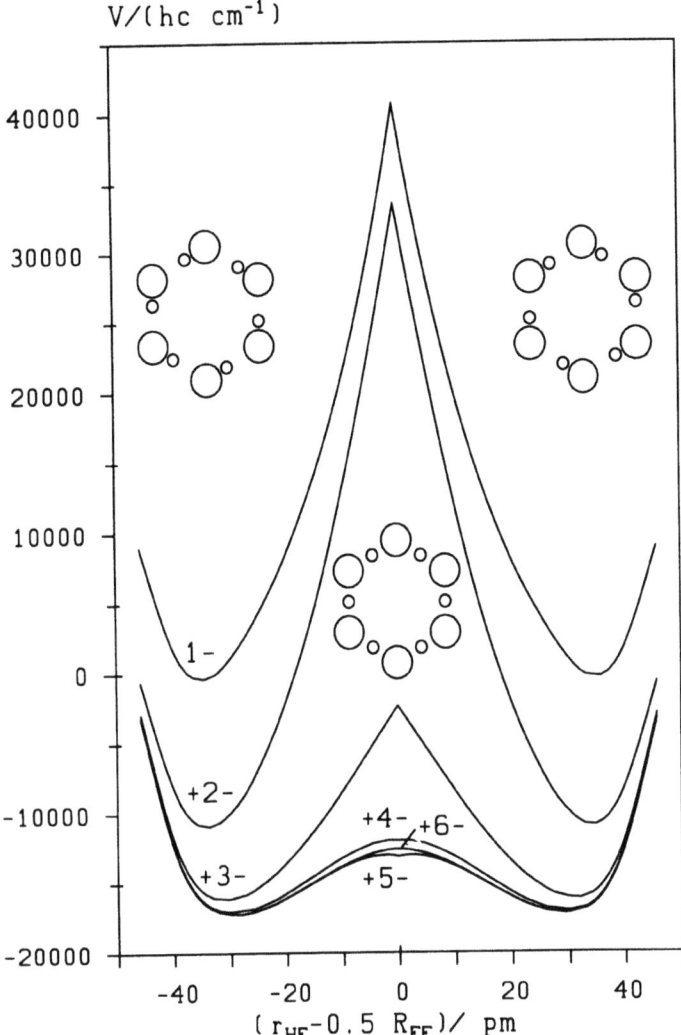

Figure 10. Many-body decomposition of planar $(HF)_6$ at the 6-311+G** B3LYP [109] level along the minimum energy path for concerted hydrogen exchange, expressed as the distance of the hydrogen from the midpoint of adjacent F atoms $r_{HF} - (R_{FF}/2)$ (with $\Theta_{HFF} = 0$). From top to bottom, the curves marked 1-, +2-, +3-, +4-, +6-, +5- represent the potential energy including up to 1-, 2-, 3-, 4-, 6- and 5-body contributions. Convergence of the many-body decomposition is seen to be slower in the region of the exchange barrier than in the hydrogen bond region, suggesting that inclusion of 4-body effects is essential in the exchange symmetrization procedure (see text).

weakly bound HF aggregates. The concepts of 'inter'-molecular and 'intra'-molecular potentials merge in this case.

6. Conclusions and outlook

We have presented the conceptual foundations and various aspects of the potential energy hypersurfaces for HF clusters. Most of these are transferable to other hydrogen bonded systems. Concerning some properties, such as the loss of monomer integrity, the importance of many-body contributions and the abundance of large clusters in the vapor phase, HF may be considered to be a particularly pronounced case. In terms of molecular complexity, it is attractively simple, the monomer consisting of only two nuclei and ten electrons. These two properties together render HF an excellent prototype system for the detailed study of hydrogen bonding.

Some of our findings should be highlighted, because of their relevance for accurate hydrogen bond potentials in general:

i) The interaction in HF clusters is far from pairwise additive. Inclusion of three-body terms recovers many of the missing structural and energetic effects in the hydrogen bond regime. Four-body terms can be useful, quantitatively.

ii) Voronoi step representation is a useful tool in the development of multidimensional PES.

iii) Quantum chemistry is a prime source of data for accurate hydrogen bond PES, provided that some quantitative discrepancies are removed by semiempirical adjustment via comparison with spectroscopic data through rigorous nuclear dynamics calculations. Among the investigated ab initio approaches, the quality ranking is clearly CCSD(T) > MP2 > B3LYP > BLYP. Since the computational effort increases strongly with the quality of these four methods, all four may be considered to be useful tools in the study of hydrogen bonding.

iv) The best available analytical pair potentials (most recently SC-2.9) are successful tools for the modelling of a large HF dimer spectroscopy and dynamics data base. In combination with an analytical three-body potential (HF3BG), reliable trimer predictions can be made. Here, rigorous comparison to experiment is more difficult than for the dimer and still limited to few available data. For larger clusters, a combination of quantum chemistry for size trends and analytical potentials for the assessment of anharmonic effects is most fruitful.

The modular structure of the many-body approach to HF cluster PES invites continued improvements at the dimer, trimer, and oligomer levels. Current and planned future developments include:

i) Construction of dipole moment surfaces in a similar spirit as described here for potential energy surfaces.

ii) Refinement of the analytical pair potential at CCSD(T) (-R12) level in order to capture subtle effects in the HF stretching manifold, which are the subject of current experiments.

iii) Empirical refinement of the three-body potential with new spectroscopic trimer data.

iv) Development of an analytical four-body potential using VSR techniques to sample the 18-d configuration space and systematic investigation of the high resolution IR spectra of $(HF)_4$, $(HF)_5$ and $(HF)_6$ in view of a comparison of experimental data with predictions from the potentials. Such experiments are still quite challenging, but should be just about in reach of current techniques.

v) Symmetrization of the 1-, 2-, 3- and 4-body potentials with respect to hydrogen exchange and studies of the quantum dynamics of these processes.

Even before these extensions are realized, it will be interesting to apply the available (1+2+3)-body approach to condensed phases and to compare it to complementary Car-Parrinello results [143]. Liquid HF is an extreme case of a dipolar, cooperatively hydrogen-bonded liquid with a characteristic delayed gas-liquid phase transition [266] and an unusually high compressibility [267, 268]. Spectroscopic accuracy in the gas phase combined with a reliable description of condensed phase properties for this prototype system in a single analytical potential energy surface is an important goal of our investigations.

We may conclude here by an overview of the recent past and near future of our understanding of the potential and dynamics of the hydrogen bond with the example of hydrogen fluoride clusters. Just about a decade ago our knowledge of even the bond dissociation energy and the thermodynamic functions for the equilibrium, i.e. the simplest questions concerning the simplest species, the HF dimer:

$$2\,HF \rightleftharpoons (HF)_2 \quad \text{equilibrium}: K_p$$

was so grossly inadequate [269], that one might justly say that there was no knowledge of relevance at all. This question was in essence solved by 1991 [32] and is certainly answered accurately today [41]. Current efforts concerning quantitative 'equilibrium dynamics', i.e. spectroscopy and thermodynamics can concentrate on the larger clusters $(HF)_n$ with $n \geq 3$ [22, 23, 25, 35, 39]. On the other hand there remains still much to be learned today about even the simplest questions on the quantitative time dependent dynamics of even the dimer, and almost nothing is known in this respect for the larger clusters:

$$2\,HF \rightleftharpoons (HF)_2 \quad \text{rate coefficients}: k$$

Our developments of accurate potential hypersurfaces should help to design, analyse and interpret current and future experiments. Adequate comparison and matching of experiment and theory can provide a general understanding for hydrogen bond structure and dynamics in this prototype system, much of which should be transferable to other hydrogen bond systems.

Acknowledgement: We acknowledge help from and discussions with our coauthors in this project, T.-K. Ha, Y. He, K. Liedl, D. Luckhaus, W. Klopper, C. Maerker, H. Müller, K. von Puttkamer, P. von Ragué Schleyer, U. Schmitt, R. Signorell, and J. Stohner as well as discussions and correspondence with D. L. Bunker, P. R. Bunker, F. Huisken, A. Karpfen, D. Nesbitt, M. Parrinello, H. F. Schaefer, and D. Truhlar. Generous computer grants from the Rechenzentrum der ETH, the CSCS and the C4 project have been essential in the construction and testing of the potential energy surfaces.

References

1. Maurice L. Huggins. 50 years of hydrogen bond theory. *Angew. Chem. Internat. Edit.*, 10:147–152, 1971.
2. Linus Pauling. *The Nature of the Chemical Bond*, $2^n d$ ed. University Press, Oxford, 1940.
3. G. C. Pimentel and A. L. McClellan. *The Hydrogen Bond.* Freeman, 1960.
4. P. Schuster, G. Zundel, and G. Sandorfy, editors. *The Hydrogen Bond*, volume 1-3. North-Holland, Amsterdam, 1976.
5. L. A. Curtiss and M. Blander. Thermodynamic properties of gas-phase hydrogen-bonded complexes. *Chem. Rev.*, 88:827–841, 1988.
6. A. D. Buckingham. Permanent and induced molecular moments and long-range intermolecular forces. *Adv. Chem. Phys.*, 12:107–142, 1967.
7. Clifford E. Dykstra. Intermolecular electrical interaction: A key ingredient in hydrogen bonding. *Accounts of Chemical Research*, 21(10):355–361, 1988.
8. Don L. Bunker. Classical trajectory methods. *Methods in Computational Physics*, 10:287–325, 1971.
9. J. N. Murrell, S. Carter, S. C. Farantos, P. Huxley, and A. J. C. Varandas. *Potential energy functions.* Wiley, N.Y., 1984.
10. Maurice Rigby, E. Brian Smith, William A. Wakeham, and Geoffrey C. Maitland. *The Forces Between Molecules.* Clarendon Press, Oxford, 1986.
11. Paul G. Mezey. *Potential Energy Hypersurfaces.* Studies in Physical and Theoretical Chemistry 53. Elsevier, 1987.
12. George C. Schatz. The analytical representation of electronic potential energy surfaces. *Reviews of Modern Physics*, 61:669–688, 1989.
13. Per-Olov Löwdin. Proton tunneling in DNA and its biological implications. *Rev. Mod. Phys.*, 35:724–732, 1963.
14. Charles L. Brooks III, Martin Karplus, and B. Montgomery Pettitt. *Proteins: A theoretical perspective of dynamics, structure, and thermodynamics*, volume 71 of *Adv. Chem. Phys.* Wiley, 1988.
15. Jeremy M. Hutson. Coupled channel methods for solving the bound-state Schrödinger equation. *Comp. Phys. Commun.*, 84:1–18, 1994.
16. Eugene S. Kryachko. On the theory of the dynamics of hydrogen bonds coupled with a bath. Part I. *J. Mol. Struct. (Theochem)*, 314:133–154, 1994.
17. Ad van der Avoird, Paul E. S. Wormer, and Robert Moszynski. From intermolecular potentials to the spectra of van der Waals molecules, and vice versa. *Chem.*

Rev., 94:1931–1974, 1994.
18. Martin Quack. Spectra and dynamics of coupled vibrations in polyatomic molecules. *Annu. Rev. Phys. Chem.*, 41:839, 1990.
19. David J. Nesbitt. High-resolution, direct infrared laser absorption spectroscopy in slit supersonic jets: intermolecular forces and unimolecular vibrational dynamics in clusters. *Annu. Rev. Phys. Chem.*, 45:367–399, 1994.
20. Tak-San Ho and Herschel Rabitz. Inversion of experimental data to extract intermolecular and intramolecular potentials. *J. Phys. Chem.*, 97:13447–13456, 1993.
21. Martin Quack and Martin A. Suhm. On hydrogen-bonded complexes: The case of $(HF)_2$. *Theor. Chim. Acta*, 93(2):61–65, 1996.
22. Katharina von Puttkamer and Martin Quack. Vibrational spectra of $(HF)_2$, $(HF)_n$ and their D-isotopomers: Mode selective rearrangements and nonstatistical unimolecular decay. *Chem. Phys.*, 139:31–53, 1989.
23. Martin A. Suhm. HF vapor. *Ber. Bunsenges. Phys. Chem.*, 99:1159–1167, 1995.
24. Peter Schuster. Intermolecular forces - An example of fruitful cooperation of theory and experiment. *Angew. Chem. Int. Ed. Engl.*, 20:546–568, 1981.
25. Martin Quack, Jürgen Stohner, and Martin A. Suhm. Vibrational dynamics of $(HF)_n$ aggregates from an ab initio based analytical (1+2+3)-body potential. *J. Mol. Struct.*, 294:33–36, 1993, and to be published.
26. Matthew J. Elrod and Richard J. Saykally. Many-body effects in intermolecular forces. *Chem. Rev.*, 94:1975–1997, 1994.
27. Katharina von Puttkamer and Martin Quack. High resolution interferometric FTIR spectroscopy of $(HF)_2$: analysis of a low frequency fundamental near 400 cm^{-1}. *Mol. Phys.*, 62(5):1047–1064, 1987.
28. Katharina von Puttkamer, Martin Quack, and Martin A. Suhm. Observation and assignment of tunnelling-rotational transitions in the far infrared spectrum of $(HF)_2$. *Mol. Phys.*, 65:1025–1045, 1988.
29. Katharina von Puttkamer, Martin Quack, and Martin A. Suhm. Infrared spectrum and dynamics of the hydrogen bonded dimer $(HF)_2$. *Infrared Phys.*, 29:535–539, 1989.
30. Martin Quack and Martin A. Suhm. Potential energy surface and energy levels of $(HF)_2$ and its D isotopomers. *Mol. Phys.*, 69:791–801, 1990.
31. Martin Quack and Martin A. Suhm. Observation and assignment of the hydrogen bond exchange disrotatory in-plane bending vibration ν_5 in $(HF)_2$. *Chem. Phys. Lett.*, 171:517–524, 1990.
32. Martin Quack and Martin A. Suhm. Potential energy surfaces, quasiadiabatic channels, rovibrational spectra, and intramolecular dynamics of $(HF)_2$ and its isotopomers from quantum Monte Carlo calculations. *J. Chem. Phys.*, 95:28–59, 1991.
33. Typographical errors in ref. [32]: In Table I, the first occurence of a_3^{222} should instead read a_3^{145} and the second occurence of a_{12} should read a_{13}/a_0^{-1}. In the polarization formula (7), 1/3 instead of 1 should be added to each of the two $\cos^2$ terms. A FORTRAN source code of the SQSBDE potential is available on request from the authors (suhm@ir.phys.chem.ethz.ch).
34. Martin Quack and Martin A. Suhm. Quasiadiabatic channels and effective transition state barriers for the disrotatory in plane hydrogen bond exchange motion in $(HF)_2$. *Chem. Phys. Lett.*, 183:187–194, 1991.
35. Martin Quack, Ulrich Schmitt, and Martin A. Suhm. Evidence for the $(HF)_5$ complex in the HF stretching FTIR absorption spectra of pulsed and continuous supersonic jet expansions of hydrogen fluoride. *Chem. Phys. Lett.*, 208:446–452, 1993.
36. Martin A. Suhm. Multidimensional vibrational quantum Monte Carlo technique using robust interpolation from static or growing sets of discrete potential energy points. *Chem. Phys. Lett.*, 214:373–380, 1993.

37. Martin A. Suhm. Reliable determination of multidimensional analytical fitting bias. *Chem. Phys. Lett.*, 223:474–480, 1994.
38. Martin Quack and Martin A. Suhm. Accurate quantum Monte Carlo calculations of the tunneling splitting in $(HF)_2$ on a 6-dimensional potential hypersurface. *Chem. Phys. Lett.*, 234:71–76, 1995.
39. David Luckhaus, Martin Quack, Ulrich Schmitt, and Martin A. Suhm. On FTIR spectroscopy in asynchronously pulsed supersonic jet expansions and on the interpretation of stretching spectra of HF clusters. *Ber. Bunsenges. Phys. Chem.*, 99:457–468, 1995.
40. Ruth Signorell, Yabai He, Holger B. Müller, Martin Quack, and Martin A. Suhm. High resolution diode laser and FTIR spectroscopy of $(HF)_n$ and its isotopomers. In John P. Maier and Martin Quack, editors, *Proceedings of the 10th International Symposium on Atomic, Molecular, Cluster, Ion, and Surface Physics*, pages 256–259. Vdf Publishers, Zürich, 1996.
41. Wim Klopper, Martin Quack, and Martin A. Suhm. A new ab initio based six-dimensional pair interaction potential for HF. *Chem. Phys. Lett.*, 261:35–44, 1996.
42. Thomas R. Dyke, Brian J. Howard, and William Klemperer. Radiofrequency and microwave spectrum of the hydrogen fluoride dimer; a nonrigid molecule. *J. Chem. Phys.*, 56(5):2442–2454, 1972.
43. David R. Yarkony, Stephen V. O'Neil, Henry F. Schaefer III, Craig P. Baskin, and Charles F. Bender. Interaction potential between two rigid HF molecules. *J. Chem. Phys.*, 60:855–865, 1974.
44. A. E. Barton and B. J. Howard. An intermolecular potential-energy surface for $(HF)_2$. *Faraday Discuss. Chem. Soc.*, 73:45–62, 1982.
45. J. T. Brobjer and J. N. Murrell. The intermolecular potential of HF. *Mol. Phys.*, 50(5):885–899, 1983.
46. A. S. Pine, W. J. Lafferty, and B. J. Howard. Vibrational predissociation, tunneling, and rotational saturation in the HF and DF dimers. *J. Chem. Phys.*, 81(7):2939–2950, 1984.
47. Michael E. Cournoyer and William L. Jorgensen. An improved intermolecular potential function for simulations of liquid hydrogen fluoride. *Mol. Phys.*, 51(1):119–132, 1984.
48. Jeffrey F. Gaw, Yukio Yamaguchi, Mark A. Vincent, and Henry F. Schaefer III. Vibrational frequency shifts in hydrogen-bonded systems: the hydrogen fluoride dimer and trimer. *J. Am. Chem. Soc.*, 106:3133–3138, 1984.
49. Daniel W. Michael, Clifford E. Dykstra, and James M. Lisy. Changes in the electronic structure and vibrational potential of hydrogen fluoride upon dimerization: A well-correlated $(HF)_2$ potential energy surface. *J. Chem. Phys.*, 81(12):5998–6006, 1984.
50. A. S. Pine and B. J. Howard. Hydrogen bond energies of the HF and HCl dimers from absolute infrared intensities. *J. Chem. Phys.*, 84(2):590–596, 1986.
51. S. Murad, K. A. Mansour, and J. G. Powles. A model intermolecular potential for hydrogen fluoride including polarizability. *Chem. Phys. Lett.*, 131(1):98–102, 1986.
52. Michael J. Redmon and J. Stephen Binkley. Global potential energy hypersurface for dynamical studies of energy transfer in HF-HF collisions. *J. Chem. Phys.*, 87(2):969–982, 1987.
53. Kirk D. Kolenbrander, Clifford E. Dykstra, and James M. Lisy. Torsional vibrational modes of $(HF)_3$: IR-IR double resonance spectroscopy and electrical interaction theory. *J. Chem. Phys.*, 88(10):5995–6012, 1988.
54. Gene C. Hancock, Donald G. Truhlar, and Clifford E. Dykstra. An analytic representation of the six-dimensional potential energy surface of hydrogen fluoride dimer. *J. Chem. Phys.*, 88(3):1786–1796, 1988.
55. Manfred Kofranek, Hans Lischka, and Alfred Karpfen. Coupled pair functional study on the hydrogen fluoride dimer. I. Energy surface and characterization of stationary points. *Chem. Phys.*, 121:137–153, 1988.

56. P. R. Bunker, Manfred Kofranek, Hans Lischka, and Alfred Karpfen. An analytical six-dimensional potential energy surface for $(HF)_2$ from ab-initio calculations. *J. Chem. Phys.*, 89(5):3002–3007, 1988.
57. David W. Schwenke and Donald G. Truhlar. A new potential energy surface for vibration-vibration coupling in HF-HF collisions. formulation and quantal scattering calculations. *J. Chem. Phys.*, 88:4800–4813, 1988.
58. G. Chałasinski, S. M. Cybulski, M. M. Szczęśniak, and S. Scheiner. Nonadditive effects in HF and HCl trimers. *J. Chem. Phys.*, 91:7048–7056, 1989.
59. P. R. Bunker, Tucker Carrington, Jr., P. C. Gomez, Mark D. Marshall, Manfred Kofranek, Hans Lischka, and Alfred Karpfen. An ab initio semirigid bender calculation of the rotation and trans-tunnelling spectra of $(HF)_2$ and $(DF)_2$. *J. Chem. Phys.*, 91(9):5154–5159, 1989.
60. W. Rijks and P. E. S. Wormer. Correlated van der Waals coefficients. II. Dimers consisting of CO, HF, H_2O, and NH_3. *J. Chem. Phys.*, 90(11):6507–6519, 1989. erratum: ibid. 92(1990),5754.
61. D. C. Dayton, K. W. Jucks, and R. E. Miller. Photofragment angular distributions for HF dimer: Scalar $J-J$ correlations in state-to-state photodissociation. *J. Chem. Phys.*, 90:2631–2638, 1989.
62. Donald G. Truhlar. The HF dimer: Potential energy surface and dynamical processes. In *Proceedings of the NATO Workshop on the dynamics of polyatomic van der Waals Complexes, NATO Ser. B 227*, pages 159–185, 1990.
63. Huai Sun and R. O. Watts. Diffusion Monte Carlo simulations of hydrogen fluoride dimers. *J. Chem. Phys.*, 92:603–616, 1990.
64. Alfred Karpfen. Ab initio studies on hydrogen bonded clusters: Structure and vibrational spectra of cyclic $(HF)_n$ complexes. *Int. J. Quantum Chem. (Quant. Chem. Symp.)*, 24:129–140, 1990.
65. Clifford E. Dykstra. Intermolecular vibrational frequencies of $(HF)_n$ and $(HCN)_n$ weak complexes by electrical molecular mechanics. *J. Phys. Chem.*, 94:180–185, 1990.
66. Per Jensen, P. R. Bunker, Alfred Karpfen, Manfred Kofranek, and Hans Lischka. An ab initio calculation of the intramolecular stretching spectra for the $(HF)_2$ and its D-substituted isotopic species. *J. Chem. Phys.*, 93:6266–6280, 1990.
67. S. C. Althorpe, D. C. Clary, and P. R. Bunker. Calculation of the far infrared spectra for $(HF)_2$, $(HCl)_2$ and $(HBr)_2$. *Chem. Phys. Lett.*, 187:345–353, 1991.
68. Mark D. Marshall, Per Jensen, and P. R. Bunker. An ab initio close-coupling calculation of the lower vibrational energies of the HF dimer. *Chem. Phys. Lett.*, 176:255–260, 1991.
69. Huai Sun, R. O. Watts, and U. Buck. The infrared spectrum and structure of hydrogen fluoride clusters and the liquid: Semiclassical and classical studies. *J. Chem. Phys.*, 96:1810–1821, 1992.
70. Martin A. Suhm, John T. Farrell, Jr., Andrew McIlroy, and David J. Nesbitt. High resolution 1.3μm overtone spectroscopy of HF dimer in a slit jet: $K_a = 0 \leftarrow 0$ and $K_a = 1 \leftarrow 0$ subbands of $v_{acc} = 2 \leftarrow 0$. *J. Chem. Phys.*, 97:5341–5354, 1992.
71. Martin A. Suhm, John T. Farrell, Jr., Stephen Ashworth, and David J. Nesbitt. High resolution infrared spectroscopy of DF trimer: A cyclic ground state structure and DF stretch induced intramolecular vibrational coupling. *J. Chem. Phys.*, 98:5985–5989, 1993.
72. E. J. Bohac and R. E. Miller. The trans-bending and F-F stretching vibrations of HF dimer in $v_{HF} = 1$: The influence of intermolecular vibrational excitation on the predissociation dynamics. *J. Chem. Phys.*, 99:1537–1544, 1993.
73. Huan-C. Chang and William Klemperer. The vibrational second overtones of HF dimer: A quartet. *J. Chem. Phys.*, 100:1–14, 1994.
74. A. Karpfen and O. Yanovitskii. Cooperativity in hydrogen bonded clusters: an improved ab initio SCF study on the structure and energetics of neutral, protonated and deprotonated chains and of neutral, cyclic hydrogen fluoride oligomers. *J. Mol.*

Struct. (THEOCHEM), 314:211–227, 1994.
75. Klaus R. Liedl, Romano T. Kroemer, and Bernd M. Rode. Hydrogen transitions between (HF)$_n$ C_{nh} structures (n=2-5) via D_{nh} transition states as models for hydrogen tunneling in hydrogen fluoride clusters. Chem. Phys. Lett., 246:455–462, 1995.
76. Martin A. Suhm and David J. Nesbitt. Potential surfaces and dynamics of weakly bound trimers: Perspectives from high resolution IR spectroscopy. Chem. Soc. Rev., 24:45–54, 1995.
77. Kirk A. Peterson and Thom H. Dunning, Jr. Benchmark calculations with correlated molecular wave functions. VII. Binding energy and structure of the HF dimer. J. Chem. Phys., 102:2032–2041, 1995.
78. Charlene L. Collins, Kenji Morihashi, Yukio Yamaguchi, and Henry F. Schaefer III. Vibrational frequencies of the HF dimer from the coupled cluster method including all single and double excitations plus perturbative connected triple excitations. J. Chem. Phys., 103:6051–6056, 1995.
79. Dong H. Zhang, Qian Wu, John Z. H. Zhang, Michael von Dirke, and Zlatko Bačić. Exact full-dimensional bound state calculations for (HF)$_2$, (DF)$_2$, and HFDF. J. Chem. Phys., 102:2315–2325, 1995.
80. Dong H. Zhang, Qian Wu, and John Z. H. Zhang. A time-dependent approach to flux calculation in molecular photofragmentation: Vibrational predissociation of HF-DF. J. Chem. Phys., 102:124–132, 1995.
81. Quian Wu, Dong H. Zhang, and John Z. H. Zhang. 6D quantum calculation of energy levels for HF stretching excited (HF)$_2$. J. Chem. Phys., 103:2548–2554, 1995.
82. Michael von Dirke, Zlatko Bacic, Dong H. Zhang, and John Z. H. Zhang. Vibrational predissociation of HF dimer in $\nu_{HF} = 1$: Influence of initially excited intermolecular vibrations on the fragmentation dynamics. J. Chem. Phys., 102:4382–4389, 1995.
83. Friedrich Huisken, Michael Kaloudis, Axel Kulcke, Curtiss Laush, and James M. Lisy. Vibrational spectroscopy of small (HF)$_n$ clusters ($n = 4 - 8$) in size-selected molecular beams. J. Chem. Phys., 103:5366–5377, 1995.
84. John T. Farrell, Jr., Martin A. Suhm, and David J. Nesbitt. Breaking symmetry with hydrogen bonds: Vibrational predissociation and isomerization dynamics in HF-DF and DF-HF isotopomers. J. Chem. Phys., 104:9313–9331, 1996.
85. David T. Anderson, Scott Davis, and David J. Nesbitt. Probing hydrogen bond potentials via combination band spectroscopy: A near-IR study of the geared bend/van der Waals stretch intermolecular modes in (HF)$_2$. J. Chem. Phys., 104:6225–6243, 1996.
86. David T. Anderson, Scott Davis, and David J. Nesbitt. Hydrogen bond spectroscopy in the near infrared: Out-of-plane torsion and antigeared bend combination bands in (HF)$_2$. J. Chem. Phys., 105:4488–4503, 1996.
87. H.-C. Chang and W. Klemperer. A phenomenological model for the vibrational dependence of hydrogen interchange tunneling in HF dimer. J. Chem. Phys., 104:7830–7835, 1996.
88. William C. Necoechea and Donald G. Truhlar. An improved potential energy surface for the degenerate rearrangement of (HF)$_2$. Chem. Phys. Lett., 248:182–188, 1996.
89. Roger L. Wilkins. Mechanisms of energy transfer in hydrogen fluoride systems. J. Chem. Phys., 67:5838–5854, 1977.
90. M. E. Coltrin, M. L. Koszykowski, and R. A. Marcus. Cross correlation trajectory study of $V-V$ energy transfer in HF-HF and DF-DF. J. Chem. Phys., 73(8):3643–3652, 1980.
91. F. A. Gianturco, U. T. Lamanna, and F. Battaglia. Vibrationally inelastic scattering and relaxation times in gaseous HF. Int. J. Quantum Chemistry, 19:217–236, 1981.

92. David W. Schwenke, Deverajan Thirumalai, and Donald G. Truhlar. Tests of the quasiclassical trajectory cross-correlation moment method against accurate quantum dynamics for $V - V$ energy transfer in HF-HF collisions. *J. Chem. Phys.*, 78:3078–3083, 1983.
93. S. Bosanac, J. T. Brobjer, and J. N. Murrell. Second virial coefficient of gaseous HF and rotational inelastic scattering for HF-HF collisions. *Mol. Phys.*, 51(2):313–322, 1984.
94. K. J. Rensberger, J. M. Robinson, and F. F. Crim. Collisional relaxation of DF($v = 1$) and HF($v = 1$) by the DF dimer. *J. Chem. Phys.*, 86(3):1340–1347, 1987.
95. Peter F. Vohralik, R. O. Watts, and Millard H. Alexander. HF-HF differential scattering cross sections. *J. Chem. Phys.*, 93:3983–4002, 1990.
96. Craig A. Taatjes and Stephen R. Leone. Laser double-resonance studies of low-temperature rotational and vibrational relaxation of HF: Rates for HF($J = 13$) + HF from 225 to 298 K and detection of HF($v = 1$) deactivation by HF clusters at 210-240 K. *J. Phys. Chem.*, 95:5870–5877, 1991.
97. G. D. Billing, V. A. Zenevich, and W. Lindinger. Semiclassical analysis of vibrational energy transfer in HF-HF and isotopic systems. I. $V - T/R$ and $V - V$ rate constants for the lowest transitions in HF-HF. *J. Chem. Phys.*, 97:3274–3281, 1992.
98. D. F. Smith. Hydrogen fluoride polymer spectrum, hexamer and tetramer. *J. Chem. Phys.*, 28:1040–1056, 1958.
99. James M. Lisy, Andrzej Tramer, Matthew F. Vernon, and Yuan T. Lee. Vibrational predissociation spectra of (HF)$_n$, $n = 2-6$. *J. Chem. Phys.*, 75(9):4733–4734, 1981.
100. P. V. Huong, J. C. Cornut, and B. Desbat. Infrared spectral profile of hydrogen fluoride diluted in deuterium fluoride. *J. Chem. Phys.*, 77(11):5406–5409, 1982.
101. R. J. Bemish, M. Wu, and R. E. Miller. Probing the dynamics of weakly bound complexes using high-resolution laser spectroscopy. *Faraday Discuss.*, 97:57–68, 1994.
102. R. J. Bemish, M. C. Chan, and R. E. Miller. Molecular control using d.c. electric fields: quenching of the tunneling in HF dimer. *Chem. Phys. Lett.*, 251:182–188, 1996.
103. B. J. Howard, T. R. Dyke, and W. Klemperer. The molecular beam spectrum and the structure of the hydrogen fluoride dimer. *J. Chem. Phys.*, 81(12):5417–5425, 1984.
104. H. S. Gutowsky, Carl Chuang, John D. Keen, T. D. Klots, and Tryggvi Emilsson. Microwave rotational spectra, hyperfine interactions, and structure of the hydrogen fluoride dimers. *J. Chem. Phys.*, 83(5):2070–2077, 1985.
105. W. J. Lafferty, R. D. Suenram, and F. J. Lovas. Microwave spectra of the (HF)$_2$, (DF)$_2$, HFDF, and DFHF hydrogen-bonded complexes. *J. Mol. Spectrosc.*, 123:434–452, 1987.
106. S. P. Belov, E. N. Karyakin, I. N. Kozin, A. F. Krupnov, O. L. Polyansky, M. Yu. Tretyakov, N. F. Zobov, R. D. Suenram, and W. J. Lafferty. Tunneling-rotation spectrum of the hydrogen fluoride dimer. *J. Mol. Spectrosc.*, 141:204–222, 1990.
107. David K. Hindermann and C. D. Cornwell. Fluorine and proton NMR study of gaseous hydrogen fluoride. *J. Chem. Phys.*, 48(5):2017–2024, 1968.
108. E. L. Mackor, C. MacLean, and C. W. Hilbers. NMR of hydrogen fluoride in the gas phase. *Recl. Trav. Chim.*, 87:655–672, 1968.
109. Christoph Maerker, Paul von Ragué Schleyer, K. R. Liedl, Tae-Kyu Ha, Martin Quack, and Martin A. Suhm. A critical analysis of electronic density functionals for structural, energetic, dynamic and magnetic properties of hydrogen fluoride clusters; to be published.
110. Lester Andrews, V. E. Bondybey, and J. H. English. FTIR spectra of (HF)$_n$ species in solid neon. *J. Chem. Phys.*, 81(8):3452–3457, 1984.
111. Lester Andrews. Fourier transform infrared spectroscopy of matrix isolated species. In J. R. Durig, editor, *Applications of FT-IR spectroscopy, Vol 18 in Vibrational*

Spectra and Structure, pages 183–216. Elsevier, 1990.
112. Lester Andrews, Stephen R. Davis, and Rodney D. Hunt. Far-infrared spectra of (HF)$_2$ and (HF)$_3$ in solid argon. Mol. Phys., 77:993–1003, 1992.
113. A. V. Nemukhin, B. L. Grigorenko, and A. V. Savin. Theoretical vibrational spectrum of (HF)$_2$ in argon matrices. Chem. Phys. Lett., 250:226–231, 1996.
114. Jay Janzen and L. S. Bartell. Electron-diffraction structural study of polymeric gaseous hydrogen fluoride. J. Chem. Phys., 50(8):3611–3618, 1969.
115. Matthew F. Vernon, James M. Lisy, Douglas J. Krajnovich, Andrzej Tramer, Hoi-Sing Kwok, Y. Ron Shen, and Yuan T. Lee. Vibrational predissociation spectra and dynamics of small molecular clusters of H$_2$O and HF. Faraday Discuss. Chem. Soc., 73:387–397, 1982.
116. A. S. Pine and W. J. Lafferty. Rotational structure and vibrational predissociation in the HF stretching bands of the HF dimer. J. Chem. Phys., 78(5):2154–2162, 1983.
117. Robert L. DeLeon and J. S. Muenter. Vibrational predissociation in the hydrogen fluoride dimer. J. Chem. Phys., 80(12):6092–6094, 1984.
118. Daniel W. Michael and James M. Lisy. Vibrational predissociation spectroscopy of (HF)$_3$. J. Chem. Phys., 85(5):2528–2537, 1986.
119. N. Halberstadt, Ph. Bréchignac, J. A. Beswick, and M. Shapiro. Theory of mode specific vibrational predissociation: The HF dimer. J. Chem. Phys., 84(1):170–175, 1986.
120. A. S. Pine and G. T. Fraser. Vibrational, rotational, and tunneling dependence of vibrational predissociation in the HF dimer. J. Chem. Phys., 89(11):6636–6643, 1988.
121. G. T. Fraser and A. S. Pine. Vibrational predissociation in the H-F stretching mode of HF-DF. J. Chem. Phys., 91:633–636, 1989.
122. E. J. Bohac, Mark D. Marshall, and R. E. Miller. Initial state effects in the vibrational predissociation of hydrogen fluoride dimer. J. Chem. Phys., 96:6681–6695, 1992.
123. Curtis Laush and James M. Lisy. Saturation predissociation spectroscopy: Vibrational transition moments of HF dimer. J. Chem. Phys., 101:7480–7487, 1994.
124. W. Strohmeier and G. Briegleb. Über den Assoziationszustand des HF im Gaszustand I, II. Z. Elektrochemie, 57:662,668, 1953.
125. E. U. Franck and F. Meyer. Fluorwasserstoff III : Spezifische Wärme und Assoziation im Gas bei niedrigem Druck. Z. Elektrochemie, 63(5):571–582, 1959.
126. Cecil E. Vanderzee and Walter W. Rodenburg. Gas imperfections and thermodynamic excess properties of gaseous hydrogen fluoride. J. Chem. Thermodynamics, 2:461–478, 1970.
127. A. D. Buckingham and Liu Fan-Chen. Differences in the hydrogen and deuterium bonds. Int. Rev. Phys. Chem., 1:253–269, 1981.
128. R. L. Redington. Nonideal associated vapor analysis of hydrogen fluoride. J. Phys. Chem., 86:552–560, 1982.
129. Richard L. Redington. Infrared absorbance of hydrogen fluoride oligomers. J. Phys. Chem., 86:561–563, 1982.
130. W. David Chandler, Keith E. Johnson, and John L. E. Campbell. Thermodynamic calculations for reactions involving hydrogen halide polymers, ions, and Lewis acid adducts. 1. Polyfluorohydrogenate(1-) anions ($H_n F_{n+1}^-$), Polyfluorohydrogen(I) cations ($H_{n+1} F_n^+$), and hydrogen fluoride polymers ((HF)$_n$). Inorg. Chem., 34:4943–4949, 1995.
131. G. K. Vasilev, E. F. Makarov, Yu. A. Chernyshov, and V. G. Yakushev. A study of the kinetic and thermodynamic characteristics of HF associates. Sov. J. Chem. Phys., 4:1515–1527, 1987.
132. Ian M. Mills. Born-Oppenheimer failure in the separation of low frequency molecular vibrations. J. Phys. Chem., 88:532–536, 1984.

133. Jon T. Hougen and Nobukimi Ohashi. Group theoretical treatment of the planar internal rotation problem in (HF)$_2$. *J. Mol. Spectrosc.*, 109:134–165, 1985.
134. Gene C. Hancock, Paul Rejto, Rozeanne Steckler, Franklin B. Brown, David W. Schwenke, and Donald G. Truhlar. Reaction-path analysis of the tunneling splitting in fluctional molecules: Application to the degenerate rearrangement of hydrogen fluoride dimer. *J. Chem. Phys.*, 85:4997–5003, 1986.
135. Edwin L. Sibert III. Corrections to the Born-Oppenheimer treatment of the tunneling splitting in the HF dimer. *J. Phys. Chem.*, 93:5022–5024, 1989.
136. G. T. Fraser. Vibrational exchange upon interconversion tunneling in (HF)$_2$ and (HCCH)$_2$. *J. Chem. Phys.*, 90:2097–2108, 1989.
137. Gene C. Hancock and Donald G. Truhlar. Reaction-path analysis of the effect of monomer excitation on the tunneling splitting of the hydrogen fluoride dimer. *J. Chem. Phys.*, 90(7):3498–3505, 1989.
138. Per Jensen, P. R. Bunker, and A. Karpfen. An ab initio calculation of the nonadiabatic effect on the tunneling splitting in vibrationally excited (HF)$_2$. *J. Mol. Spectrosc.*, 148:385–390, 1991.
139. Dong H. Zhang and John Z. H. Zhang. Photofragmentation of HF dimer: Quantum dynamics studies on ab initio potential energy surfaces. *J. Chem. Phys.*, 99:6624–6633, 1993.
140. William C. Necoechea and Donald G. Truhlar. A converged full-dimensional calculation of the vibrational energy levels of (HF)$_2$. *Chem. Phys. Lett.*, 224:297–304, 1994. Erratum: Vol. 231 (1994), 125–126.
141. David M. Bishop, Janusz Pipin, and Bernard Kirtman. Effect of vibration on the linear and nonlinear optical properties of HF and (HF)$_2$. *J. Chem. Phys.*, 102:6778–6786, 1995.
142. Michael L. Klein and Ian R. McDonald. Structure and dynamics of associated molecular systems. I. Computer simulation of liquid hydrogen fluoride. *J. Chem. Phys.*, 71(1):298–308, 1979.
143. Ursula Röthlisberger and Michele Parrinello. preprint.
144. Changyin Zhang, David L. Freeman, and J. D. Doll. Monte Carlo studies of hydrogen fluoride clusters: Cluster size distributions in hydrogen fluoride vapor. *J. Chem. Phys.*, 91:2489–2497, 1989.
145. R. O. Watts. Infrared spectroscopy of large clusters. In G. Scoles, editor, *The chemical physics of atomic and molecular clusters. Proceedings of the international school of physics 'Enrico Fermi', Course CVII*, pages 271–329. North-Holland, 1990.
146. Geerd H. F. Diercksen and Wolfgang P. Kraemer. SCF MO LCGO studies on hydrogen bonding: the hydrogen fluoride dimer. *Chem. Phys. Lett.*, 6(5):419–422, 1970.
147. Peter A. Kollman and Leland C. Allen. Hydrogen bonded dimers and polymers involving hydrogen fluoride, water, and ammonia. *J. Am. Chem. Soc.*, 92(4):753–759, 1970.
148. Janet E. DelBene and J. A. Pople. Theory of molecular interactions. II. Molecular orbital studies of HF polymers using a minimal slater-type basis. *J. Chem. Phys.*, 55(1):2296–2299, 1971.
149. T. Aoyama and H. Yamakawa. Non-empirical calculations of HF oligomers. *Chem. Phys. Lett.*, 60:326–328, 1979.
150. Hans Lischka. Ab initio calculations on intermolecular forces. III. Effect of electron correlation on the hydrogen bond in the HF dimer. *J. Am. Chem. Soc.*, 96(15):4761–4766, 1974.
151. J. D. Dill, L. C. Allen, W. C. Topp, and J. A. Pople. A systematic study of the nine hydrogen-bonded dimers involving NH_3, OH_2, and HF. *J. Am. Chem. Soc.*, 97(25):7220–7226, 1975.
152. Anton Beyer and Alfred Karpfen. Ab initio studies on hydrogen bonded chains. II. Equilibrium geometry and vibrational spectra of the bent chain of hydrogen

fluoride molecules. *Chem. Phys.*, 64:343–357, 1982.
153. Alfred Karpfen, Anton Beyer, and Peter Schuster. Ab initio studies on clusters of polar molecules. Stability of cyclic versus open–chain trimers of hydrogen fluoride. *Chem. Phys. Lett.*, 102(4):289–291, 1983.
154. Reinhart Ahlrichs, Peter Scharf, and Claus Erhardt. The coupled pair functional (CPF). A size consistent modification of the CI(SD) based on an energy functional. *J. Chem. Phys.*, 82(2):890–898, 1985.
155. Shiyi Liu, Daniel W. Michael, Clifford E. Dykstra, and James M. Lisy. The stabilities of the hydrogen fluoride trimer and tetramer. *J. Chem. Phys.*, 84(9):5032–5036, 1986.
156. Michael J. Frisch, Janet E. DelBene, J. Stephen Binkley, and Henry F. Schaefer III. Extensive theoretical studies of the hydrogen-bonded complexes $(H_2O)_2$, $(H_2O)_2H^+$, $(HF)_2$, $(HF)_2H^+$, F_2H^-, and $(NH_3)_2$. *J. Chem. Phys.*, 84(4):2279–2289, 1986.
157. Alan E. Reed, Frank Weinhold, Larry A. Curtiss, and David J. Pochatko. Natural bond orbital analysis of molecular interactions: Theoretical studies of binary complexes of HF, H_2O, NH_3, N_2, O_2, F_2, CO, and CO_2 with HF, H_2O, and NH_3. *J. Chem. Phys.*, 84(10):5687–5705, 1986.
158. Pavel Hobza, Bohdan Schneider, Petr Čársky, and Rudolf Zahradník. The superposition error problem: The $(HF)_2$ and $(H_2O)_2$ complexes at the SCF and MP2 levels. *J. Mol. Struct. (Theochem)*, 138:377–385, 1986.
159. Zdzislaw Latajka and Steve Scheiner. Structure, energetics and vibrational spectra of H-bonded systems. Dimers and trimers of HF and HCl. *Chem. Phys.*, 122:413–430, 1988.
160. Małgorzata M. Szczęśniak and Grzegorz Chałasinski. Anisotropy of correlation effects in hydrogen-bonded systems: The HF dimer. *Chem. Phys. Lett.*, 161:532–538, 1989.
161. I. Røeggen. An analysis of hydrogen-bonded systems: $(HF)_2$, $(H_2O)_2$ and $H_2O \cdots HF$. *Mol. Phys.*, 70(3):353–376, 1990.
162. P. R. Bunker, Per Jensen, Alfred Karpfen, Manfred Kofranek, and Hans Lischka. An ab initio calculation of the stretching energies for the HF dimer. *J. Chem. Phys.*, 92:7432–7440, 1990.
163. Stanislaw Rybak, Bogumil Jeziorski, and Krzysztof Szalewicz. Many-body symmetry-adapted perturbation theory of intermolecular interactions. H_2O and HF dimers. *J. Chem. Phys.*, 95:6576–6601, 1991.
164. Janet E. DelBene and Isaiah Shavitt. A theoretical study of the neutral, protonated, and deprotonated trimers of HF and HCl. *J. Mol. Struct. (Theochem)*, 234:499–508, 1991.
165. D. Heidrich, W. Kliesch, and W. Quapp. *Properties of chemically interesting potential energy surfaces.* Springer (Berlin), 1991.
166. Andrew Komornicki, David A. Dixon, and Peter R. Taylor. Concerted hydrogen atom exchange between three HF molecules. *J. Chem. Phys.*, 96:2920–2925, 1992.
167. Stephen C. Racine and Ernest R. Davidson. Electron correlation contribution to the hydrogen bond in $(HF)_2$. *J. Phys. Chem.*, 97:6367–6372, 1993.
168. D. Heidrich, Nicolaas J. R. van Eikema Hommes, and Paul von Ragué Schleyer. Ab initio models for multiple-hydrogen exchange: Comparison of cyclic four- and six-center systems. *J. Comput. Chem.*, 14:1149–1163, 1993.
169. Zdzislaw Latajka and Yves Bouteiller. Application of density functional methods for the study of hydrogen-bonded systems: The hydrogen fluoride dimer. *J. Chem. Phys.*, 101:9793–9799, 1994.
170. Janet E. DelBene and Isaiah Shavitt. Basis-set effects on computed acid-base interaction energies using the Dunning correlation-consistent polarized split-valence basis sets. *J. Mol. Struct. (Theochem)*, 307:27–34, 1994.
171. Masanori Tachikawa and Kaoru Iguchi. Nonadditivity effects in the molecular interactions of H_2O and HF trimers by the symmetry-adapted perturbation theory.

J. Chem. Phys., 101:3062–3072, 1994.
172. A. Karpfen and O. Yanovitskii. Structure and vibrational spectra of neutral, protonated and deprotonated hydrogen bonded polymers: an ab initio SCF study on chain-like hydrogen fluoride clusters. J. Mol. Struct. (THEOCHEM), 307:81–97, 1994.
173. M. H. Alexander and Andrew E. DePristo. Fitting an ab initio HF-HF potential surface. J. Chem. Phys., 65(11):5009–5016, 1976.
174. Frank H. Stillinger. Polarization model representation of hydrogen fluoride for use in gas- and condensed phase studies. Int. J. Quant. Chem., 14:649–657, 1978.
175. William H. Press, Brian P. Flannery, Saul A. Teukolsky, and William T. Vetterling. Numerical Recipes - The Art of Scientific Computing. Cambridge University Press, 1986.
176. Wilfred F. van Gunsteren and Herman J. C. Berendsen. Computer simulation of molecular dynamics: Methodology, applications, and perspectives in chemistry. Angew. Chem. Int. Ed. Engl., 29:992–1023, 1990.
177. Saroj K. Nayak and Ramakrishna Ramaswamy. Solid $\rightleftharpoons$ liquid transition in model $(HF)_n$ clusters. Mol. Phys., 89:809–817, 1996.
178. Warren T. Zemke, William C. Stwalley, John A. Coxon, and Photos G. Hajigeorgiou. Improved potential energy curves and dissociation energies for HF, DF and TF. Chem. Phys. Lett., 177:412–418, 1991.
179. Jonathon K. Gregory and David C. Clary. Quantum simulation of weakly bound complexes using direct ab initio points. Chem. Phys. Lett., 237:39–44, 1995.
180. Ronald C. Cohen and Richard J. Saykally. Multidimensional intermolecular potential surfaces from vibration-rotation-tunneling (vrt) spectra of van der Waals complexes. Ann. Rev. Phys. Chem., 42:369–392, 1991.
181. Klaus Stark and Hans-Joachim Werner. An accurate multireference configuration interaction calculation of the potential energy surface for the $F+H_2 \rightarrow HF+H$ reaction. J. Chem. Phys., 104:6515–6530, 1996.
182. Zdzislaw Latajka, Yves Bouteiller, and Steve Scheiner. Critical assessment of density functional methods for study of proton transfer processes. $(FHF)^-$. Chem. Phys. Lett., 234:159–164, 1995.
183. F. A. Gianturco, E. Buonomo, E. Semprini, F. Stefani, and Amadeo Palma. Ab initio potential energy function for the dynamics of the fluoronium ion. Int. J. Quantum Chemistry, 47:335–373, 1993.
184. R. Car and M. Parrinello. Unified approach for molecular dynamics and density-functional theory. Phys. Rev. Lett., 55:2471–2474, 1985.
185. Bernd Hartke and Emily A. Carter. Ab initio molecular dynamics with correlated molecular wave functions: Generalized valence bond molecular dynamics and simulated annealing. J. Chem. Phys., 97:6569–6578, 1992.
186. Toshio Asada and Suehiro Iwata. Hybrid procedure of ab initio molecular orbital calculation and Monte Carlo simulation for studying intracluster reactions: applications to $Mg^+(H_2O)_n$, $(n = 1 - 4)$. Chem. Phys. Lett., 260:1–6, 1996.
187. A. D. Becke. A multicenter numerical integration scheme for polyatomic molecules. J. Chem. Phys., 88:2547–2553, 1988.
188. Chengteh Lee, Weitao Yang, and Robert G. Parr. Development of the Colle-Salvetti correlation-energy formula into a functional of the electron density. Phys. Rev. B, 37:785–789, 1988.
189. Axel D. Becke. Density-functional thermochemistry. III. The role of exact exchange. J. Chem. Phys., 98:5648–5652, 1993.
190. Peter Botschwina. Anharmonic potential-energy surfaces, vibrational frequencies and infrared intensities calculated from highly correlated wavefunctions. J. Chem. Soc., Faraday Trans. 2, 84:1263–1276, 1988.
191. M. J. D. Powell. Radial basis functions for multivariable interpolation: A review. In J. C. Mason and M. G. Cox, editors, Algorithms for approximation, pages 143–167. Clarendon Press, Oxford, 1987.

192. Tae-Kyu Ha, David Luckhaus, and Martin Quack. Spectrum and dynamics of the CH chromophore in CD_2HF: II. Ab initio calculations of the potential and dipole moment functions. *Chem. Phys. Lett.*, 190:590–598, 1992.
193. Donald W. Marquardt. An algorithm for least-squares estimation of nonlinear parameters. *J. Soc. Indust. Appl. Math.*, 11(2):431–441, 1963.
194. Clifford E. Dykstra. Weak interaction potentials of large clusters developed from small cluster information. *J. Phys. Chem.*, 94:6948–6956, 1990.
195. Atsuyuki Okabe, Barry Boots, and Kokichi Sugihara. *Spatial Tesselations. Concepts and Applications of Voronoi Diagrams.* Wiley, Chichester, 1992.
196. Josef Ischtwan and Michael A. Collins. Molecular potential energy surfaces by interpolation. *J. Chem. Phys.*, 100:8080–8088, 1994.
197. Michael A. Collins. The interface between electronic structure theory and reaction dynamics by reaction path methods. *Adv. Chem. Phys.*, XCIII:389–453, 1996.
198. Joel M. Bowman, Joseph S. Bittman, and Lawrence B. Harding. Ab initio calculations of electronic and vibrational energies of HCO and HOC. *J. Chem. Phys.*, 85:911–921, 1986.
199. Trygve Helgaker, Einar Uggerud, and Hans Jorgen Aa. Jensen. Integration of the classical equations of motion on ab initio molecular potential energy surfaces using gradients and Hessians: application to translational energy release upon fragmentation. *Chem. Phys. Lett.*, 173:145–150, 1990.
200. Wlodzislaw Duch and Geerd H. F. Diercksen. Neural networks as tools to solve problems in physics and chemistry. *Comp. Phys. Communications*, 82:91–103, 1994.
201. Thomas B. Blank, Steven D. Brown, August W. Calhoun, and Douglas J. Doren. Neural network models of potential energy surfaces. *J. Chem. Phys.*, 103:4129–4137, 1995.
202. David F. R. Brown, Mark N. Gibbs, and David C. Clary. Combining ab initio computations, neural networks, and diffusion Monte Carlo: An efficient method to treat weakly bound molecules. *J. Chem. Phys.*, 105:7597–7604, 1996.
203. U. Niesar, G. Corongiu, M.-J. Huang, M. Dupuis, and E. Clementi. Preliminary observations on a new water-water potential. *Int. J. Quantum Chem. (Symposium)*, 23:421–443, 1989.
204. Ilya G. Kaplan, Ruben Santamaria, and Octavio Novaro. Non-additive forces in atomic clusters. The case of Ag_n. *J. Mol. Spectrosc.*, 84:105–114, 1994.
205. Alberto De Santis and Dario Rocca. Evidence of nonadditive many-body terms in the water potential. *J. Chem. Phys.*, 105:7227–7230, 1996.
206. J. D. Cruzan, M. G. Brown, Kun Liu, L. B. Braly, and R. J. Saykally. The far-infrared vibration-rotation-tunneling spectrum of the water tetramer-d8. *J. Chem. Phys.*, 105:6634–6644, 1996.
207. Bretta F. King and Frank Weinhold. Structure and spectroscopy of $(HCN)_n$ clusters: Cooperative and electronic delocalization effects in C-H···N hydrogen bonding. *J. Chem. Phys.*, 103:333–347, 1995.
208. Alfred Karpfen. Case studies on cooperativity in hydrogen-bonded clusters and polymers: neutral chain-like and cyclic clusters of hydrogen fluoride and hydrogen cyanide and charged defects on chain-like hydrogen fluoride oligomers. In S. Scheiner, editor, *Molecular Interactions - From van der Waals to Strongly Bound Complexes*, in press, Wiley, New York.
209. Richard M. Adams and J. J. Katz. New variable thickness infrared cell and the infrared spectra of HF, DF, H_2O and D_2O. *J. Opt. Soc. Amer.*, 46:895–898, 1956.
210. Bernard Desbat and Pham V. Huong. Structure of liquid hydrogen fluoride studied by infrared and Raman spectroscopy. *J. Chem. Phys.*, 78(11):6377–6383, 1983.
211. M. Deraman, J. C. Dore, J. G. Powles, J. H. Holloway, and P. Chieux. Structural studies of liquid hydrogen fluoride by neutron diffraction. I. Liquid DF at 293 K. *Mol. Phys.*, 55:1351–1367, 1985.
212. J. S. Kittelberger and D. F. Hornig. Vibrational spectrum of crystalline HF and

DF. *J. Chem. Phys.*, 46(8):3099–3108, 1967.
213. Kazuhiko Honda, Kazuo Kitaura, and Kichisuke Nishimoto. Theoretical study of structure and thermodynamic properties of liquid hydrogen fluoride. *Bull. Chem. Soc. Japan*, 65:3122–3134, 1992.
214. M. P. Allen and D. J. Tildesley. *Computer Simulation of Liquids*. Clarendon Press, Oxford, 1989.
215. Martin Quack. Molecular quantum dynamics from high resolution spectroscopy and laser chemistry. *J. Mol. Struct.*, 292:171–195, 1993.
216. Wim Klopper. Limiting values for Møller-Plesset second-order correlation energies of polyatomic systems: A benchmark study on Ne, HF, H_2O, N_2, and He$\cdots$He. *J. Chem. Phys.*, 102:6168–6179, 1995.
217. Y. Bouteiller. Electronic and vibrational properties of hydrogen bonded complexes in the gas phase. *Trends in Chemical Physics*, 1:277–302, 1991.
218. Grzegorz Chałasinski and Małgorzata M. Szczęśniak. Origins of structure and energetics of van der Waals clusters from ab initio calculations. *Chem. Rev.*, 94:1723–1765, 1994.
219. M. Kertesz, J. Koller, and A. Azman. Ab initio crystal orbital treatment of hydrogen fluoride (HF) chains. *Chem. Phys. Lett.*, 36(5):576–579, 1975.
220. Ernest R. Davidson and Subhas J. Chakravorty. A possible definition of basis set superposition error. *Chem. Phys. Lett.*, 217:48–54, 1994.
221. A. J. Perkins. The refractive index of anhydrous hydrogen fluoride. *J. Phys. Chem.*, 68:654–655, 1964.
222. J. S. Muenter. Polarizability anisotropy of hydrogen fluoride. *J. Chem. Phys.*, 56:5409–5412, 1972.
223. Juan J. Novoa, Marc Planas, and Myung-Hwan Whangbo. A numerical evaluation of the counterpoise method on hydrogen bond complexes using near complete basis sets. *Chem. Phys. Lett.*, 225:240–246, 1994.
224. Frans B. van Duijneveldt, Jeanne G. C. M. van Duijneveldt-van de Rijdt, and Joop H. van Lenthe. State of the art in counterpoise theory. *Chem. Rev.*, 94:1873–1885, 1994.
225. Werner Kutzelnigg and Wim Klopper. Wave functions with terms linear in the interelectronic coordinates to take care of the correlation cusp. I. General theory. *J. Chem. Phys.*, 94:1985–2001, 1991.
226. Eugene S. Kryachko and Eduardo V. Ludeña. *Energy Density Functional Theory of Many-Electron Systems*. Kluwer, Dordrecht, 1990.
227. Michael Springborg. On solitonic defects in hydrogen-bonded $(HF)_x$. *Chem. Phys. Lett.*, 195:143–155, 1994.
228. Jozef Noga, Werner Kutzelnigg, and Wim Klopper. CC-R12, a correlation cusp corrected coupled-cluster method with a pilot application to the Be_2 potential curve. *Chem. Phys. Lett.*, 199:497–504, 1992.
229. Martin A. Suhm and Robert O. Watts. Quantum Monte Carlo studies of vibrational states in molecules and clusters. *Physics Reports*, 204:293–329, 1991.
230. J. B. Anderson. A random-walk simulation of the Schrödinger equation: H_3^+. *J. Chem. Phys.*, 63:1499–1503, 1975.
231. William A. Lester, Jr. and Brian L. Hammond. Quantum Monte Carlo for the electronic structure of atoms and molecules. *Ann. Rev. Phys. Chem.*, 41:283–311, 1990.
232. David M. Ceperley and Lubos Mitas. Quantum Monte Carlo methods in chemistry. *Adv. Chem. Phys.*, XCIII:1–38, 1996.
233. James B. Anderson. Exact quantum chemistry by Monte Carlo methods. In S. R. Langhoff, editor, *Quantum Mechanical Electronic Structure Calculations with Chemical Accuracy*, pages 1–45. Kluwer Academic Publishers, Dordrecht, 1995.
234. Shyn-Yi Leu and Chung-Yuan Mou. Floating spherical Gaussian orbitals based quantum Monte Carlo method in molecular electronic calculations. *J. Chem. Phys.*, 101:5910–5918, 1994.

235. Arne Lüchow and J. B. Anderson. Accurate quantum Monte Carlo calculations for hydrogen fluoride and the fluorine atom. *J. Chem. Phys.*, 105:4636–4640, 1996.
236. J. N. Huffaker, Majid Karimi, and Loc Binh Tran. Accurate analytic vibrational potential and dipole functions for HF. *J. Mol. Spectrosc.*, 124:393–406, 1987.
237. John A. Coxon and Photos G. Hajigeorgiou. Isotopic dependence of Born-Oppenheimer breakdown effects in diatomic hydrides: The $B^1\Sigma^+$ and $X^1\Sigma^+$ states of HF and DF. *J. Mol. Spectrosc.*, 142:254–278, 1990.
238. Daniel Kastler. Théorie quantique de la molécule HF. *Journal de Physique*, 50:556–572, 1953.
239. S. Peyerimhoff. Berechnungen am HF-Molekül. *Z. f. Naturforschung A*, 18:1197–1204, 1963.
240. Vladimir Bondybey, Peter K. Pearson, and Henry F. Schaefer III. Theoretical potential energy curves for OH, HF^+, HF, HF^-, NeH^+ and NeH. *J. Chem. Phys.*, 57:1123–1128, 1972.
241. Richard N. Sileo and Terrill A. Cool. Overtone emission spectroscopy of HF and DF: Vibrational matrix elements and dipole moment function. *J. Chem. Phys.*, 65:117–133, 1976.
242. Hans-Joachim Werner and Pavel Rosmus. Theoretical dipole moment functions of the HF, HCl, and HBr molecules. *J. Chem. Phys.*, 73:2319–2328, 1980.
243. Jan M. L. Martin and Peter R. Taylor. Basis set convergence for geometry and harmonic frequencies. Are h-functions enough? *Chem. Phys. Lett.*, 225:473–479, 1994.
244. Mehmet Şimşek and Zeynel Yalçin. Generalized Pöschl-Teller potential. *J. Math. Chem.*, 16:211–215, 1994.
245. Peter A. Kollman and Leland C. Allen. Theory of the hydrogen bond: Ab initio calculations on hydrogen fluoride dimer and the mixed water hydrogen fluoride dimer. *J. Chem. Phys.*, 52(10):5085–5093, 1970.
246. William J. Jorgensen and Michael E. Cournoyer. An intermolecular potential function for the hydrogen fluoride dimer from ab initio 6-31G computations. *J. Am. Chem. Soc.*, 100(16):4942–4945, 1978.
247. William L. Jorgensen. Basis set dependence of the structure and properties of liquid hydrogen fluoride. *J. Chem. Phys.*, 70(12):5888–5897, 1979.
248. Michael L. Klein, Ian R. McDonald, and Séamus F. O'Shea. An intermolecular force model for $(HF)_2$. *J. Chem. Phys.*, 69(1):63–66, 1978.
249. Lise Lotte Poulsen, Gert D. Billing, and J. I. Steinfeld. Temperature dependence of HF vibrational relaxation. *J. Chem. Phys.*, 68:5121–5127, 1978.
250. Richard L. Redington and Delphia F. Hamill. Infrared matrix isolation spectra of the H_2F_2 dimer. *J. Chem. Phys.*, 80(6):2446–2461, 1984.
251. David W. Schwenke and Donald G. Truhlar. Converged close coupling calculations for $V-V$ energy transfer: $2HF(v=1) \rightarrow HF(v=2)+HF(v=0)$. *Theor. Chim. Acta*, 69:175, 1986.
252. G. D. Billing. Semiclassical calculation of cross sections for vibration-rotation energy transfer in HF-HF collisions. *J. Chem. Phys.*, 84:2593–2603, 1986.
253. Claudio Amovilli and Roy McWeeny. Perturbation calculation of molecular interaction energies: an example, HF-HF. *Chem. Phys. Lett.*, 128(1):11–17, 1986.
254. W. A. Sokalski, A. H. Lowrey, S. Roszak, V. Lewchenko, J. Blaisdell, P. C. Hariharan, and Joyce J. Kaufman. Nonempirical atom-atom potentials for main components of intermolecular interaction energy. *J. Comp. Chem.*, 7:693–700, 1986.
255. George C. Berend and Ronald L. Thommarson. Vibrational relaxation of HF and DF. *J. Chem. Phys.*, 58(8):3203–3208, 1973.
256. Scott Davis, David T. Anderson, and David J. Nesbitt. Plucking a hydrogen bond: A near IR study of all four intermolecular modes in $(DF)_2$. *J. Chem. Phys.*, 105:6645–6664, 1996.
257. Dong H. Zhang and John Z. H. Zhang. Total and partial decay widths in vibrational predissociation of HF dimer. *J. Chem. Phys.*, 98:5978–5981, 1993.

258. Kersti Hermansson. Redshifts and blueshifts of OH vibrations. *Int. J. Quantum Chem.*, 45:747–758, 1993. erratum: Vol. 47, 175 (1993).
259. B. I. Stepanov. Interpretation of the regularities in the spectra of molecules forming the intermolecular hydrogen bond by the predissociation effect. *Nature*, 157:808, 1946.
260. Jonathon K. Gregory and David C. Clary. Tunneling dynamics in water tetramer and pentamer. *J. Chem. Phys.*, 105:6626–6633, 1996.
261. Tiffany R. Walsh and David J. Wales. Rearrangements of the water trimer. *J. Chem. Soc., Faraday Trans.*, 92:2505–2517, 1996.
262. M. W. Johnson, E. Sandor, and E. Arzi. The crystal structure of deuterium fluoride. *Acta Cryst. B*, 31:1998–2003, 1975.
263. H. L. Frisch and E. Wasserman. Chemical topology. *J. Am. Chem. Soc.*, 83:3789–3795, 1961.
264. Friedrich Huisken, Elena G. Tarakanova, Andrei A. Vigasin, and Gregory V. Yukhnevich. Vibrational frequency shifts and thermodynamic stabilities of $(HF)_n$ isomers ($n = 4 - 8$). *Chem. Phys. Lett.*, 245:319–325, 1995.
265. Martin Quack. Molecular infrared spectra and molecular motion. *J. Mol. Struct.*, 347:245–266, 1995.
266. M. E. van Leeuwen and B. Smit. What makes a polar liquid a liquid? *Phys. Rev. Lett.*, 71(24):3991–3994, 1993.
267. Robert T. Lagemann and C. Harry Knowles. Velocity of compressional waves in liquid hydrogen fluoride and some thermodynamic properties derived therefrom. *J. Chem. Phys.*, 32:561–564, 1960.
268. N. Karger, T. Vardag, and H.-D. Lüdemann. p, T-dependence of self-diffusion in liquid hydrogen fluoride. *J. Chem. Phys.*, 100:8271–8276, 1994.
269. M. W. Chase, Jr., C. A. Davies, J. R. Downey, Jr., D. J. Frurip, R. A. McDonald, and A. N. Syverud. JANAF thermochemical tables. *J. Phys. Chem. Ref. Data*, 14:Suppl. No. 1, 1985.

One–Electron Pictures of Electronic Structure: Propagator Calculations on Photoelectron Spectra of Aromatic Molecules

J.V. Ortiz, V.G. Zakrzewski and O. Dolgounitcheva
Department of Chemistry
University of New Mexico
Albuquerque, New Mexico 87131-1096
United States of America

Current Address:
Department of Chemistry
Kansas State University
Manhattan, Kansas 66506-3701
United States of America

1 Introduction

Hartree–Fock [1] and density functional [2, 3] theories offer one–electron pictures of molecular electronic structure. The former has been the standard by which the concept of electron correlation has been defined [4] and its errors therefore are known as correlation effects. Density functional theories provide a variety of exchange and correlation approximations that can be incorporated into orbital eigenvalue equations. The widespread application of methods based on the Kohn–Sham equations [5] attests to the appeal of the one–electron paradigm.

Electron propagator theory [6-12] presents an alternative one–electron picture. Residues and poles of the electron propagator yield Dyson orbitals and electron binding energies that generalize the canonical orbitals and orbital energies of Hartree–Fock theory. Electron correlation may be systematically included through improvements to the self–energy: an energy–dependent, nonlocal potential.

After a general discussion of the physical content of the electron propagator, poles and residues are displayed in the uncorrelated case. Several mathematical devices, including superoperators, inner projection and partitioning, conduce to a computationally useful form of the electron propagator. Relationships with Hilbert space methods, such as configuration interaction and equation–of–motion, coupled–cluster theory are clarified. The Dyson equation is derived subsequently and perturbative formulae for the self–energy are obtained. Several computational techniques, including pole search methods, semidirect contractions and symmetry adaptation, enable efficient evaluation of electron binding energies. Quasiparticle approximations, a set of conservative, but useful extensions of Koopmans's theorem, are presented. Applications on a variety of aromatic molecules illustrate the utility of electron propagator theory in predicting photoelectron spectra and in interpreting the results in terms of one–electron concepts.

2 The Electron Propagator

The physical content of the electron propagator resides chiefly in its poles (energies where singularities lie) and residues (coefficients of the terms responsible for the singularities) [13]. In its spectral form, the r,s element of the electron propagator (or, one-electron Green function) matrix is

$$G_{rs}(E) \equiv \langle\langle a_r^\dagger; a_s \rangle\rangle =$$

$$\lim_{\eta \to 0} \{ \sum_n \frac{\langle N|a_r^\dagger|N-1,n\rangle\langle N-1,n|a_s|N\rangle}{E + E_n(N-1) - E_0(N) - i\eta} + \sum_m \frac{\langle N|a_s|N+1,m\rangle\langle N+1,m|a_r^\dagger|N\rangle}{E - E_m(N+1) + E_0(N) + i\eta} \}. \quad (1)$$

The limit with respect to η is taken because of integration techniques required in a Fourier transform from the time-dependent representation. Indices r and s refer to general, orthonormal spin-orbitals, $\phi_r(x)$ and $\phi_s(x)$, respectively, where x is a space–spin coordinate. Matrix elements of the corresponding field operators, $a_r^\dagger$ and a_s, depend on the N-electron reference state, $|N\rangle$, and final states with N±1 electrons, labelled by the indices m and n. The propagator matrix is energy-dependent; poles occur when E equals an ionization energy, $E_0(N) - E_n(N-1)$, or an electron affinity, $E_0(N) - E_m(N+1)$. Corresponding residues, such as $\langle N|a_r^\dagger|N-1,n\rangle\langle N-1,n|a_s|N\rangle$ or $\langle N|a_s|N+1,m\rangle\langle N+1,m|a_r^\dagger|N\rangle$, are related to the Feynman-Dyson amplitudes (FDAs), where

$$U_{r,n} = \langle N-1,n|a_r|N\rangle \quad (2)$$

or

$$U_{r,n} = \langle N+1,n|a_r^\dagger|N\rangle. \quad (3)$$

FDAs suffice for constructing Dyson orbitals (DOs) for ionization energies, where

$$\phi_n^{Dyson,IE}(x) = \sum_r \phi_r(x) U_{r,n}, \quad (4)$$

and for electron affinities, where

$$\phi_n^{Dyson,EA}(x) = \sum_r \phi_r(x) U_{r,n}. \quad (5)$$

In the former case, the DO is related to initial and final state wavefunctions via

$$\phi_n^{Dyson,IE}(x_1) = \int \Psi_N(x_1, x_2, x_3, \ldots, x_N) \Psi^*_{N-1,n}(x_2, x_3, x_4, \ldots, x_N) dx_2 dx_3 dx_4 \cdots dx_N, \quad (6)$$

while for electron affinities,

$$\phi_n^{Dyson,EA}(x_1) = \int \Psi_N^*(x_2, x_3, x_4, \ldots, x_{N+1}) \Psi_{N+1,n}(x_1, x_2, x_3, \ldots, x_{N+1}) dx_2 dx_3 dx_4 \cdots dx_{N+1}. \quad (7)$$

Many kinds of transition probabilities depend on DOs. Photoionization cross sections, σ^{PI}, are proportional to the absolute squares of matrix elements between DOs and continuum orbitals, or

$$\sigma^{PI} = \kappa^{PI} |\langle \phi^{Dyson} | \hat{T} \phi^{Continuum} \rangle|^2, \quad (8)$$

where κ^{pI} is a constant and T is a transition operator describing the interaction between electrons and the radiation field [14]. DOs also are useful in computing cross sections for various electron scattering processes [15, 16].

Canonical molecular orbitals (MOs) are a special case of DOs, where Ψ_N is chosen to be a Hartree–Fock wavefunction and the final states are approximated by the frozen–orbital, single–determinant wavefunction assumed in Koopmans's theorem [17]. It is possible to improve the DOs through insertion of configuration interaction (CI) wavefunctions in equation 1. This need not be done, for propagator approaches to quantum chemistry eschew evaluation of many–electron wavefunctions and energies [18]. Instead, direct paths are sought to experimentally relevant quantities: transition energies and probabilities.

3 The Uncorrelated Case

A treatment of the unperturbed Møller–Plesset Hamiltonian [19],

$$H_0 = \sum_p \epsilon_p a_p^\dagger a_p, \tag{9}$$

in the canonical spin–orbital basis displays the simple structure of the electron propagator matrix. Here, the ground state is a determinantal wavefunction,

$$|N\rangle = \prod_p a_p^\dagger |vacuum\rangle. \tag{10}$$

and the ground state energy is a sum of occupied spin–orbital energies,

$$\langle N|H_0 N\rangle = \sum_p \epsilon_p n_p. \tag{11}$$

In the reference state, there are N spin–orbitals (i,j,k,...) with occupation numbers, n_p, equal to one; the remaining spin–orbitals (a,b,c,...) have occupation numbers equal to zero. Final states with N±1 electrons have different occupation numbers from the reference state. Total energy differences corresponding to poles are

$$E_{pole} = E_{n_a=1}(N+1) - E_0(N) = \epsilon_a \tag{12}$$

for electron affinities and

$$E_{pole} = E_0(N) - E_{n_i=0}(N-1) = \epsilon_i \tag{13}$$

for ionization energies. Because all states are single determinant wavefunctions, FDAs are especially simple:

$$\langle N+1, n_a=1|a_q^\dagger|N\rangle = \delta_{aq} \tag{14}$$

$$\langle N-1, n_i=0|a_p|N\rangle = \delta_{ip}. \tag{15}$$

Koopmans's theorem is recovered and DOs equal canonical orbitals.

4 Superoperator Theory

A concise, but flexible, formulation of propagator theory employs superoperators in deriving approximations from equation 1. Multiplying the latter expression by E implies that

$$E\langle\langle a_r^\dagger; a_s\rangle\rangle = \lim_{\eta\to 0}\{\sum_n \langle N|a_r^\dagger|N-1,n\rangle\langle N-1,n|a_s|N\rangle(1-\frac{E_n(N-1)-E_0(N)-i\eta}{E+E_n(N-1)-E_0(N)-i\eta})+$$

$$\sum_m \langle N|a_s|N+1,m\rangle\langle N+1,m|a_r^\dagger|N\rangle(1-\frac{-E_m(N+1)+E_0(N)+i\eta}{E-E_m(N+1)+E_0(N)+i\eta})\}. \tag{16}$$

Because the bras and kets are eigenstates of the Hamiltonian, H, such that,

$$H|N\rangle = E_0(N)|N\rangle \tag{17}$$

and

$$H|N\pm 1,m\rangle = E_m(N\pm 1)|N\pm 1,m\rangle, \tag{18}$$

there exists a relationship between the electron propagator and a more complicated propagator in which the field operator a_s has been replaced by $[a_s, H]$:

$$E\langle\langle a_r^\dagger; a_s\rangle\rangle = \langle N|[a_r^\dagger, a_s]_+|N\rangle +$$

$$\lim_{\eta\to 0}\{\sum_n \frac{\langle N|a_r^\dagger|N-1,n\rangle\langle N-1,n|[a_s,H]|N\rangle}{E+E_n(N-1)-E_0(N)-i\eta} + \sum_m \frac{\langle N|[a_s,H]|N+1,m\rangle\langle N+1,m|a_r^\dagger|N\rangle}{E-E_m(N+1)+E_0(N)+i\eta}\}. \tag{19}$$

Equation 19 may be rewritten as

$$E\langle\langle a_r^\dagger; a_s\rangle\rangle = \langle N|[a_r^\dagger, a_s]_+|N\rangle + \langle\langle a_r^\dagger; [a_s, H]\rangle\rangle. \tag{20}$$

This result is but a special case of a more general relation,

$$E\langle\langle \mu^\dagger; \nu\rangle\rangle = \langle N|[\mu^\dagger, \nu]_+|N\rangle + \langle\langle \mu^\dagger; [\nu, H]\rangle\rangle, \tag{21}$$

where μ and ν are, in general, products of field operators. A chain of equations, including

$$E\langle\langle a_r^\dagger; [a_s, H]\rangle\rangle = \langle N|[a_r^\dagger, [a_s, H]]_+|N\rangle + \langle\langle a_r^\dagger; [[a_s, H], H]\rangle\rangle \tag{22}$$

and

$$E\langle\langle a_r^\dagger; [[a_s, H], H]\rangle\rangle = \langle N|[a_r^\dagger, [[a_s, H], H]]_+|N\rangle + \langle\langle a_r^\dagger; [[[a_s, H], H], H]\rangle\rangle \tag{23}$$

may be forged. A series expression for the electron propagator thereby follows, where

$$\langle\langle a_r^\dagger; a_s\rangle\rangle = \frac{1}{E}\langle N|[a_r^\dagger, a_s]_+|N\rangle + \frac{1}{E^2}\langle N|[a_r^\dagger, [a_s, H]]_+|N\rangle +$$

$$\frac{1}{E^3}\langle N|[a_r^\dagger, [[a_s, H], H]]_+|N\rangle + \frac{1}{E^4}\langle N|[a_r^\dagger, [[[a_s, H], H], H]]_+|N\rangle + \cdots. \tag{24}$$

Through introduction of superoperators and a corresponding metric, the series may be represented more compactly [20]. Superoperators act on field operator products, X, where the number of annihilators exceeds the number of creators by one. The identity superoperator, $\hat{I}$, and the Hamiltonian superoperator, $\hat{H}$, are defined by

$$\hat{I} X = X \tag{25}$$

and

$$\hat{H} X = [X, H]_-, \tag{26}$$

respectively. The superoperator metric, defined by

$$(\mu|\nu) = \langle N|[\mu^\dagger, \nu]_+|N\rangle, \tag{27}$$

depends on the choice of the N–electron reference state, $|N\rangle$. Consideration of ionization energy and electron affinity poles in a single propagator leads to the anticommutator, $\mu^\dagger \nu + \nu \mu^\dagger$, contained in the metric definition. Had this discussion considered either the first or second terms in equation 1, the anticommutator would have been abandoned in favor of $\mu^\dagger \nu$ or $\nu \mu^\dagger$, respectively. (See Section 6.) With this notation, one may write

$$\langle\langle a_r^\dagger; a_s \rangle\rangle = \frac{1}{E}(a_r|a_s) + \frac{1}{E^2}(a_r|\hat{H} a_s) + \frac{1}{E^3}(a_r|\hat{H}^2 a_s) + \frac{1}{E^4}(a_r|\hat{H}^3 a_s) + \cdots \tag{28}$$

A more succinct statement is

$$\langle\langle a_r^\dagger; a_s \rangle\rangle \equiv G_{r,s}(E) = (a_r|(E\hat{I} - \hat{H})^{-1} a_s). \tag{29}$$

Thus the matrix elements of the electron propagator are related to field operator products arising from the superoperator resolvent, $(E\hat{I} - \hat{H})^{-1}$, that are evaluated with respect to $|N\rangle$. In this sense, electron binding energies and DOs are properties of the reference state.

5 Projection and Partitioning Techniques

Instead of considering polynomial expressions in $\hat{H}$ and E or other approaches to evaluating the matrix elements of the superoperator resolvent, an inner projection [21] is employed. In matrix notation, equation 29 is rewritten as

$$\mathbf{G}(E) = (\mathbf{a}|(E\hat{I} - \hat{H})^{-1}\mathbf{a}), \tag{30}$$

where

$$\mathbf{a} = \begin{bmatrix} a_1 \\ a_2 \\ a_3 \\ \vdots \\ a_K \end{bmatrix}, \tag{31}$$

and K is the dimension of the spin–orbital basis. After the inner projection,

$$\mathbf{G}(E) = (\mathbf{a}|\mathbf{u})(\mathbf{u}|(E\hat{I} - \hat{H})\mathbf{u})^{-1}(\mathbf{u}|\mathbf{a}), \tag{32}$$

where u is the vector of all X field operator products. An inverse matrix instead of an inverse superoperator is considered henceforth.

If u is partitioned into the primary space, a, and an orthogonal space, f, such that

$$(a|a) = 1, \qquad (33)$$

$$(f|f) = 1, \qquad (34)$$

$$(a|f) = 0, \qquad (35)$$

and

$$(f|a) = 0, \qquad (36)$$

then the partitioned form of the propagator matrix,

$$G(E) = \begin{bmatrix} (a|a) & (a|f) \end{bmatrix} \begin{bmatrix} (a|(E\hat{I} - \hat{H})a) & (a|(E\hat{I} - \hat{H})f) \\ (f|(E\hat{I} - \hat{H})a) & (f|(E\hat{I} - \hat{H})f) \end{bmatrix}^{-1} \begin{bmatrix} (a|a) \\ (f|a) \end{bmatrix}, \qquad (37)$$

reduces to

$$G(E) = \begin{bmatrix} 1 & 0 \end{bmatrix} \begin{bmatrix} E1 - (a|\hat{H}a) & -(a|\hat{H}f) \\ -(f|\hat{H}a) & E1 - (f|\hat{H}f) \end{bmatrix}^{-1} \begin{bmatrix} 1 \\ 0 \end{bmatrix}. \qquad (38)$$

Poles of the propagator therefore occur at values of E that are equal to eigenvalues, ω, of the superoperator Hamiltonian matrix:

$$\omega_n \begin{bmatrix} U_{a,n} \\ U_{f,n} \end{bmatrix} = \begin{bmatrix} (a|\hat{H}a) & (a|\hat{H}f) \\ (f|\hat{H}a) & (f|\hat{H}f) \end{bmatrix} \begin{bmatrix} U_{a,n} \\ U_{f,n} \end{bmatrix} \qquad (39)$$

or

$$U\omega = \hat{H}U. \qquad (40)$$

In the new basis of operators,

$$G(E) = \begin{bmatrix} 1 & 0 \end{bmatrix} [U(E1 - \omega)^{-1}U^\dagger] \begin{bmatrix} 1 \\ 0 \end{bmatrix}. \qquad (41)$$

Residues corresponding to the nth electron binding energy, ω_n, defined by

$$Res(\omega_n) = \lim_{E \to \omega_n} G_{rs}(E)(E - \omega_n), \qquad (42)$$

are obtained from

$$\lim_{E \to \omega_n} G_{rs}(E)(E - \omega_n) = U_{r,n}U^*_{s,n}. \qquad (43)$$

The DO corresponding to the same pole is

$$\phi_n^{Dyson} = \sum_r \phi_r U^*_{r,n}. \qquad (44)$$

Equations 35 and 36 imply that contributions from the secondary sector of the eigenvectors, U_f, do not appear in the residues.

Because this route to poles and residues requires only solutions of equation 40, the usual matrix diagonalization techniques characteristic of CI calculations may be applied [22, 23]. The chief conceptual difference is that operators, not configurations, form the basis. For ionization energies, operators corresponding to virtual (particle or p) orbitals and shakeon (two particle, one hole or 2p–h) processes may contribute to the eigenvector, U, in addition to the usual CI-like operators for occupied (hole or h) orbitals and shakeup (two hole, one particle or 2h–p) processes that generate (N–1)–electron states in Hilbert space when operating on a reference configuration. Electron affinity operators also have h, p, 2h-p and 2p-h constituents.

6 Relationships to Hilbert Space Theories

The relationship of propagator methods to Hilbert space theories such as configuration interaction [24] or the so-called coupled-cluster equation of motion [25] method can be clarified by separating the ionization energy and electron affinity components of the electron propagator. Let

$$G_{pq}^+(E) = \lim_{\eta \to \infty} \{ \sum_u \frac{\langle N|a_p|N+1,u\rangle \langle N+1,u|a_q^\dagger|N\rangle}{E - E_u(N+1) + E_0(N) + i\eta} \} \tag{45}$$

and

$$G_{pq}^-(E) = \lim_{\eta \to \infty} \{ \sum_v \frac{\langle N|a_q^\dagger|N-1,v\rangle \langle N-1,v|a_p|N\rangle}{E + E_v(N-1) - E_0(N) - i\eta} \}, \tag{46}$$

so that

$$G_{pq}(E) = G_{pq}^+(E) + G_{pq}^-(E). \tag{47}$$

An alternative notation facilitates comparisons with the preceding Sections:

$$G_{pq}^\pm(E) \equiv \langle\langle a_p; a_q^\dagger \rangle\rangle_E^\pm. \tag{48}$$

Making use of the same identities that were used deriving equation 20, one discovers that

$$E\langle\langle a_p; a_q^\dagger \rangle\rangle_E^+ = \langle N|a_p a_q^\dagger|N\rangle + \langle\langle [a_p, H]; a_q^\dagger \rangle\rangle_E^+ \tag{49}$$

and

$$E\langle\langle a_p; a_q^\dagger \rangle\rangle_E^- = \langle N|a_q^\dagger a_p|N\rangle + \langle\langle [a_p, H]; a_q^\dagger \rangle\rangle_E^-. \tag{50}$$

Note that the first term on the right side of equation 49 and its counterpart in equation 50 contain one field operator product instead of the anticommutator that occurs in the analogous term of equation 20. The series expression for each propagator is

$$\langle\langle a_p; a_q^\dagger \rangle\rangle_E^+ = E^{-1}\langle N|a_p a_q^\dagger|N\rangle + E^{-2}\langle N|[a_p, H]a_q^\dagger|N\rangle + E^{-3}\langle N|[[a_p, H], H]a_q^\dagger|N\rangle + \cdots \tag{51}$$

for the electron affinity part and is

$$\langle\langle a_p; a_q^\dagger \rangle\rangle_E^- = E^{-1}\langle N|a_q^\dagger a_p|N\rangle + E^{-2}\langle N|a_q^\dagger[a_p, H]|N\rangle + E^{-3}\langle N|a_q^\dagger[[a_p, H], H]|N\rangle + \cdots \tag{52}$$

for the ionization energy part. The superoperator expression of this result is

$$\langle\langle a_p; a_q^\dagger\rangle\rangle_E^\pm = (a_q|(E\hat{I} - \hat{H})^{-1}a_p)^\pm, \tag{53}$$

where

$$(X|Y)^+ = \langle N|YX^\dagger|N\rangle \tag{54}$$

and

$$(X|Y)^- = \langle N|X^\dagger Y|N\rangle. \tag{55}$$

After performing the inner projection,

$$\langle\langle a_p; a_q^\dagger\rangle\rangle_E^\pm = (a_q|\mathbf{h})^\pm(\mathbf{h}|(E\hat{I} - \hat{H})\mathbf{h})^{\pm^{-1}}(\mathbf{h}|a_p)^\pm, \tag{56}$$

it is clear that poles occur when

$$det\{E(\mathbf{h}|\mathbf{h})^\pm - (\mathbf{h}|\mathbf{h}H)^\pm + (\mathbf{h}|H\mathbf{h})^\pm\} = 0. \tag{57}$$

For ionization energies,

$$\begin{aligned}0 &= det\{E\langle N|\mathbf{h}^\dagger\mathbf{h}|N\rangle - \langle N|\mathbf{h}^\dagger\mathbf{h}H|N\rangle + \langle N|\mathbf{h}^\dagger H\mathbf{h}|N\rangle\}\\ &= det\{[E - E_0(N)]\langle N|\mathbf{h}^\dagger\mathbf{h}|N\rangle + \langle N|\mathbf{h}^\dagger H\mathbf{h}|N\rangle\}\end{aligned} \tag{58}$$

and for electron affinities,

$$\begin{aligned}0 &= det\{E\langle N|\mathbf{h}\mathbf{h}^\dagger|N\rangle - \langle N|\mathbf{h}H\mathbf{h}^\dagger|N\rangle + \langle N|H\mathbf{h}\mathbf{h}^\dagger|N\rangle\}\\ &= det\{[E + E_0(N)]\langle N|\mathbf{h}\mathbf{h}^\dagger|N\rangle - \langle N|\mathbf{h}H\mathbf{h}^\dagger|N\rangle\}.\end{aligned} \tag{59}$$

Solutions of the usual secular equations for the final states yield poles shifted by $E_0(N)$.

Approximate configuration interaction calculations can be defined by a reference state wavefunction, $|N\rangle$, and truncations of the operator manifold, $\mathbf{h}$. Alternatively, one can choose a reference configuration for the ground state and obtain the same poles with a suitably defined operator manifold. Eigenvectors will differ and residues therefore will depend on the choice of reference state. For example, when a Hartree–Fock reference configuration is chosen, a complete separation of ionization energy and electron affinity operator manifolds can be effected: only h, 2hp, 3h2p, ... operators contribute to $\mathbf{h}$ in the ionization energy case and only p, 2ph, 3p2h, ... operators contribute in the electron affinity case. A balanced treatment of electron binding energies requires a judicious choice of operator manifolds in initial and final states. Koopmans's theorem is a special case where a Hartree–Fock initial state is paired with h and p final states. Singles and doubles configuration interaction for the N-electron state might be coupled with final state treatments that include all operators up to 3h2p or 3p2h. A somewhat different approach is taken in certain coupled–cluster methods where the final state Hamiltonian matrices of equations 58 and 59 are written as

$$\langle HF|\mathbf{h}^\dagger H\mathbf{h}e^T|HF\rangle = \langle HF|\mathbf{h}^\dagger He^T\mathbf{h}|HF\rangle \tag{60}$$

and

$$\langle HF|\mathbf{h}H\mathbf{h}^\dagger e^T|HF\rangle = \langle HF|\mathbf{h}He^T\mathbf{h}^\dagger|HF\rangle, \tag{61}$$

respectively. The same operator manifolds that apply to the Hartree–Fock reference, configuration interaction methods are used and these operators commute with e^T. In effect, one performs a configuration interaction calculation with an effective Hamiltonian, He^T.

7 The Dyson Equation

Partitioning the operator manifold can lead to efficient strategies for finding poles and residues. In equation 38, only the upper left block of the inverse matrix is relevant. After a few elementary matrix manipulations, a convenient form of the inverse propagator matrix emerges, where

$$\mathbf{G}^{-1}(E) = E\mathbf{1} - (\mathbf{a}|\hat{H}\mathbf{a}) - (\mathbf{a}|\hat{H}\mathbf{f})\left[E\mathbf{1} - (\mathbf{f}|\hat{H}\mathbf{f})\right]^{-1}(\mathbf{f}|\hat{H}\mathbf{a}). \tag{62}$$

Because

$$(a_r|\hat{H}a_s) = h_{rs} + \sum_{tu}(rs||tu)\rho_{tu}, \tag{63}$$

where ρ is the one-electron density matrix, it is possible to separate the correlated and uncorrelated contributions to the $(\mathbf{a}|\hat{H}\mathbf{a})$ block. This block consists of elements of a generalized Fock matrix. In the canonical MO basis,

$$(a_r|\hat{H}a_s) = \epsilon_r \delta_{rs} + \sum_{tu}(rs||tu)\rho_{tu}^c, \tag{64}$$

where the correlation contribution to the one-electron density matrix is ρ^c and

$$\rho = \rho^{HF} + \rho^c. \tag{65}$$

Elements of the zeroth order inverse propagator matrix are

$$[\mathbf{G}_0^{-1}(E)]_{rs} = (E - \epsilon_r)\delta_{rs}. \tag{66}$$

(The poles equal Koopmans's theorem results.) The inverse propagator matrix and its zeroth order counterpart therefore are related through

$$\mathbf{G}^{-1}(E) = \mathbf{G}_0^{-1}(E) - \mathbf{\Sigma}(\infty) - \mathbf{\Sigma}'(E) \tag{67}$$

where

$$\mathbf{\Sigma}(\infty)_{rs} = (a_r|\hat{H}a_s)_{correlation} = \sum_{tu}(rs||tu)\rho_{tu}^c \tag{68}$$

and

$$\mathbf{\Sigma}'(E) = (\mathbf{a}|\hat{H}\mathbf{f})\left[E\mathbf{1} - (\mathbf{f}|\hat{H}\mathbf{f})\right]^{-1}(\mathbf{f}|\hat{H}\mathbf{a}). \tag{69}$$

Corrections to the zeroth order propagator in equation 67 are gathered together in a term known as the self-energy matrix, $\mathbf{\Sigma}(E)$. The Dyson equation may be written as

$$\mathbf{G}^{-1}(E) = \mathbf{G}_0^{-1}(E) - \mathbf{\Sigma}(E). \tag{70}$$

In the self-energy matrix, there are energy-independent terms and energy-dependent terms:

$$\mathbf{\Sigma}(E) = \mathbf{\Sigma}(\infty) + \mathbf{\Sigma}'(E). \tag{71}$$

In the limit of $|E| \to \infty$, $\mathbf{\Sigma}(E)$ approaches its energy-independent component, $\mathbf{\Sigma}(\infty)$.

At a pole energy,
$$det\{\mathbf{G}(E)\} \to \infty. \tag{72}$$
Poles may be found by requiring that
$$det\{\mathbf{G}^{-1}(E)\} = 0. \tag{73}$$
The Dyson equation may be expressed in terms of a generalized Fock matrix, $\mathbf{F}$, where
$$F_{rs} = (a_r|\dot{H}a_s), \tag{74}$$
according to
$$\mathbf{G}^{-1}(E) = E\mathbf{1} - \mathbf{F} - \mathbf{\Sigma}'(E). \tag{75}$$
Now let the inverse propagator matrix's eigenvalues and eigenvectors be written as
$$\mathbf{G}^{-1}(E)\mathbf{U}(E) = \mathbf{U}(E)\omega(E), \tag{76}$$
so that
$$\mathbf{G}^{-1}(E) = \mathbf{U}(E)\omega(E)\mathbf{U}^{\dagger}(E). \tag{77}$$
Therefore,
$$det\{\mathbf{G}^{-1}(E)\} = det\{\omega(E)\}. \tag{78}$$
When E is equal to a pole, E_{pole}, at least one element of the diagonal matrix $\omega(E)$ vanishes. For this zero eigenvalue case,
$$\mathbf{G}^{-1}(E_{pole})\mathbf{U}(E_{pole}) = 0\mathbf{U}(E_{pole}), \tag{79}$$
or
$$\{E_{pole}\mathbf{1} - \mathbf{F} - \mathbf{\Sigma}'(E_{pole})\}\mathbf{U}(E_{pole}) = 0\mathbf{U}(E_{pole}). \tag{80}$$
It follows that
$$\{\mathbf{F} + \mathbf{\Sigma}'(E_{pole})\}\mathbf{U}(E_{pole}) = E_{pole}\mathbf{U}(E_{pole}). \tag{81}$$
The canonical form of the Hartree–Fock equations is recovered in the last expression through neglect of the energy-dependent part of the self-energy matrix and correlation terms in $\mathbf{F}$. The eigenvalues of this equation are electron binding energies and the eigenvectors provide the FDAs. When correlation terms are restored, the poles contain corrections to Koopmans's theorem and the DO of equation 44 has contributions from more than one canonical orbital. The DO is subject to the usual Coulomb and exchange potentials, but the underlying density matrix may have correlation contributions. Dynamical correlation is included in the energy-dependent self-energy term, $\mathbf{\Sigma}'(E)$.

8 Perturbative Self–Energies

The partitioning of the operator space leading to the Dyson equation has informed the order analysis of correlation corrections. The second order self–energy matrix,
$$\mathbf{\Sigma}^{(2)}(E) = (\mathbf{a}|\dot{H}\mathbf{f}_3)^{(1)} \left[E\mathbf{1} - (\mathbf{f}_3|\dot{H}\mathbf{f}_3)^{(0)}\right]^{-1} (\mathbf{f}_3|\dot{H}\mathbf{a})^{(1)}, \tag{82}$$

follows from zeroth and first order choices for superoperator blocks defined in terms of the a and $\mathbf{f}_3$ (2hp and 2ph) operator manifolds. (Superscripts in parentheses denote the order through which a superoperator matrix element is evaluated.) Upon retention of first order terms in $(\mathbf{f}_3|\hat{H}\mathbf{f}_3)$, third and higher order contributions to the self-energy matrix are generated. Third order terms also are created by perturbative improvements in the ground state averages. The third order self-energy matrix contains contributions of both types and may be expressed as

$$\Sigma^{(3)}(E) = \Sigma^{(2)}(E) + (\mathbf{a}|\hat{V}\mathbf{a})^{(3)} +$$

$$(\mathbf{a}|\hat{H}\mathbf{f}_3)^{(2)}\left[E\mathbf{1} - (\mathbf{f}_3|H_0\mathbf{f}_3)^{(0)}\right]^{-1}(\mathbf{f}_3|\hat{H}\mathbf{a})^{(1)} + (\mathbf{a}|\hat{H}\mathbf{f}_3)^{(1)}\left[E\mathbf{1} - (\mathbf{f}_3|\hat{H}_0\mathbf{f}_3)^{(0)}\right]^{-1}(\mathbf{f}_3|\hat{H}\mathbf{a})^{(2)} +$$

$$(\mathbf{a}|\hat{H}\mathbf{f}_3)^{(1)}\left[E\mathbf{1} - (\mathbf{f}_3|H_0\mathbf{f}_3)^{(0)}\right]^{-1}(\mathbf{f}_3|\hat{V}\mathbf{f}_3)^{(1)}\left[E\mathbf{1} - (\mathbf{f}_3|\hat{H}_0\mathbf{f}_3)^{(0)}\right]^{-1}(\mathbf{f}_3|\hat{H}\mathbf{a})^{(1)}. \qquad (83)$$

9 Quasiparticle Methods

Results on valence ionization energies and electron affinities of closed-shell molecules generally indicate that off-diagonal elements of the self-energy matrix in the canonical basis are small and have a negligible effect on poles and DOs. Quasiparticle approximations explicitly neglect these matrix elements and, as a consequence of equations 44 and 81, constrain the DOs to be equal to canonical orbitals. The associated pole search becomes especially easy, for the zeroes of the diagonal elements of the Dyson equation can be found by solving

$$E = \epsilon_p + \Sigma_{pp}(E). \qquad (84)$$

The usual initial guess, $\epsilon_p + \Sigma_{pp}(\epsilon_p)$, equals the second or third order perturbation estimates [26] when second or third order approximations, respectively, of $\Sigma_{pp}(E)$ are made. Reinsertion of new energies into $\Sigma_{pp}(E)$ generally leads to rapid convergence.

9.1 OVGF Methods

Simple measures for estimating fourth and higher order terms have been applied widely. Once a third order pole is found, various terms are rearranged to make these estimates. The A version of the Outer Valence Green's Function (OVGF) methods [12], for example, corrects canonical orbital energies with

$$\Sigma_{rr}^{(A,OVGF)}(E) = \Sigma_{rr}^{(2)}(E) + (1 + X_r)^{-1}\Sigma_{rr}^{(3)}(E) \qquad (85)$$

where E is chosen to be the third order pole and

$$X_r = -2\frac{\{(\mathbf{a}|\hat{H}\mathbf{f}_3)^{(2)}\{E\mathbf{1} - (\mathbf{f}_3|\hat{H}\mathbf{f}_3)^{(0)}\}^{-1}(\mathbf{f}_3|\hat{H}\mathbf{a})^{(1)}\}_{rr}}{\{(\mathbf{a}|\hat{H}\mathbf{f}_3)^{(1)}\{E\mathbf{1} - (\mathbf{f}_3|\hat{H}\mathbf{f}_3)^{(0)}\}^{-1}(\mathbf{f}_3|\hat{H}\mathbf{a})^{(1)}\}_{rr}}. \qquad (86)$$

The B approximation is similar, but separates the energy-dependent 2p-h and 2h-p contributions through

$$\Sigma_{rr}^{(B,OVGF)}(E) = \Sigma_{rr}^{(2)}(E) + (\mathbf{a}|\hat{V}\mathbf{a})_{rr}^{(3)}$$
$$+ (1 + X_r^{2h-p})^{-1}\Sigma_{rr}^{(3,2h-p)}(E) + (1 + X_r^{2p-h})^{-1}\Sigma_{rr}^{(3,2p-h)}(E), \qquad (87)$$

where

$$X_r^{2h-p} = -2\frac{\{(a|\hat{H}f_{2h-p})^{(2)}\{E1 - (f_{2h-p}|\hat{H}f_{2h-p})^{(0)}\}^{-1}(f_{2h-p}|\hat{H}a)^{(1)}\}_{rr}}{\{(a|\hat{H}f_{2h-p})^{(1)}\{E1 - (f_{2h-p}|\hat{H}f_{2h-p})^{(0)}\}^{-1}(f_{2h-p}|Ha)^{(1)}\}_{rr}} \quad (88)$$

and

$$X_r^{2p-h} = -2\frac{\{(a|\hat{H}f_{2p-h})^{(2)}\{E1 - (f_{2p-h}|\hat{H}f_{2p-h})^{(0)}\}^{-1}(f_{2p-h}|\hat{H}a)^{(1)}\}_{rr}}{\{(a|\hat{H}f_{2p-h})^{(1)}\{E1 - (f_{2p-h}|\hat{H}f_{2p-h})^{(0)}\}^{-1}(f_{2p-h}|Ha)^{(1)}\}_{rr}}. \quad (89)$$

Note that the constant terms are not scaled in this approximation. These schemes function best when the third order numerators in equations 86, 88 and 89 are small compared to the second order denominators. For other cases, the C recipe often is preferable. The self-energy correction is

$$\Sigma_{rr}^{(C,OVGF)}(E) = \Sigma_{rr}^{(2)}(E) + (1 + X_r^C)^{-1}\Sigma_{rr}^{(3)}(E) \quad (90)$$

where

$$X_r^C = \frac{X_r^{2h-p}\Sigma_{rr}^{(3,2h-p)}(E) + X_r^{2p-h}\Sigma_{rr}^{(3,2p-h)}(E)}{\Sigma_{rr}^{(3,2h-p)}(E) + \Sigma_{rr}^{(3,2p-h)}(E)}. \quad (91)$$

This procedure is implemented in a widely distributed program [27] for closed–shell and unrestricted [28, 29] Hartree–Fock orbitals.

The utility of quasiparticle approximations was tested recently through a comparison of third order and OVGF results for 25 ionization energies of ten closed–shell molecules with H, C, N, O and F nuclei [30]. Calculations were performed with a correlation–consistent, triple ζ basis [31]. Root mean square errors with respect to experiment in eV are: 1.25 for Koopmans's theorem, 0.67 for third order, 0.27 for OVGF–A, 0.33 for OVGF–B and 0.32 for OVGF–C. After employment of the OVGF selection criteria [30], the root mean square error is 0.25 eV. These results indicate that third order, quasiparticle estimates are useful for semiquantitative purposes, such as correctly ordering final states. Once third order results are available, OVGF estimates are very efficient tools for making assignments and predicting ionization energies.

9.2 Partial Third Order Theory

The usual choice of superoperator metric starts from a Hartree–Fock wavefunction plus perturbative corrections:

$$(Y|Z) = \langle HF|(1+T^\dagger)[Y^\dagger, Z]_+(1+T)|HF\rangle \quad (92)$$

where

$$T = T_2^{(1)} + T_1^{(2)} + T_2^{(2)} + T_3^{(2)} + T_4^{(2)} + \cdots. \quad (93)$$

The level of excitation in $T_e^{(f)}$ is indicated by the subscript, e, and the order is defined by the superscript, f. For example, second order triple excitations are represented by $T_3^{(2)}$. Coupled-cluster parametrizations of this metric suggest an alternative form:

$$(Y|Z) = \langle HF|e^{-T}[Y^\dagger, Z]_+e^T|HF\rangle. \quad (94)$$

This choice produces asymmetric superoperator matrices. A simplified final form for the self–energy matrix that does not require optimization of cluster amplitudes is sought presently; the approximation

$$e^T \approx 1 + T_2^{(1)} \quad (95)$$

therefore is made.

With this choice. several third order terms that appeared with the usual metric are eliminated. The new self-energy matrix in third order is asymmetric and is expressed by

$$\Sigma(E) = (\mathbf{a}|\dot{H}\mathbf{f}_3)^{(1)}\{E\mathbf{1} - (\mathbf{f}_3|H\mathbf{f}_3)^{(0)}\}^{-1}(\mathbf{f}_3|\dot{H}\mathbf{a})^{(1)} + (\mathbf{a}|\dot{H}\mathbf{f}_3)^{(1)}\{E\mathbf{1} - (\mathbf{f}_3|\dot{H}\mathbf{f}_3)^{(0)}\}^{-1}(\mathbf{f}_3|\dot{H}\mathbf{a})^{(2)}$$
$$+ (\mathbf{a}|\dot{H}\mathbf{f}_3)^{(1)}\{E\mathbf{1} - (\mathbf{f}_3|\dot{H}\mathbf{f}_3)^{(0)}\}^{-1}(\mathbf{f}_3|\dot{V}\mathbf{f}_3)^{(1)}\{E\mathbf{1} - (\mathbf{f}_3|\dot{H}\mathbf{f}_3)^{(0)}\}^{-1}(\mathbf{f}_3|\dot{H}\mathbf{a})^{(1)}. \quad (96)$$

Note that energy-independent terms in the third order self-energy matrix are not retained.

Two observations suggest additional economies. First, numerical results for ionization energies show that third order, 2p-h terms in equation 96 are small relative to their 2h-p counterparts [32]. Terms arising from these operators are important in second order, however. Second, evaluation of the third order 2p-h terms requires electron repulsion integrals with four virtual indices. Because of the large number of these integrals that typically is generated. their storage is often avoided through semidirect algorithms [33]. Contractions involving integrals with four virtual indices remain the bottleneck in third order quasiparticle theory.

Neglect of third order, 2p-h terms produces this self-energy matrix:

$$\Sigma(E)_{pq} = \frac{1}{2}\sum_{iab}\frac{\langle pi||ab\rangle\langle ab||qi\rangle}{E+\epsilon_i-\epsilon_a-\epsilon_b} + \frac{1}{2}\sum_{aij}\frac{\langle pa||ij\rangle W_{qaij}}{E+\epsilon_a-\epsilon_i-\epsilon_j} + \frac{1}{2}\sum_{aij}\frac{U_{paij}(E)\langle ij||qa\rangle}{E+\epsilon_a-\epsilon_i-\epsilon_j}, \quad (97)$$

where i,j,k are occupied indices, a,b,c are virtual indices, p,q are general indices,

$$W_{qaij} = \langle qa||ij\rangle + \frac{1}{2}\sum_{bc}\frac{\langle qa||bc\rangle\langle bc||ij\rangle}{\epsilon_i+\epsilon_j-\epsilon_b-\epsilon_c} + (1-P_{ij})\sum_{bk}\frac{\langle qk||bi\rangle\langle ba||jk\rangle}{\epsilon_j+\epsilon_k-\epsilon_a-\epsilon_b} \quad (98)$$

and

$$U_{paij}(E) = -\frac{1}{2}\sum_{kl}\frac{\langle pa||kl\rangle\langle kl||ij\rangle}{E+\epsilon_a-\epsilon_k-\epsilon_l} - (1-P_{ij})\sum_{bk}\frac{\langle pb||jk\rangle\langle ak||bi\rangle}{E+\epsilon_b-\epsilon_j-\epsilon_k}. \quad (99)$$

This partial third order expression has been designated by the abbreviation P3.

Comparison of the self-energy matrix elements of equation 97 with older, related methods [12, 22] reveals the advantages of the P3 approximation. OVGF results [12] and the third order quasiparticle calculations on which they are based [9] require the evaluation of the diagonal elements of the self-energy matrices given in equation 82 (second order terms) and equation 83 (third order terms). The second term (the third-order, energy-independent or constant diagrams) and the third term in equation 83 are omitted in P3 theory. Furthermore, the 2p-h operators included in the fourth and fifth terms of equation 83 are dropped in P3 theory as well. Another difference between P3 and OVGF is the absence of scaling factors, such as those in equations 85 through 91, in P3.

In the P3 quasiparticle propagator, only diagonal elements of the self-energy matrix are retained. The first contraction in equation 98 is the most demanding, for it has an arithmetic scaling factor of O^2V^3. This step also requires electron repulsion integrals with one occupied and three virtual indices. This intermediate is energy-independent, however, and must be evaluated once only for each ionization energy of interest.

A quasiparticle implementation of the P3 method was applied to 19 ionization energies of six closed-shell molecules with a correlation-consistent, triple ζ basis [32]. Comparisons with results of more costly propagator techniques were made by applying the same basis sets. The average absolute errors in eV are:

1.34 for Koopmans's theorem, 0.25 for OVGF-B and 0.19 for P3. The P3 procedure is clearly the best for the lowest ionization energies of closed–shell molecules, for it exhibits accuracy at least as good as that of other methods, superior arithmetic scaling and no need for electron repulsion integrals with four virtual indices.

10 Computational Techniques

10.1 Pole Search Methods

Pole searches based on the Dyson equation usually converge rapidly with respect to E. Starting from equation 70, one can insert a matrix expression for the unperturbed inverse propagator matrix,

$$\mathbf{G}_0^{-1}(E) = E\mathbf{1} - \epsilon, \tag{100}$$

where ϵ is the diagonal matrix of canonical orbital energies. Because E is a pole when $\mathbf{G}^{-1}(E)$ has a vanishing eigenvalue, one searches for E such that E is an eigenvalue of $\Gamma(E)$, where

$$\Gamma(E) = \epsilon + \Sigma(E). \tag{101}$$

With the second and third order self-energy matrices, it is possible to evaluate derivatives of the eigenvalues of $\Gamma(E)$ with respect to E. This information is then used in a Newton procedure for determining the next guess [34]. If

$$E'\mathbf{C}(E_{old}) = \Gamma(E_{old})\mathbf{C}(E_{old}), \tag{102}$$

then the new guess is

$$E_{new} = E_{old} - \frac{E' - E_{old}}{\mathbf{C}^\dagger(E_{old})\frac{d\Sigma(E)}{dE}|_{E=E_{old}}\mathbf{C}(E_{old}) - 1}. \tag{103}$$

Iterations are continued until an eigenvalue of $\Gamma(E)$ is within 10^{-5} atomic units of E. Convergence seldom requires more than three iterations for final states with relatively minor contributions from the f operator manifold. Another advantage of this technique is its applicability to any choice of E.

The normalized DO from equation 102,

$$\psi^{Dyson}(x) = \sum_r \phi_r(x) C_r, \tag{104}$$

satisfies

$$\langle \psi^{Dyson} | \psi^{Dyson} \rangle = 1 \tag{105}$$

provided $\mathbf{C}^\dagger \mathbf{C} = 1$. The normalization factor, $\sqrt{P}$, occurring in

$$\phi^{Dyson}(x) = \sqrt{P} \psi^{Dyson} \tag{106}$$

is related to the pole strength, P, such that

$$P = \left[1 - \mathbf{C}^\dagger(E_{pole}) \frac{d\Sigma(E)}{dE} |_{E=E_{pole}} \mathbf{C}(E_{pole}) \right]^{-1}. \tag{107}$$

Because

$$P_n = \sum_r |U_{r,n}|^2 \qquad (108)$$

for the nth final state, the pole strength is an index of the primary operator space's contribution to the U vector of equation 40. The term pole strength is founded on the proportionality of the cross sections in equation 8 to P when ψ^{Dyson} replaces φ^{Dyson} according to equation 106. Final states with large shakeup or shakeon character have pole strengths closer to zero than one. When $\Sigma(E)$ is neglected, P equals unity for each Koopmans final state.

10.2 Semidirect Contractions

Among the chief obstacles to applying quasiparticle methods to larger molecules are the storage and retrieval of electron repulsion integrals. A semidirect strategy has been adopted, where only some classes of these integrals are calculated and stored [33]. Several intermediates are evaluated in order to improve the efficiency of the pole search algorithms.

Evaluation of the energy-independent $(a|\dot{V}a)^{(3)}$ term in the third order self-energy matrix of equation 83 is preceded by calculation of the second-order density matrix, $\rho^{c^{(2)}}$. This step requires contractions with O^2V^3 arithmetic operations (where O is the number of occupied spin-orbitals and V is the number of virtual spin-orbitals) and electron repulsion integrals with up to three virtual indices. A contraction resembling the assembly of the Fock matrix from the density matrix follows according to equation 68. All elements of the constant part of the self-energy matrix may be calculated prior to energy iterations in this way.

For the last term in equation 83, the intermediate

$$(f_3|\hat{V}f_3)^{(1)} \left[E\mathbf{1} - (f_3|\hat{H}_0 f_3)^{(0)} \right]^{-1} (f_3|\hat{H}a)^{(1)} \qquad (109)$$

is required. The most difficult contribution is

$$X^{C_i}_{iab}(E) = \sum_{cd} \frac{\langle ab|cd\rangle \langle pi|cd\rangle}{E + \epsilon_i - \epsilon_c - \epsilon_d} \qquad (110)$$

Explicit use of the $\langle ab|cd\rangle$ integrals can be circumvented. If one denotes

$$W_{icd}(E) = \frac{\langle pi|cd\rangle}{(E + \epsilon_i - \epsilon_c - \epsilon_d)}, \qquad (111)$$

then a transformation to the AO basis (see Ref. [35]),

$$W'_{i\mu\nu}(E) = \sum_{cd} C_{\mu c} C_{\nu d} W_{icd}(E), \qquad (112)$$

may be applied, where C is the matrix of MO coefficients. W' may be contracted with AO integrals generated as needed using

$$Y^{C_i}_{ip\sigma}(E) = \sum_{\mu\nu} W'_{i\mu\nu}(E)\langle \rho\sigma|\mu\nu\rangle. \qquad (113)$$

X^{C_1} is calculated using

$$X_{iab}^{C_1} = \sum_{\rho\sigma} C_{\rho a} C_{\sigma b} Y_{i\rho\sigma}^{C_1}(E). \qquad (114)$$

At the price of performing three fifth power contractions instead of one, evaluation and storage of $\langle ab|cd\rangle$ integrals is avoided. The worst of these contractions scales as OB^4, where B is the dimension of the AO basis; the other two contractions have OV^2B^2 scaling factors. Because these contractions must be repeated for each value of E, they are the computational bottleneck for OVGF and third order calculations. P3 calculations of ionization energies do not require these intermediates.

The remaining terms in equation 83 require evaluation of an intermediate that does not depend on E. For closed–shell reference states, this vector is given by

$$X_{aij}^{C_4} = \sum_{bc} \frac{(2\langle ij|bc\rangle - \langle ij|cb\rangle)\langle pa|bc\rangle}{(\epsilon_i - \epsilon_j + \epsilon_b + \epsilon_c)}. \qquad (115)$$

For the electron affinity case, where p is a virtual index, this contraction requires electron repulsion integrals with four virtual indices. Similar techniques that are applied to the X^{C_1} intermediate are applied here as well. The calculation of X^{C_4} for all virtual p is done once and only those terms which are related to states under consideration are stored.

When only ionization energies from the occupied MOs are required, it is possible to avoid transformation and storage of integrals with one occupied and three virtual indices, $\langle ia|bc\rangle$. There are two terms to which these integrals contribute. One of them occurs in the second order density matrix required for the constant self–energy term:

$$\rho_{cj}^{A_3} = \sum_{iab} \frac{\langle ic|ab\rangle(2\langle ij|ab\rangle - \langle ij|ba\rangle)}{(\epsilon_i + \epsilon_j - \epsilon_a - \epsilon_b)}. \qquad (116)$$

The other occurs in the X^{C_4} intermediate. The $\langle ia|bc\rangle$ integrals are contracted with the first order wave-function coefficients such that

$$Y_{ijkc} = \sum_{ab} \frac{\langle kc|ab\rangle(2\langle ij|ab\rangle - \langle ij|ba\rangle)}{(\epsilon_i + \epsilon_j - \epsilon_a - \epsilon_b)}. \qquad (117)$$

Contributions to $\rho_{cj}^{A_3}$ require an additional summation:

$$\rho_{cj}^{A_3} = \sum_i Y(ijic). \qquad (118)$$

10.3 Integral Transformation and Storage

The integral transformation program of Gaussian 94 [27] has direct and semi–direct capabilities and the ability to omit certain types of integrals, such as $\langle ab|cd\rangle$, from the transformation. A modified version now stores only those integrals which are non–zero by symmetry. The necessary symmetry assignment of MOs with respect to the closest abelian group is obtained from the direct SCF program of Gaussian 94.

A simple ordering for symmetry–compressed integral lists stores integrals immediately after the transformation to the canonical orbital basis and is implemented in the present codes. Loops over four indices form the usual upper or lower triangle of transformed integrals; integrals that are zero by symmetry are skipped. Auxiliary vectors are introduced to calculate the location of a particular integral from the four canonical indices. This algorithm to locate integrals was first proposed for the case of a lower triangular form and then modified for the upper triangle case [36]. Storage of the integrals in the upper triangular form is more convenient because the block of $(ab|cd)$ integrals may be omitted without other significant changes in the location procedure. As long as the procedure to locate a particular integral is known, the creation of the symmetry-compressed upper triangle of integrals, or of the reduced triangle without the $(ab|cd)$ block, presents no difficulties.

After these temporary lists are generated, the programs sort the integrals into $(ij|kl)$, $(ij|ka)$, $(ij|ab)$ and other buckets that are useful in performing contractions. These buckets do not take advantage of permutational symmetry in order to facilitate reorderings of indices and syntheses of related intermediates.

Another feature of these lists is that the orbitals are reordered first by representations and then according to ascending orbital energies within each representation. Each orbital p is described by an index of its representation, P_r, and by the relative index within the representation, P_{rel}. The vectors for one-to-one correspondence,

$$(P_r, P_{rel}) \Longleftrightarrow (p), \tag{119}$$

for the description of a given orbital are easily built. If the first of the orbitals belongs to the representation Γ_p, then the remaining three orbitals in non-zero $\langle pq|rs \rangle$ are subject to the following restriction,

$$\Gamma_q \otimes \Gamma_r \otimes \Gamma_s = \Gamma_p. \tag{120}$$

Similarly one may write

$$\Gamma_p \otimes \Gamma_q = \Gamma_r \otimes \Gamma_s \tag{121}$$

and

$$\Gamma_p \otimes \Gamma_q \otimes \Gamma_r = \Gamma_s. \tag{122}$$

These relations indicate what kind of restrictions exist on the possible values of the representation indices in the inner loops of actual calculations where loops over the orbitals within each of the representations are running inside loops over representations.

We assume the extreme right index in four-index arrays is running first and the extreme left is running last. The lengths of the batches of integrals $\langle p[q|rs]\rangle$ in square brackets are equal for all p in Γ_p. It is possible to introduce a pair of vectors $A3(8)$ and $L3(8)$ (the dimension, eight, is the maximum number of representations) which set the entry points ($A3$) and lengths ($L3$) for the batches of integrals, $[q|rs]\rangle$. The entry point into the batch of $[q|rs]\rangle$ integrals is given by the expression

$$Entry = A3(P_r) + (P_{rel} - P_s) * L3(P_r), \tag{123}$$

where P_s is equal to 1 in each of the representations for every occupied MO and is equal to the index of the first virtual in each of the representations for every virtual MO.

For a particular pair of p and q values, the lengths of the $|[rs]\rangle$ batches may be established in a similar way. These batch lengths remain the same for all pairs where p transforms according to Γ_p and q transforms according to Γ_q. Two square matrices, $A2(8,8)$ and $L2(8,8)$, contain information about entry points and lengths of $|[rs]\rangle$ integral batches. The entry point to the batch $|[rs]\rangle$ for any of the orbitals from representations Γ_p and Γ_q is given by the formula

$$Entry = A3(P_r) + (P_{rel} - P_s) * L3(P_r) + A2(P_r, Q_r) + (Q_{rel} - Q_s) * L2(P_r, Q_r). \tag{124}$$

This procedure facilitates retrieval of $[q|rs])$ or $|[rs])$ batches from disk and also simplifies contractions over the last two or three indices.

An element's location in the batch $|[rs])$ is determined by introducing the array $IS(Nsym, N, M)$, where $Nsym$ is the index of representations and N and M are the ranges of the indices r and s without symmetry. The direct product of the representations of the first pair of indices determines the direct product of the second pair. The array $IS(Nsym, N, M)$ for the batch $|[rs])$ may be built by running loops over the representation index, R_r, R_{rel}, and S_{rel}. (S_r is determined from the direct product of the first index of IS and Γ_r.) Values of R_r, R_{rel}, S_r and S_{rel} are mapped onto r and s. The consecutive number generated in these loops is the value of the IS element.

These tools for locating matrix elements are used with binary sort routines to perform sorts and transpositions of indices. These operations are often required to form precursors for various contractions.

11 Interpretations of Photoelectron Spectra

Quasiparticle methods discussed above are applied presently to photoelectron spectra (PES) of molecules exhibiting one or more six-membered, aromatic rings. DOs remain equal to canonical orbitals, but the ionization energies contain correlation corrections arising from the self-energy matrix. Contours of the DOs are generated with MOLDEN [37].

11.1 Benzene

With the cc-pVDZ [31] basis, the vertical ionization energies of benzene were calculated with C-C and C-H bond lengths of 1.397 and 1.084 Å, respectively [32]. The P3 and OVGF values are close to each other, but there are large differences with the Koopmans's theorem (KT) results. Agreement with experiment [38] is generally satisfactory, although the average absolute error for P3, 0.23 eV, is somewhat lower than its OVGF counterpart, 0.32 eV. Basis set improvements are likely to reduce these errors. The largest discrepancies with experiment occur for the second ionization energy. The presence of a low-lying π^* orbital probably enhances the importance of shakeup contributions to this final state.

DOs for benzene are displayed in Fig. 1. The first degenerate cationic state corresponds to ionization from two π MOs. The second state, also degenerate, belongs to σ MOs delocalized over C-C and C-H bonding regions. The third ionization occurs from a π level with a_{2u} symmetry. The remaining states correspond to ionizations from σ MOs.

Benzene Ionization Energies (eV)

State	KT	OVGF	P3	Expt. [38]
$^2E_{1g}$	9.066	8.912	9.105	9.25
$^2E_{2g}$	13.357	11.939	12.026	11.53
$^2A_{2u}$	13.552	12.176	12.134	12.38
$^2E_{1u}$	15.898	14.257	14.242	13.98
$^2B_{2u}$	16.712	14.574	14.774	14.86
$^2B_{1u}$	17.459	15.786	15.582	15.46
$^2A_{1g}$	19.179	17.222	17.112	16.84

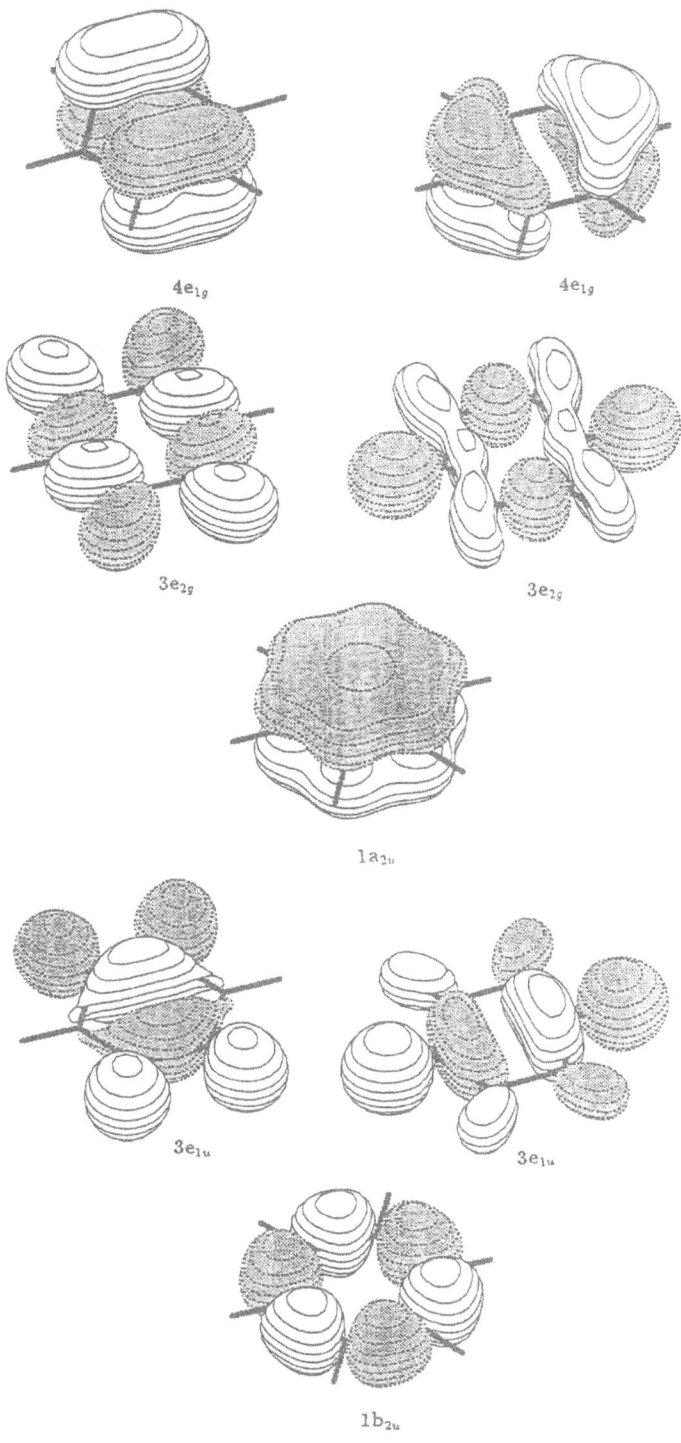

11.2 Chlorobenzene

Ionization energies and pole strengths (PSs) obtained with aug-cc-pVDZ and cc-pVTZ basis sets [31] for the six lowest cationic states are listed below [39]. These basis sets provide good quantitative agreement with experimental ionization energies. The best results are produced by the cc-pVTZ basis set. While the P3 method tends to overestimate the lowest ionization energy, it predicts the same order of final states as OVGF. All pole strengths exceed 0.8; therefore, the perturbative arguments underlying the OVGF and P3 approximations are very likely to be valid. The Koopmans description of final states is qualitatively correct. P3 calculations take only 40 minutes of CPU time and 2.75 hours of wall-clock time for 12 states of chlorobenzene with the cc-pVTZ basis set on an IBM RS6000/550 computer with 128 Mb of memory and 4 Gb of available external disk storage.

The first two bands originate from splitting of benzene's degenerate highest occupied MOs (HOMOs): $e_{1g} \rightarrow b_1 + a_2$. For chlorobenzene, the HOMO consists of a benzene π_3 pattern with significant admixtures from chlorine p_π orbitals (Fig. 2). Antibonding character between Cl and the adjacent carbon explains the relative instability (low ionization energy) of this one-electron state. The second band in the PES refers to $1a_2$, a benzene π_2 orbital that has negligible chlorine contributions. It can be seen from the plot in Fig. 2 that this MO remains almost unchanged when compared to its benzene counterpart even though this ionization energy is larger than benzene's first ionization energy, 9.25 eV [38]. This contrast is probably due to depletion of electron density in the ring produced by the Cl substituent. Each of the first two bands displays vibrational structure, in agreement with the π delocalization seen in the two HOMOs The third band ($9b_2$) originates from the Cl p orbital which is parallel to the plane of the ring ($n_{||}$); the corresponding MO also has a weak C-Cl antibonding interaction. Ionization from $3b_1$ pertains to the fourth band. Here, a Cl p orbital perpendicular to the ring plane ($n_\perp$) is the largest contributor, but extensive delocalization into the meta and para carbons is seen. Sharp, intense peaks corroborate the chiefly nonbonding character of these two lone pair MOs, but each band has a high energy vibrational feature that discloses the presence of some C-Cl antibonding character [38]. Bands 5 and 6 originate from splitting of a degenerate set of benzene σ orbitals: $e_{2g} \rightarrow a_1 + b_2$. The fifth band corresponds to ionization from $15a_1$. This component of the degenerate e_{2g} benzene set gains significant Cl contributions from s and p_σ AOs. The second MO component, $8b_2$, is slightly stabilized by the Cl substituent through an in-plane π interaction. The splitting between the two is less than 0.1 eV in our calculations, but experimental works report a considerable splitting of 0.6-0.7 eV. PES display extended vibrational structure from 12.2 to 13.3 eV [38, 40] and a thorough understanding of this energy region requires a treatment of final state vibronic interactions.

Agreement with experiment is reasonable for all basis sets and approximations used, except for the fifth final state due to the reasons given above. Closer agreement with experiment obtains for the larger basis set, except for the first two states with the P3 approximation.

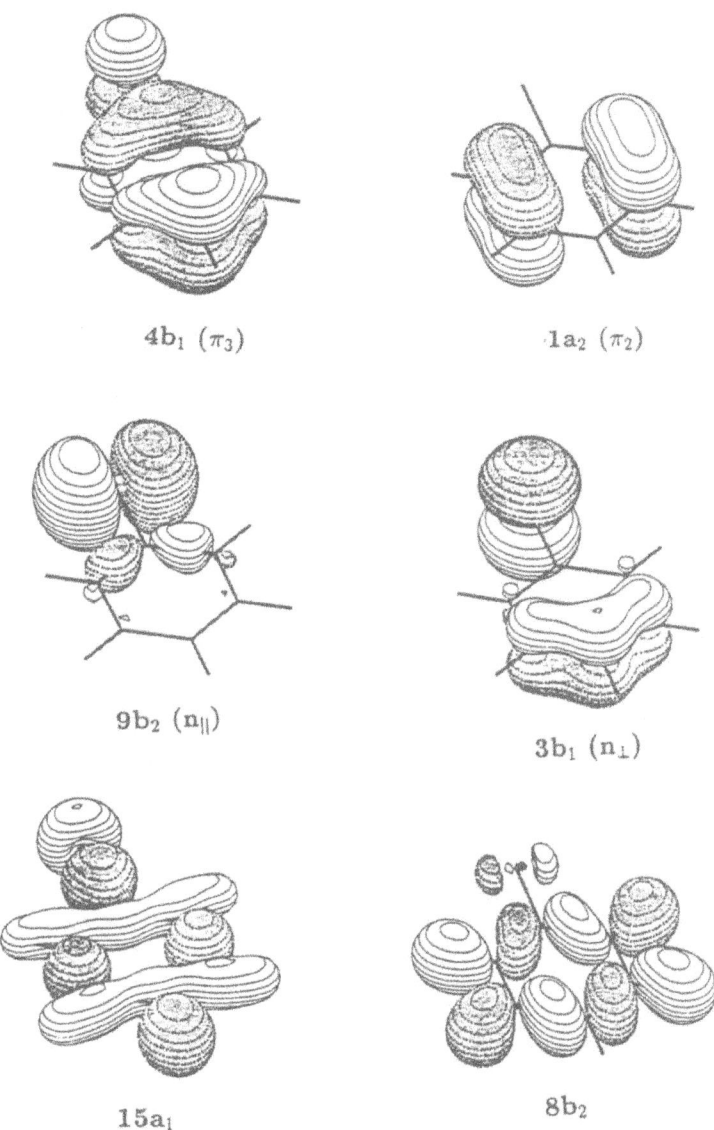

$4b_1$ (π_3) $1a_2$ (π_2)

$9b_2$ ($n_\parallel$) $3b_1$ ($n_\perp$)

$15a_1$ $8b_2$

Chlorobenzene Ionization Energies (eV)

Final State	aug-cc-pVDZ			cc-pVTZ			Expt. [38, 40]
	KT	OVGF PS	P3 PS	KT	OVGF PS	P3 PS	
$^2B_1$	9.24	9.02 0.90	9.15 0.90	9.20	9.12 0.90	9.31 0.89	9.10, 9.07
$^2A_2$	9.65	9.60 0.89	9.65 0.88	9.61	9.60 0.89	9.81 0.88	9.69, 9.64
$^2B_2$	12.41	11.47 0.90	11.13 0.89	12.34	11.44 0.90	11.26 0.89	11.32, 11.31
$^2B_1$	12.82	11.91 0.88	11.65 0.87	12.78	11.88 0.87	11.79 0.86	11.69, 11.67
$^2A_1$	13.96	12.68 0.90	12.67 0.89	13.90	12.61 0.90	12.81 0.89	12.26
$^2B_2$	14.13	12.73 0.90	12.70 0.89	14.09	12.69 0.90	12.86 0.89	12.9, 12.98

11.3 Dichlorobenzenes

OVGF and P3 results on the three dichlorobenzene isomers [41] produce average discrepancies with respect to experiments [38, 40, 42] that are approximately 0.1 eV. The aug-cc-pVDZ basis set provides a good compromise of accuracy and efficiency.

11.3.1 *Para*-Dichlorobenzene

Because all pole strengths exceed 0.85 for the states presented here, the qualitative validity of the Koopmans description of ionization energies is confirmed. The first two bands originate from the splitting of benzene's degenerate HOMOs: $e_{1g} \rightarrow b_{1g} + b_{2g}$. For dichlorobenzene, the HOMO consists of a benzene π_3 pattern (Fig. 1) with significant admixtures from chlorine p_π orbitals. (See Fig. 3.) Antibonding character on four bonds (two C-Cl and two C-C bonds) explains the relative instability (low ionization energy) of this one-electron state. The second band in the PES refers to $1b_{1g}$, a benzene π_2 orbital that has negligible chlorine contributions. As was the case with chlorobenzene, this MO remains almost unchanged when compared to its benzene counterpart even though the ionization energy is larger than benzene's first ionization energy, 9.25 eV [38]. Each of the first two bands displays vibrational structure, in agreement with the π delocalization seen in the two HOMOs. The third band originates from the Cl lone pairs ($n_\perp$) which are perpendicular to the plane of the ring. Next, two bands (4 and 5) of the spectra pertain to ionization from MOs $6b_{2u}$ and $5b_{3g}$. The evident similarity of these two MOs explains why the bands were considered to have the same energy in Ref. [40] and to differ by only 0.15 and 0.14 eV in Ref.s [38] and [42], respectively. Three closely spaced, but sharp, peaks are compatible with the present results for bands 3-5. The sixth band pertains to ionization from the $2b_{2g}$ MO which consists of Cl p orbitals perpendicular to the plane of the ring ($n_\perp$). This MO also has significant π C-Cl bonding character. (It was considered to be another Cl lone pair orbital perpendicular to the ring in Ref. [42]). Some vibrational structure is present for this feature as well.

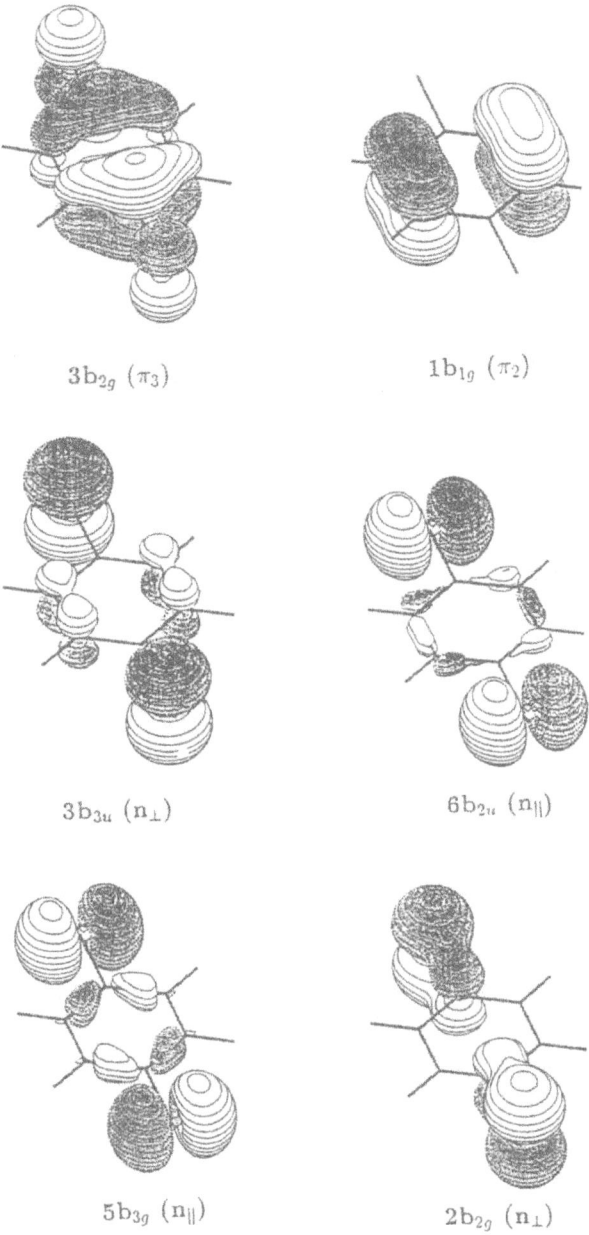

$3b_{2g}$ (π_3) $1b_{1g}$ (π_2)

$3b_{3u}$ ($n_\perp$) $6b_{2u}$ ($n_\parallel$)

$5b_{3g}$ ($n_\parallel$) $2b_{2g}$ ($n_\perp$)

Para–Dichlorobenzene Ionization Energies (eV)

Final State	KT	OVGF	P3	Expt. [38, 40, 42]
$^2B_{2g}$	9.28	8.90	9.08	8.94–8.97
$^2B_{1g}$	10.06	9.90	9.99	9.84
$^2B_{3u}$	12.47	11.53	11.26	11.37
$^2B_{2u}$	12.59	11.61	11.28	11.46–11.50
$^2B_{3g}$	12.74	11.74	11.43	11.50–11.64
$^2B_{2g}$	14.07	13.04	12.79	12.65, 12.78

11.3.2 *Meta*–Dichlorobenzene

Our results for *meta*–dichlorobenzene are in complete qualitative agreement with the assignments in the text and in Table II of Ref. [42]. Koopmans's theorem predicts the correct order of states for this isomer. The HOMO, $3a_2$, is formed from the π_2 (instead of π_3 as was stated in Ref. [42]) MO of benzene and chlorine p_π orbitals (see Fig. 4). $4b_1$ is formed from the π_3 of benzene and smaller contributions from Cl p_π orbitals. This MO has more C–Cl antibonding character compared to the second HOMO in the *para* isomer; the gap between this final state and the lowest cation state is smaller in the *meta* isomer. In the PES, the first two bands display extensive vibrational structure [38]. The third band represents ionization from Cl lone pairs ($13b_2$, $n_{||}$) parallel to the ring plane. It is more destabilized than its *para* counterpart, $5b_{3g}$. $3b_1$ ($n_\perp$) contains Cl lone pairs perpendicular to the ring. It also has a π bonding contribution from three nonadjacent C atoms, but is not as destabilized by C–Cl antibonding interactions as the $3b_{3u}$ MO of the *para* isomer. The fifth MO ($17a_1$, $n_{||}$) represents Cl lone pairs parallel to the plane of the ring and is more stable than its *para* counterpart, $6b_{2u}$. Three sharp, closely spaced features [38] correspond to states 3, 4 and 5. $2a_2$ ($n_\perp$) contains chiefly Cl lone pairs perpendicular to the ring that are significantly delocalized onto the adjacent carbons in a π bonding relationship. A nearly identical ionization energy obtains in the analogous *para* case, $2b_{2g}$.

Meta–Dichlorobenzene Ionization Energies (eV)

Final State	KT	OVGF	P3	Expt. [38, 40, 42]
$^2A_2$	9.46	9.10	9.28	9.14–9.28
$^2B_1$	9.82	9.49	9.69	9.65–9.70
$^2B_2$	12.58	11.59	11.28	11.46, 11.47
$^2B_1$	12.61	11.67	11.44	11.56, 11.58
$^2A_1$	12.80	11.81	11.49	11.70–11.73
$^2A_2$	13.96	12.96	12.70	12.77, 1280

11.3.3 *Ortho*–Dichlorobenzene

In the case of *ortho*–dichlorobenzene, the order of cationic states 1–6 is the same as in Ref. [42] except for two pairs of states. States 4 and 5 (MOs $3b_1$ and $16a_1$) are switched by correlation corrections. Experimental results from Refs [38] and [42] do not distinguish between these states, but in Ref. [40], the gap is 0.05 eV.

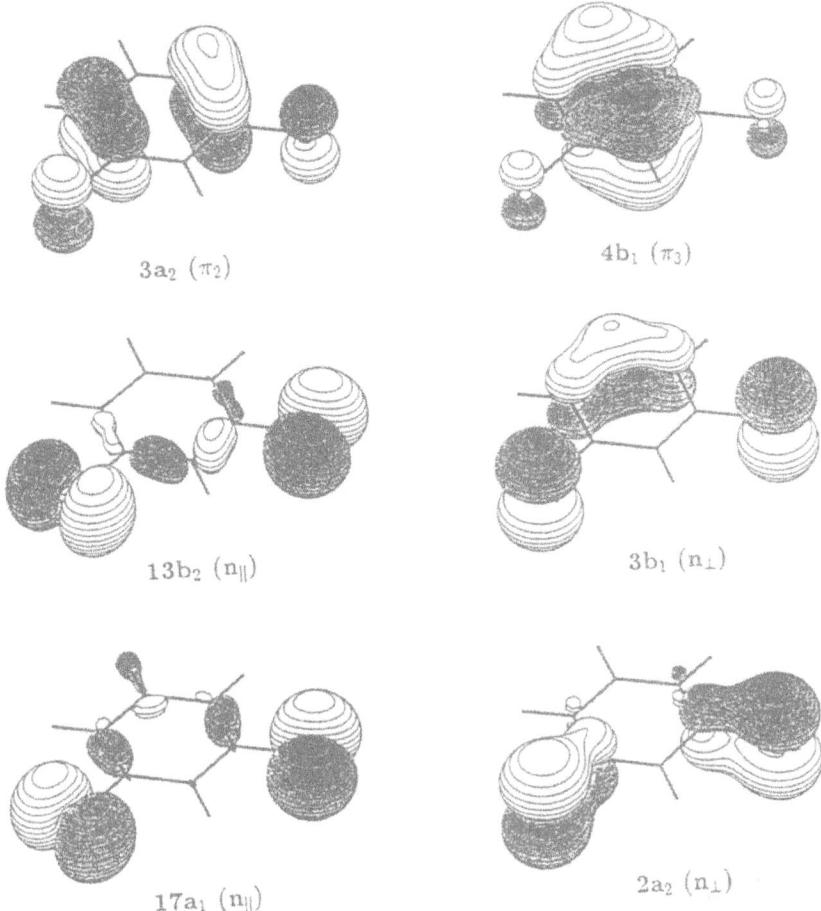

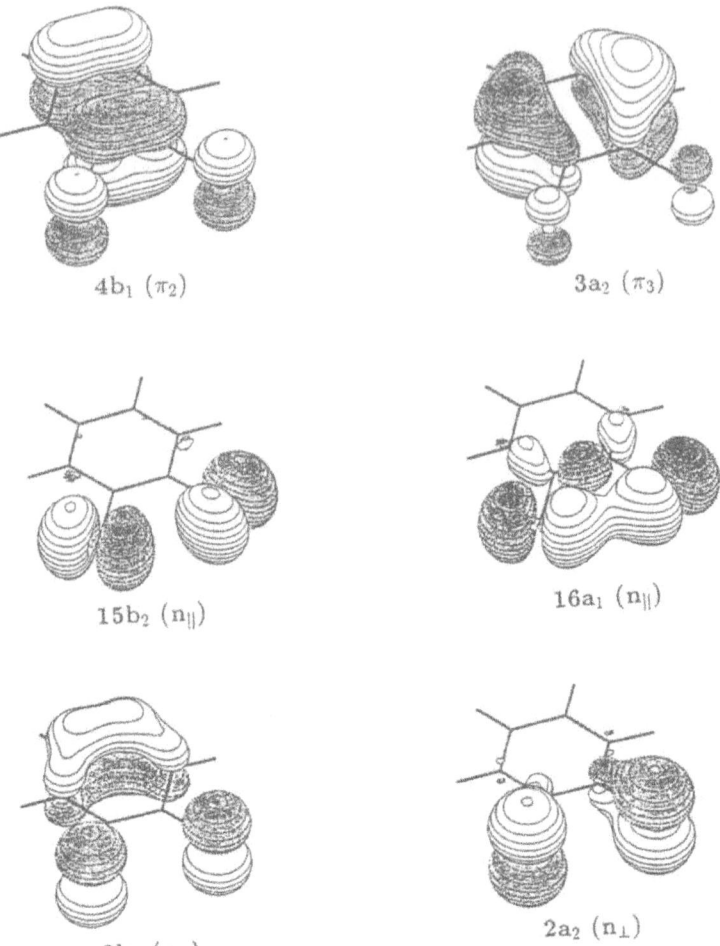

4b₁ (π₂)

3a₂ (π₃)

15b₂ (n∥)

16a₁ (n∥)

3b₁ (n⊥)

2a₂ (n⊥)

Fig. 5 shows that the HOMO, $4b_1$, is composed of the benzene π_2 orbital with an antibonding contribution from Cl π orbitals. $3a_2$ is chiefly a π_3 benzene orbital with a small contribution from Cl π orbitals. Note that closer Cl–Cl contacts produce the opposite ordering from the *meta* case, but similar ionization energies. Extensive vibrational structure appears in the first two bands of the PES [38]. The third final state's MO ($15b_2$, $n_{||}$) is almost entirely Cl lone pairs parallel to the ring plane. This final state is destabilized with respect to its *meta* counterpart and corresponds to a simple, sharp peak [38]. In the fourth MO ($16a_1$, $n_{||}$), Cl lone pairs parallel to the ring constructively interfere, but are destabilized by antibonding interactions with adjacent carbon contributions. The fifth band's MO, $3b_1$ ($n_\perp$), contains Cl lone pairs perpendicular to the plane of the ring and extensive delocalization onto all four nonadjacent carbons. Experimental and calculated ionization energies are higher for this case than for the corresponding *meta* MO, $3b_1$. Final states 4 and 5 contribute to the large peak observed at 11.77 eV [38]. $2a_2$ displays C–Cl bonding and is less stable than its *meta* counterpart with the same label.

Ortho–Dichlorobenzene Ionization Energies (eV)

Final State	KT	OVGF	P3	Expt. [38, 40, 42]
$^2B_1$	9.40	9.04	9.24	9.08–9.24
$^2A_2$	9.72	9.41	9.61	9.63–9.65
$^2B_2$	12.28	11.33	10.97	11.23–11.26
$^2A_1$	12.95	11.88	11.61	11.70–11.77
$^2B_1$	12.94	11.95	11.77	11.75–11.77
$^2A_2$	13.40	12.44	12.13	12.37, 12.38

11.4 Azabenzenes

P3 and OVGF calculations have been performed for the heterocyclic molecules known as azabenzenes [43]. One–electron levels with π or nitrogen lone pair character are energetically close and assignments require quantitative agreement with experiment. Each of these molecules has a six-membered ring with one (pyridine), two (pyridazine, pyrimidine, pyrazine), three (s-triazine) or four (s-tetrazine) nitrogen atoms replacing CH groups. Nitrogen atoms in pyridazine, pyrimidine and pyrazine are *ortho*, *meta* and *para*, respectively, in relation to each other. Carbon and nitrogen sites alternate in the D_{3h} structure of s–triazine. Carbon atoms are *para* with respect to each other in s–tetrazine.

11.4.1 Pyridine

Assignments of the first two final states have been disputed extensively. PES display a feature with two sharp maxima at 9.60 and 9.75 eV, followed by a sharp peak at 10.51 eV [38]. Post-SCF calculations agree that the Koopmans ordering ($a_2 < b_1 < a_1$) is incorrect and that correlation corrections sharply decrease the ionization energy for the $^2A_1$ final state [44–48].

The present calculations on pyridine indicate that a definitive ordering of the first two states must await the application of more precise correlation methods and larger basis sets. There is not much doubt, however, that the third state is $^2B_1$ and corresponds to ionization from the benzene–like π_3 MO shown in Fig. 6. P3 and OVGF results place the $^2A_1$ (ionization from nitrogen lone pair) and $^2A_2$ (ionization

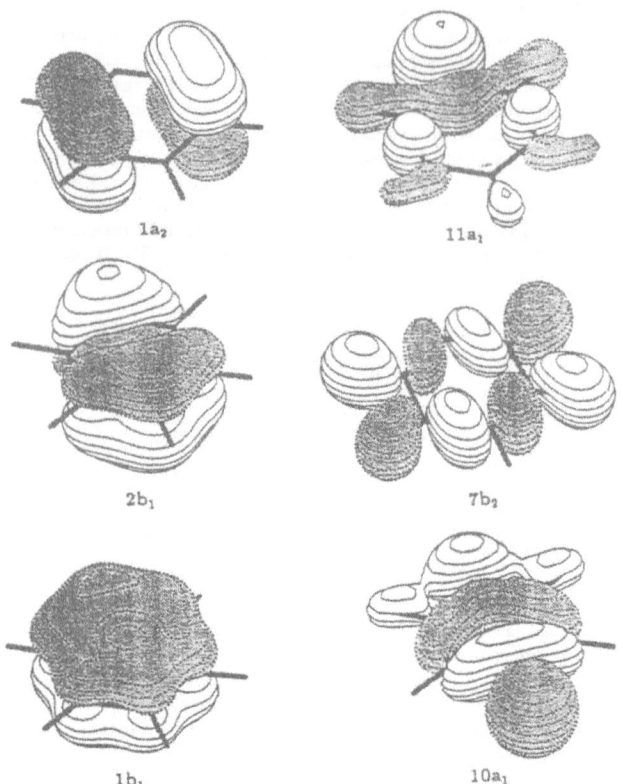

from a benzene–like π_2 orbital) final states within 0.1 eV of each other. Because the average mean errors with these methods are 0.19 and 0.25 eV, respectively, for representative closed-shell molecules [30, 32], these calculations cannot settle the controversy. When the basis set is improved to cc-pVTZ, a reversal of ordering occurs, but the energy separation is only 0.02 eV. Absolute changes in the ionization energies produced by the basis set improvement for the first three states range from 0.27 to 0.37 eV.

To settle the ordering of the first two final states, correlation methods that are accurate to within 0.1 eV of experiment will be needed. Basis sets of greater than triple ζ plus double polarization will required to demonstrate saturation. Even at this level, a consideration of vibronic coupling may be required to interpret PES with higher resolution.

Pole strengths for each of the final states exceed 0.85 with an exception of the second $^2B_1$ state, where the pole strength is only 0.80. This state corresponds to ionization from a π_1 $1b_1$ MO which is delocalized over all six atoms of the aromatic ring. It is likely that higher levels of theory will be needed to identify shakeup states with appreciable pole strengths occurring near 13.1 eV. In contrast, the first four states have pole strengths of 0.88 or higher and the Koopmans description retains greater qualitative validity. Correlation corrections are much smaller for π final states than for σ.

Pyridine Ionization Energies (eV)

Final State	cc-pVDZ			cc-pVTZ		Expt. [38]
	KT	OVGF	P3	KT	P3	
$^2A_2$	9.42	9.34	9.56	9.46	9.83	9.60
$^2A_1$	11.20	9.44	9.48	11.28	9.85	9.75
$^2B_1$	10.42	10.12	10.31	10.48	10.60	10.51
$^2B_2$	14.11	12.70	12.78	14.17	13.03	12.61
$^2B_1$	14.79	13.33	13.29	14.85	13.54	13.1
$^2A_1$	15.64	13.83	13.86	15.69	14.11	13.8

11.4.2 Pyridazine

Large correlation corrections cause the lowest $^2B_2$ final state to lie below the lowest π-hole final state, $^2A_2$. The corresponding $8b_2$ MO displays much nitrogen lone pair character, but delocalization into σ bonding regions is present as well. (See Fig. 7.) P3 calculations with a cc-pVTZ basis obtain ionization energies of 9.28 and 10.85 eV, respectively. Therefore, the first two ionization bands in the spectrum correspond to electron detachment from an antisymmetric combination of adjacent nitrogen lone pairs ($8b_2$) and a π_3 benzene-type MO ($1a_2$).

A broad feature centered at 9.3 eV has an integrated intensity that is approximately equal to a relatively sharp peak centered at 10.5–10.6 eV [49]. Approximately twice the integrated intensity belongs to an asymmetric feature with a maximum at 11.3 eV. At the cc-pVTZ, P3 level of theory, the $^2B_1$ and $^2A_1$ final states lie at 11.29 and 11.66 eV. This evidence suggests that two final states contribute to the third feature in the PES. These correspond to ionizations from π_2 (MO $2b_1$) and a symmetric nitrogen lone pair combination (MO $10a_1$), respectively. Delocalization into σ bonding regions is important in the latter MO. Poles strengths of 0.88–0.89 for the first four final states indicate that the Koopmans description is qualitatively valid.

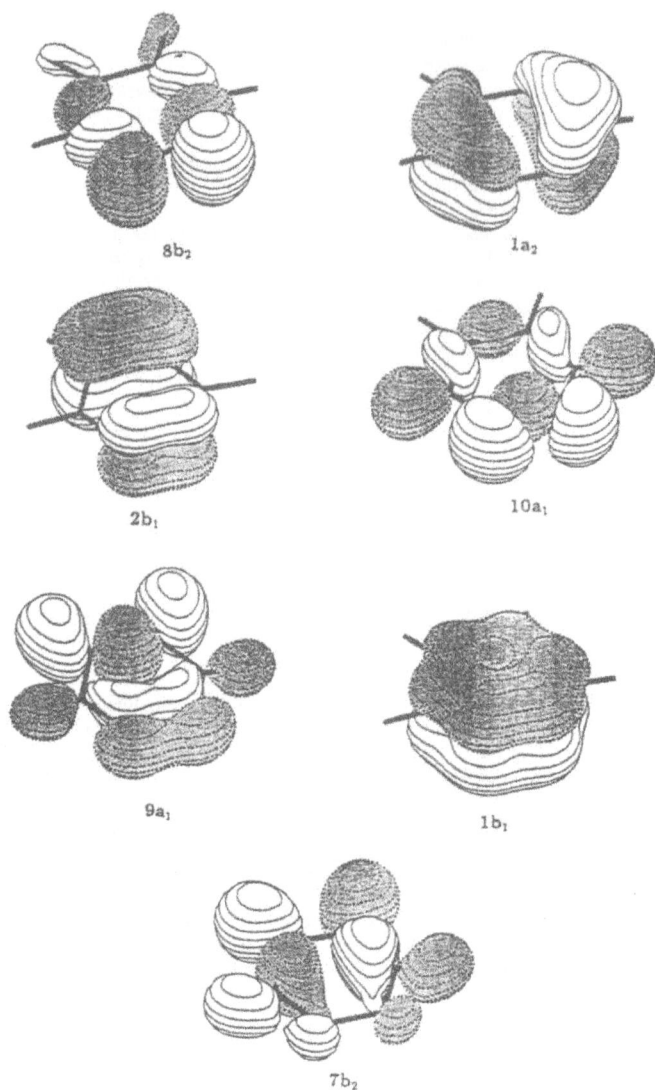

Two higher states lie within 0.1 eV of each other according to P3 calculations. The last final state with a π hole lies at 14.55 eV, but the pole strength is only 0.80. It is likely that there are many $^2B_1$ shakeup states with appreciable intensity that are close in energy. The nearby $^2A_1$ state at 14.52 eV has a pole strength of 0.87. The corresponding $9a_1$ MO is similar to $10a_1$, but has less nitrogen lone pair character. A broad feature between 13.4 and 14.3 eV appears with a maximum at 13.8 eV and a shoulder at 14.2 eV [49]. While the two final states calculated here contribute in this energy range, it is not possible to state which is lower.

P3, cc-pVTZ calculations predict a $^2B_2$ final state at 14.88 eV with a 0.87 pole strength. A prominent maximum lies at 14.8 eV [49]. The Koopmans-hole final state corresponding to a delocalized σ $7b_2$ MO is among the chief contributors to this spectral feature.

Pyridazine Ionization Energies (eV)

Final State	cc-pVDZ			cc-pVTZ		Expt. [49]
	KT	OVGF	P3	KT	P3	
$^2B_2$	10.73	8.77	8.88	10.82	9.28	9.3
$^2A_2$	10.47	10.35	10.59	10.50	10.85	10.5
$^2B_1$	11.03	10.77	11.02	11.08	11.29	11.3
$^2A_1$	13.01	11.20	11.32	13.08	11.66	11.3
$^2A_1$	16.11	14.19	14.26	16.15	14.52	13.8
$^2B_1$	15.88	14.32	14.31	15.94	14.55	14.2
$^2B_2$	16.33	14.61	14.67	16.36	14.88	14.8

11.4.3 Pyrimidine

Correlation effects cause a complete reordering of the first four final states. In the lowest cationic state, $^2B_2$, there is a much larger correlation correction than in the lowest π_3 state, $^2B_1$. The $7b_2$ MO displays large lone pair amplitudes, but delocalization into the σ framework is clear in Fig. 8. Similar arguments apply to the next lone pair state, $^2A_1$, with respect to the nearby π_2 state, $^2A_2$. Agreement with experiment [50] (see also [51]) is within 0.2 eV in each case. Discrepancies between cc-pVDZ and cc-pVTZ results are 0.3–0.4 eV. The experimental feature at 11.2 eV appears as structure in a larger peak whose maximum occurs near 11.5 eV. Calculated energy separations between the $^2A_1$ and $^2A_2$ states obtained with the P3 method are between 0.25 and 0.30 eV and provide a reasonable account of this spectral feature. The corresponding OVGF, cc-pVDZ results produce a separation that is less than 0.1 eV. Because the pole strengths for all four states exceed 0.87, the Koopmans picture of the final states is qualitatively valid.

A broad feature with a maximum at 14.5 eV has a pronounced shoulder at 13.9 eV [50]. Two final states are calculated to lie at 14.73 ($^2B_2$) and 14.56 ($^2A_1$) eV; both states have pole strengths equal to 0.88 and correspond to ionizations from delocalized σ MOs, $6b_2$ and $10a_1$, respectively. In addition, there is a $^2B_1$ π_1 final state at 14.56 eV with a 0.81 pole strength. It is therefore likely that the spectral feature in question contains contributions from more than one $^2B_1$ final state.

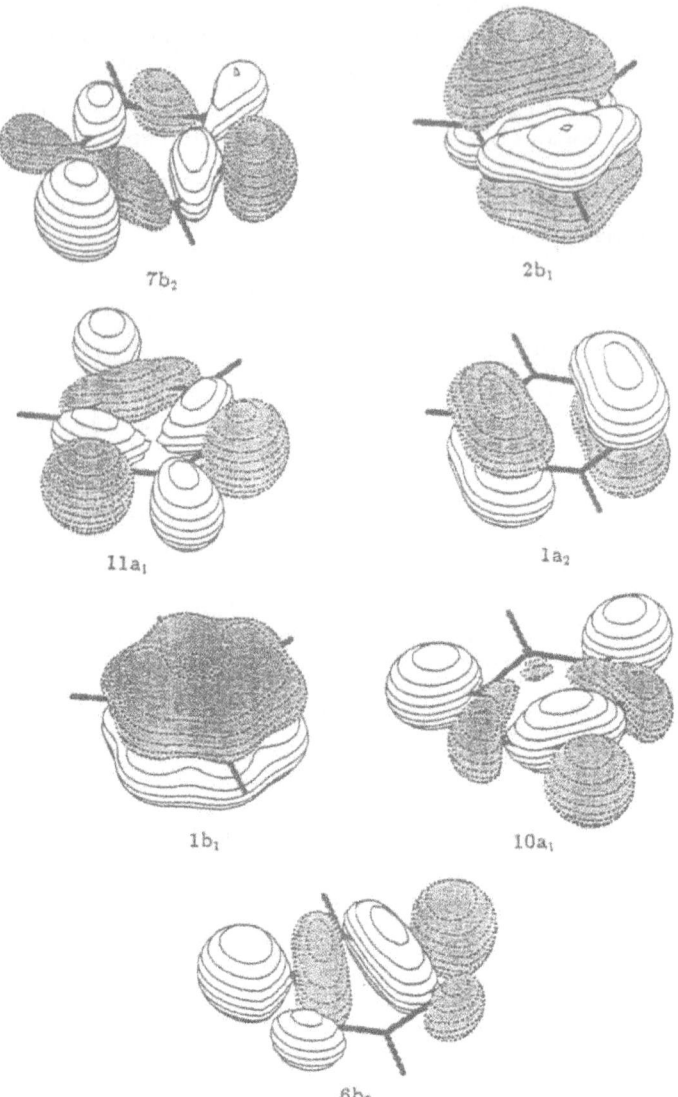

Pyrimidine Ionization Energies (eV)

Final State	cc-pVDZ			cc-pVTZ		Expt. [50]
	KT	OVGF	P3	KT	P3	
$^2B_2$	11.16	9.44	9.48	11.25	9.86	9.7
$^2B_1$	10.26	10.14	10.37	10.32	10.64	10.5
$^2A_1$	12.76	10.99	10.98	12.85	11.33	11.2
$^2A_2$	11.49	11.08	11.27	11.58	11.58	11.5
$^2B_1$	15.80	14.32	14.29	15.88	14.56	13.9
$^2A_1$	15.96	14.27	14.29	16.02	14.56	14.5
$^2B_2$	16.19	14.40	14.46	16.25	14.73	14.5

11.4.4 Pyrazine

Larger correlation corrections again obtain for final states with lone pair holes as opposed to π holes. For pyrazine, this trend produces the novel result that the lowest ionization energy, to ^2A_g, corresponds to an in-phase combination of N lone pairs. (See Fig. 9.) After the lowest π state, $^2B_{3g}$, there follows a state, $^2B_{3u}$, corresponding to the out-of-phase combination of N lone pairs. Note that both $6a_g$ and $5b_{3u}$ exhibit significant σ framework amplitudes in addition to their lone pair features. The next is another π state, $^2B_{2g}$. All of these states have pole strengths in excess of 0.87; the Koopmans description of the final states is qualitatively reasonable. Four clearly identified peaks correspond to these final states [52]. Agreement with experiment [52] is within 0.3 eV for all states except for the lowest state, where an anomalously large error of 0.4 eV is found when the cc-pVTZ basis is used. (A larger value for the lowest ionization energy, 9.6 eV, is reported elsewhere [51]). A more thorough analysis of conical intersections [53] has shown the necessity of treating nonadiabatic dynamics in this case. The P3 cc-pVDZ result, 9.79 eV, can therefore be regarded as a reasonable estimate for the vertical ionization energy.

Peaks at higher energies display larger discrepancies with calculated results. Pole strengths are above 0.83 for all remaining states, except for $^2B_{1u}$, where the pole strength is only 0.79. The latter state corresponds to the deepest π hole. Shakeup character is relatively prominent in this final state. These data suggest that some additional states with appreciable pole strength are present above 14 eV.

Pyrazine Ionization Energies (eV)

Final State	cc-pVDZ			cc-pVTZ		Expt. [52]
	KT	OVGF	P3	KT	P3	
^2A_g	10.97	9.23	9.40	11.08	9.79	9.4
$^2B_{3g}$	9.76	9.76	10.02	9.79	10.29	10.2
$^2B_{3u}$	13.48	11.16	11.22	13.54	11.57	11.4
$^2B_{2g}$	11.90	11.50	11.71	11.99	12.01	11.7
$^2B_{1g}$	15.04	13.62	13.72	15.11	13.98	13.3
$^2B_{1u}$	15.75	14.22	14.20	15.82	14.44	14.0
$^2B_{2u}$	16.63	14.99	15.01	16.68	15.23	15.0

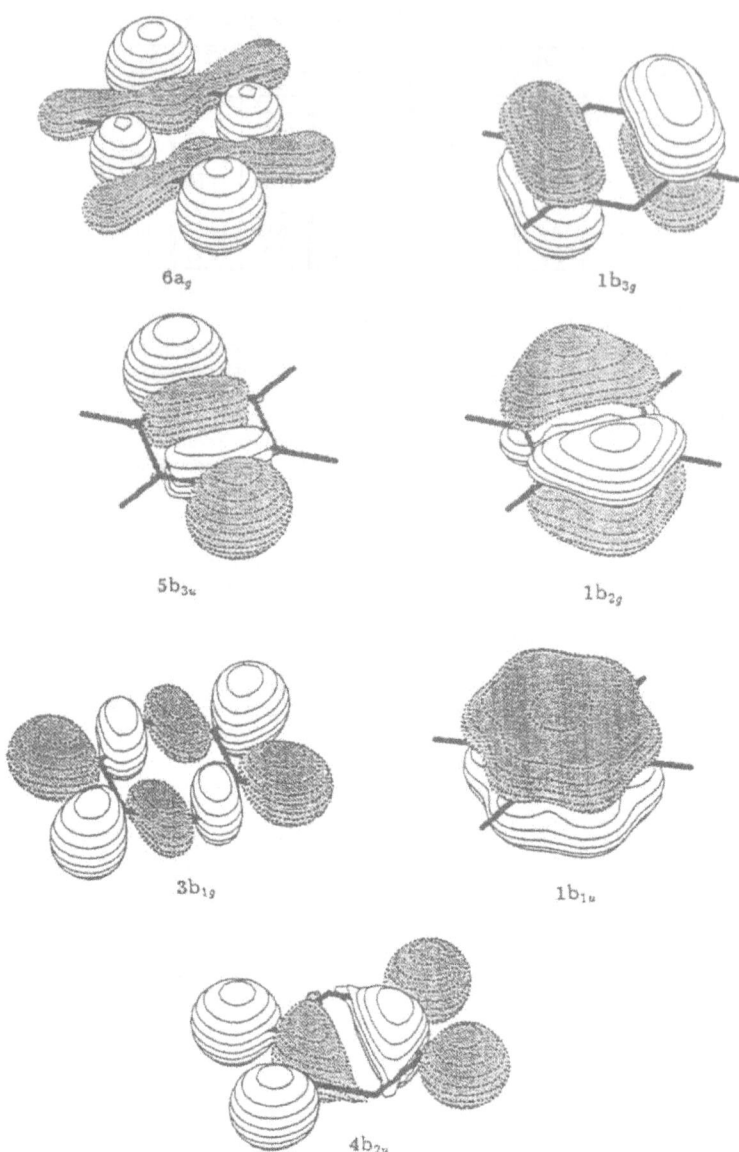

11.4.5 S-triazine

The order of final states for s-triazine is treated correctly by Koopmans's theorem, but the propagator calculations apply large quantitative corrections. For the $^2E'$ final state, a lowering of 1.4 eV obtains in P3, cc-pVTZ calculations. In the lowest π final state, $^2E''$, the same level of theory produces only a minor increase in the vertical ionization energy with respect to the Koopmans's theorem result. For the nondegenerate lone pair hole ($^2A'_1$ final state), correlation corrections exceed 2 eV, but are not large enough to affect the order of states. Agreement with experiment [54] is within 0.2 eV for these three states, which correspond to well-separated peaks of roughly equal intensity. All three pole strengths are above 0.85. Fig. 10 shows that the three lone pair MOs exhibit delocalization into σ bond regions.

A broad peak [54] has several maxima between 14.7 and 15.6 eV. In addition, there is a slowly descending hump that persists until 16.0 eV. The next two calculated states, $^2E'$ at 15.43 eV and $^2A''_2$ at 15.59 eV, contribute to this spectral feature. In accord with the pattern established by the other azabenzenes presently under consideration, the first ionization energy pertaining to σ bonding in the ring is overestimated. Despite having the lowest pole strength, 0.82, the $^2A''_2$ ionization energy is calculated to within 0.1 eV of the experimental estimate.

S–Triazine Ionization Energies (eV)

Final State	cc-pVDZ			cc-pVTZ		Expt. [54]
	KT	OVGF	P3	KT	P3	
$^2E'$	11.84	10.15	10.16	11.94	10.55	10.4
$^2E''$	11.95	11.57	11.80	12.06	12.10	12.0
$^2A'_1$	15.52	13.35	13.18	15.59	13.52	13.3
$^2E'$	16.75	15.07	15.15	16.83	15.43	14.7
$^2A''_2$	16.78	15.33	15.31	16.88	15.59	15.6

11.4.6 S-tetrazine

A complete rearrangement of final states is produced by electron correlation. The lowest ionization energy is assigned correctly by Koopmans's theorem, but correlation corrections over 1 eV are necessary to achieve quantitative agreement with experiment. The $^2B_{3g}$ final state corresponds to a nitrogen lone pair hole. Correlation effects indicate that the next state, $^2B_{1u}$, is related to another lone pair hole. A π-hole final state, $^2B_{2g}$, follows. A spectral feature at 12.1 eV with a shoulder at 11.9 eV [55] can be attributed to the $^2B_{2g}$ and $^2B_{1u}$ states, respectively. In the P3, cc-pVTZ results, the spacing between the two states is approximately 0.2 eV. The pattern of final states repeats with two lone pair states and another π state: ^2A_g, $^2B_{2u}$ and $^2B_{1g}$. All pole strengths are above 0.85 and confirm the qualitative validity of the Koopmans description of the first six final states. Lone pair MOs retain considerable C–C σ bonding character. (See Fig. 11.) Discrepancies between cc-pVDZ and cc-pVTZ results with the P3 method show that fortuitously good agreement may obtain for some final states with the smaller basis. The deepest π state calculated here again has an anomalously low pole strength, 0.79. This result indicates that another final state with $^2B_{3u}$ symmetry has an appreciable pole strength.

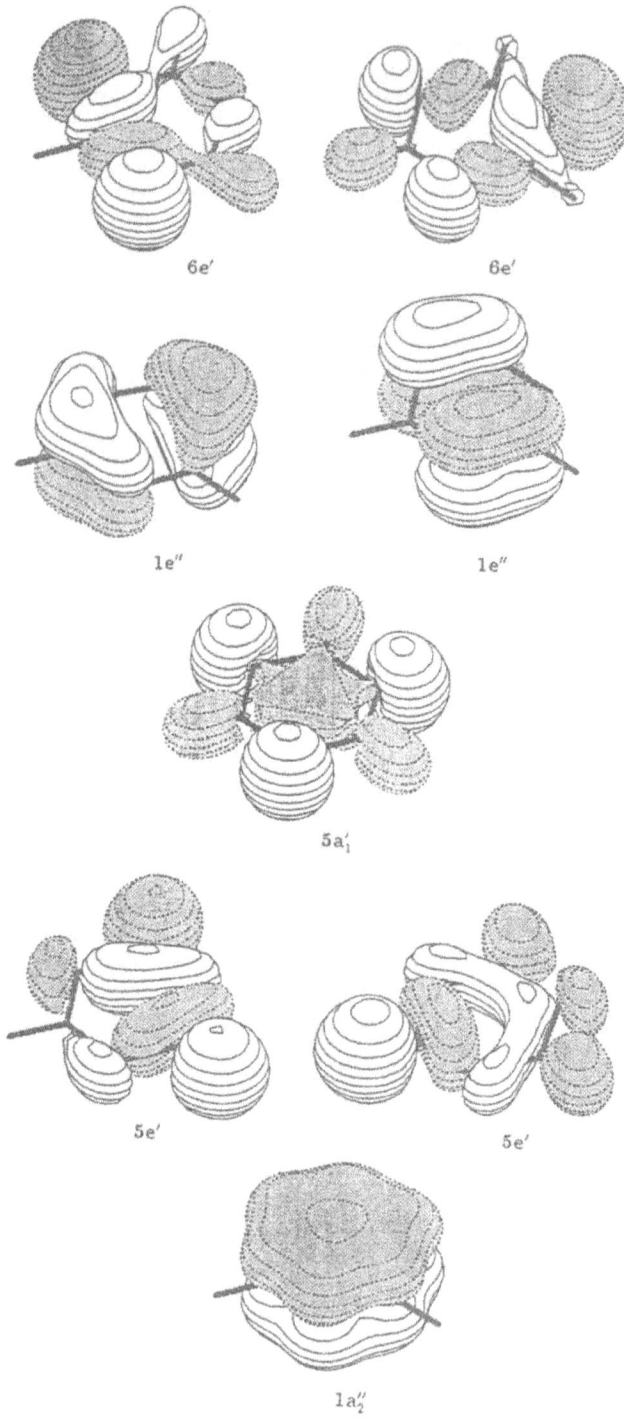

ONE-ELECTRON PICTURES OF ELECTRONIC STRUCTURE

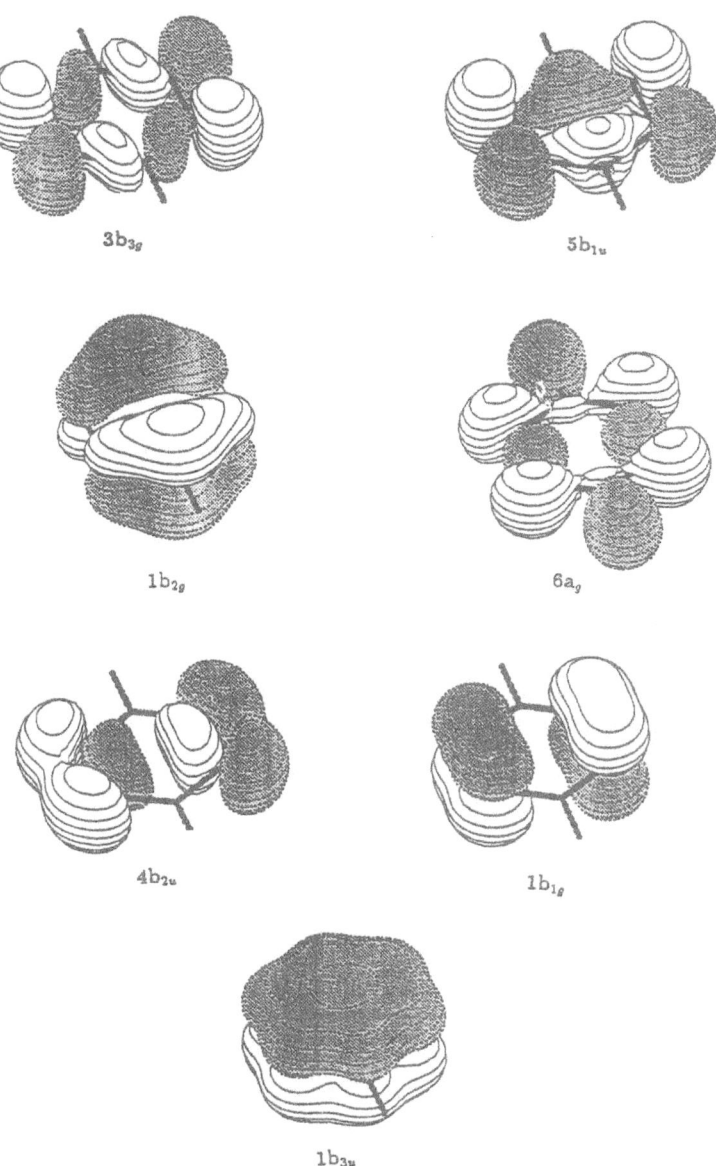

$3b_{3g}$

$5b_{1u}$

$1b_{2g}$

$6a_g$

$4b_{2u}$

$1b_{1g}$

$1b_{3u}$

S–Tetrazine Ionization Energies (eV)

Final State	cc-pVDZ			cc-pVTZ		Expt. [55]
	KT	OVGF	P3	KT	P3	
$^2B_{3g}$	10.99	9.03	9.33	11.08	9.72	9.7
$^2B_{1u}$	14.22	11.78	11.90	14.26	12.23	11.9
$^2B_{2g}$	11.86	11.81	12.17	11.87	12.40	12.1
^2A_g	14.40	12.67	12.90	14.47	13.21	12.8
$^2B_{2u}$	15.46	12.93	13.15	15.51	13.48	13.3
$^2B_{1g}$	13.71	13.26	13.60	13.76	13.88	13.5
$^2B_{3u}$	17.90	16.25	16.32	17.95	16.53	15.8

11.5 Polyacenes

A numbering scheme for nuclei is given in Fig. 12. Vertical ionization energies calculated with the cc-pVDZ basis are considered here [56], together with experimental data on anthracene, phenanthrene and naphthacene.

11.5.1 Anthracene

Ionization energies predicted for the $^2B_{2g}$ final state are somewhat lower than the experimental results. P3 is slightly closer to experiment than OVGF. The corresponding DO, the $2b_{2g}$ HOMO, has four nodal planes and its largest positive and negative amplitudes reside on C_9 and C_{10}. (See Fig. 13.) This nonbonding relationship is mixed with bonding patterns that are present between the 1-2, 3-4, 5-6 and 7-8 pairs of carbons. A sharp peak with two high-energy, vibrational satellites occurs in each reported spectrum [57-60]. The orbital amplitudes correspond very well with the preferred sites for electrophilic aromatic substitution, carbons 9 and 10 in the central ring [61].

The second band represents ionization from $2b_{1g}$, an MO which is approximately equal to an antisymmetric combination of π_2 benzene orbitals on each of the exterior rings. Somewhat larger amplitudes occur in the interior positions: 4a, 9a, 8a and 10a. As in the case of the lowest final state, the P3 approximation gives better agreement with the experimental results. Another sharp peak with several vibrational satellites is found in the spectra [57-60].

The third band occurs due to ionization from $1a_u$, but this MO is composed chiefly of π_3 benzene fragments on the exterior rings with opposite phases. In this case, a distortion reduces the amplitudes near carbons 4a, 9a, 8a and 10a. The spectra display a sharp feature and a high-energy shoulder.

The fourth and the fifth ionization bands overlap in the experimental spectra and there are some uncertainties in assignment of their maxima [59, 62, 63]. P3 values are closer to those described in Ref. [63] than the other experimental reports. The fourth band corresponds to ionization from the $1b_{2g}$ MO. This orbital is built from the middle ring's π_3 system with π components from the outer rings. Amplitudes on carbon atoms 2, 3, 6 and 7 have negligible importance. Only two nodal planes are present. Nearby in energy is the $^2B_{3u}$ final state. In the corresponding MO, a benzene π_1 pattern on the inner ring interferes destructively with four-center bonding lobes spread over carbons 5, 6, 7 and 8 and over carbons 1, 2, 3 and 4. The net effect is to locate the largest contributions on carbons 2, 3, 6 and 7. In the reported spectra, there are two closely-spaced maxima followed by a broad shoulder. Two final electronic states contribute to this feature's enhanced area.

ONE-ELECTRON PICTURES OF ELECTRONIC STRUCTURE

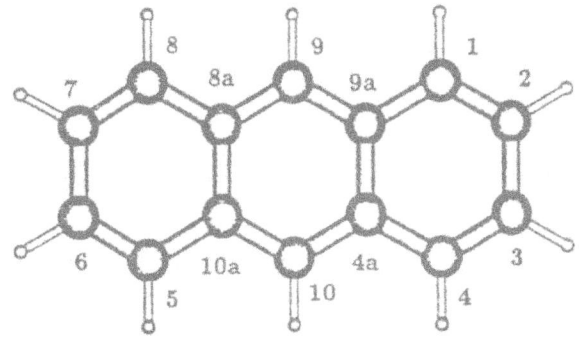

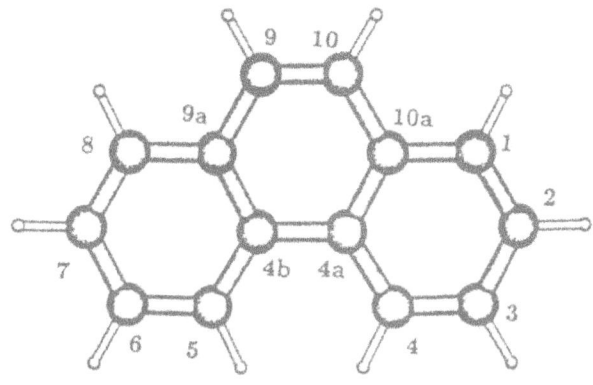

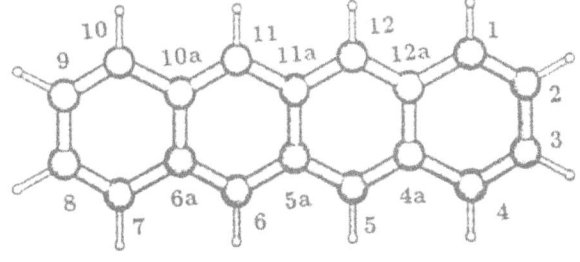

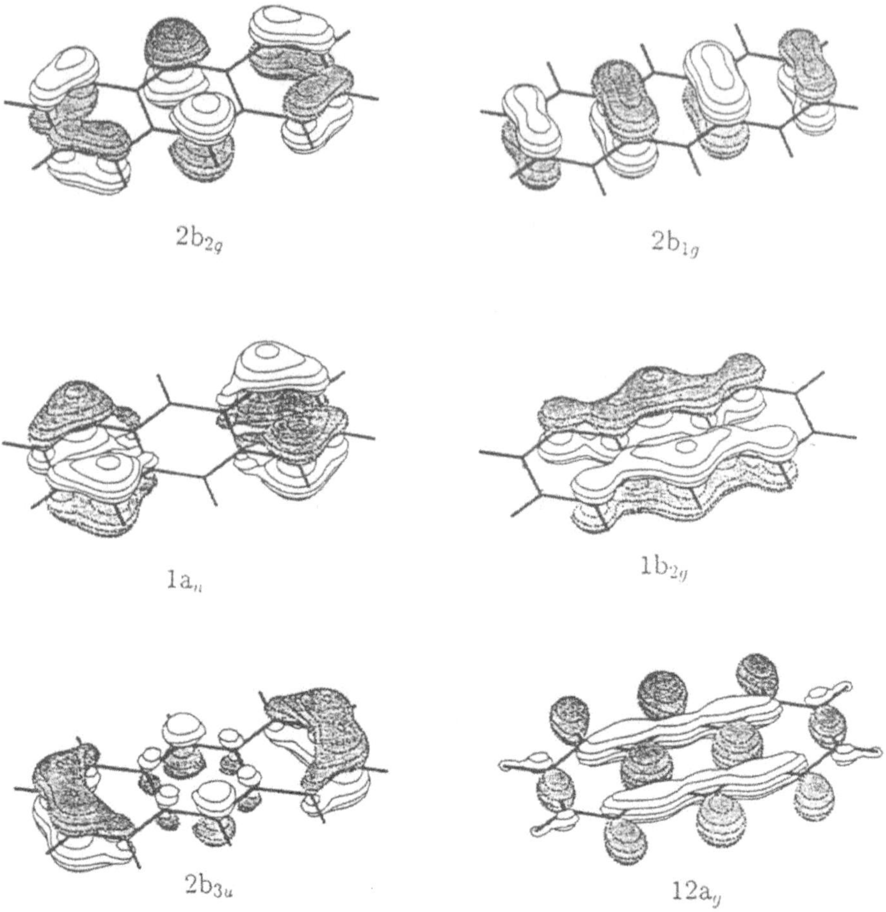

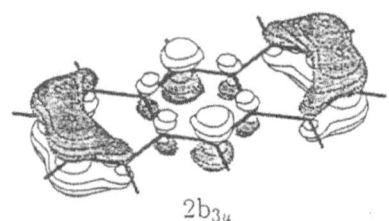
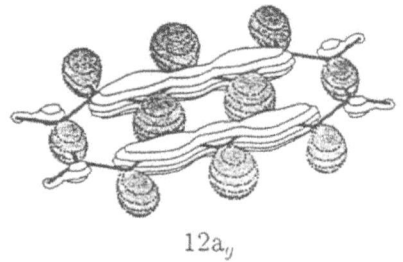

$2b_{2g}$ $2b_{1g}$

$1a_u$ $1b_{2g}$

$2b_{3u}$ $12a_g$

These five ionic states are the only ones assigned in the experimental articles, although He I spectra are sometimes depicted in their entirety [57, 58, 60]. The next ionization represents the onset of σ-states.

Anthracene Ionization Energies (eV)

Final State	KT	OVGF PS	P3 PS	Expt. [56]
$^2B_{2g}$	7.13	7.17 0.89	7.31 0.88	7.40–7.47
$^2B_{1g}$	8.41	8.27 0.89	8.46 0.87	8.52–8.57
^2A_u	9.45	9.07 0.87	9.09 0.86	9.16–9.23
$^2B_{2g}$	11.00	10.13 0.86	10.21 0.85	10.13–10.26
$^2B_{3u}$	11.28	10.31 0.84	10.34 0.82	10.21–10.45
$^2A_{1g}$	12.78	11.14 0.89	11.28 0.88	

11.5.2 Phenanthrene

The first peak in the experimental spectrum is resolved very well and relates to ionization from $4b_1$. OVGF and P3 energies are in very good agreement with experimental values. The HOMO of phenanthrene is extensively delocalized, but the most prominent π-bonding contributions are on carbons 9 and 10. (See Fig. 14.) Other bonding lobes are found between the following pairs of carbons: 1-10a, 3-4, 5-6 and 8-9a. In each of these cases, some polarization to the exterior carbons takes place. As is the case with anthracene, the distribution of this orbital is compatible with observed patterns in electrophilic aromatic substitution at sites 9 and 10 [61].

The second peak in the spectrum corresponds to $3a_2$. There are some ambiguities in assignment of this band [57,60–64]. Brogli and Heilbronner [62] describe it as an unresolved double band, but Hush et al., while having a maximum at 8.06 eV, still report a vertical ionization energy of 8.30 eV in order to obtain consistency with their scaled, INDO ionization energies [60]. The value of Ref. [65], 8.28 eV, corresponds to a peak that is less intense than another occurring near 8.10 eV. P3 results are in especially good agreement with the latter figure and also agree closely with the experimental range, 8.06–8.15 eV, reported elsewhere [57, 60, 63, 66]. The $3a_2$ MO is a π system with 4 nodal planes. Two, three-center π bond lobes at carbons 4, 4a and 10a and at 5, 4b and 9a and two, two-center bond lobes between 1 and 2 and between 7 and 8 have alternating phases.

A sharp peak with a high-energy vibrational feature has been assigned to a $^2A_2$ final state. Four, two-center π lobes dominate the pertinent orbital, $2a_2$, which appears to be an antisymmetric combination of two benzene π_3 motifs with slight distortions toward the center of the molecule. P3 results are again in excellent agreement with experiment.

The next spectral peak also is adequately described at the P3 level. In the corresponding MO, $3b_1$, there are nodal planes between four-center π bond lobes at carbons 4, 4a, 4b and 5 and two, two-center

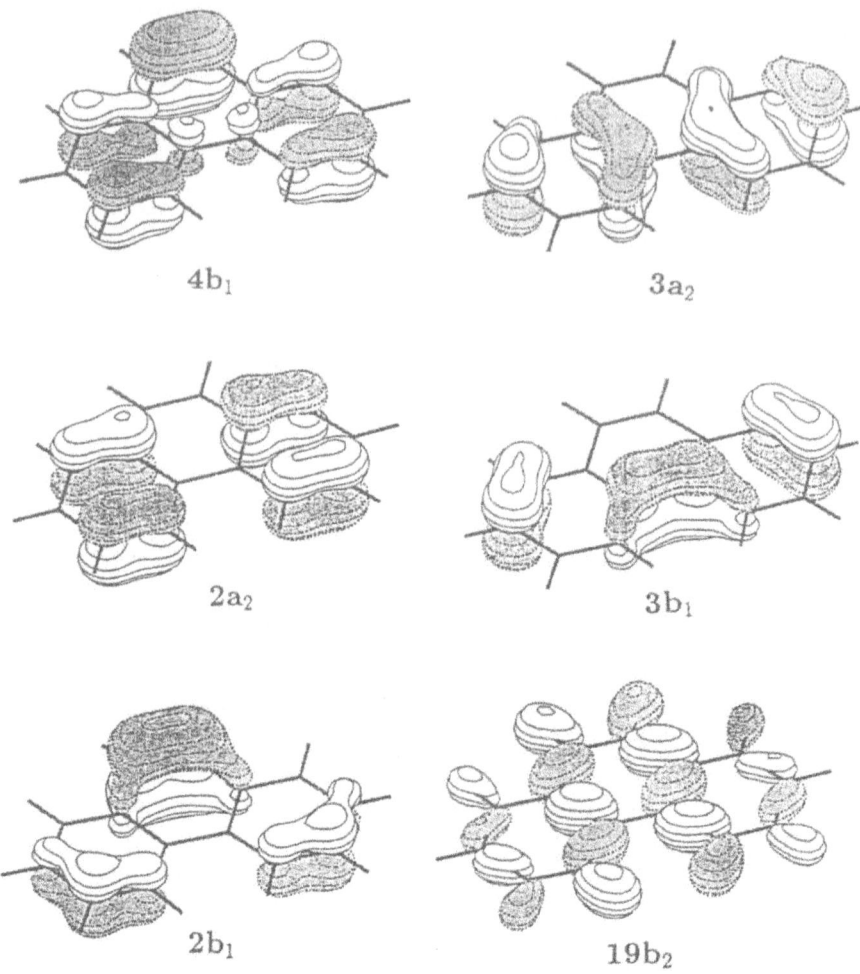

π bond lobes at carbons 1 and 2 and at carbons 7 and 8. This feature is sharp, but with a series of high-energy vibrational structures.

The last peak in the experimental spectrum that was assigned previously is situated at 10.53-10.64 eV. Calculated ionization energies are in closer agreement with the higher values. The MO corresponding to this ionization, $2b_1$, is concentrated on the C_9-C_{10} bond, although there are definite contributions from three-center π lobes at carbons 2, 3 and 4 and at carbons 5, 6 and 7.

The higher energy region of the experimental spectrum [57, 60, 65] consists of a series of closely-spaced peaks. Two distinct features at 10.53 and 10.71 eV have been observed in Ref. [65]. Because of the relatively low pole strength of our calculated, P3 state at 10.72 eV (0.83), it is likely that a shakeup $^2B_1$ state is present. It is not possible, however, to discern which of the two final states has greater shakeup character.

The first σ-hole final state is reported to lie at 11.03 eV [65]. OVGF and P3 results are in good agreement. The corresponding $19b_2$ orbital displays a pattern of σ bond functions with alternating phases.

Phenanthrene Ionization Energies (eV)

Final State	KT	OVGF PS	P3 PS	Expt. [56]
$^2B_1$	7.78	7.72 0.89	7.85 0.88	7.85-7.92
$^2A_2$	8.04	7.95 0.89	8.11 0.87	8.10-8.30
$^2A_2$	9.49	9.12 0.88	9.18 0.87	9.25-9.28
$^2B_1$	10.35	9.81 0.87	9.84 0.86	9.85-9.90
$^2B_1$	11.70	10.71 0.85	10.72 0.83	10.53-10.64
$^2B_2$	12.73	11.08 0.89	11.23 0.88	11.03

11.5.3 Naphthacene

The first ionization band is located at 6.97-7.04 eV and contains a sharp peak with well-resolved vibrational structure [58]. P3 agrees with experiment more closely than OVGF. Five nodal planes are present in the corresponding $2a_u$ orbital. Electron density is localized mostly at carbon sites 5, 6, 11 and 12. (See Fig. 15.)

Contradictory assignments have been made for the overlapping second and third ionization bands. Besides the sharp peak appearing at ~8.4 eV, one finds a low-energy shoulder and a small, high-energy peak. Our P3 calculations give 8.30 and 8.39 eV for the $^2B_{3u}$ and $^2B_{2g}$ states, respectively. The spacing of the states suggests that the shoulder and the principal peak should be assigned to $^2B_{3u}$ and $^2B_{2g}$ final states, respectively, with the high energy peak corresponding to a vibrational satellite.

For the next three states, agreement between our propagator results and experimental data is very close. Corresponding MOs ($1a_u$, $2b_{1g}$ and $1b_{2g}$) become more bonding in character and exhibit more many-center lobes. Each of the associated bands has extensive vibrational structure.

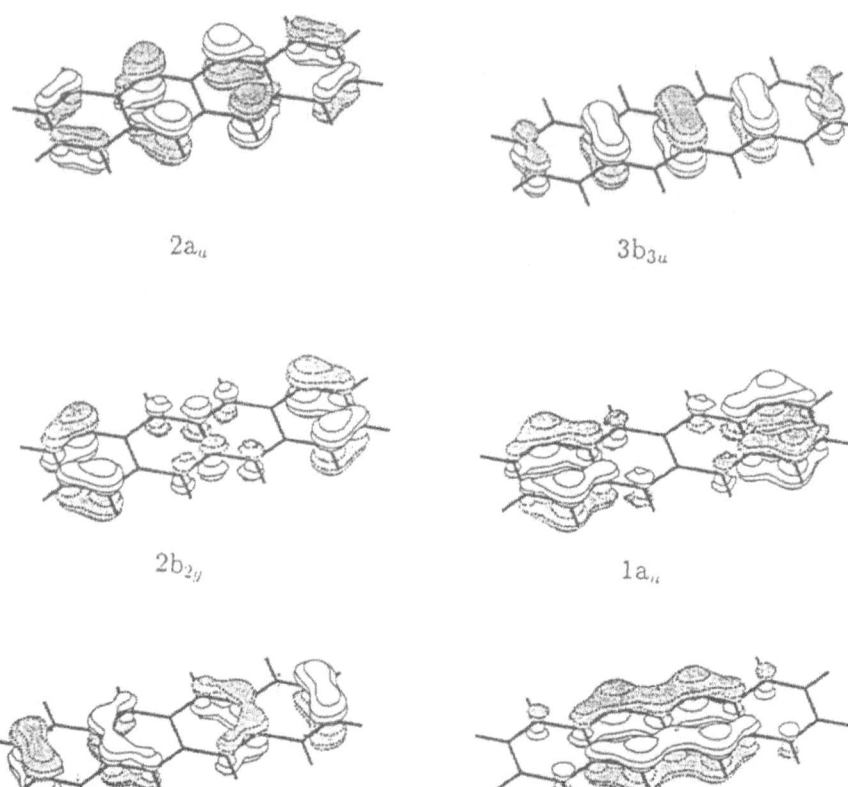

$2a_u$

$3b_{3u}$

$2b_{2g}$

$1a_u$

$2b_{1g}$

$1b_{2g}$

Naphthacene Ionization Energies (eV)

Final State	KT	OVGF PS	P3 PS	Expt. [56]
^2A_u	6.59	6.65	6.83	6.97–7.01
		0.88	0.87	
$^2B_{3u}$	8.25	7.99	8.30	8.41–8.44
		0.88	0.86	
$^2B_{2g}$	8.64	8.27	8.39	8.41–8.63
		0.87	0.86	
^2A_u	10.28	9.57	9.62	9.56–9.60
		0.86	0.85	
$^2B_{1g}$	10.65	9.77	9.83	9.70–9.75
		0.83	0.86	
$^2B_{2g}$	11.29	10.30	10.33	10.25–10.26
		0.85	0.84	

11.6 Borazine

Borazine is isoelectronic with benzene and displays an alternating pattern of BH and NH groups in a D_{3h} framework. P3 and OVGF calculations were performed with a cc-pVDZ basis [31] and the results are listed below [32]. Because all pole strengths are between 0.88 and 0.92, the final states are qualitatively similar to their Koopmans descriptions. For borazine, various assignments of final states have been made [67–70]. The lowest ionization energy corresponds to a π MO that is more localized on the nitrogens than on the borons. The next state has σ symmetry and has a pattern of B-H lobes with appreciable B-N contributions. (See Fig. 16.) A π MO is associated with the third state. For the π ionization energies, both correlated predictions are in close agreement with experiment. P3 is somewhat closer to experiment than OVGF for the σ final states. A σ MO with almost pure B-H lobes corresponds to the fourth state. It is likely that there are $^2E'$ and $^2A'_2$ contributions to the peak at 14.76 eV [68], for the area under it is greater than the area under the peaks that have been assigned to the first two, degenerate final states. The latter state corresponds to a σ MO delocalized over all B-N bonds. Note that correlation effects change the order of these two final states.

Borazine Ionization Energies (eV)

State	KT	OVGF	P3	Expt. [68]
$^2E''$	11.007	10.103	9.913	10.14
$^2E'$	12.671	11.785	11.551	11.42
$^2A''_2$	13.961	12.877	12.604	12.83
$^2A'_1$	14.982	14.130	13.830	13.84
$^2E'$	16.411	15.216	14.882	14.76
$^2A'_2$	16.475	14.918	14.660	

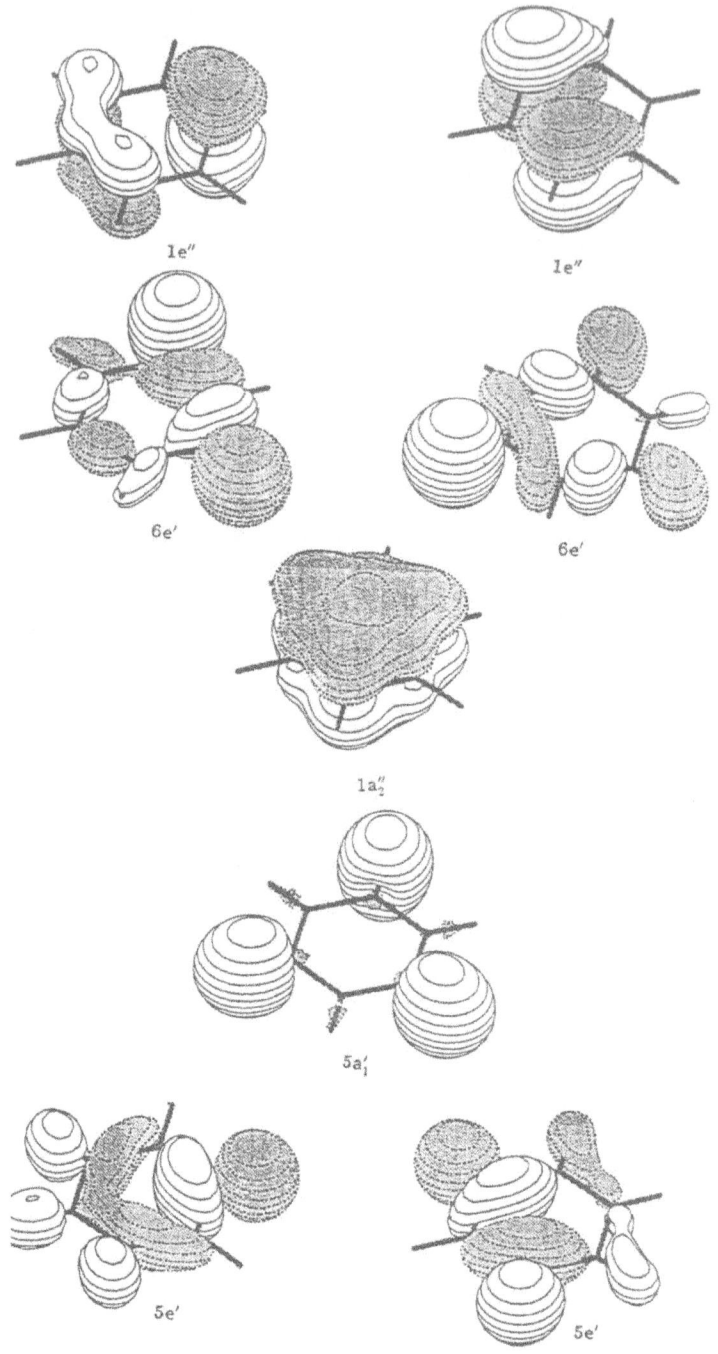

11.7 TCNQ

A cc-pVDZ basis set augmented with s and p diffuse functions was used for calculations on tetracyanoquinone (TCNQ) [71]. The experimental PES is well-resolved and enables clear assignments of spectral features [72]. The lowest cationic state corresponds to $3b_{3u}$, a delocalized π orbital with significant contributions from all nonhydrogen atoms except for the cyano carbons. (See Fig. 17.) There is good agreement between calculated and experimental values. For the second ionization energy, the $2b_{1g}$ MO is essentially identical to one of benzene's degenerate HOMOs. P3 and OVGF ionization energies are much larger than the lowest ionization energy of benzene, 9.25 eV [38]. Another delocalized π MO, $2b_{2g}$, pertains to the third calculated ionization energy. There is a broad band between 11 and 12 eV [72], with local maxima we have estimated to reside at 11.3 and 11.5 eV.

TCNQ Ionization Energies (eV)

Final State	KT	OVGF PS	P3 PS	Expt. [72]
$^2B_{3u}$	9.83	9.42	9.81	9.61
		0.88	0.87	
$^2B_{1g}$	12.08	11.24	11.62	11.3 (est.)
		0.88	0.86	
$^2B_{2g}$	12.45	11.55	11.97	11.5 (est.)
		0.85	0.83	
$^2B_{3g}$	13.92	12.91	13.16	12.68
		0.86	0.84	
$^2B_{2u}$	13.93	12.96	13.20	12.90
		0.86	0.84	
^2A_u	14.27	13.24	13.49	13.30
		0.86	0.84	
$^2B_{1g}$	14.28	13.25	13.50	13.38
		0.86	0.84	
$^2B_{3g}$	15.21	13.25	13.64	
		0.88	0.87	
$^2B_{1u}$	14.34	13.26	13.55	13.38
		0.87	0.85	
^2A_g	14.49	13.44	13.71	13.53
		0.86	0.84	
$^2B_{3u}$	14.79	13.65	13.96	
		0.86	0.84	

The fourth band corresponds to ionization from $9b_{3g}$, an MO which is built from C-N π bonding lobes. A distinct phase pattern of similar lobes occurs in the next MO, $10b_{2u}$.

Electron correlation compresses the next four states into a narrow energy range and yields a different order of final states. A broad, incompletely resolved band from 12 to 14 eV contains contributions from two MOs, $1a_u$ and $1b_{1g}$, which display combinations of C-N π lobes perpendicular to the nuclear plane. Correlation again induces a reordering of final states in this energy region. The $8b_{3g}$ MO exhibits σ

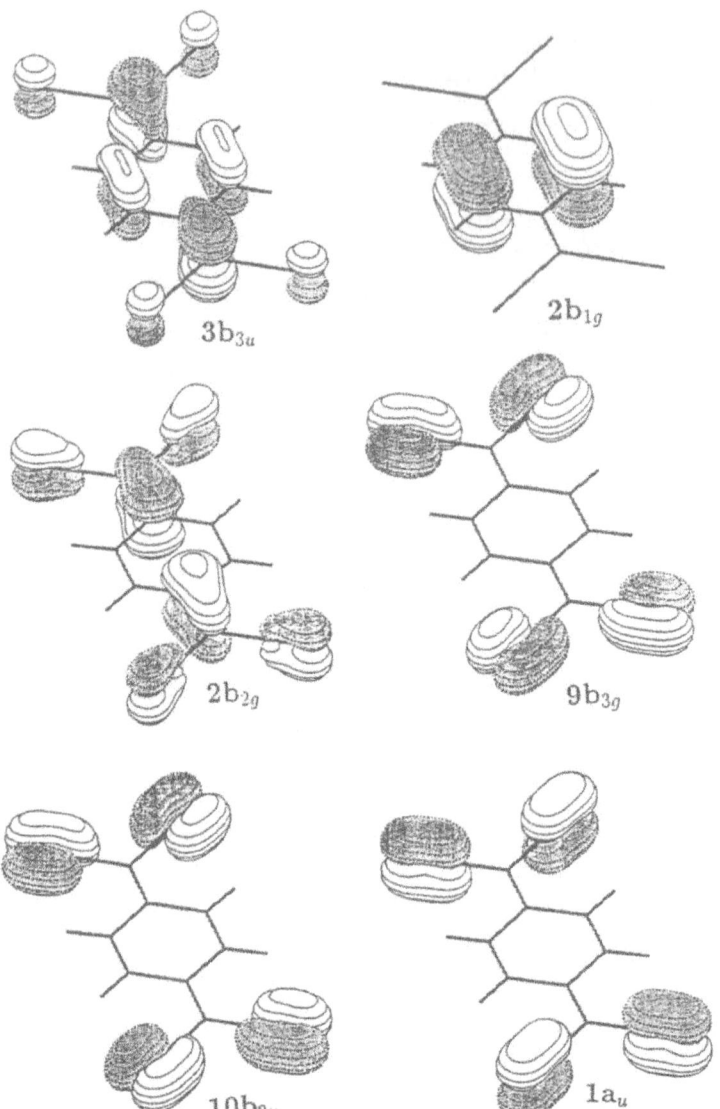

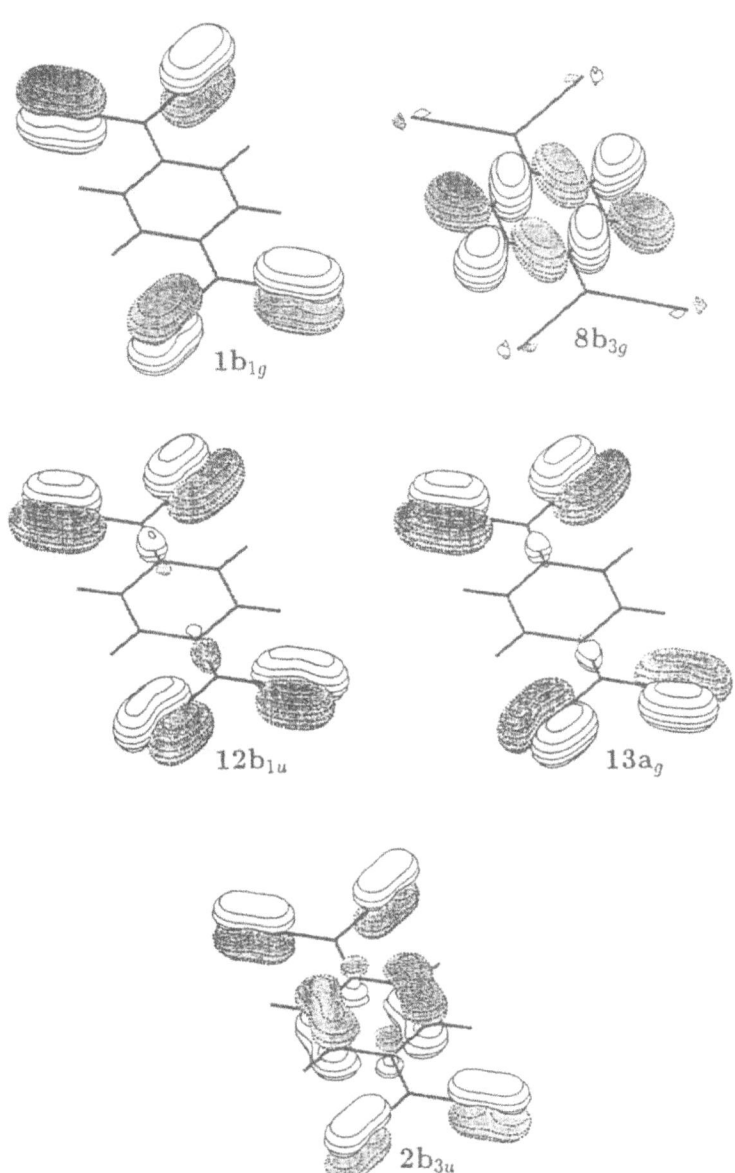

bonding patterns in the ring. It is followed by states corresponding to two MOs with CN π interactions in the nuclear plane, $12b_{1u}$ and $13a_g$.

12 Conclusions

The working equations of electron propagator theory resemble the one-electron eigenvalue problems that arise in Hartree-Fock theory and in the Kohn-Sham approach to density functional theory. DOs and electron binding energies are the correlated generalizations of canonical orbitals and their energies. These solutions of the Dyson equation are subject to a nonlocal, energy-dependent potential that accounts for relaxation and correlation corrections to the familiar Coulomb and exchange terms. Retention of the ionization energy and electron affinity components of the electron propagator distinguishes this approach from Hilbert space theories, which can be rederived through separate consideration of these two parts.

Calculation of third order terms in the self-energy matrix suffices for useful quasiparticle models of valence ionization energies. Efficient implementations of the P3 and OVGF methods require fifth power contractions and eliminate the storage of electron repulsion integrals with four virtual indices. Practical calculations on molecules with up to four fused aromatic rings have been realized with these schemes.

Quasiparticle models associate uncorrelated DOs (canonical orbitals) with correlated electron binding energies. These methods produce accurate assignments of photoelectron final states in molecules where a consistent treatment of electron correlation is essential. Association of DOs to correlated ionization energies facilitates comparisons between related molecules and provides pictorial explanations of trends.

13 Acknowledgments

This work was supported by the National Science Foundation under grant CHE-9321434, the Petroleum Research Fund under grant 29848-AC6 and Gaussian, Incorporated.

References

[1] See, for example, R. McWeeny, *Methods of Molecular Quantum Mechanics*, Academic Press, New York (1989).

[2] R.G. Parr and W. Yang. *Density Functional Theory of Atoms and Molecules*, Oxford University Press. Oxford (1989).

[3] E.S. Kryachko and E.V. Ludeña, *Energy Density Functional Theory of Many Electron Systems*, Kluwer Academic, Boston (1990).

[4] P.O. Löwdin, Adv. Chem. Phys. *2*, 207 (1959).

[5] W. Kohn and L.J. Sham, Phys. Rev. *140*, A1133 (1965).

[6] J. Linderberg and Y. Öhrn, *Propagators in Quantum Chemistry*, Academic Press, New York (1973).

[7] B. Pickup and O. Goscinski, Mol. Phys. *26* 1013 (1973).

[8] L.S. Cederbaum and W. Domcke, Adv. Chem. Phys. *36*, 206 (1977).

[9] J. Simons, Theor. Chem. Adv. Persp. *3*, 1 (1979).

[10] M.F. Herman, K.F. Freed and D.L. Yeager, Adv. Chem. Phys. *48*, 1 (1981).

[11] Y. Öhrn and G. Born, Adv. Quant. Chem. *13*, 1 (1981).

[12] W. von Niessen, J. Schirmer and L.S. Cederbaum, Comput. Phys. Rep. *1*, 57 (1984).

[13] The residue corresponding to an electron propagator pole, E_{pole}, is defined by $Res(E_{pole}) = \lim_{E \to E_{pole}} G_{pq}(E)(E - E_{pole})$.

[14] H.A. Bethe and E.E. Salpeter, *Quantum Mechanics of One and Two Electron Atoms*, Academic Press, New York (1957).

[15] G. Csanak, H.S. Taylor and R. Yaris, in *Advances in Atomic and Molecular Physics*, Vol. 7, D.R. Bates and I. Esterman, Editors, Academic Press, New York (1971).

[16] M.A. Coplan, J.H. Moore and J.P. Doering, Rev. Mod. Phys. *66*, 985 (1994).

[17] T. Koopmans, Physica *1*, 104 (1933).

[18] In multiconfigurational electron propagator theory, considerable effort is expended in preparing correlated reference wavefunctions; see the following paper and references therein: J.T. Golab and D.L. Yeager, J. Chem. Phys. *87*, 2925 (1987). In equation of motion, coupled-cluster methods, a ground state optimization creates an effective Hamiltonian that is used in the ionization energy or electron affinity portions of the electron propagator. Some leading references are J.F. Stanton and J. Gauss, J. Chem. Phys. *101*, 8938 (1994) and M. Nooijen and J.G. Snijders, Int. J. Quantum Chem. *48*, 15 (1993).

[19] C. Møller and C.R. Plesset, Phys. Rev. *46*, 618 (1934).

[20] O. Goscinski and B. Lukman, Chem. Phys. Lett. *7*, 573 (1970).

[21] P.O. Löwdin, Phys. Rev. *139*, 357 (1965).

[22] J.V. Ortiz, J. Chem. Phys. *99*, 6716 (1993).

[23] J. Baker and B.T. Pickup, Chem. Phys. Lett. *76*, 537 (1980).

[24] I. Shavitt in *Modern Theoretical Chemistry*, H.F. Schaefer III, Editor, Plenum, New York (1977).

[25] See the following and references therein: J.F. Stanton and J. Gauss, J. Chem. Phys. *101*, 8938 (1994); M. Nooijen and J.G. Snijders, Int. J. Quantum Chem. *48*, 15 (1993).

[26] D.P. Chong, F.G. Herring and D. McWilliams, J. Chem. Phys. *61*, 78 (1974).

[27] Gaussian 94, M.J. Frisch, G.W. Trucks, H.B. Schlegel, P.M.W. Gill, B.G. Johnson, M.A. Robb, J.R. Cheeseman, T. Keith, G.A. Petersson, J.A. Montgomery, K. Raghavachari, M.A. Al-Laham, V.G. Zakrzewski, J.V. Ortiz, J.B. Foresman, J. Cioslowski, B.B. Stefanov, A. Nanayakkara, M. Challacombe, C.Y. Peng, P.Y. Ayala, W. Chen, M.W. Wong, J.L. Andres, E.S. Replogle, R. Gomperts, R.L. Martin, D.J. Fox, J.S. Binkley, D.J. Defrees, J. Baker, J.P. Stewart, M. Head-Gordon, C. Gonzalez and J.A. Pople, Gaussian, Inc., Pittsburgh PA, 1995.

[28] J.V. Ortiz, Int. J. Quant. Chem., Quant. Chem. Symp. *23*, 321 (1989).

[29] J.S. Lin and J.V. Ortiz, Chem. Phys. Lett. *171*, 197 (1990).

[30] V.G. Zakrzewski, J.V. Ortiz, J.A. Nichols, D. Heryadi, D.L. Yeager and J.T. Golab, Int. J. Quant. Chem. *60*, 29 (1996).

[31] T.H. Dunning, J. Chem. Phys. *90*, 1007 (1989).

[32] J. V. Ortiz, J. Chem. Phys. *104*, 7599 (1996).

[33] V.G. Zakrzewski and J.V. Ortiz, Int. J. Quant. Chem., Quant. Chem. Symp. *28*, 23 (1994); V.G. Zakrzewski and J.V. Ortiz, Int. J. Quant. Chem. *53*, 583 (1995).

[34] G.D. Purvis and Y. Öhrn, J. Chem. Phys. *60*, 4063 (1974).

[35] J.A. Pople, J.S. Binkley and R. Seeger, Int. J. Quant. Chem., Quant. Chem. Symp. *10*, 1 (1976).

[36] W. von Niessen, private communication; V.G. Zakrzewski, unpublished work.

[37] G. Schaftenaar, MOLDEN, CAOS/CAMM Center, The Netherlands (1991).

[38] K. Kimura, S. Katsumata, Y. Achiba, T. Yamazaki and S. Iwata, *Handbook of HeI Photoelectron Spectra of Fundamental Organic Molecules*, Halsted Press, New York (1981) and references therein.

[39] V.G. Zakrzewski and J.V. Ortiz, J. Mol. Struct. (*Theochem*), in press.

[40] B. Ruščič, L. Klasinc, A. Wolf and J.V. Knop, J. Phys. Chem. *85*, 1486 (1981).

[41] V.G. Zakrzewski and J.V. Ortiz, J. Phys. Chem. *100*, 13979 (1996).

[42] S. Fujisawa, I. Oonishi, S. Masuda, K. Ohno and Y. Harada, J. Phys. Chem. *95*, 4250 (1991).

[43] J.V. Ortiz and V.G. Zakrzewski, J. Chem. Phys. *105*, 2762 (1996).

[44] W. von Niessen, G.H.F. Diercksen and L.S. Cederbaum, Chem. Phys. *10*, 345 (1975).

[45] W. von Niessen, W.P. Kraemer and G.H.F. Diercksen, Chem. Phys. *41*, 113 (1979).

[46] G.F. Tantardini and M. Simonetta, Int. J. Quant. Chem. *20*, 705 (1981).

[47] O. Kitao and H. Nakatsuji, J. Chem. Phys. *88*, 4913 (1988).

[48] I.C. Walker, M.H. Palmer and A. Hopkirk, Chem. Phys. *141*, 365 (1989).

[49] L. Åsbrink, C. Fridh, B.Ö. Jonsson and E. Lindholm, Int. J. Mass Spectrom. and Ion. Phys. *8*, 229 (1972).

[50] L. Åsbrink, C. Fridh, B.Ö. Jonsson and E. Lindholm, Int. J. Mass Spectrom. and Ion. Phys. *8*, 215 (1972).

[51] R. Gleiter, E. Heilbronner and V. Hornung, Helv. Chim. Acta *55*, 255 (1970).

[52] C. Fridh, L. Åsbrink, B.Ö. Jonsson and E. Lindholm, Int. J. Mass Spectrom. and Ion. Phys. *8*, 101 (1972).

[53] L. Seidner, W. Domcke and W. von Niessen, Chem. Phys. Lett. *205*, 117 (1993).

[54] C. Fridh, L. Åsbrink, B.Ö. Jonsson and E. Lindholm, Int. J. Mass Spectrom. and Ion. Phys. *8*, 85 (1972).

[55] C. Fridh, L. Åsbrink, B.Ö. Jonsson and E. Lindholm, Int. J. Mass Spectrom. and Ion. Phys. *9*, 485 (1972).

[56] V.G. Zakrzewski, O. Dolgounitcheva and J.V. Ortiz, J. Chem. Phys., in press and references therein.

[57] R. Boschi, J. N. Murrell and W. Schmidt, Discuss. Faraday Soc. *54*, 116 (1972).

[58] P. A. Clark, F. Brogli and E. Heilbronner, Helv. Chim. Acta *55*, 1415 (1972). Also see F. Brogli and E. Heilbronner, Angew. Chem. *84*, 551 (1972).

[59] D. G. Streets and T. A. Williams, J. Electron Spectrosc. Relat. Phenom. *3*, 71 (1974).

[60] N. S. Hush, A. S. Cheung and P. R. Hilton, J. Electron Spectrosc. Relat. Phenom. *7*, 385 (1975).

[61] A. Streitwieser and C. H. Heathcock, *Introduction to Organic Chemistry*, Macmillan Publishing Co., New York (1981).

[62] F. Brogli and E. Heilbronner, Theor. Chim. Acta *26*, 289 (1972).

[63] W. Schmidt, J. Chem. Phys. *66*, 828 (1977).

[64] D.G. Streets and G.P. Caesar, Mol. Phys. *26*, 1037 (1973).

[65] B. Ruščić, B. Kovač, L. Klasinc, and H. Güsten, Z. Naturforsch. *33a*, 1006 (1974).

[66] R. Boschi, E. Clar and W. Schmidt, J. Chem. Phys. *60*, 4406 (1974).

[67] D.C. Frost, F.G. Herring, C.A. McDowell and I.A. Stenhouse, Chem. Phys. Lett. *5*, 291 (1970).

[68] D.R. Lloyd and N. Lynaugh, Phil. Trans. Roy. Soc. (London) *A268*, 97 (1970).

[69] J. Kroner, D. Proch, W. Fuss and H. Bock, Tetrahedron *28*, 1585 (1972).

[70] W.P. Anderson, W.D. Edwards, M.C. Zerner and S. Canuto, Chem. Phys. Lett. *88*, 185 (1982).

[71] V.G. Zakrzewski, O. Dolgounitcheva and J.V. Ortiz, J. Chem. Phys. *106*, 5872 (1996).

[72] I. Ikemoto, K. Samizo, T. Fujikawa, K. Ishii, T. Ohto and H. Kurada, Chem. Lett. 785 (1974).

SHAPE IN QUANTUM CHEMISTRY

Paul G. Mezey

Mathematical Chemistry Research Unit,
Department of Chemistry and
Department of Mathematics and Statistics
University of Saskatchewan,
110 Science Place,
Saskatoon, SK, Canada, S7N 5C9

1. INTRODUCTION

Among the conceptual trends involving novel approaches in quantum chemistry, the study of three-dimensional shape properties of molecules, in particular, the shape properties of molecular electron density distributions has become an important component of both theoretical research and computational applications. The fundamental quantum chemical concept of shape is an intriguing one, involving questions of the role of the Heisenberg uncertainty relation, and the problem of localizability in nonrelativistic and relativistic quantum mechanics. Nevertheless, strong motivation for the exploration of the fundamentals of the molecular shape concept is provided by the practical advantages of detailed shape representations and shape analysis using appropriate quantum chemical models in applications to molecular similarity studies in molecular engineering, pharmaceutical drug design, toxicological risk assessment, and in a wide area of biochemical applications.

Another important shape problem of quantum chemistry involves the multidimensional shape of potential energy hypersurfaces. These shapes determine the reactivities, and the reaction mechanisms of all chemical species (including various stable and unstable conformers) of each molecule with a common stoichiometry. This common stoichiometry specifies the nuclear configuration space and the associated (internal) coordinates which define the potential energy hypersurface.

These two shape problems show some common features, in particular, both can be studied using discrete models derived from group-theoretical approaches of algebraic topology. In this contribution, some of the relevant conceptual aspects of the quantum chemical shape problem are reviewed.

The quantum chemical concept of molecular shape has strong connections to the fundamentals of quantum mechanics, involving the Heisenberg Uncertainty relation, localizability, and the structure of relativistic space-time. On the basis of the particle-wave duality of matter, it is arguable whether any shape concept analogous to the macroscopic shape concept is applicable on the molecular level, and a thought-provoking series of papers[1-4] by Woolley has clearly focused attention on the fact that the most common shape concepts applied in chemistry are primarily classical, requiring a thorough re-investigation within a quantum mechanical context.

One of the main questions is localizability in a quantum mechanical sense, since the conventional concept of molecular shape involves the classical mechanical concept of localization within the three-dimensional space. Unfortunately, the classical concept of localization of dynamic objects like molecules cannot be generalized directly to quantum mechanics. The Heisenberg uncertainty relation places constraints on the simultaneous determination of the position and momentum of various contributors to molecular shapes.

In nonrelativistic quantum mechanics, one may view spatial localization of a particle as the limit for a sequence of monotonically shrinking volumes; this sequence, however, is associated with a sequence of increasing uncertainties for the momentum of the particle. Nevertheless, within the framework of nonrelativistic quantum mechanics, a sharp position variable is meaningful if one considers an isolated particle, even if this localization implies loss of momentum information. However, for a complex system generated by the formal combination of several subsystems, the generalization of the procedure of shrinking volumes can no longer be carried out in isolation, and, inevitably, non-classical features appear.

One faces even more cumbersome problems if localization is considered within the framework of relativistic quantum mechanics. A procedure of generating a sequence of monotonically shrinking volumes is not in general Lorentz-invariant. In particular, one volume sequence that converges to zero volume in one Lorentz frame, as desired, may show entirely different convergence behavior in another Lorentz frame,[5] leading to no formal localization at all.

Within a relativistic model, if the rest frame of a particle is to be specified, then the uncertainty in the momentum of the particle implies uncertainty in the rest frame. In turn, the momentum uncertainties affecting the rest frames imply that the question whether two events can or cannot be considered simultaneous also involves some uncertainty. Even if one assumes to start at some t=0 time with a localized relativistic particle within a bounded region of the three-dimensional space, it has been shown that this particle instantaneously "spreads" over the space, and for any later time t > 0, causality is not maintained.[6]

These are severe conceptual problems, and the approaches to overcome these difficulties involve the reformulation of some fundamental quantum mechanical models in phase space.[7-12]

These reformulations also imply some modifications of the fundamental classical concepts of shape. Nevertheless, in chemistry at least, classical analogies for shape representations appear to work remarkably well, as long as one regards molecules as nonrigid, fuzzy objects, and treats them using primarily topological tools. The proposition that *quantum mechanical molecules are not geometrical but topological entities,* has been exposed in some detail,[13-19] in the context of representations of the nuclear configuration space of molecules and the associated potential energy hypersurfaces. These representations involved fuzziness for both electron densities (a very natural condition, see, e.g., refs. 20,21) and nuclear arrangements.

In fact, an application of the fuzzy set methodology[22-24] to a more general quantum mechanical framework[25-32] appears to offer a well-justified model that incorporates both the constraints of the Heisenberg uncertainty relation and the general wave-particle duality aspects. The consequences of applying this and analogous approaches to chemistry are of importance; molecules are fuzzy, topological entities and not geometrical objects in the classical, macroscopic sense.[14-19,33,34]

These realizations naturally lead to a re-evaluation of two conventional chemical concepts: molecular shape and reaction mechanism, where the latter can be viewed as a general shape problem of potential energy surfaces. In the analyses reviewed in this contribution, the molecular shape problem is treated using the *algebraic shape groups of molecular electron densities*, relying on the homology groups of algebraic topology, whereas the shape problem of potential energy hypersurfaces is treated using homotopy groups of algebraic topology, leading to the *energy-dependent fundamental groups of reaction mechanisms*.

In both cases, a modern chapter of mathematics, algebraic topology, offers both a novel conceptual framework as well as practical, computational techniques which properly reflect the fundamental, quantum mechanical nature of molecules, subject to the Heisenberg uncertainty relation. These approaches are based on the following conceptual foundation: *quantum mechanical molecules are not geometrical but topological entities,* and lead to new approaches towards an answer to the fundamental questions of chemists: *what are molecules, and what are chemical reactions?*

2. MOLECULAR SHAPE: THE SHAPE GROUP METHODS.

Whereas in chemistry the traditional shape concepts have often been associated with various molecular orbital representations, the focus of current molecular shape analysis is on electron densities, as motivated by the following observation: *electron density is reality, whereas a wavefunction is only a formal "square-root" of reality.* Within the Born-Oppenheimer approximation, a Molecular Iso-Density Contour (MIDCO) surface $G(K,a)$ is defined for any fixed nuclear configuration K in terms of the electronic density $\rho(K,r)$:

$$G(K,a) = \{\, r : \rho(K,r) = a \,\}, \tag{1}$$

that is, as the collection of all points r of the three-dimensional space where the density is equal to a given threshold value a. The corresponding *density domain*, denoted by DD(K,a), is defined as the point set that includes all points of the MIDCO G(K,a) as well as all the points within its interior:

$$DD(K,a) = \{ \, r : \rho(K,r) \geq a \, \}. \qquad (2)$$

Whereas there are infinitely many MIDCOs (one for each value of the continuous electron density threshold parameter a), and there are infinitely many density domains for each nuclear configuration K of the molecule, there are only a finite number of topologically different bodies of density domains. That is, there is only a finite number of equivalence classes, and these classes are used for a rather general description of chemical bonding,[35] as well as for a quantum chemical definition of chemical functional groups.[36-39]

If the density threshold value a is regarded as the coordinate along a fourth dimension, augmenting the three spatial dimensions, as has been defined earlier,[40] then the density domains and the associated MIDCOs can be regarded as lower-dimensional projections of the four dimensional density representation. This approach also leads to a new family of molecular similarity measures.[35] In fact, MIDCOs are two-dimensional surfaces embedded in the ordinary three-dimensional space.[37]

The shape groups are algebraic groups describing shape properties. Whereas in molecules the nuclear point symmetry is an important (although rather limited) aspect of shape, the shape groups are not related to point symmetry groups. The shape groups of molecules are the homology groups of truncated MIDCOs, where the truncation is determined by local shape properties, for example, by local curvature properties of the MIDCOs.[42-45]

Local curvature properties are usually specified in terms of shape domains, such as locally convex, concave, or saddle-type regions of MIDCOs relative to some reference object T, for example, a tangent sphere of some specified curvature.[43] A reference object T, tangent to the MIDCO at some point **r**, may fall locally on the outside, on the inside of a MIDCO, or it may cut into the given MIDCO surface G(K,a) within any small neighborhood of point **r**. Based on such a characterization with reference to T, the points of the MIDCO are classified into three types, that is, the MIDCO G(K,a) is subdivided into local shape domains of types D_2, D_0, and D_1. These domains are referred to as locally convex, locally concave, and locally saddle-type shape domains, respectively, relative to the tangent object T. Since a typical MIDCO G(K,a) is an orientable surface, and a tangent sphere T may osculate to G(K,a) either from the inside or the from the outside of G(K,a), one distinguishes negative or positive reference curvature values b. In a detailed shape analysis a continuum of b values of the reference curvature parameter is considered.

Each value of the reference curvature b provides a complete partitioning of the MIDCO G(K,a) and leads to a family of local shape domains D_2, D_0, and D_1. All D_μ domains of a specified type μ, for example, all the locally convex domains D_2 relative to reference curvature b, can be excised from the MIDCO surface G(K,a), leading to a truncated contour surface G(K,a,μ). This truncated surface encodes some essential shape information of the original MIDCO surface G(K,a) that can be detected and identified by simple topological methods. If the truncation is repeated for MIDCOs of a whole range of reference curvature values b, then a detailed shape analysis of the original, non-truncated MIDCO surface G(K,a) is obtained. Note that within the entire range of possible reference curvature values b there are only a finite number of topologically different truncated MIDCOs $G(K,a,\mu)$. This simplifies the shape analysis considerably. The homology groups of these truncated surfaces are topological invariants. The ranks of the homology groups are the Betti numbers, and these are the numerical topological invariants used for shape characterization.

The formal dimensions p of the electron density shape groups $HP_\mu(a,b)$ for each pair of parameter values a and b are zero, one, and two, and the numerical shape codes are generated by the lists of the $bP_\mu(a,b)$ Betti numbers of these shape groups. For each truncation type μ, reference curvature b and density threshold a of a given MIDCO $G(K,a)$, three shape groups, $H^0_\mu(a,b)$, $H^1_\mu(a,b)$, and $H^2_\mu(a,b)$ are determined. These shape groups collectively describe the essential shape information of the MIDCO $G(K,a)$. The corresponding three Betti numbers are denoted by $b^0_\mu(a,b)$, $b^1_\mu(a,b)$, and $b^2_\mu(a,b)$, respectively. The most important characterization is obtained in terms of the one-dimensional shape groups $H^1_\mu(a,b)$ and their Betti numbers $b^1_\mu(a,b)$, associated with a truncation type $\mu = 2$.

A two-dimensional (a,b)-map describes the distribution of the Betti numbers of various shape groups within the ranges of parameters a and b. A positive b value for reference curvature indicates a tangent sphere of radius 1/b placed on the exterior side of the MIDCO surface, whereas a negative b value indicates a tangent sphere of radius 1/b placed on the interior side of the MIDCO $G(K,a)$. A numerical *shape code* is defined in terms of the (a,b)-parameter map molecule M.

For practical, computational considerations, a grid of a and b parameter values is taken within some interval $[a_{min}, a_{max}]$ of density thresholds a and some interval $[b_{min}, b_{max}]$ of reference curvature values b, where logarithmic scales are used. For negative values of the curvature parameter b, the log|b| values are considered. Typically, the ranges of [0.001, 0.1 a.u.] (a.u. = atomic unit) and [-1.0, 1.0] are considered for the density threshold values a and the curvature b of the test spheres, respectively. A 41 × 21 logarithmic grid is usually sufficient for a detailed enough representation. A matrix $M(a,b)$ is defined by the values of the Betti numbers at the grid points (a,b) that represents a numerical shape code of the shape of the fuzzy electron density of the molecule M.

The numerical shape code matrices $M(a,b)$ are used for the evaluation of a numerical shape similarity measure between molecules. Denote the total number of elements in $M(a,b)$ as

$$t = n_a n_b, \qquad (3)$$

where n_a and n_b are the number of grid divisions for parameters a and b, respectively. If $m[M(a,b),A, M(a,b),B]$ denotes the number of matches between corresponding elements in two shape code matrices $M(a,b),A$ and $M(a,b),B$ of two molecules, A and B, then a numerical shape similarity measure s(A,B) can be defined as

$$s(A,B) = m[M(a,b),A, M(a,b),B] / t, \qquad (4)$$

This similarity measure has been applied to a numerical evaluation of the similarities of molecular shape features in several molecular families, leading to useful shape-property correlations.[46-52]

With the introduction of the Additive Fuzzy Density Fragmentation (AFDF) methods based on the Mulliken-Mezey scheme,[35,53-56] the Shape

Group method became applicable to the study of both the global and local shape features of the electron densities of macromolecules. The AFDF methods serve as the basis for the Molecular Electron Density Loge Assembler (MEDLA) method,[57-63] the Adjustable Density Matrix Assembler (ADMA) techniques,[54,56,64-66] and the ADMA-FORCE approach to macromolecular forces.[65]

3. THE SHAPE OF POTENTIAL ENERGY HYPERSURFACES: THE FUNDAMENTAL GROUP OF REACTION MECHANISMS.

The shape of potential energy hypersurfaces has a fundamental influence on the distribution of molecules and their stable conformations and transition structures, all belonging to a given stoichiometry. Each family of stable nuclear arrangements (stabilized by the associated electron distribution) can be thought of as a formal basin, a "catchment region" on the potential energy hypersurface, taken as the collection of all those nuclear arrangements which relax to a common critical point on the hypersurface. In this model, one assumes that the relaxation occurs infinitely slowly along some concerted path of the nuclear rearrangement. One important shape problem of potential energy hypersurfaces is the distribution of these catchment regions. Alternatively, one may consider various energy bounds and the associated level sets (analogous to flooded areas) of the potential energy hypersurface. In this context, the shape of the potential energy hypersurfaces also determines the reactivities and reaction mechanisms of various species of the given stoichiometry.

The following concepts and notations are used: the nuclear configuration space M is a metric space[67] that belongs to the given collection of nuclei (to the given stoichiometry), and the nuclear arrangements (also referred to as nuclear configurations) are denoted by K. Since energetic considerations are of major importance in the study of reaction mechanisms,[13] we restrict the investigations to reaction paths and reaction mechanisms that fall below some energy bound A. Various upper bounds for energy may be considered; each upper bound A defines a *level set* F(A) of the nuclear configuration space M:

$$F(A) = \{ K : E(K) < A \} . \qquad (5)$$

A level set may be disconnected, as well as multiply connected; that is, it may have several separate parts, and it may also contain "holes".

Within the algebraic topological model,[68-71] formal reaction paths p are considered as mappings from the unit interval I,

$$I = [0, 1] \qquad (6)$$

into a level set F(A) of the configuration space M

$$p: I \rightarrow F(A). \qquad (7)$$

For a precise treatment, a continuous parametrization is introduced: as a mapping, the path p assigns parameter values x from the unit interval I to points p(x) of the level set F(A). In this context, the path p itself is *not* a curve in F(A), but a continuous *function*, where the *image* of p is a curve in the level set F(A).

A common image may be obtained by two different assignments, p1 and p2, of the x parameter values to the *same* set of points of F(A); nevertheless, these different assignments are regarded as different paths, p1≠p2.

The point p(0) of the image (that is, the point assigned to the parameter value x=0) is called the origin (or the beginning) and the point p(1) of the image (that is, the point assigned to the parameter value x=1) is called the extremity (or the end) of the path p.

A path p is not necessarily a classical trajectory, in fact, a path p is not constrained by any classical mechanical restriction.[68-71] In a quantum mechanical model no sharply defined lines within the nuclear configuration space M is meaningful, and instead of individual paths, *families* of topologically related paths are used to represent reactions.[68-71]

The *inverse path* p^{-1} of path p involves the same image, that is, the same curve in F(A) as the original path p, but the direction of the inverse path is reversed, hence, the origin and extremity are interchanged. The inverse path p^{-1} of path p is defined as

$$p^{-1}(x) = p(1-x). \qquad (8)$$

A path where the entire image is a single point K of the level set F(A) is called a *constant path* :

$$p: I \to K \in F(A). \qquad (9)$$

Special paths are obtained if the two endpoints coincide. A *closed path* or a *loop* is a path p with coincident origin and extremity:

$$p(0) = p(1) = K \in F(A). \tag{10}$$

Loop paths provide important tools for the description of reaction mechanisms using topological methods, such as homotopy groups,[68-71] discussed in the next section.

It is possible to provide a family of paths with an algebraic structure if a formal *product path* is defined for two paths in the following sense: the product of two paths is the first path continued by the second one, whenever the endpoint of the first path coincides with the beginning of the second path. Such a product p_3 of two paths, p_1 and p_2, exists if and only if the extremity of the first path p_1 coincides with the origin of the second path p_2:

$$p_1(1) = p_2(0). \tag{11}$$

This product path p_3 can be denoted by $p_1 \, p_2$:

$$p_3 = p_1 \, p_2, \tag{12}$$

where the parametrization of the product path p_3 is chosen as

$$p_3(x) = p_1(2x) \quad \text{if } 0 \le x \le 1/2, \tag{13}$$

and

$$p_3(x) = p_2(2x-1) \quad \text{if } 1/2 \le x \le 1. \tag{14}$$

These definitions serve as the basis for the topological concepts used for the construction of an algebraic structure for quantum chemical reaction mechanisms[68-71] within the potential energy hypersurface model: *homotopy, homotopical equivalence,* and *homotopy equivalence classes.* Additional details of the topological background can be found in the mathematical literature.[72-74]

Two paths, q_1 and q_2, with images within a level set F(A) are *homotopic relative to their endpoints,* (in short, *homotopic*), if they have common origin as well as common extremity and if they *can be continuously deformed into one another within the level set* F(A) while keeping their endpoints fixed:

$$q_1(0) = q_2(0), \qquad (15)$$

$$q_1(1) = q_2(1), \qquad (16)$$

$$q_1 \text{ is continuously deformable into } q_2. \qquad (17)$$

The fact that path q_1 is homotopic to q_2 is denoted by

$$q_1 \sim q_2. \qquad (18)$$

Each family of paths which are homotopic to one another and are confined to the actual level set F(A) form a *homotopy equivalence class (relative to fixed endpoints),* or in short, a *homotopy class* of paths of level set F(A). The homotopy class of all paths homotopic to some path p is denoted by [p],

$$[p] = \{p': p' \sim p\}. \qquad (19)$$

The homotopy equivalence classes of formal reaction paths have been proposed to represent reaction mechanisms.[68-71] Whereas the classes defined above involve the geometrical restriction of fixed endpoints of these paths, this restriction can be eliminated. This leads to a fully topological model of reaction mechanisms and an associated group theoretical structure.[68-71] Whereas individual reaction paths are not strictly quantum mechanical, the topological model of homotopy classes of paths is compatible with quantum mechanics.[13]

As an illustration of the importance of energetic considerations, as manifested in the level set approach, consider a two-dimensional level set $F(A)$ and the images of four paths within $F(A)$, q_1, q_2, q_3, and q_4. Assume that paths q_1, q_2 and q_3 have the same endpoints. Also assume that there is a high-energy domain above energy A that is missing from the interior of level set $F(A)$, where path q_3 leads along one side of the missing domain, whereas paths q_1 and q_2 lead along the other side of the missing domain. Whereas paths q_1 and q_2 are continuously deformable into one another, that is, q_1 and q_2 are homotopically equivalent, $q_1 \sim q_2$, the path q_3 cannot be deformed into either of q_1 and q_2, as long as the allowed deformations are confined to the level set $F(A)$. That is, q_3 is not homotopic to q_1 and q_2. The paths q_1 and q_2 belong to a common homotopy equivalence class $[q_1]$, whereas q_3 belongs to a different homotopy equivalence class $[q_3]$ of the level set $F(A)$.

It is important to realize, however, that q_3 and q_1 may become homotopically equivalent if the energy bound A of the level set is raised to a new value A' that lies above the missing mountain top that separates these paths within the original level set $F(A)$. If one considers a new level set $F(A')$ of such a higher threshold A', then all three paths q_1, q_2 and q_3 may become homotopically equivalent, that is, all three paths may become associated with a common homotopy equivalence class of the paths of the new, larger level set F(A').

If the endpoints of the fourth path q_4 do not coincide with the endpoints of the first three paths q_1, q_2 and q_3, then path q_4 necessarily belongs to a different homotopy class $[q_3]$ of paths for all values of the thresholds A" of level sets F(A").

Whereas the mismatch of endpoints of paths is a common occurrence, it is still possible to construct an algebraic structure for such "endpoint-constrained" reaction paths. This algebraic structure, the *fundamental groupoid of paths* within a level set F(A) forms an important step in the construction of a more important algebraic structure, the *fundamental group of reaction mechanisms* within a level set F(A), that is no longer constrained by geometrical conditions on the endpoints of paths.

Following the earlier conventions,[13] the lower case notation p is used for reaction paths if $p(0) \neq p(1)$ is allowed, whereas the upper case notation P is used only for loop paths (also called cycles) for which $P(0) = P(1)$. The notation **P** is used for the family of all paths with images within a selected level set F(A) of the potential energy hypersurface E(K).

For each path $p \in \mathbf{P}$, two mappings, L* and R* are defined, called the *left* and *right unit paths* of reaction path $p \in \mathbf{P}$, respectively. The first of these is given as

$$L^*: \mathbf{P} \to \mathbf{P}, \qquad L^*(p) = q \in \mathbf{P}, \qquad (20)$$

where q is a constant path,

$$q(I) = p(0) \in F(A). \qquad (21)$$

The second mapping is given by

$$R^*: \mathbf{P} \to \mathbf{P}, \qquad R^*(p) = q' \in \mathbf{P}, \tag{22}$$

where q' is a constant path,

$$q'(I) = p(1) \in F(A). \tag{23}$$

The left unit $L^*(p)$ assigns the constant path q at the origin $p(0)$ of p to the path p, whereas the right unit $R^*(p)$ assigns the constant path q' at the extremity $p(1)$ of p to the path p.

For any loop path P

$$L^*(P) = R^*(P). \tag{24}$$

These left and right units L^* and R^* are useful for a concise representation of the condition $p_1(1) = p_2(0)$ for the existence of a product path $p_1 p_2$ generated by paths $p_1, p_2 \in \mathbf{P}$:

$$R^*(p_1) = L^*(p_2). \tag{25}$$

For any two paths $p, p' \in [p]$ from the same homotopy equivalence class $[p] = \{p': p' \sim p, p', p \in \mathbf{P}\}$ relative to endpoints

$$L^*(p) = L^*(p'), \tag{26}$$

and

$$R^*(p) = R^*(p') \qquad (27)$$

must hold.

Take the family $\Pi(F(A))$ of all homotopy classes [p] of the level set F(A),

$$\Pi(F(A)) = \{ [p_\alpha]: p_\alpha \in \mathbf{P} \}. \qquad (28)$$

Since the left units and also the right units are common within each homotopy class [p], two mappings, L and R can be defined on the set Π:

$$L: \Pi \to \Pi, \qquad L([p_\alpha]) = [L^*(p_\alpha)] \in \Pi, \qquad (29)$$

and

$$R: \Pi \to \Pi, \qquad R([p_\alpha]) = [R^*(p_\alpha)] \in \Pi. \qquad (30)$$

The classes L([p]) and R([p]) so defined are called the *left unit* and the *right unit* of the homotopy class [p], respectively.

If the condition

$$R([p_1]) = L([p_2]) \qquad (31)$$

is fulfilled for the right unit $R([p_1])$ of a homotopy class $[p_1]$ and the left unit $L([p_2])$ of a homotopy class $[p_2]$, then the *product* $[p_1][p_2]$ of the two homotopy classes $[p_1]$ and $[p_2]$ of paths is defined as

$$[p_1][p_2] = [p_1 p_2] \in \Pi, \tag{32}$$

and is interpreted as the homotopy class which contains the products of paths from the homotopy classes $[p_1]$ and $[p_2]$.

This product, if it exists, is unique and does not depend on the choice of reaction paths $p_1, p_2 \in P$, representing equivalence classes $[p_1], [p_2] \in \Pi$, as it is implied by the homotopy equivalence within each class.

The set Π of all homotopy classes of the set P of all reaction paths on the level set $F(A)$, with respect to mappings L and R, and the product defined above fulfill the following conditions (i) - (vi) of a *groupoid*:

(i) If the symbol $\circ$ denotes the composition of mappings (one mapping followed by the other, in the order from right to left), then for units L and R

$$L \circ L = L = R \circ L, \tag{33}$$

and

$$L \circ R = R = R \circ R \tag{34}$$

hold.

(ii) The products $L([p])[p]$ and $[p]R([p])$ exist for each class $[p] \in \Pi$, and fulfill the relations

$$L([p])[p] = [p] = [p]R([p]) \in \Pi. \tag{35}$$

(iii) The products $L([p])L([p])$ and $R([p])R([p])$ exist for each class $[p] \in \Pi$ and fulfill the idempotency relations

$$L([p])L([p]) = L([p]) \in \Pi, \tag{36}$$

$$R([p])\,R([p]) = R([p]) \in \Pi. \tag{37}$$

(iv) For any two classes $[p_1], [p_2] \in \Pi$ fulfilling the condition for product

$$L([p_1][p_2]) = L([p_1 p_2]) = L([p_1]), \tag{38}$$

and

$$R([p_1][p_2]) = R([p_1 p_2]) = R([p_2]) \tag{39}$$

hold. If the condition

$$L([p_3]) = R([p_2]) \tag{40}$$

also holds for some homotopy class $[p_3] \in \Pi$, then the products

$$([p_1]\,[p_2])\,[p_3] \in \Pi, \tag{41}$$

and

$$[p_1]([p_2][p_3]) \in \Pi \qquad (42)$$

also exist.

(v) The product of homotopy classes of paths, if it exists, is *associative*, that is, if the products (41) and (42) exist, then they are equal,

$$([p_1][p_2])[p_3] = [p_1]([p_2][p_3]). \qquad (43)$$

The parentheses can be omitted and one may write $[p_1][p_2][p_3]$ for this product.

Note that associativity is not in general assured for the products $(p_1p_2)p_3$ and $p_1(p_2p_3)$ of paths p_1, p_2, and p_3 from the homotopy classes $[p_1]$, $[p_2]$, and $[p_3]$, respectively. In fact, even if the paths $(p_1p_2)p_3$ and $p_1(p_2p_3)$ are homotopically equivalent, $(p_1p_2)p_3 \sim p_1(p_2p_3)$, they are different paths, $(p_1p_2)p_3 \neq p_1(p_2p_3)$, as long as p_1, p_2, and p_3 are not constant paths. By contrast, the equivalence classes $[p_1]$, $[p_2]$, and $[p_3]$ have stronger algebraic properties.

(vi) Since a unique inverse path p^{-1} exists for each path $p \in \mathbf{P}$, there exists a *unique inverse* class $[p]^{-1}$ for every homotopy class $[p] \in \Pi$:

$$[p]^{-1} = [p^{-1}] \in \Pi. \qquad (44)$$

Here the following relations must hold for the pair [p], [p]$^{-1}$:

$$L([p]) = R([p]^{-1}) \qquad (45)$$

and

$$R([p]) = L([p]^{-1}). \qquad (46)$$

The set Π of homotopy classes is called the *fundamental groupoid* of reaction paths of level set F(A).[68-71]

The fundamental groupoid $\Pi(F(A))$ of paths of level set F(A) has many unit elements, and $\Pi(F(A))$ does not have the closure property: for arbitrary pairs of elements the product is not required to exist. These disadvantages limit the practical applications of this algebraic structure. However, a related algebraic structure, the fundamental group of reaction mechanisms (FRM group), avoids these problems.

For an arbitrary point $K_o \in F(A)$ one may consider the following subset $\Pi'_1(F(A),K_o)$ of groupoid $\Pi(F(A))$:

$$\Pi'_1(F(A),K_o) = \{[P]: P(0) = P(1) = K_o,\ P \in [P],\ [P] \in \Pi(F(A))\ \}. \qquad (47)$$

The elements of this set $\Pi'_1(F(A),K_o)$ are the homotopy classes of all the loop paths P with the common endpoint $K_o \in F(A)$. As a consequence of the common endpoint K_o, all left and right units, all inverses, and all possible pairwise products of these homotopy classes exist and are necessarily elements of the same subset $\Pi'_1(F(A),K_o)$, that is, for every

$$[P], [P_1], [P_2] \in \Pi'_1(F(A),K_o), \qquad (48)$$

the following relations hold:

$$L([P]), R([P]) \in \Pi'_1(F(A),K_o), \qquad (49)$$

$$[P_1]^{-1} = [P_1^{-1}] \in \Pi'_1(F(A),K_o), \qquad (50)$$

and

$$[P_1][P_2] \in \Pi'_1(F(A),K_o). \qquad (51)$$

In particular, the closure property of product within this set follows from the fact that in the family $\Pi'_1(F(A),K_o)$ of loop paths P with the common endpoint K_o, each path can be continued by any other such path, hence, the product exists for any pairwise combination of these paths. Consequently, the product also exists for any pairwise combination of the homotopy classes [P] of these paths.

In general, if some subset π of a groupoid Π has properties (48) - (51), then this subset π is called a *stable subset* of the groupoid Π. Furthermore, π is called a *subgroupoid* of the original groupoid Π, if the mappings left unit L and right unit R are restricted to this stable subset π.

Take the equivalence class $[P_o]$ within $\Pi'_1(F(A),K_o)$, where $[P_o]$ contains the element constant path P_o at point K_o,

$$P_o \in [P_o]. \qquad (52)$$

Since

$$P_0(I) = K_0, \quad (53)$$

$$L([P]) = R([P]) = [P_0] \quad (54)$$

follows for *every* homotopy class $[P] \in \Pi'_1(F(A),K_0)$, that is, the left unit mapping L and the right unit mapping R, if restricted to the subgroupoid $\Pi'_1(F(A),K_0)$, are *constant maps*. Consequently, the unit element $[P_0]$ of subgroupoid $\Pi'_1(F(A),K_0)$ *exists* and is *unique*.

Besides the existence of a unique *unit element* and the *closure* property within subgroupoid $\Pi'_1(F(A),K_0)$, the existence of *inverse* and *associativity* of the product are also guaranteed, since these two properties are inherited from the groupoid $\Pi(F(A))$. Consequently, the subgroupoid $\Pi'_1(F(A),K_0)$ is a *group*, that may be regarded as a *subgroup* of the original groupoid $\Pi(F(A))$.

These additional properties, satisfying the conditions for a group, greatly enhance the chemical relevance of the subgroupoid $\Pi'_1(F(A),K_0)$.

For each arcwise connected level set $F(A)$ and interior point $K_0 \in F(A)$, the group $\Pi_1(F(A),K_0)$ is called the *fundamental group* of level set $F(A)$ at point $K_0 \in F(A)$.

If within an arcwise connected level set $F(A)$ one regards the fundamental group $\Pi'_1(F(A),K_0)$ as an abstract group, its algebraic structure is independent of the choice of point K_0. Evidently, if $F(A)$ is arcwise connected, then there must exist within $F(A)$ some path R leading from point $K_0 \in F(A)$ to any other point $K_1 \in F(A)$. This implies that by simply attaching a "detour" from K_0 to K_1 and back, any loop path P with both extremities at point K_0 can be extended into a homotopically equivalent loop path PRR^{-1} with both extremities at point K_1. Consequently, the choice of the actual point K_0 does not affect the algebraic structure. In fact, the two concrete groups $\Pi_1(F(A), K_0)$ and $\Pi_1(F(A), K_1)$ are isomorphic:

$$\Pi_1(F(A), K_0) \approx \Pi_1(F(A), K_1). \qquad (55)$$

If one is interested only in the algebraic structure of the fundamental group of any arcwise connected level set $F(A)$, the reference to point K_0 can be omitted. The abstract fundamental group $\Pi_1(F(A))$ is a fully topological algebraic structure of homotopy equivalence classes of reaction paths within the arcwise connected level set $F(A)$.

Individual reaction paths are not compatible with quantum mechanics. A reaction mechanism does not imply a precise, geometrical path, in fact, a reaction mechanism can be thought of as a formal *reaction itinerary,* where some details of individual paths are irrelevant, and the only relevant properties are the main, invariant features of the chemical transformation. A reaction itinerary, that is, a reaction mechanism, can be represented by a homotopy equivalence class of paths.

For any arcwise connected level set $F(A)$, the fundamental group $\Pi_1(F(A))$ is the algebraic structure of all loop reaction mechanisms below energy A. This model avoids any classical mechanical reference to formal, fixed nuclear configurations, and to individual, classical reaction paths.

Only loop reaction mechanisms are involved directly in the fundamental group $\Pi_1(F(A))$. However, any reaction mechanism within a level set $F(A)$ can be extended into a loop reaction mechanism and any reaction mechanism that requires activation energy less than the threshold A is a segment of some loop mechanism within the level set $F(A)$. Consequently, the family $\Pi_1(F(A))$ of all loop mechanisms, in fact, represents all reaction mechanisms within the level set $F(A)$.

For a finite energy bound A, the fundamental group $\Pi_1(F(A))$ of reaction mechanisms on an arcwise connected level set $F(A)$ is a finitely generated free group. A finite family of generator reaction mechanisms

$$[P_1], [P_2], \ldots, [P_m] \qquad (56)$$

can serve as free generators for this group. The generator reaction mechanisms are the chemically most important fundamental mechanisms within the arcwise connected level set $F(A)$.

Any fundamental reaction mechanism $[P] \in \Pi_1(F(A))$ is a product of some of the generator mechanisms:

$$[P] = [P_{i_1}]^{\alpha_{i_1}} [P_{i_2}]^{\alpha_{i_2}} \cdots [P_{i_j}]^{\alpha_{i_j}} \cdots [P_{i_k}]^{\alpha_{i_k}}, \qquad (57)$$

where the exponents are chosen as $\alpha_{i_k} = \pm 1$ and repetitions are allowed for each generator mechanism P_{i_j}.

The fundamental groups of reaction mechanisms are not in general commutative; the products $[P_1][P_2]$ and $[P_2][P_1]$ may represent two different reaction mechanisms.

4. AN ANALOGY BETWEEN SHAPE GROUPS AND THE FUNDAMENTAL GROUPS OF REACTION MECHANISMS.

There is an interesting analogy between the fundamental groups of reaction mechanisms for a stoichiometric family of molecules and the shape groups of electron densities of individual molecules. The level set F(A) depends on the threshold value A, consequently, the fundamental group $\Pi_1(F(A))$ of reaction mechanisms also depends on the energy bound A. Since $\Pi_1(F(A))$ is a discrete algebraic structure, this dependence is not continuous. The changes in the fundamental group provides an energy dependent description of reaction mechanisms. These changes in the group structure of $\Pi_1(F(A))$ are analogous to the changes of shape groups of electron density contour surfaces of an individual molecule as the function of the electron density threshold a. These changes are also discontinuous, and in both families of groups, there are only a finite number of groups. Finite numbers of shape descriptors, whether shape groups, or fundamental groups of reaction mechanisms, provide a discrete representation of shape: shape of individual molecules or the shapes of potential energy hypersurfaces of stoichiometric families of molecules and molecular conformations.

SUMMARY

Two shape representations, the shape groups for molecular electron densities, and the fundamental group of reaction mechanisms describing the shape of potential energy hypersurfaces are reviewed and compared.

REFERENCES

1. R.G. Woolley, *Adv. Physics,* **25**, 27 (1976).
2. R.G. Woolley, B.T. Sutcliffe, *Chem. Phys. Letters,* **45**, 393 (1977).
3. R.G. Woolley, *Chem. Phys. Letters,* **55**, 443 (1978).
4. R.G. Woolley, *J. Am. Chem. Soc.,* **100**, 1073 (1978).
5. S. Twareque Ali, *La rivista del Nuovo Cimento,* **8**, 1 (1985).
6. G.C. Hegerfeldt, and S.N.M. Ruijsenaars, *Phys. Rev. D.,* **22**, 377 (1980).
7. S.T. Ali and E. Prugovecki, *Physica,* **89A**, 501 (1977).
8. F.E. Schroeck, Jr., "Measures with Minimum Uncertainty of Non-Commutative Algebras with Application to Measurement Theory in Quantum Mechanics", in *Mathematical Foundations of Quantum Theory,* A. Marlow, ed., Academic Press, New York, 1978, pp. 299-327.
9. F.E. Schroeck, Jr., *J. Math. Phys.,* **30**, 2078 (1989).
10. D.M. Healy, Jr. and F.E. Schroeck, Jr., *J. Math. Phys.* **36**, 453 (1995).
11. F.E. Schroeck, Jr., *Int. J. Theor. Phys.*, **33**, 157 (1994).
12. F.E. Schroeck, Jr., *Quantum Mechanics on Phase Space* (Fundamental Theories of Physics, Volume 74), Kluwer Academic Publishers, Dordrecht, 1996.
13. P.G. Mezey, *Potential Energy Hypersurfaces,* Elsevier, Amsterdam, 1987.
14. P.G. Mezey, *Theor. Chim, Acta,* **60**, 97 (1981).
15. P.G. Mezey, *Int. J. Quantum Chem., Quantum Biol. Symp.,* **8**, 185 (1981).
16. P.G. Mezey, *Theor. Chim. Acta,* **60**, 409 (1982).
17. P.G. Mezey, *Theor. Chim. Acta,* **60**, 133 (1982).

18. P.G. Mezey, "The Topological Model of Non-Rigid Molecules and Reaction Mechanisms" in *Symmetries and Properties of Non-rigid Molecules: A Comprehensive Survey,* J. Maruani, and J. Serre, Eds., Elsevier, Amsterdam, 1983.

19. P.G. Mezey, *Can. J. Chem.,* **61**, 956 (1983).

20. R.G. Parr, and W. Yang, *Density Functional Theory of Atoms and Molecules*, Clarendon Press, Oxford, 1989.

21. E.S. Kryachko, and E.V. Ludena, *Density Functional Theory of Many-Electron Systems*, Kluwer, Dordrecht, 1989.

22. L.A. Zadeh, *Inform. Control,* **8**, 338 (1965).

23. L.A. Zadeh, *J. Math. Anal. Appl.,* **23**, 421 (1968).

24. P.P. Wang, and S.K. Chang, Eds., *Fuzzy Sets,* Plenum, New York, 1980.

25. E. Prugovecki, *Quantum Mechanics in Hilbert Space,* Academic Press, New York, 1971.

26. E. Prugovecki, *Found. Phys.,* **4**, 9 (1974).

27. E. Prugovecki, *Found. Phys.,* **5**, 557 (1975).

28. E. Prugovecki, *J. Phys. A,* **9**, 1851 (1976).

29. S.T. Ali and H.D. Doebner, *J. Math. Phys.,* **17**, 1105 (1976).

30. E. Prugovecki, *J. Math. Phys.,* **17**, 517 (1976).

31. S.T. Ali and E. Prugovecki, *J. Math. Phys.,* **18**, 219 (1977).

32. J.A. Brooke and E. Prugovecki, preprint.

33. P.G. Mezey, *J. Chem. Phys.,* **78**, 6182 (1983).

34. P.G. Mezey, *J. Mol. Struct. Theochem,* **103**, 81 (1983) (Volume dedicated to Nobel Laureate Prof. K. Fukui).

35. P.G. Mezey, "Density Domain Bonding Topology and Molecular Similarity Measures". In K. Sen, ed., *Topics in Current Chemistry,* Vol. **173**, *Molecular Similarity,* Springer-Verlag, Heidelberg, 1995.

36. P.G. Mezey, *J. Chem. Inf. Comp. Sci.*, **32**, 650 (1992).
37. P.G. Mezey, *Shape in Chemistry: An Introduction to Molecular Shape and Topology*, VCH Publishers, New York, 1993.
38. P.G. Mezey, *Canad. J. Chem.*, **72**, 928 (1994). (Special issue dedicated to Prof. J. C. Polanyi).
39. P.G. Mezey, "Functional Groups in Quantum Chemistry", in *Advances in Quantum Chemistry,* **27**, 163 (1996).
40. P.G. Mezey, "Topological Quantum Chemistry". In H. Weinstein, and G. Naray-Szabo, eds., *Reports in Molecular Theory,* CRC Press, Boca Raton, 1990.
41. Z. Zimpel, and P.G. Mezey, *Int. J. Quantum Chem.*, **59**, 379 (1996).
42. P.G. Mezey, *Int. J. Quant. Chem. Quant. Biol. Symp.*, **12**, 113 (1986).
43. P.G. Mezey, *J. Comput. Chem.*, **8**, 462 (1987).
44. P.G. Mezey, *Int. J. Quantum Chem., Quant. Biol. Symp.*, **14**, 127 (1987).
45. P.G. Mezey, *J. Math. Chem.*, **2**, 325 (1988).
46. G.A. Arteca, and P.G. Mezey, *Chem. Phys.*, **161**, 1 (1992).
47. P.D. Walker, G.A. Arteca, and P.G. Mezey, *J. Comput. Chem.*, **14**, 1172 (1993).
48. P.D. Walker, G.M. Maggiora, M.A. Johnson, J.D. Petke, and P.G. Mezey, *J. Chem. Inf. Comp. Sci.*, **35**, 568 (1995).
49. P.D. Walker, P.G. Mezey, G.M. Maggiora, M.A. Johnson, and J.D. Petke, *J. Comput. Chem.*, **16**, 1474 (1995).
50. G.A. Heal, P.D. Walker, M Ramek, and P.G. Mezey, *Canad. J. Chem.*, **74**, 1660 (1996).
51. P.G. Mezey, Z. Zimpel, P. Warburton, P.D. Walker, D.G. Irvine, D.G. Dixon, and B. Greenberg, *J. Chem. Inf. Comp. Sci.*, **36**, 602 (1996).

52. P.G. Mezey, Z. Zimpel, P. Warburton, P.D. Walker, D.G. Irvine, D.G. Dixon, and B. Greenberg, to be published.

53. P.G. Mezey, "Methods of Molecular Shape-Similarity Analysis and Topological Shape Design". In P.M. Dean, ed., *Molecular Similarity in Drug Design*, Chapman & Hall - Blackie Publishers, Glasgow, U.K., 1995.

54. P.G. Mezey, "Shape Analysis of Macromolecular Electron Densities", *Structural Chem.*, **6**, 261 (1995).

55. P.G. Mezey, "Fuzzy Measures of Molecular Shape and Size", in D.H. Rouvray, ed., *Fuzzy Logic in Chemistry,* Academic Press, San Diego, 1997.

56. P.G. Mezey, "Local Shape Analysis of Macromolecular Electron Densities", in J. Leszczynski, ed. *Computational Chemistry: Reviews and Current Trends,* Vol. 1, World Scientific Publ., Singapore, 1996.

57. P.D. Walker, and P.G. Mezey, *J. Am. Chem. Soc.,* **115**, 12423 (1993).

58. P.D. Walker, and P.G. Mezey, *J. Am. Chem. Soc.,* **116**, 12022 (1994).

59. P.D. Walker, and P.G. Mezey, *Canad. J. Chem.,* **72**, 2531 (1994).

60. P.D. Walker, and P.G. Mezey, *J. Math. Chem.,* **17**, 203 (1995).

61. P.D. Walker, and P.G. Mezey, *J. Comput. Chem.,* **16**, 1238 (1995).

62. P.D. Walker, and P.G. Mezey, *Program MEDLA 93* (Mathematical Chemistry Research Unit, University of Saskatchewan, Saskatoon, Canada, 1993).

63. S. Borman, *Chem. & Eng. News,* **73**, 29 (1995).

64. P.G. Mezey, *J. Math. Chem.,* **18**, 141 (1995).

65. P.G. Mezey, *Int. J. Quantum Chem.,* in press.

66. P.G. Mezey, *Program ADMA 95* (Mathematical Chemistry Research Unit, University of Saskatchewan, Saskatoon, Canada, 1995).

67. P.G. Mezey, *Int.J. Quant. Chem.*, **26**, 983 (1984).

68. P.G. Mezey, *Int.J. Quant. Chem. Symp.*, **18**, 77 (1984).

69. P.G. Mezey, *Theor. Chim. Acta,* **67**, 43 (1985).

70. P.G. Mezey, *Theor. Chim. Acta,* **67**, 91 (1985).

71. P.G. Mezey, *Theor. Chim. Acta,* **67**, 115 (1985).

72. E.H. Spanier, *Algebraic Topology*, McGraw-Hill, New York, 1966.

73. M. Greenberg, *Lectures on Algebraic Topology,* Benjamin, New York, 1967.

74. J. Vick, *Homology Theory,* Academic Press, New York, 1973.

INDEX

Ab initio theory	428	Damped dispersion series	341
adiabatic path analysis	434	density functional	1, 342, 429
alias function	65	dichlorobenzene	486
alternation	72	dipole function	430
analytical representation of PES	420	dipole moment	88, 435
angular expansion	435	dipole polarization	436
angular matrix	65	distance norm	421
angular momentum representation	6	DNA	415
anharmonic effect	444	driven harmonic oscillator	176
anthracene	502	Dyson equation	473
atomic cell orbital	14	Dyson orbital	466
azabenzenes	492ff		
		Effective pair potential	425
Baker-Campbell-Hausdorff relation	167	electric polarization field	98
barrier height	442	electron correlation	428
basis alias function	70	electron propagator	470
basis-set-superposition error	346, 429	eliminative reduction	323
basis space	64	equivalent set	60
benzene	483	exchange-correlation density	
binding energy	434	functional	2
borazine	509	exchange operator	245
Born approximation	95	exchange tunneling process	436
Born-Oppenheimer approximation	139, 427	extended Hückel method	136
Brandow diagram	243		
		Factorization theorem	252
Car-Parrinello technique	419	Feynman-Dyson amplitude	466
"catchment region"	528	Fock operator	245
causality	522	four-body contribution	425
chlorobenzene	484	four-body potential	449
closed path	530	frequency shift	418
cluster binding energy	427	full CI method	240
cluster operator	240	full potential theory	4
coherent state	166	functional group	92
common stoichiometry	520	fundamental group of reaction	
concerted hydrogen exchange	446	mechanisms	523
concerted hydrogen transfer	440	fuzzy set	522
configuration space	520		
constant map	542	Gauge transformation	97
constant path	530	generalized Poeschl-Teller oscillator	431
correlation effects	239	global PES	418
Coulomb gauge	120	Green function	4, 26
Coulomb operator	245		
Coulomb self-energy	53	Harmonic approximation	444
counterpoise correction	436	harmonic oscillator	171
coupled-cluster method	240, 472	Heisenberg-Weyl algebra	150
coupled pair functional	432	Heitler-London-Slater-Pauling	
critical point	528	valence bond method	383
cubic hybrid	73	Hellmann-Feynman theorem	207

Helmholtz equation	4
HF cluster	425
HF oligomer	440
HF pair potential	435
HF stretching frequency	445
$(HF)_n$	415
Hilbert space theories	471
high resolution IR	416
homology group	524
Hund-Mulliken method	383
hybrid functional	429
hydrogen bond	415
hydrogen bonding libration	444
hydrogen bond interconversion	437
hydrogen bond rearrangement	440
hydrogen fluoride	415
hydrogen fluoride dimer	417, 435
hydrogen fluoride trimer	437
hydrogen exchange	446
hydrogen motion	419
Idealized observation	329
inverse path	530
intermediate array	272
interpolation scheme	422
IR band strength	445
IR spectroscopy	416
isomeric structure	442
Kohn-Rostoker variational principle	15
Kohn-Sham equations	343
Koopmans' theorem	467
Korringa-Kohn-Rostoker method	2
Kramers-Heisenberg formula	115
Legendre polynomial	431
Lie algebra	146
linked diagram	243
linked diagram theorem	242
Lippmann-Schwinger integral equation	4
liquid water	415
local curvature	419
local density approximation	344
localization	76
localized hybrid	60
localized state	2
local spin-density approximation	2

Lorentz frame	521
Many-body decomposition	422
many-body expansion	423
many-body interaction	427
many-body perturbation theory	239
materialism	327
Maxwell equation	98
meta-dichlorobenzene	488
microscopic rigidity	330
MP2-R12 approach	432
molecular shape	320, 520
molecular similarity	522
molecular structure	91, 319
Møller-Plesset perturbation operator	239
Monte Carlo random walk	422
Morse oscillator	431
muffin-tin model	2
Mulliken population analysis	136
multidimensional PES	419
multiple scattering theory	1
multipolar expansion	346
Naphthacene	507
Octahedral group	63
octahedral hybrid	72
operator alias function	80
ortho-dichlorobenzene	488
Pair potential	432
pairwise interaction potential	431
para-dichlorobenzene	486
Pauli principle	228, 436
permutational symmetry	227
perturbation operator	245
phase space	522
phenanthrene	505
physicalism	327
physical observable	98
Poisson equation	42
polarizability	435
pole search	478
pole strength	478
polyacenes	502
potential energy surface	338, 416
Power-Zienau-Woolley transformation	100

product path	531	Tetracyanoquinone	511
puckering	441	tetrahedral hybrid	59
pyrazine	498	three-body potential	436
pyridazine	494	time-dependent Schrödinger	
pyridine	492	equation	155
pyrimidine	496	transition state	127
		tunneling barrier	434
Quadratic CI approach	241	tunneling splitting	434
quadrature	86	two-point fluctuation function	113
quantum Monte Carlo	417		
		Uncertainty relation	320
Reductionism	321	unlinked diagram	243
reference-state density functional		unperturbed Hamiltonian	245
theory	3, 45		
realism	319, 327	Van der Waals complex	337
renormalization term	248	variational cellular method	15
resolvent	246	velocity-dipole formula	107
rigid monomer approximation	418	Verlet algorithm	198
ring strain	441	vibration-rotation tunneling	338
ring structure	441	Voronoi step representation	421
rotational constant	427		
Röntgen current	108	Wave operator	244
R12 approach	429	weak binding condition	109
		weight	85
Sample space	66		
scalar product	64	Zero point energy	418
Schlosser-Marcus variational			
principle	16		
Schrödinger algebra	147		
Schrödinger equation	4, 145, 226, 245		
self-energy matrix	473		
self-interaction correction	52		
semidirect methods	480		
semiglobal PES	419		
separability	329		
shape group	523		
shape similarity measure	527		
simulated annealing	212		
Slater determinant	232		
spectral form	466		
square hybrid	69		
stationary point	438		
s-tetrazine	500		
strength	59		
s-triazine	500		
superoperator	468		
supersonic jet expansion	443		
surface integral formalism	10		
surface matching theorem	8		
symmetry adaptation	481		

GPSR Compliance
The European Union's (EU) General Product Safety Regulation (GPSR) is a set of rules that requires consumer products to be safe and our obligations to ensure this.

If you have any concerns about our products, you can contact us on

ProductSafety@springernature.com

In case Publisher is established outside the EU, the EU authorized representative is:

Springer Nature Customer Service Center GmbH
Europaplatz 3
69115 Heidelberg, Germany

www.ingramcontent.com/pod-product-compliance
Ingram Content Group UK Ltd.
Pitfield, Milton Keynes, MK11 3LW, UK
UKHW022229230426
12048UKWH00016BA/1153